D.E. Richardson P. van Oosterom (Eds.)

Advances in Spatial Data Handling

10th International Symposium on Spatial Data Handling

Organising Committee:
Dianne E. Richardson (Director, SDH 2002)
Brian Klinkenberg (co-chair)
Liza Charbonneau
Craig Dillabaugh

Programme Committee:
Peter van Oosterom and Henk Ottens (co-chairs)

Sponsors:
Canada Centre for Remote Sensing, Natural Resources Canada
Canadian Institute for Geomatics, Ottawa, Canada
Department of Geodesy, Delft University of Technology, The Netherlands

Springer
Berlin
Heidelberg
New York
Barcelona
Hong Kong
London
Milan
Paris
Tokyo

Dianne E. Richardson Peter van Oosterom (Eds.)

Advances in Spatial Data Handling

10th International Symposium on Spatial Data Handling

With 248 Figures and 39 Tables

Springer

Dr. Dianne E. Richardson
Canada Centre for Remote Sensing
588 Booth Street
Ottawa, Ontario K1A 0Y7, Canada
e-mail: dianne.richardson@ccrs.nrcan.gc.ca

Professor Dr. Peter van Oosterom
Delft University of Technology
Faculty of Civil Engineering and Geosciences
Department of Geodesy, Section GIS-Technology
Thijsseweg 11
2629 JA Delft, The Netherlands
e-mail: P.J.M.vanOosterom@geo.tudelft.nl

ISBN 3-540-43802-5 Springer-Verlag Berlin Heidelberg New York

Cataloging-in-Publication Data applied for

Die Deutsche Bibliothek - CIP-Einheitsaufnahme
Advances in spatial data handling : 10th International
Symposium on Spatial Data Handling / Dianne E. Richardson ; Peter van
Oosterom (ed.). - Berlin ; Heidelberg ; New York ; Barcelona ; Hongkong ;
London ; Mailand ; Paris ; Tokyo : Springer, 2002
ISBN 3-540-43802-5

Springer-Verlag Berlin Heidelberg New York
a member of BertelsmannSpringer Science+Business Media GmbH

http://www.springer.de

Printed in Germany

Production: PRO EDIT GmbH, Heidelberg, Germany
Cover Design: Erich Kirchner, Heidelberg, Germany
Typesetting: Camera-Ready by the Editors

Printed on acid-free paper SPIN: 10858277 30/3130/Di 5 4 3 2 1 0

Foreword

This book, entitled *Advances in Spatial Data Handling,* is a compendium of papers resulting from the International Symposium on Spatial Data Handling (SDH), held in Ottawa, Canada, July 9-12, 2002. The SDH conference series has been organised as one of the main activities of the International Geographical Union (IGU) since it was first started in Zurich in 1984.

In the late 1990's the IGU Commission of Geographic Information Systems was discontinued and a study group was formed to succeed it in 1997. Much like the IGU Commission, the objectives of the Study Group are to create a network of people and research centres addressing geographical information science and to facilitate exchange of information. The International Symposium on Spatial Data Handling, which is the most important activity of the IGU Study Group, has, throughout its 18 year history been highly regarded as one of the most important GIS conferences in the world.

In the spring of 2000, IGU and the International Society for Phtogrammetry and Remote Sensing (ISPRS) established an agreement to facilitate collaboration between the two organisations. In so doing the present collaborative symposium was planned to foster closer relations and provide opportunities for greater interaction among the larger geomatics communities especially among those involved in research, development, applications or management of geo-spatial data and information. The Canadian Institute for Geomatics, which hold an annual geomatics conference, was invited to host their 95th Annual Conference jointly with the 2002 SDH and ISPRS Commission IV symposium. Collectively, some 279 papers and 56 posters were submitted to the combined event "Joint International Symposium on Geospatial Theory, Processing and Application".

This 10th International SDH symposium presents a set of papers that demonstrate a maturing in geographical information science (GISc). Looking back over close to two decades since the SDH was started, it is clear that a larger community now accepts GISc. International standards such as those put forward by the International Organisation for Standards (ISO) now include topology operators first developed by Egenhofer and Clementini, the results of which are available in commercial products. Other issues that are still found in current research endeavours have undergone remarkable progress such as integration of heterogeneous data through spatial and semantic operators, spatial-temporal modelling for real time object assessments in mobile GIS, application of methodologies for on-demand dynamic web mapping. Advances resulting from Internet access are especially far reaching, such as the research undertaken in interoperability, search and access protocols for data discovery, and visualisation and analysis of distributed and heterogeneous sources. Both the more traditional topics and emerging issues can be found within the SDH, presented as solid contributions to the scientific literature.

Of the many remaining challenges under the general topic of spatial data handling, 12 representative categories were identified as essential elements that require in-depth research and development and that reflect the purview of the IGU science objectives. These topics were selected as a means of focussing the symposium and in an effort to bring together research efforts. These are the topics under which the paper submissions appear in the table of contents.

The SDH is a refereed conference. Full papers were submitted by authors and reviewed by a minimum of three experts in the field, and in addition, were reviewed by both editors. The criteria for the paper review involved an assessment of the originality, its contribution to existing knowledge, soundness of the methodology, and the overall scientific quality and presentation. Papers were then ranked accordingly. Of the 59 papers that were submitted to the SDH, only 40 have been accepted for publication. A number of other papers submitted to SDH have been presented at the conference, as they were interesting and innovative, but have not been published in the volume at hand, as they did not meet the publication requirements for this particular compendium.

Dianne Richardson and Peter van Oosterom
Ottawa, April 13, 2002

Acknowledgements

Each person involved in organising the Spatial Data Handling Symposium balanced existing workloads, ongoing commitments, and the contribution of many long hours to the conference arrangements. The first words of thanks go to Craig Dillabaugh and Liza Charbonneau (Canada Centre for Remote Sensing), for their ever cheerful contributions in compiling the proceedings and for their help in every aspect of the symposium. Appreciation is also extended to Brian Klinkenberg, co-chair of the organising committee and to Peter van Oosterom and Henk Ottens Co-chairs of the Programme Committee. Appreciation is also extended to the Programme Committee who is responsible for the quality of the overall programme. Finally, thanks also to Elfriede Fendel of the Delft University of Technology, for proof reading a selection of papers, and to Marguerite Trindade for her excellent design work, portions of which have been used for the cover of this book.

Dianne Richardson
Ottawa, April 13, 2002

Table of Contents

Data Mining

Databases and Interoperability

DEM, Terrain and Landscape Modelling

Generalisation

Integration and Fusion

Spatial and Temporal Data Modelling and Analysis

Spatial Reasoning

Topographic and Thematic Digital Processing

Uncertainty, Accuracy and Error

Urban Modelling and Applications

Visualisation

Web-based Mapping

Author Index

Programme Committee

Co-chairs
Peter van Oosterom and Henk Ottens

Henri	Aalders	Robert	Laurini
Dave	Abel	Pal	Lavai
Tinghua	Ai	David	Mark
Mike	Barry	Bela	Markus
Nicholas	Chrisman	Martien	Molenaar
Eliseo	Clemetini	Esa	Mononen
Leila	De Floriani	Ferjan	Ormeling
Paulino	Di Felice	Wanning	Peng
Geoffrey	Edwards	Donna	Peuquet
Max	Egenhofer	Tom	Poiker
Peter	Fischer	Wilko	Quak
Pip	Forer	Anna	Raus
Randolph	Franklin	Hanan	Samet
Barend	Gehrels	Tapani	Sarjakoski
Chris	Gold	Monika	Sester
Mike	Goodchild	Jack	Snoeyink
Stephen	Guptill	Vladimir	Tikunov
Paul	Hardy	Thierry	Toutin
John	Herring	Marc	van Krefeld
Angela	Ionita	Jan	van Roessel
Chris	Jones	Edward	Verbree
Wolfgang	Kainz	Rob	Weibel
Brian	Klinkenberg	Mike	Worboys
Menno-Jan	Kraak	Anthony	Yeh

A Decision Tree for Multi-Layered Spatial Data

Nadjim Chelghoum[1], Karine Zeitouni[1], and Azedine Boulmakoul[2]

PRISM Laboratory, University of Versailles, 45 avenue des Etats-Unis, 78035 Versailles Cedex, France, Nadjim.Chelghoum@prism.uvsq.fr, Karine.Zeitouni@prism.uvsq.fr[1]
LIST Laboratory, Science and Techniques Faculty of Mohammedia, B.P. 146 Mohammedia, Morocco, boul@uh2m.ac.ma[2]

Abstract

Spatial data mining fulfils real needs of many geomatic applications. It allows the geomatics community to take advantage of the growing availability of geographically referenced data and benefit from this rich information resource. This article addresses spatial data classification for using decision trees. A new method called SCART which differs from conventional decision trees by considering the specifics of geographical data, namely their organisation in thematic layers, and their spatial relationships is proposed. SCART is an extension of CART methods in two ways. On the one hand, the algorithm considers several thematic layers as in the so-called relational data mining area, and on the other hand, it extends discriminating criteria to address concerns about the neighbourhood. As such, the algorithm determines which combination of attribute values and spatial relationships of neighbouring objects provide the best criterion.
Keywords: spatial data mining, classification rules, decision tree, spatial relationship, spatial database

1 Requirements

The growing development of automatic mapping results in the production of large spatial databases. More and more applications require access to large data volumes, however, the complexity and size of these databases exceed our capacity to effectively analyse them. It thus seems appropriate to develop and apply techniques in automatic knowledge extraction through processes referred to as data mining.

The domain of interest for this particular paper is in traffic risk analysis (Huguenin 00). Traffic risk analysis requires the identification of road safety problems in an effort to propose appropriate safety measures. This project aims at identifying relevant risk models to help in various traffic safety tasks. The risk assessment is based on an analysis of information about previous injury accidents collected by police forces. Currently, however, , this analysis has been based on statistics with no consideration about the various spatial relationships that are associated with the accidents. This work aims at identifying risky road sections and analysing and explaining those risks with respect to the geographic context.

The risk analysis presented in this study combines accident information with thematic information about the road networks, the population census, the buildings, and other geographic neighbourhood detail. The paper presents details about the classification task and builds a decision tree that integrates the spatial features of the thematic layer in this case, accident information. Through the decision tree along with the spatial assessment of accidents, one can explain and predict the danger of roads by their geographic context.

Thus, it appears that decision trees can be effectively extended through an integrated assessment of the properties of neighbouring objects. As such potential exists for the development of explanations about analysed phenomena. Two technical problems arise:

1. Neighbouring objects could belong to thematic layers other than the theme analysed.. Yet decision trees consider only one table (theme attributes) where each row represents a learning example. In these cases , a multi-table decision tree is needed;
2. Many definitions of neighbours exist, giving rise to confusion. Indeed, a spatial relationship could be topologic when the objects touch each other, or metric when they are close. In this case, each separating distance represents a particular spatial relationship. Consequently, multi-layered spatial decision trees are more than a multi-table decision tree. The multi-layered spatial decision tree should support the automatic filtering of the multiple and even infinite number of spatial criteria.

The remainder of this paper is organised as follows: section 2 gives a summary of the state of the art in spatial data mining; section 3 presents the proposed method and specifies the algorithm and section 4 gives the prototype architecture, test results, and discusses the implementation issue. Our conclusions are presented in section 5.

2 Background

This section links this work to general research in spatial data mining, highlights the support of spatial relationships and describes other works on decision trees.

2.1 Spatial Data Mining

The goal of spatial data mining is to discover hidden knowledge from spatial databases by combining spatial and non-spatial properties. The spatial data mining methods are usually an extension of those used in conventional data mining (Fayyad 96). Spatial data mining consists of two functions (Zeitouni 00a). The first function addresses a spatial phenomenon by exploring data, for example identifying risky zones by viewing the spatial distribution of the accident locations. The second function explains or even predicts the phenomena while looking for some association or relationship with properties of the geographic environment. For instance, accidents could be "explained" by the state of the road or the surrounding urban density. The spatial classification clarifies these explanatory methods.

2.2 Spatial Relationships

As emphasised above, the main considerations in spatial data mining is that it considers the spatial relationships among objects (Egenhofer 93). Unlike the relational data model, spatial relationships are implicit. Computing them requires many spatial join operations, which can be computationally burdensome. .

In a recent article, (Zeitouni 00b) a method to simplify this process using a secondary structure has been presented. This structure is called spatial join index (SJI), and is an extension of the well-known join indices introduced by (Valduriez 87) in the relational database framework. It pre-computes the exact spatial relationships between objects of two thematic layers. As shown in Fig. 1, a SJI is a secondary table that references matching objects from thematic layers R and S and stores their spatial relationships. In case this relationship is topological (such as

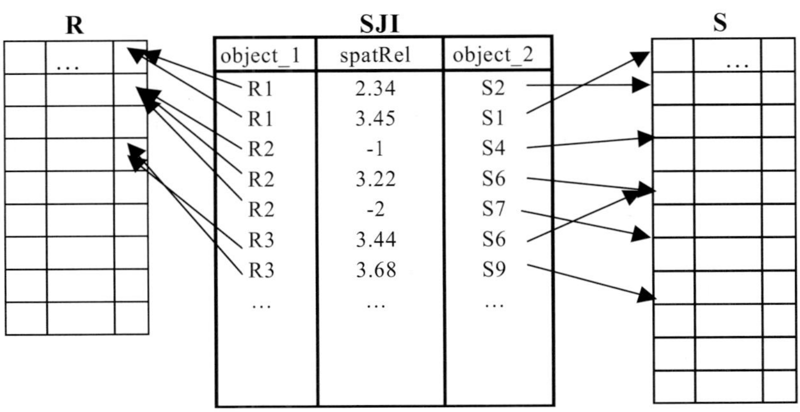

Fig. 1. Spatial join index

inclusion, adjacency or overlapping), the *spatRel* attribute will contain a negative code, such as *(R2, S4)*. Otherwise, it will store the exact distance value. For performance reasons, the distance calculation is limited to a given useful perimeter.

Unlike join indices for a relational model, this extension optimises join operations for multiple criteria (topological and metric). Indeed, objects will match if *spatRel* fulfils certain criteria. This corresponds to a basic relational selection on the SJI table.

2.3 Spatial Decision Tree

A decision tree is an hierarchical knowledge structure that corresponds to a sequence of decision rules. This method aims to determine which attributes (called explanatory) or which criteria of these attributes provide the best distribution of the actual dataset regardless of the given attribute values (called classes). The tree is built recursively by testing and applying subdivision criteria on a training dataset. The test for criteria is based on statistical computation of entropy or information gain. Subdivision criteria are determined at attribute level in the ID3 (Quinlan 86) method while they operate on attribute values in the CART method (Breiman 84). The decision rule sequences are composed of criteria of tree paths starting from the root to the leaves. The main advantage of this technique is its simplicity for decision-makers and people who are not well versed in the complexities of the data analysis domain. It may be less powerful, however, in terms of quality of prediction, than some of the more complex tools such as neural networks.

As emphasised above, unlike conventional decision trees, a spatial decision tree uses data from several tables. One approach consists in using predicate logic instead of attribute values. However, this approach requires that all relational data be transformed into a predicate set. Recently, a new field called relational data mining has been developed. It addresses notably the extension of decision trees for multiple relational tables (Knobbe 99). This more recent method, however, does not solve the problem of spatial relationship determination.

Ester et al. (Ester 97) proposes an algorithm dealing with spatial databases based on ID3. They use the concept of a neighbourhood graph to represent the spatial relationships. This algorithm considers the properties of neighbouring objects in addition to those of the actual object. In the traffic accident example, each object could have many neighbours (e.g., an accident could be near a school and a bus stop). As a result, spatial criteria are not sufficiently discriminating and thus the segmentation may be incorrect. Moreover, this method is limited to only one given relationship. Finally, it does not support the concept of thematic layers which is an essential component in geographical databases. An additional classification method has been proposed in (Koperski 98). In this case, data are first generalised, then all "attribute = value" are transformed into logic predicates. Such transformations are computationally costly and are limited to a few spatial relationships. In previous work, a two-step solution was implemented to address some of the above shortcomings (Zeitouni 01). The first step computes the spatial join between the target object collection and other themes, while the second step builds a conventional decision tree on the join result. Since spatial criteria are a many-to-many relationship, join operations could result in some target objects being duplicated and give rise to being classified into incorrect classes. As in (Ester 97), the results were shown to be problematic. .

3 The Proposed Method

The proposed classification algorithm is based on two ideas. The first is the utilisation of the spatial join index presented in the section 2.2 and the second is the adaptation of relational data mining methods.

Since the spatial join index formalises neighbourhood links within thematic layers and represents them using relational tables, the classification can directly use the relational schema instead of a predicate set. Indeed, the method uses a target table, the join index tables, and neighbour tables describing other theme attributes. The algorithm details are given in section 3.2 below.

This approach is an extension of the CART method (Breiman 84) that we call SCART and includes the concept of Spatial CART. The information gain is computed using the Twoing expression. The difference with CART is that a node may be partitioned according to a criterion resulting from neighbouring objects, which may have a particular spatial relationship with the target objects. To avoid duplications, the right son of a node is defined as the complement of the left son (*right_son* = *node* – *left_son*). The originality of our method, regardless of relational decision trees, is to precisely qualify the neighbourhood relationship. Thus, computing the information gain combines the neighbours' attributes and their distance or their topological relationships with target objects.

3.1 The Method Concepts

Information gain: This is a measurement used to split a node in the CART algorithm. This measurement relates the gain of class homogeneity in case the node splits according to a particular criterion – such as (*attribute* = *value*) or (*attribute* < *value*). The "best" split would be the one maximising the information gain. A number of formulas exist for information gain such as Gini. The proposed algorithm uses the Twoing indice that is more suitable for multi-class target attributes.

Saturation conditions: These are the criteria under which the node split terminates. Usually the user specifies these criteria.. The node split is stopped when all objects in the node are in the same target attribute class. In this case, the node is referred to as a pure node. It will also stop when no criteria exist that improves the information gain. The other possible criteria may be a minimal occupation of the node, a maximal depth of the tree or a threshold value for the information gain.

Node encoding: Since the decision tree is binary, an encoding technique has been adopted to identify each tree node. The root has a value code of 1. A node code is then defined recursively by:

$$left\_son\_code = 2 * father\_code$$

and

$$right\_son\_code = 2 * father\_code + 1$$

Assignment procedure: A decision tree is a progressive partitioning technique. In the splitting process, objects will be assigned to a left or right son. We propose a virtual representation of partitions by dynamic assignment of target objects to a node (a leaf of the tree). The object will be assigned to the node code.

3.2 Algorithm of a Spatial Decision Tree

The following provides details about the algorithm.

Input parameters:

- Target_table: the analysed objects (i.e. the analysed thematic layer),
- Neighbor_table: thematic layer objects (neighbors of analysed objects),
- Spatial_join_index: the join index table,
- Target_attribute: the attribute to predict (i.e. class labels),
- Predictive_attributes: attributes from a target table or neighbour table that could be used to predict the target attribute,
- Saturation_condition: condition under which the split is considered invalid..

Output:

A binary decision tree

```
Step 1:
Initially, assign all target objects to the root (i.e. to node number 1)
Step 2:
Best_gain = 0
For each predictive_attribute
    For each attribute_value
      If the predictive_attribute belongs to the target_table
              Info_gain = compute information gain        -- as in CART
      Else If the predictive_attribute belongs to neighbor_table
            For each spatial relationship spatRel
                Info_Gain = Compute information gain for the split criterion
```

```
                    "exists neighbours (by mean of spatRel) having such
                    attribute_value".
              If Info_gain > Best_gain
                    Save the split criterion
  -- The retained split criterion is the one maximising the information gain for all
   predictive attributes.
  Step 3:
  If the current leaf is not saturated
        Perform the node split
        Assign its objects to the left son or to the right son accordingly.
  Step 4:
  Replace the current node by the following according to the code number.
  Iterate step 2 and step 3.
  --The algorithm will stop when all the leaves of the lower level are saturated.
```

Fig. 2. SCART Algorithm

SCART is an extension of CART in two ways. First it uses several tables and attributes of a complex relational database. Second, it may combine some attribute values in the split criterion (neighbour predictive attributes and spatial relationships). For simplification, this description is limited here to one neighbour table.

As an example, the target table may be the accident thematic layer; neighbour table may be the building thematic layer; the target attribute may be the accident gravity or the involved category; a predictive attribute may belong to the target table such as speed, or to the neighbour table such as the building category. Note that when the split condition uses an attribute of another thematic layer, the semantics of the partitioning is somewhat different. It means that the existence of neighbouring objects with such neighbourhood relationships, fulfils a condition such as (*attribute compared_to value*) and gives the best information gain along with the best partitioning of the actual node.

4 Implementation and Discussion

This method has been implemented and tested on real data sets for an application in road transport safety. An example of the results is given in Fig. 3. It classifies accidents according to the involved categories (pedestrians, two wheels – bicycles and motorcycles – or others – vehicles –). As shown here, the first split criterion relates the closeness of a "walkway" within a particular distance (100 m). This criterion leads to more pedestrian accident categories. The right son is partitioned again into the left part close to schools where the pedestrian rate increases and the vehicle rate decreases, and conversely for the right son. The third level shows a case where the algorithm chooses a predictive attribute belonging to the target table. The last leaf on the bottom of the tree could be interpreted such as "when accidents are far from "walkways", schools and administration, then they involve fewer vulnerable categories such as pedestrians and two wheels.

This implementation was made using the Oracle 8i DBMS and the Java language environment (see Fig. 4). It allows a first validation of the proposed method. More work is required to validate this method at two levels. An operational level of validation is required and needs domain expert input both for the procedures and for prototyping. Additionally, tests need to be extended to other datasets and other geographical areas. Finally, a performance evaluation and optimisation are necessary especially since large volumes of data may effect the behaviour of the algorithm.

Some optimisation techniques have already been implemented such as the direct object reference (ROWID in Oracle). Other techniques have been considered such as reducing the scan of tables by prior implementation of join operations and database schema transformation. Zhe, (98) explores fast joins and will be assessed in a forthcoming study.

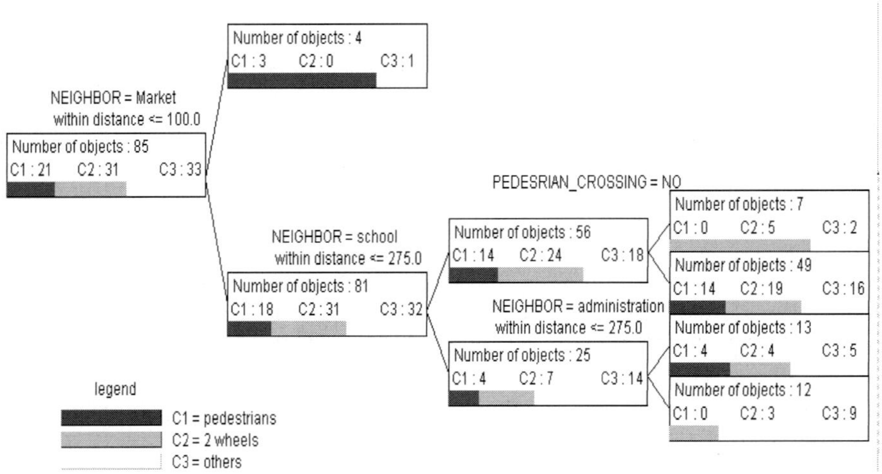

Fig. 3. A spatial decision tree example

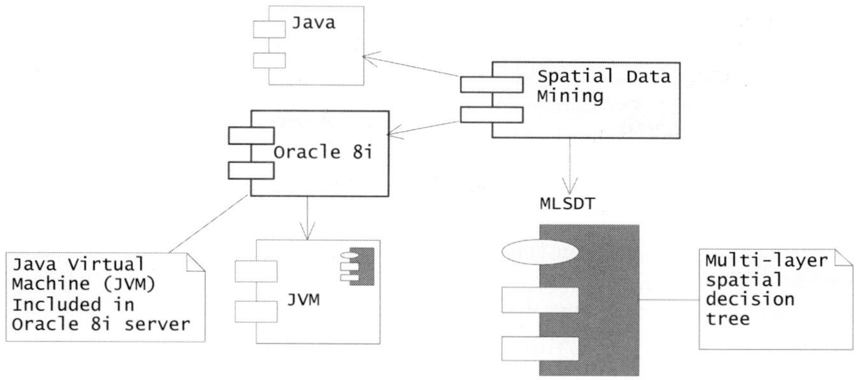

Fig. 4. The software architecture

5 Conclusion and Future Work

This article proposes SCART, a new classification method for spatial data. Two main requirements have been considered in effectively using spatial data in decision trees: multiple layers and automatic filtering, have been briefly assessed and the calculation of neighbourhood relationships have been considered. The intent is to facilitate the use of all spatial relationships through a relational table, and then use and extend the relational data mining methodology.

The spatial join index is merely a correspondence table between two relational tables and in this was corresponds to relational data mining methods. In exploring SCART, it is apparent that it supports the classification of spatial objects according to both their attributes and their neighbours' attributes. It also determines the relevant neighbourhood relationship. Moreover, the organisation of thematic layers has been completely integrated.

From a spatial data mining stand point, the general approach of representation of the spatial relationships as tabular data is very promising. This *a priori* structure could be used in other methods such as spatial clustering or spatial association rules.

Research investigations are required for the algorithmic performance and optimisation, however, other decision tree methods that are disk oriented also require further assessment (Mehta 96), (Gehrke 98). Their application needs to be assessed in an effort to improve the algorithm cost relative to large volumes of data. The second orientation will be the extension to spatio-temporal data and multimedia data that also have complex structures.

References

Breiman L, Friedman JH, Olshen RA, Stone CJ (1984) Classification and Regression Trees. Ed: Wadsworth & Brooks. Montery, California

Egenhofer MJ, Sharma J (1993) Topological Relations Between Regions in R2 and Z2. In: 5th International Symposium, SSD'93. Singapore, Springer-Verlag, pp 316-331
Ester M, Kriegel HP, Sander J (1997) Spatial Data Mining: A Database Approach. In: Proceedings of 5th Symposium on Spatial Databases, Berlin, Germany
Fayyad et al (1996) Advances in Knowledge Discovery and Data Mining. AAAI Press / MIT Press
Gehrke J, Ramakrishnan R, Ganti V (1998) RainForest – A Framework for Fast Decision Tree Construction on Large Datasets. In: Proceedings of the 24 the Annual International Conference on Very Large Data Bases (VLDB). New York, pp 416 - 427
Huguenin-Richard F (2000) Approche géographique des accidents de la circulation : proposition de modes opératoires de diagnostic, application au territoire de la métropole lilloise. Ph.D. thesis, Université de Franche-Comté
Knobbe AJ, Siebes A, Wallen V, Daniel MG (1999) Relational Decision Tree Induction. In: Proceedings of PKDD' 99. Prague, Czech Republic
Koperski K, Han J, Stefanovic N (1998) An Efficient Two-Step Method for Classification of Spatial Data. In: Proceedings of International Symposium on Spatial Data Handling (SDH'98). Vancouver, Canada, pp 45-54
Mehta M, Agrawal R, Rissanen J (1996) SLIQ: A Fast Scalable Classifier for Data Mining. In: Proc. of Int. Conf. On Extending Database Technology (EDBT'96). Avignon, France, pp 18-32
Quinlan JR (1986) Induction of Decision Trees. Machine Learning 1: 82 - 106
Valduriez P (1987) Join indices. ACM Transactions on Database Systems 12(2): 218-246
Zeitouni K (2000) A Survey on Spatial Data Mining Methods Databases and Statistics Point of Views. In: IRMA 2000, Information Resources Management Association International Conference, Data Warehousing and Mining. Anchorage, Alaska.
Zeitouni K, Yeh L, Aufaure MA (2000) Join indices as a tool for spatial data mining. In: International Workshop on Temporal, Spatial and Spatio-Temporal Data Mining, Lecture Notes in Artificial Intelligence, no. 2007. Lyon, France, Springer, pp 102-114
Zeitouni K, Chelghoum N (2001) Spatial Decision Tree - Application to Traffic Risk Analysis. In: ACS/IEEE International Conference on Computer Systems and Applications, Beirut, Lebanon
Zhe L, Kenneth AR (1998) Fast joins using join indices. The VLDB Journal 8:1-24

Computational and Visual Support for Geographical Knowledge Construction: Filling in the Gaps Between *Exploration* and *Explanation*

Mark Gahegan and Boyan Brodaric[1]

GeoVISTA Center, Department of Geography, The Pennsylvania State University, University Park, PA 16802. Tel: +1-814-865-2612, Fax: +1-814-863-7943, mng1@psu.edu, bmb184@psu.edu

Abstract

Although many different types of data mining tools have been developed for geographic analysis, the broader perspective of geographic knowledge discovery—the stages required and their computational support—have been largely overlooked. This paper describes the process of knowledge construction as a number of inter-related activities and the support of these activities in an integrated visual and computational environment, GeoVISTA *Studio*. Results are presented showing examples of each stage in the knowledge construction process and a summary of the inter-relationships between visualisation, computation, representation and reasoning is provided.
Keywords: knowledge discovery, data mining, visualisation, machine learning, abduction

1 Introduction

Despite enormous efforts in quantification, our understanding of many of the Earth's systems remains *non-axiomatic*; the systems are 'open' and consequently it is not possible to deduce all outcomes from known laws and rules. Geographic science must therefore adopt a manner that encourages the creation or uncovering of new knowledge (Baker, 1996; Takatsuka and Gahegan, 2001). For this reason alone—and completely uncoupled from concerns about increasing data volumes—it is vital that knowledge discovery methods can be brought successfully to bear on problems across geography and the wider geosciences.

---

[1] Also from Geological Survey of Canada.

The foundation of this paper is the relationship that knowledge construction and discovery activities have with the different approaches used for scientific inference; it is via an understanding of this relationship that we can categorise the kinds of knowledge that can be discovered or learned (epistemology), and thus begin to comprehend the roles required of domain experts and computational tools in the knowledge discovery process. Building on this foundation, the goal of the work described here is to facilitate the knowledge construction process in geography (or geo-sciences in general), by providing better computational support and closer integration of various exploratory visual and computational methods.

1.1 Background: Data Mining Knowledge Discovery to Date

Many different styles of spatio-temporal knowledge discovery have been proposed (see Roddick and Spiliopoulou, 1999 for a wide-ranging bibliography), from the entirely computational to visually-led methods (e.g. Openshaw *et al.*, 1990; Koperski and Han, 1995; Knorr and Ng, 1996; Ester *et al.*, 1996, 1998; Gahegan *et al.*, 2001). Automated methods currently available concentrate on mining categories, clusters, outliers and other kinds of patterns that might occur in data, rather than on developing these patterns into knowledge structures. Examples are the Self Organising Map (Kohonen, 1997) and the AutoClass (Cheeseman and Stutz, 1996) and AutoClust (Estivill-Castro and Lee, 2001) packages.

Parallel efforts across many disciplines, including statistics, machine learning, databases and information visualisation have emphasized different aspects of the knowledge discovery process (e.g. Agrawal *et al.*, 1993; Gehrke *et al.*, 1999; Glymour *et al.*, 1997; Rainsford and Roddick, 1999; Haslett *et al.*, 1991; MacEachren *et al.*, 1999). Within this spectrum, the various roles played by the expert and the machine differ greatly, sometimes with little thought to the different abilities that each has to offer (Valdez-Perez, 1999). The nature of the results obtained also differ, from probabilistic rules in computational form to deep insights gained by the expert from visual displays that cannot easily be shared or formalized. See Table 1 for a summary.

A specifically geographically-oriented overview of data mining and knowledge discovery has been recently developed as an emerging theme by the *University Consortium on Geographic Information Science* (UCGIS) (Yuan *et al.*, 2001) and is available via http://www.ucgis.org/emerging/. The role played by geographic visualisation in supporting knowledge discovery activities is specifically tackled in a related research initiative by the *International Cartographic Association* (ICA) and reported by Gahegan *et al.* (2001). Section 3 below discusses the tasks and roles in the knowledge discovery process in more detail, but first we begin by examining some of the problems encountered by data mining and knowledge discovery methods applied in the geographical domain.

Table 1. Summary of the different discipline perspectives on data mining and knowledge discovery activities. Contributed by the author to Yuan *et al.* (2001).

	Databases	Statistics	Machine learning	Visualisation
Finding	Association rules	Local pattern analysis and global inferential tests	Neural networks, decision trees	Exploratory visualisation, Visual data mining
Reporting	Rule lists	Summary statistics, significance and power	Likelihood estimation, information gain	A stimulus within the visual domain
Representing	Schema update, metadata	Fitted statistical models, local or global	Rules, graphs, functions,	Shared between the scene and the observer
Validating	Weak significance testing	Statistical significance tests	Learning followed by verification	Human subjects testing.
Optimising	Reducing computationa l complexity	Data reduction and stratified sampling strategies	Stochastic search, gradient ascent methods	Hierarchical and adaptive methods, grand tours

2 Problems With Geographical Knowledge Discovery

2.1 What Constitutes Discovery?

Scientific discovery draws on a wide range of techniques, often simultaneously, in the search for new insights and theories (e.g. Hanson, 1958; Popper, 1959; Langley, 2000). Within this complex process, different forms of inference are required.

Philosophically, the process of actually discovering something new is closely tied to a form of reasoning called *abduction* (Psillos, 2000; Peirce, 1878). Abduction is the simultaneous act of uncovering some structure within the data and producing a hypothesis with which to explain it. The structure uncovered is necessarily unknown at the outset, while the theoretic explanation may be drawn from what is already known or may involve an expansion or a reshaping of existing knowledge. The importance of abductive reasoning in the geosciences is argued convincingly by Baker (1996).

Practically, Fayadd *et al.* (1996) also point to the development of knowledge via a number of stages: data selection, pre-processing, transformation, data mining and interpretation/evaluation that progressively refine a large dataset to the point where it makes sense to propose object structures and relationships.

2.2 Why is Geographical Knowledge Discovery Unique and Difficult?

Geography is an integrative discipline, so data necessarily spans a wide range of perspectives and interests, from the social to the physical and all points in between. Arising from this complex mix of perspectives, and coupled with a growing infrastructure for gathering information, the following problems arise.

1. **Data volume**. Like many disciplines where data mining is applied, geography is rich in data. Knowing which portions of a dataset to analyse, and which to ignore, becomes problematic.
2. **Complexities caused by data gathering and sampling**. Although data are available in increasing volume, it is still often the case that we must resort to surrogates or aggregates for the phenomena of interest, rather than direct measurements (Openshaw, 1984; Yuan *et al.*, 2001).
3. **Complexities caused by local relationships**. Earth systems are so intrinsically interconnected that it is difficult to isolate an analysis conducted on some part of a system from the effects of other unmodelled aspects. The outcome often appears in statistical form as heteroskedasticity.
4. **Complexities associated with the domain itself**. Interesting and relevant signals in data are often entirely hidden by stronger patterns that must first be removed. For example, the cyclic nature of many geographical systems (daily, annual, sunspot) impose a heavy signal on data that can overshadow more localised variance (Roddick and Lees, 2001).
5. **Lack of appropriate methods.** While the existing techniques described in section 1.1 are useful for exploring a dataset, they fail to offer the explicit connection to theory or explanation that characterises abduction (section 2.1). The best that can be managed in a computational setting is a kind of low-level explanation offered in the language of the underlying feature-space, not in a higher form as domain knowledge.
6. **Difficulty in formalising the geographic domain**. There is, as yet, no universally accepted conceptual model of geography (e.g. Goodchild, 1992), and the models that are currently implemented in commercial GIS vary significantly one from another, often in quite fundamental, philosophical ways. This leads to three distinct problems: (**a**) data are often intrinsically non-commensurate, they cannot be directly compared or combined; (**b**) it is difficult to apply formal geographical knowledge to the process of knowledge discovery, since such knowledge is not readily available; (**c**) when new knowledge is uncovered it is difficult to represent that knowledge formally—there is nowhere to put it!

Additional details of some of these problems are presented by Yuan *et al.* (2001) and Miller and Han, (2001). All six of these problems speak to the need to bring domain knowledge to bear on the knowledge discovery process. As described above (point 6), the lack of a formal conceptual model for geographic information, models and processes, presents a formidable barrier to the automation of knowledge discovery and precludes the use of computationally-based abduction. As Psillos (2000) argues: "*The more conceptually adequate a model of abduction becomes, the less computationally tractable it is*".

In the absence of formal mechanisms for representing and applying domain knowledge, many researchers have modified the problem to focus on ways to engage the human as a direct node in the problem-solving process, rather than simply the consumer of the results. Using visually-led approaches, an abductive task is performed collaboratively between the observer and the visualisation. The stimulus to abduction—patterns in the visual displays—are observed as a consequence of the way the data are presented and the way the observer perceives and comprehends them. The simultaneous task of hypothesis generation is also similarly split, the mappings used to visualise the data may imply a hypothesis and an observer may generate one or more theories to explain the observed structure.

An additional problem is that, with few exceptions, the tools currently available operate in isolation, typically performing a single function (e.g. clustering, classification), or providing a view onto the data from a single perspective (e.g. scatterplot, parallel coordinate plot). By doing so, they implicitly assume that problems in science can be isolated to a single conceptual 'plane', which, when correctly understood and represented, can be fixed to form the basis on which further science can be constructed. However, it is often the case that we must experiment across more than one level simultaneously, particularly where complex situations and under-constrained theory present multiple alternatives that must be evaluated (Baker, 1999). Take, for example, the case of eco-regions: to be useful, an eco-region must not only neatly summarise many complex environmental dimensions, it must also form a useful basis for further analysis.

We thus regard *knowledge discovery* or *knowledge construction* within the geospatial sphere as a *developmental process*, with meaning being progressively constructed and refined through a series of pre-processing and interpretative steps (e.g. Fayyad *et al.*, 1996; MacEachren, *et al.*, 1999; Valdez-Perez, 1999; Risch *et al.*, 1997; Ribarsky *et al.*, 1999; Wong, 1999). Current systems lack adequate tools for supporting this process.

3 A Software Laboratory for Knowledge Construction

What we envisage is a computing environment where a user can move seamlessly between exploring data, constructing elements to represent observed structures, applying these structures operationally, assessing their performance and communicating findings. This has led us to construct GeoVISTA *Studio*, which aims to encompass this entire spectrum of activities in an integrated manner (Gahegan *et al.*, 2002). A technical description of *Studio* has been previously reported (Takatsuka and Gahegan, 2001) so will not be repeated here. In short, it is a visual programming environment, which allows users to quickly design, test and refine strategies to explore and analyse geospatial data. Functionality is encapsulated in JavaBeans that support a range of activities, from visualising high dimensional feature spaces, applying neural networks and traditional statistical analysis tools through to mapping outcomes. Examples of the use of *Studio*, and its support for the process of discovery, appear in the following sub-section.

3.1. Scientific Activities in Support of Human-Directed Discovery

Although there is no consensus on any one scientific method, there are several prevalent activities proposed both by philosophers (e.g., Feyerband, 1975; Hanson, 1958; Kuhn, 1962; Popper, 1959) and geographers alike (e.g. Harvey, 1969). A subset of these activities is clearly pertinent to geographical knowledge discovery, beginning with exploratory activities from which concepts are synthesised then woven together into models or theories that can be evaluated and presented to others. Such activities are empirically supported by psychologists (Feist and Gorman, 1998; Zimmerman, 2000), and some are even implemented computationally (Langley, 2000; Shrager and Langley, 1990; Thagard, 1988), though largely without geographical focus.

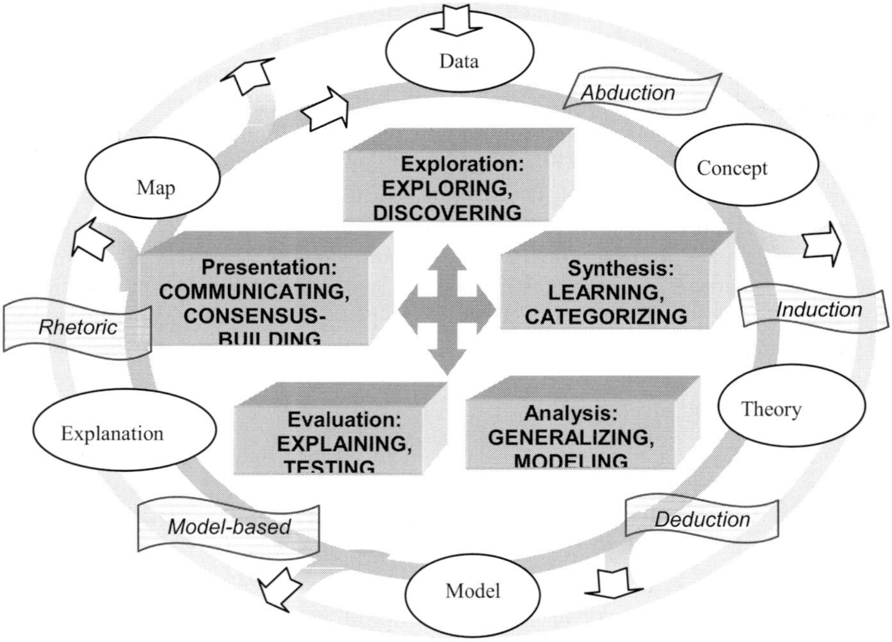

Fig. 1. A generalised framework for geoscientific discovery consisting of exploratory, synthesis, analysis, evaluation, and presentation activities. These respectively correlate with the evolution of features, concepts, theories and models, explanations, and presentations.

Fig. 1 presents an overview of some of the key aspects of the discovery process. It is depicted as a cycle, or a spiral, and this is apt because we can then envisage the negative outcomes causing the cycle to begin again and the positive outcomes (such as a validated theory described by a map) as being then becoming the data that is fed into the next iteration of the cycle at a more abstract level. Thus meaning is constructed in successive layers, each one supported by those below it (Popper, 1959). However, in practice, all stages are connected and may proceed in almost any order and simultaneously, as the connecting boxes in the interior of the diagram attempt to show. Further descriptions of the main stages follow.

Exploration: involves selecting which data, features (i.e. attributes), and feature weightings, are appropriate and significant, based on what is known and the prevailing scientific climate. It includes orienting human and other sensors to an environment by making preliminary observations. These observations can be notoriously subjective, being influenced by existing knowledge and various social pressures; moreover, this activity may be guided by perceptual and other implicit mechanisms that are difficult to express or even be conscious of (Shrager, 1990). Exploration is associated with scientific discovery inasmuch as it provides a stimulus for hypothesis generation and grounds for abduction; i.e. exploration involves selecting the data for which explanations are to be sought (abductively).

Fig. 2 shows two exploratory views provided by *Studio* onto a feature-space of state-level, socio-demographic data of the USA: a dynamic map and a clustering tool (a self-organising neural map or SOM). These and other tools provide the user with a number of different perspectives onto the data, ranging from the geographic distribution of variables (the map), to the clustering of places based on their similarity in feature space (SOM). The user can explore connections between these views via the linking and brushing tools (Cleveland and McGill, 1988; Buja *et al.*, 1996; Hardisty *et al.*, 2001) that *Studio* provides. A number of more traditional methods are also included here, such as the re-ordering of the feature space using correlation analysis and principal components. When used in conjunction with the visualisation tools, these allow the user to discard attributes that offer little or no additional information.

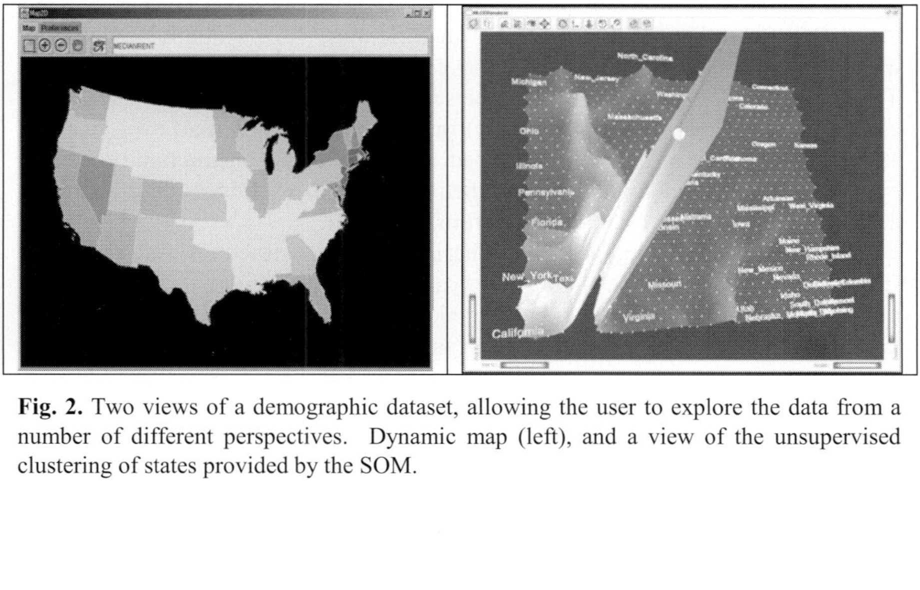

Fig. 2. Two views of a demographic dataset, allowing the user to explore the data from a number of different perspectives. Dynamic map (left), and a view of the unsupervised clustering of states provided by the SOM.

Synthesis: involves building concepts and taxonomies, or revising them, from selected features and data. These taxonomies then form the basis for terminology used in various explanatory structures, such as logical theories, mathematical laws or even text narratives. Synthesis is therefore closely aligned with the process of categorisation (in cognitive science) and classification (in computation), in which concepts (or classes) are explicitly induced from data. Induction is an important mechanism here, as classes are often determined by recognising recurring patterns in select feature dimensions of specific example data.

The following *Studio* examples (Fig. 3) show experiments in the development of categories for describing a complex forest habitat. The Parallel Coordinate Plot (PCP) is used, along with tools that allow user-driven groupings of the data to be defined and imposed. Categories are constructed visually by recolouring ranges of data across one or more dimensions, then synthesised inductively via a Learning Vector Quantisation (LVQ) classifier.

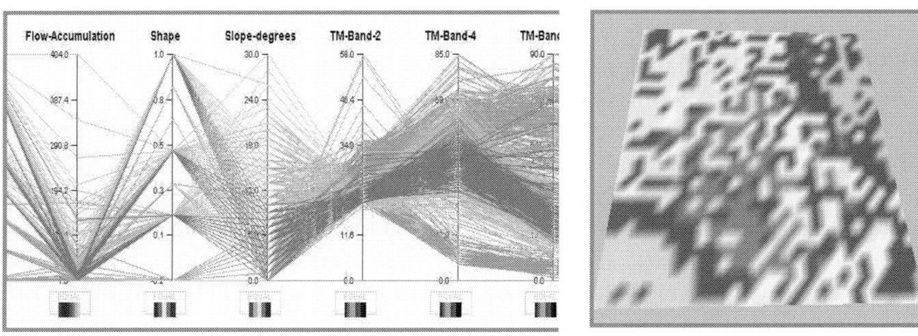

Fig. 3. Five candidate categories constructed in a PCP then learned using LVQ

Analysis: involves, on the one hand, using the established taxonomic framework and given data to develop general explanatory structures for concept behaviour and structure. On the other hand, it involves developing statements about how data are specifically related to each other (e.g. spatially, temporally, thematically, causally, etc.) and to the explanatory structures they exemplify. Taken together, these two aspects can be seen to form a model for the data; which can be either formal (e.g. probabilistic) or informal (e.g. text narrative, Suppes, 1960; 1962). Once a model is developed, the knowledge construction process in complete. Evaluation strategies must then establish its usefulness and reliability, and these strategies are often deduced from the model.

For example, the outcomes of synthesising various structures and categories (from the previous example) into themes can be used in the more traditional forms of GIS analysis (e.g. overlay) to build models of phenomena such as hydrology or landcover change (Fig. 4), and to suggest strategies for their evaluation.

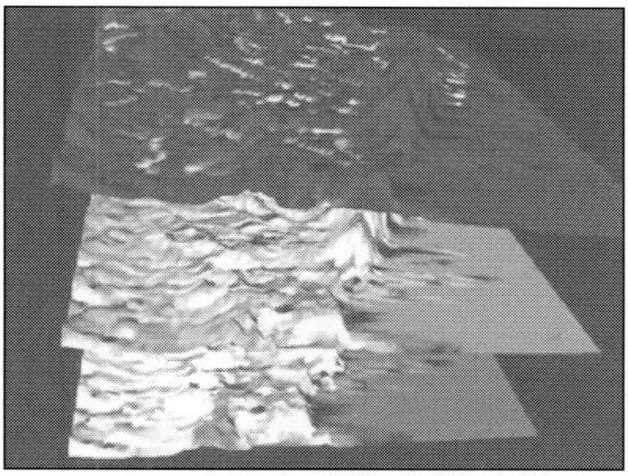

Fig. 4. Outcomes of synthesis form the inputs to analysis

Evaluation: involves testing the developed model against the validation data, or possibly against other models. It specifically requires data regularities to be explained by the model, using standards acceptable within a discipline. Philosophically and logically, unsuccessfully disconfirming the conceptual model is more informative than confirming it (Popper, 1959), a notion exploited by successful scientists who confirm early in their work, while model-building, but seek to disconfirm later as models evolve and settle (Feist & Gorman, 1998). The reasoning performed in evaluation is often model-based in that the behavioural, structural and logical-mathematical aspects of models are first proposed and then used as constraints on reasoning when testing models against the data.

Table 2 shows one form of evaluation where conflicts in meaning within a geological map are quantified; this evaluation can also be portrayed visually.

Table 2. Evaluation of semantic conflict between geologic mappers

Data Type	X	Y	r1	r2	Taxonomic Semantic Conflict (# sites)	Operational Semantic Conflict (# sites)	Total Semantic Conflict (% sites)	m Mean Conflict Distance	r Feature Space Radius	m/r
Structure	4	4	3	1	0	41	2.7 %	86.4	1635.45	0.0528
Lithology	7	7	5	2	2	155	10.61 %	158.38	40910.7	0.0038
Combined	10	10	8	2	1	137	10.43 %	133.50	40911.2	0.0036

Presentation: involves communicating with the external community to build consensus. This explicitly engages the social aspects of doing science (Kuhn, 1962), and possesses deep rhetorical components related to the holistic impact of narrative (Baker 1999; Ricoeur, 1985). Maps, and other visual devices in general, have always been an important knowledge construct in geography, and their rhetorical nature is well documented (e.g. Harley, 1989). Their transformation into digital products provides opportunity for elaborating communication methods between knowledge producers (e.g. map-based group collaboration and decision making systems (Jankowski *et al.*, 2001)), and between producers and consumers (e.g. digital libraries, NRC, 1999).

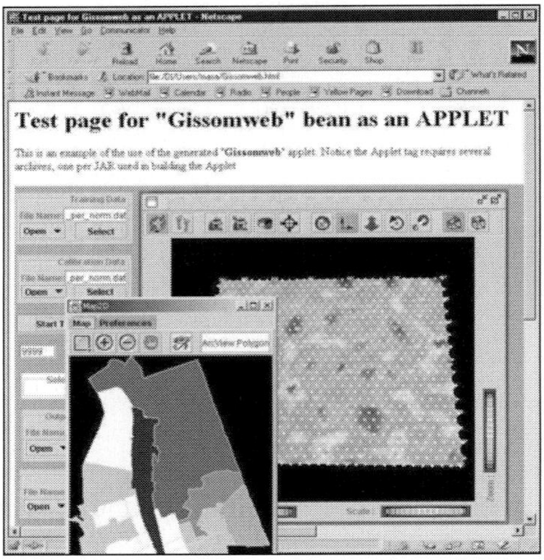

Fig. 5. Java Applet encapsulating the entire analysis of gentrification activities in Harrisburg, PA, that can be readily shared with others (Takatsuka and Gahegan, 2001)

The example in Fig. 5 shows a Java Applet, automatically created by *Studio* for web deployment, directly from an analysis of urban gentrification activities. It contains the entire process by which the analysis was constructed and is readily explorable by other researchers who can then accept or refute the findings.

4 Summary and Conclusions

The different knowledge construction activities are summarised in Table 3 according to the visualisation and computational techniques that can be used to support them via *Studio*. The dominant form of inference employed in each stage is also shown, as are the various representation devices required to support the activities. (The latter are not yet explicitly supported in *Studio* but are the subject of current and future work.)

Table 3. Knowledge discovery activities related to computational, visual, representation and reasoning issues

Scientific Activity	Visualisation	Computation	Representation		Reasoning
			Object	Structure	
Exploration	PCP, scatterplot, iconographic displays	SOM, k-means, clustering methods, GAM	Feature	Dataset	Abductive
Synthesis	Interactive visual classification, PCP	Machine learning, maximum likelihood, decision tree	Concept	Taxonomy	Inductive
Analysis	Scene composition, visual overlay	Statistical analysis	Rule Model	Theory	Deductive
Evaluation	Uncertainty visualisation	Statistical testing, M-C simulation	Inference	Explanation	Model-based
Presentation	Maps, charts, Reports, etc.	Web mapping, digital libraries, multi-media hypermaps, collaboratories	Document (e.g. Map)	Library (e.g. Atlas)	Rhetorical

Although presented in a coherent sequence above, these activities are generally thought to be mutually affective as depicted in Fig. 1. Of particular note is the fact that fundamental knowledge emergence is explicit within exploratory abductive activity. Fig. 1 also summarises the key representation and reasoning elements, which should act as a first order requirements statement for computational scientific systems. In addition to this, and following from the description in Sections 2 and 3, a geographical knowledge construction environment should:

1. Offer a variety of simultaneous views onto the data to gain alternative perspectives.
2. Allow different conceptual structures to be imposed onto the data (such as categories and relations) to prompt the generation of useful hypotheses.
3. Provide quick evaluation strategies for findings to assess their utility and viability in terms of existing theory, and means to revise proposed structures in the light of this evaluation, with little or no inertia.
4. Include higher-level analysis tools with which to formulate models based on knowledge construction outcomes, to test both the utility of the findings, and the correctness of the models themselves.
5. Incorporate tools to extract re-usable knowledge gained then represent it and communicate it to human experts (Gains, 1996).

6. Provide a packaging mechanism so that the entire knowledge discovery process can be shared and independently validated or refuted.

GeoVISTA *Studio* is able to touch on all of these issues to a limited extent at present, but at the time of writing concentrates on computational and visual support for the early discovery activities of exploration and synthesis. As such it can be seen as an environment within which to construct the categories and concepts that existing GIS make such heavy use of (but provide little support for their creation). *Studio* is available for no-cost download from http://www.geovistastudio.psu.edu/jsp/tryit.jsp, and we are actively seeking contributions of functionality (in the form of Java Beans) from other researchers.

Of the many challenges that remain, perhaps most difficult among them is to connect higher level, geographical domain knowledge with the tools used to search for potentially interesting anomalies or regularities within data. This higher level knowledge is required to support the abductive inference that converts data artifacts into useful domain knowledge (Sowa, 1999), and without it the structures uncovered are likely to remain unused. Taking a long-term perspective, it is unclear as to whether abduction might ultimately become fully automated, though such progress is highly unlikely in the short term. The current focus is thus to engage the head-knowledge of the domain expert as effectively as possible, using a variety of visual and computational tools brought to bear in a highly co-coordinated fashion.

Acknowledgements

The support of NSF for this work (under Digital Government grant EIA-9983445) is gratefully acknowledged.

References

Agrawal R, Imielinski T, Swami A (1993) Mining association rules between sets of items in large databases. ACM SIGMOD, pp 207-216

Baker V (1996) Hypotheses and geomorphological reasoning. In: Rhoads, BL and Thorn, CE (eds) The scientific nature of geomorphology. Wiley, New York, pp 57-86

Baker V (1999) Geosemiosis. Geological Society of America Bulletin 111(5): 633-645

Brodaric B, Gahegan M (2001) Learning Geoscience Categories In Situ: Implications for Geographic Knowledge Representation. In: Proceedings, ACM-GIS 2001: The Ninth Association for Computing Machinery International Symposium on Advances in Geographic Information Systems. Atlanta, GA, Nov. 9-10, 2001

Buja A, Cook D, Swayne D (1996) Interactive high-dimensional data visualization. Journal of Computational and Graphical Statistics 5(1):78-99

Cheeseman P, Stutz J (1996) Bayesian Classification: Theory and results. In: Fayyad U, Piatetsky-Shapiro G, Smyth P, Uthurusamy R (eds) Advances in Knowledge Discovery and Data Mining, Cambridge, MA: AAAI/MIT Press, pp 153-189

Cleveland WS, McGill ME (1988) Dynamic Graphics for Statistics. Wadsworth & Brookes/Cole, Belmont, California, USA

Ester M, Kriegel H-P, Sander J, Xu X (1996) A Density Based Algorithm for Discovering Clusters in Large Spatial Databases with Noise. In: Proc. 2nd International Conference on Knowledge Discovery and Data Mining (KDD-96). pp 226-231

Ester M, Kriegel H-P, Sander J (1998) Algorithms for characterization and trend detection in spatial databases. In: Proc. 4th International Conference on Knowledge Discovery and Data Mining (KDD'98). New York, USA, pp 44-50

Estivill-Castro V, Lee I (2001) Argument free clustering for large spatial point-data sets via boundary extraction from Delaunay diagram. Computers, Environment and Urban Systems (in press)

Fayyad U, Piatetsky-Shapiro G, Smyth P (1996) From data mining to knowledge discovery in databases. AI Magazine, Fall 1996, pp 37-54

Feist GJ, Gorman ME (1998) The psychology of science: review and integration of a nascent discipline. Review of general psychology 2(1): 3-47

Feyerband P (1975) Against Method. London: Verso

Gahegan M, Wachowicz M, Harrower M, Rhyne TM (2001) The integration of geographic visualization with knowledge discovery in databases and geocomputation. Cartography and Geographic Information Systems (special issue on the ICA research agenda) 28(1): 29-44

Gahegan M, Takatsuka M, Wheeler M, Hardisty F (2002) GeoVISTA Studio: a geocomputational workbench. Computers, Environment and Urban Systems (in press)

Gains BR (1996) Transforming Rules and Trees into Comprehensible Knowledge Structures. In: Fayyad U, Piatetsky-Shapiro G, Smyth P, Uthurusamy R (eds) Advances in Knowledge Discovery and Data Mining. AAAI/MIT Press, Cambridge, MA

Gehrke J, Ganti V, Ramrakrishnan R, Loh W-Y (1999) BOAT—Optimistic decision tree construction. In: Proc. SIGMOD 1999, ACM Press: New York, pp 169-180

Glymour C, Madigan D, Pregibon D, Smyth P (1997) Statistical themes and lessons for data mining. Journal of Data Mining and Knowledge Discovery 1:11-28

Goodchild MF (1992) Geographical data modeling. Computers and Geosciences 18(4): 401-408

Hanson N (1958) Patterns of discovery. Cambridge University Press, Cambridge

Hardisty F, MacEachren AM, Gahegan M, Takatsuka M, Wheeler M (2001) Cartographic Animation in Three Dimensions: Experimenting with the Scene Graph. In: 20th International Cartographic Conference (Beijing, China, August 6-10, 2001)

Harley JB (1989) Deconstructing the map. Cartographica 26(2): 1-20

Haslett J, Bradley R, Craig P, Unwin A, Wills G (1991) Dynamic graphics for exploring spatial data with application to locating global and local anomalies. The American Statistician 45(3): 234-242

Jankowski P, Andrienko N, Andrienko G (2001) Map-centred exploratory approach to multiple criteria spatial decision making. International Journal of Geographical Science 15(2): 101-127

Kohonen T (1997) Self-organizing maps. Berlin, New York

Knorr EM, Ng RT (1996) Finding aggregate proximity relationships and commonalities in spatial data mining. IEEE Transactions on Knowledge and Data Engineering 8(6): 884-897

Koperski K, Han J (1995) Discovery of Spatial Association Rules in Geographic Information Databases. In: Proc. 4th International Symposium on Large Spatial Databases (SSD95). Maine, pp 47-66

Kuhn TS (1962) The structure of scientific revolutions. University of Chicago Press, Chicago

Langley P (2000) The computational support of scientific discovery. Int. Journal of Human-Computer Studies 53: 393-410

MacEachren AM, Wachowicz M, Edsall R, Haug D, Masters R (1999) Constructing knowledge from multivariate spatio-temporal data: integrating geographical visualization with knowledge discovery in database methods. International Journal of Geographic Information Science 13(4): 311-334

Miller HJ, Han J (2001) Geographic data mining and knowledge discovery: An overview. In: Miller HJ and Han J (eds) Geographic Data Mining and Knowledge Discovery. Taylor and Francis, London, pp 3-32

NRC (1999) National Research Council Distributed Geolibraries: Spatial Information Resources, Summary of Workshop, Washington, DC: National Academy Press

Openshaw S (1984) The modifiable areal unit problem. CATMOG 38, (Norwich: Geo Abstracts)

Openshaw S, Cross A, Charlton M (1990) Building a Prototype Geographical Correlates Machine. International Journal of Geographical Information Systems 4(4): 297-312

Peirce CS (1878) Deduction, induction and hypothesis. Popular Science Monthly 13: 470-482

Popper K (1959) The logic of scientific discovery. Basic Books: New York

Psillos S (2000) Abduction: Between conceptual richness and computational complexity. In: Flach PA, Kakas AC (eds) Abduction and Induction. Dordrecht: Kluwer, pp 59-74

Rainsford CP, Roddick JF (1999) Database issues in knowledge discovery and data mining. Australian Journal of Information Systems 6(2): 101-128

Ribarsky W, Katz J, Holland A (1999) Discovery visualization using fast clustering. IEEE Computer Graphics and Applications. September/October 1999, pp 32-39

Ricoeur P (1985) Time and Narrative. McLaughlin K, Pellauer D (trans.), University of Chicago Press, Chicago

Risch JS, Rex DB, Dawson ST, Walters TB, May RA, Moon BD (1997) The STARLIGHT information visualization system. In: IEEE Proceedings, International Conference on Information Visualization '97. pp 42-49

Roddick JF, Lees B (2001) Paradigms for spatial and spatio-temporal data mining. In: Miller HJ, Han J (eds) Geographic Data Mining and Knowledge Discovery. Taylor and Francis, London, pp 33-50

Roddick JF, Spiliopoulou M (1999) A bibliography of temporal, spatial and spatio-temporal data mining research [online]. SIGKDD Explorations. 1(1) Available from: http://www.cis.unisa.edu.au/~cisjfr/STDMPapers/

Shrager J (1990) Commonsense Perception and the psychology of theory formation. In: Shrager J, Langley P (eds.) Computational Models of Scientific Discovery and Theory Formation. Morgan Kaufman, San Mateo, pp 437-470

Shrager J, Langley P (1990) (eds) Computational Models of Scientific Discovery and Theory Formation. Morgan Kaufman, San Mateo

Sowa JF (1999) Knowledge Representation: Logical, Philosophical, and Computational Foundations. Brooks/Cole, Pacific Grove, CA

Suppes P (1960) A comparison of the meaning and uses of models in mathematics and the empirical sciences. In: Suppes P (ed) Studies in the Methodology and Foundations of Science. Reidel, Dordrecht

Suppes P (1962) Models of Data. In: Nagel E, Suppes P and Tarski A (eds) Logic, Methodology and the Philosophy of Science: proceedings of the 1960 International Congress. Stanford University Press, Stanford, CA, pp 252-61

Takatsuka M, Gahegan M (2001) GeoVISTA Studio: A codeless visual programming environment for geoscientific data analysis and visualization. Computers and Geosciences (In Press)

Thagard P (1988) Computational philosophy of science. MIT Press, Cambridge, Mass

Valdez-Perez RE (1999) Principles of human computer collaboration for knowledge discovery in science. Artificial Intelligence 107(2): 335-346

Wong PC (1999) Visual data mining. IEEE Computer Graphics and Applications 19(5): 20-21

Yuan M, Buttenfield B, Gahegan M, Miller H (2001) Geospatial Data Mining and Knowledge Discovery [online]. A UCGIS White Paper on Emergent Research Themes. Available from: http://www.ucgis.org/emerging/

Zimmerman C (2000) The development of scientific reasoning skills. Developmental Review 20: 99-149

Polygonization of Point Clusters through Cluster Boundary Extraction for Geographical Data Mining

Ickjai Lee and Vladimir Estivill-Castro

School of Electrical Engineering and Computer Science, The University of Newcastle, NSW 2308, Australia, ijlee@cs.newcastle.edu.au, vlad@cs.newcastle.edu.au

Abstract

Interpretability and usability of clustering results are of fundamental importance. A linear time method for transforming point clusters into polygons is explored. This method automatically translates a point data layer into a space filling layer where clusters are identified as some of the resulting regions. The method is based on robustly identifying cluster boundaries in point data. The cluster polygonization process analyses the distribution of intra-cluster edges and the distribution of inter-cluster edges in Delaunay Triangulations. It approximates shapes of clusters and suggests polygons of clusters. The method can then be applied to display choropleth maps of point data without a reference map or to identify associations in the spatial dimension for geographical data mining.
Keywords: clustering, Delaunay triangulation, geographical data mining, cluster boundaries, cluster polygonization

1 Introduction

Geographic Information Systems (GIS) analyse real world phenomena with many data layers. Each layer captures some unique feature. Fast data gathering process results in data-rich environments with many layers that exceed the capability of human analysis (Miller and Han, 2001). Thus, sophisticated geographical data mining tools become necessary for handling hundreds of different themes that may contain thousands of data points. Among many other data mining techniques, clustering is one of the most popular and frequently used approaches for finding undetected or unexpected patterns of spatial concentrations residing in large databases. That is, clustering provides answers for "where" and suggests leads into "why" for post-clustering explorations. Thus, clustering is seen as a starting point for a series of knowledge discovery processes using extensive spatial databases. However, post-clustering processes have attracted less attention than clustering itself despite their importance.

Identifying shapes (boundaries) of clusters is an intuitive way of reasoning about clusters. If shapes of clusters of a set P of points match with a particular feature data, then clusters exhibit a high association with the feature. Generally however, it is difficult to extract the shape and as yet, there is no method that can be considered as an absolute winner since most approaches are application-dependent (Okabe *et al.*, 1999). The convex hull $CH(P)$ of P is one of the simplest methods to extract a shape. Since $CH(P)$ is the smallest convex set containing P, it is an appropriate choice. In addition, it is unique for P. This is useful for data mining where parameter tuning is laborious and time consuming. It is too crude a method however, to capture shape details about P unless the points are all arranged in a convex shape (in which , the vertices of $CH(P)$ coincide with P, *but do not help in summarising P).*). The α-shape (Edelsbrunner *et al.*, 1983) (this is a generalisation of $CH(P)$) overcomes the crudeness of $CH(P)$. The value of the real number α controls the desired level of detail. The set of α values leads to a whole family of shapes capturing from crude to fine shapes of P. Boundary Shape Matching (BSM) (Knorr *et al.*, 1997) uses the α-shape to explore associations among layers. However, several problems arise when the α-shape is used for detecting spatial cluster boundaries in data-rich environments.

1. It is difficult to determine the best value α that produces neither too crude nor too fine a shape (Knorr *et al.*, 1997).
2. Several trial-and-error steps for tuning the value of α are necessary. In data-rich environments, this is laborious and time consuming.
3. If P contains i clusters, then we need to tune α for as many as i times to find the best shape of each cluster. Since clusters in P require different values for α, the independent manipulation of values of α for clusters belonging to the same layer not only increases required time but incorporates human bias.
4. The α-shape of a cluster $c \subset P$ is not affected by the distribution of points $p \in P \backslash c$. Clusters are truly the result of the characteristics in the entire distribution sampled by P. Thus, although a point $p \in P \backslash c$ does not belong to the cluster c, it has some effect on the shape of c.

In this paper, we propose an automatic boundary extraction process for clusters in point data and extend it to cluster polygonization. In contrast to the α-shape approach, our approach minimises the need for parameter tuning. Instead, it derives the shape of a cluster c from both the spread of points and the spread of points in $P \backslash c$. The proposed argument-free approach is well-suited for data-rich environments. It approximates shapes of clusters using the Delaunay Triangulation. Once the Delaunay Triangulation is constructed, the boundary extraction process requires $O(n)$ time. The algorithm can be used as post-processing for Short-Long clustering (Estivill-Castro and Lee, 2000). Short-Long clustering is based on the Delaunay Triangulation as an underlying graph and detects various types of spatially aggregated concentrations.

The remainder of this paper is organised as follows. Section 2 details our boundary extraction and polygonization processes and analyses their complexity. We discuss experimental results with both synthetic and real datasets in Section 3. Section 4 provides three applications of cluster polygonization. Finally, the last section draws concluding remarks.

2 Extracting Cluster Boundaries

Clustering quality is of central concern when clustering is used in mining for associations. Although several spatial clustering methods have been proposed within the data mining and the GIS disciplines, most peak-inference clustering methods are based on global thresholds (Ester *et al.*, 1996; Openshaw, 1987). Thus, these approaches fail to detect some spatially interesting groups.

Fig. 1 illustrates one such example. This example exhibits regions with data concentration in arbitrary shapes. These regions can be interpreted as clusters of different densities. In particular, a region ("8-like" shape) surrounding two high density regions ("ball-like" shapes) shows higher density in the study region. However, peak-inference clustering may be unable to detect this distinction without careful and laborious tuning of the global thresholds. This is because the "8-like" shape is nearby the spherical clusters. When considered as clusters, the intra-cluster distance of the sparse "8-like" shape is greater than the inter-cluster distance between the "8-like" shape and the high density "ball-like" regions. Thus, peak-inference clustering methods report three high density clusters (2 balls and the helix region at the bottom) when global parameters are set for high density regions (refer to Fig. 1(a)). Alternatively, the parameters may be set for low density regions, and in such settings the output is two clusters: one large cluster (for the merging of the "8-like" region and the two "ball-like" regions) and one high density cluster for the helix region below (refer to Fig. 1(b)). The analyst would have to contrast these two results to infer the "8-like" region that is also of interest, and only after consuming time on parameter tuning and experimentation. Fig. 1(a) and (b) demonstrate this effect with DBSCAN (Ester *et al.*, 1996). The DBSCAN approach requires two global thresholds *MinPts* and *Eps*. The value of *Eps* defines a spherical neighbourhood for each point. Fixing the value of *MinPts* to 4 and then tuning the other threshold *Eps* as originally suggested for the method, we see that DBSCAN detects three high density clusters with *Eps* = 50, but leaves the sparse "8-like" region undetected as depicted in Fig. 1(a). As we increase the value of *Eps*, some points within the "8-like" are merged with the "ball-like" clusters, breaking the "8-like" shape into many meaningless sub-clusters. Finally, when the value of *Eps* is large enough so all the points within the "8-like" shape are in the same cluster, then the "8-like" and the "ball-like" shapes are all merged into a single cluster as shown in Fig. 1(b). Since this unsatisfactory behaviour occurs with this simple visual example, we cannot expect peak-inference clustering to be effective in scenarios that are more complex. Clustering methods by partitioning (Ng and Han, 1994) utilise some objective function to find spatial groups. Although these methods are suitable for facility location

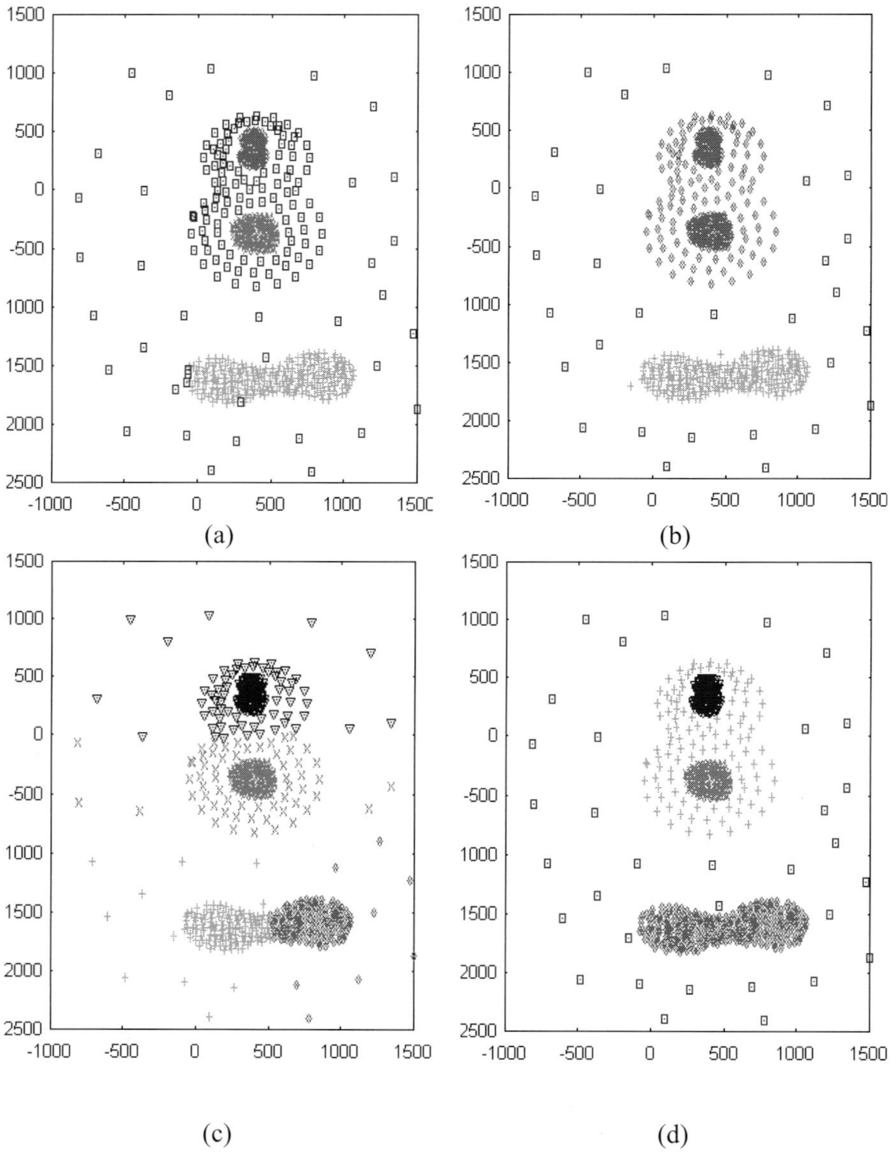

Fig. 1. Clustering with different approaches with 600 data points (Cluster-1: θ, Cluster-2: +, Cluster-3: ×, Cluster-4: ∇, Noise: ⊡), **(a)** DBSCAN with *Eps* = 50 and *MinPts* = 4 (3 clusters), **(b)** DBSCAN with *Eps* = 150 and *MinPts* = 4 (2 clusters), **(c)** Partitioning clustering approaches (4 clusters), **(d)** Short-Long clustering (4 clusters)

problems, their results are convex clusters. Thus, they are less informative for cluster reasoning. Fig. 1(c) shows 4 clusters detected by medoid-basedpartitioning. Not only is the “8-like” region assigned to two clusters, but there seems to be a degree of heterogeneity in the density of the proposed clusters.

Recently, the Short-Long criteria for edge analysis in proximity graphs result in a clustering method (Estivill-Castro and Lee, 2000) overcoming drawbacks of traditional clustering methods. This approach detects possible boundaries of clusters, and as such, it is able to identify various types of spatial concentrations. Fig. 1(d) demonstrates that the Short-Long criteria find the four regions of similar concentration, despite the complexity of the “8-like” shape around the balls. Clustering for massive datasets should remain efficient and effective while minimising the number of user-supplied arguments for clustering. Efficient clustering algorithms that demand tuning of several user-supplied arguments for their best result remain unsuitable for mining vast amounts of data. Finding best values for arguments is expensive in terms of time efficiency since this necessitates several trial-and-error steps or pre-processing. Short-Long clustering is a solid candidate for data mining and post-clustering analysis, since it needs $O(n\log n)$ time, produces quality clusters and requires minimum tuning of only one parameter. For these reasons, Short-Long clustering is used as a basis for our post-clustering analysis in this paper. However, our method for polygonization extends in a straightforward manner to other clustering approaches.

2.1 The Short-Long Clustering Criteria

Short-Long clustering belongs to the family of graph-based clustering since it is based on the Delaunay Triangulation $DT(P)$ of P. In graph-based clustering, we first build a proximity graph G. The graph has node-set $N(G)$ and edge-set $E(G)$. An undirected edge $e \in E(G)$ has associated with it a set of two nodes $\{v, w\}$ while a directed edge has associated with it a pair of nodes $(v, w) \in N(G) \times N(G)$. Here, nodes represent data points and edges connect pairs of nodes to model spatial proximity, interaction or adjacency. Clustering operations are then performed on the graph. Graph-based clustering removes edges connecting points in different clusters based on a certain criterion function and computes connected components of the remaining graph $G' \subset G$ to represent clusters.

For every node $v_i \in N(DT(P))$, Short-Long clustering classifies incident edges into globally short, long or other edges. Then, it removes globally long edges when analysis reveals they are inter-cluster links and eliminates short edges and other edges when analysis reveals they are links (chains or bridges) between clusters. The notions of $ShortEdges(v_i)$, $LongEdges(v_i)$ and $OtherEdges(v_i)$ for edges incident to a node v_i are defined as follows.

$$ShortEdges(v_i) = \{ e_{vi,vj} : |e_{vi,vj}| < LocalMean(v_i) \ \ m \times MeanStDev(P)\}, \quad (1)$$

$$LongEdges(v_i) = \{ e_{vi,vj} : |e_{vi,vj}| > LocalMean(v_i) + m \times MeanStDev(P)\}, \quad (2)$$

$$OtherEdges(v_i) = AdjEdges(v_i) - (LongEdges(v_i) \cup ShortEdges(v_i)), \quad (3)$$

where $AdjEdges(v_i)$ denotes the set of edges incident to v_i in $DT(P)$, $LocalMean(v_i)$ is the mean length of $AdjEdges(v_i)$, $MeanStDev(P)$ is the average of the standard deviation of lengths of edges in $AdjEdges(v_i)$, and m is a control value. Criteria for $ShortEdges(v_i)$, $LongEdges(v_i)$ and $OtherEdges(v_i)$ are not static, but rather dynamic over the study region, since $LocalMean(v_i)$ varies with v_i. This dynamic nature of the Short-Long criteria overcomes the static nature of peak-inference clustering and partition-based clustering methods.

2.2 Cluster Polygonization Algorithm

Receiving a set of points with cluster identifier for each point is not enough for post-clustering analysis. We need to obtain the boundaries of those clusters. There are many possibilities for where the boundary might be, and following is a fast and accurate way of resolving this problem. The following technique is considered to be better than computing convex hulls or α-shapes for mining large spatial databases. In the following, the assumption is made that a dataset P is provided and each point $p \in P$ is labelled with a cluster identifier $ID(p)$. This is the input to the entire process. An illustration of a cluster polygonization process appears in Fig. 2.

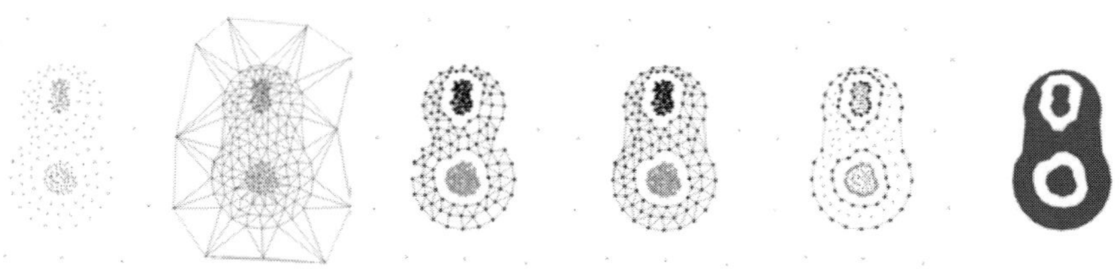

(a) (b) (c) (d) (e) (f)

Fig. 2. Polygonization through cluster boundary extraction ($\|P\| = 250$), **(a)** A set P of points, **(b)** $DT(P)$, **(c)** 3 clusters found by Short-Long clustering, **(d)** Intra-cluster edges after Phase 1 (inter-cluster edges are not shown), **(e)** After Phase 2 (the boundary extraction process), **(f)** After Phase 3 (the polygonization process)

Fig. 2(c) depicts clusters after Short-Long clustering. Fig. 2(d) displays only intra-cluster edges that approximate shapes of clusters (note that, inter-cluster edges are ignored).. Directed cluster boundaries that represent cluster regions are illustrated in Fig. 2(e). Sets of cyclic lists of edges, that are in counterclockwise ordering, are external cluster boundaries and definine inner areas as cluster regions, while sets of cyclic lists of edges, that are in clockwise ordering, are internal cluster boundaries defining holes within corresponding cluster regions. Fig. 2(f) depicts derived cluster regions. Details of the cluster polygonization process are as follows. Phase 0 will compute the Delaunay Triangulation $DT(P)$ of P (this phase is omitted when using Short-Long clustering). Since the Delaunay Triangulation is a planar graph embedded in the plane, in what follows we may refer to points $p \in P$ as nodes, and to edges $e = \{p_a, p_b\}$ as sides of the Delaunay triangles. Phase 1 labels each Delaunay edge in $DT(P)$ with either intra-cluster edge or inter-cluster edge based on cluster identifiers provided by a clustering method (in this case Short-Long clustering). Intra-cluster edges are those that expand endpoints with the same cluster identifier while inter-cluster edges are those that connect endpoints in different clusters. That is, two endpoints of intra-cluster edge $e = \{p_a, p_b\}$ satisfy $ID(p_a)$ equals to $ID(p_b)$. Phase 2 extracts and orients boundary edges. Each intra-cluster edge e is analysed. Because we have a triangulation (a planar subdivision into triangles), each edge in $DT(P)$ belongs to at least one Delaunay triangle or at most to two in $DT(P)$. For each intra-cluster edge e after Phase 1, the intra-cluster edge analysis proceeds as follows. Let the edge $e = \{p_a, p_b\}$ be an intra-cluster edge in $DT(P)$.

- Simple case: The intra-cluster edge $e = \{p_a, p_b\}$ is in only one Delaunay triangle (external Delaunay edge or hull edge (Okabe *et al.*, 1999)).

  1. Sub-case: Triangle in a cluster --- The third node p_c of the triangle $<p_a, p_b, p_c>$ has the same cluster identifier as nodes p_a and p_b. Then, edge e is labelled as a boundary edge and oriented (because $DT(P)$ is a planar embedding) such that p_c is on its left (the interior of the triangle) and the exterior is on its right. Fig. 3(a) illustrates this sub-case.
  2. Sub-case: No triangle --- The third node p_c of the triangle $<p_a, p_b, p_c>$ does not have the same cluster identifier as p_a and p_b. Then, e is not labelled as a boundary edge (it will be removed). That is, e is not placed in the output of this phase because there is no area (region or triangle) of the cluster of p_a and p_b delimited by e. Fig. 3(b) illustrates this sub-case.

- Complex case: The edge $e = \{p_a, p_b\}$ is shared by two Delaunay triangles with opposing nodes p_c and p_d (internal Delaunay edge (Okabe *et al.*, 1999)).

  1. Sub-case: Two triangles in a cluster --- The third nodes p_c and p_d of triangles $<p_a, p_b, p_c>$ and $<p_a, p_b, p_d>$ are both in the same connected component (cluster) as p_a and p_b. It follows that e is removed and it will not be included in the output as a boundary edge since clearly it is inside the quadrilateral $<p_a, p_c, p_b, p_d>$. Of course, his quadrilateral is inside the cluster. Fig. 4(a) illustrates this subcase.
  2. Sub-case: Triangle in a cluster --- One of the third nodes p_c or p_d is in the cluster of p_a and p_b but the other is not. Without loss of generality, we assume p_c is in the cluster of p_a and p_b (recall that $ID(p_a) = ID(p_b)$) while p_d is not (i.e. $ID(p_d) \neq ID(p_a)$). Then, edge e is a boundary edge since the triangle $<p_a, p_b, p_c>$ is in the cluster but the triangle $<p_a, p_b, p_d>$ is not. Then, the boundary edge e is oriented (because $DT(P)$ is a planar embedding) in such a way that p_c is on its left (the interior) and p_d (the exterior) is on its right. Fig. 4(b) and (c) illustrate this sub-case.
  3. Sub-case: No triangle --- Both, p_c and p_d are not in the same cluster as p_a and p_b. Again e is removed. There is no area of the cluster of p_a and p_b, thus, e can not be in the boundary. Fig. 4(d) illustrates this sub-case.

(a) (b)

Fig. 3. The two sub-cases of the simple case with $p_a = 1$, $p_b = 2$ and $p_c = 3$ (only intra-cluster edges are drawn), **(a)** triangle in a cluster, **(b)** no triangle

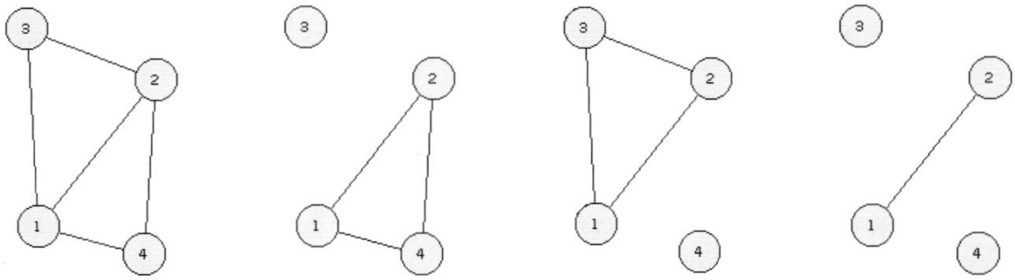

(a) (b) (c) (d)

Fig. 4. The sub-cases of the complex case. Here $p_a = 1$, $p_b = 2$, $p_c = 3$ and $p_d = 4$ (only intra-cluster edges are shown), **(a)** illustration of the first complex sub-case, two triangles in a cluster, **(b)** and **(c)** are the second sub-case triangle in a cluster, **(d)** no triangle.

After the intra-cluster edge analysis, the output of this phase is a set of lists of oriented edges in no particular order (see Fig. 2(e)). Phase 3 (polygonization) constructs a list of edges such that traversing the list of edges corresponds to navigating along the boundary of a cluster. Because of the orientation of the previous phase, the interior of the polygon will be on the left while the exterior will be on the right. The list will be circular, eventually returning to the same node. A polygon with holes will be made of several of these lists. We now prove that cluster polygonization is always possible from the output of Phase 2.

Lemma 2.2.1 *For every node p_a attached to an oriented edge e in the output of Phase 2, the following holds.*

- The degree of p_a is even.
- The *in*degree of p_a equals the *out*degree of p_a.
- It is possible to alternate the incoming edges and the outgoing edges either clockwise or counterclockwise in the planar embedding representing the output of Phase 2.

Proof. Because we started from a triangulation and the output of Phase 2 is edges who belong to one and exactly one triangle in the cluster, the output of Phase 2 is equivalent to a union of disjoint triangles (the triangles share edges but not their interiors). Note that, when Phase 2 removes intra-cluster edges, it performs the union of two regions that are the union of triangles and thus the new region is the union of triangles. The point p_a must belong to a triangle in the union because it is attached to an oriented edge e. Let T be the sequence of triangles clockwise around the point p_a. This sequence T of triangles corresponds to a sequence of edges E. These edges are all incident to p_a with e oriented. Without loss of generality, assume e is an incoming edge and that the sequence $E = \langle e, e_1, \ldots, e_k \rangle$ is the sequence of edges in the triangulation incident to p_a when travelling clockwise around p_a. Because e is incoming, the triangle determined by $\langle e, p_a, e_1 \rangle$ must be exterior. If e_1 is oriented, then it must be outgoing and the triangle $\langle e_1, p_a, e_2 \rangle$ must be interior. If e_1 is not oriented, it is because the next triangle $\langle e_1, p_a, e_2 \rangle$ is exterior. Continuing in this way, we see that the triangles in T alternate between a few that are interior and a few that are exterior. In any case, when the triangles swap from interior to exterior, the edge incident to p_a is incoming and when they swap back the edge must be outgoing. The sequence of triangles around node $p_{a,,}$ however is finite, so when completing the clockwise traversal we see that the lemma is satisfied.

The previous lemma proves that the graph after Phase 2 is Eulerian and it is now a matter of traversing the graph within the planar embedding to extract the circular (cyclic) lists of edges (equivalently nodes) that constitute the polygons.

2.3 Polygonization Requires Linear Time

For a set P of two dimensional points, the number of edges in $DT(P)$ is linear in P (Okabe *et al.*, 1999). Thus, storing $DT(P)$ requires linear space. To compute the connected components of a graph is linear in the sum of both number of edges and number of nodes. Thus, Phase 1 requires $O(n)$ time where n is the number of nodes in $DT(P)$. The boundary extraction process in Phase 2 tests if every intra-cluster edge e is a boundary edge. It performs constant work for each edge. Thus, this is linear as well. Therefore, polygonization of clusters requires linear time for Short-Long clustering. However, if other clustering methods do not use the Delaunay Triangulation as an underlying proximity graph, then we need to compute $DT(P)$, which requires $O(n\log n)$ time.

3. Performance Evaluation

We present results from experiments with synthetic datasets and real datasets that illustrate the robustness of our approach. In Short-Long clustering, we use $m = 1$ as the default control value for all datasets. In addition, in all experiments, clusters whose sizes are less than 1% of the total number of points are considered as noise.

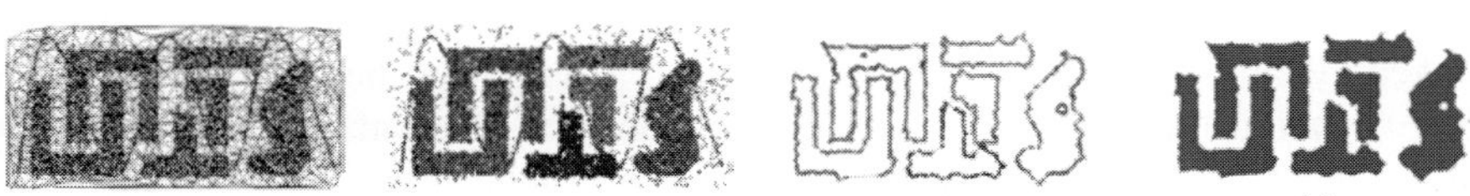

(a) (b) (c) (d)

Fig. 5. Synthetic dataset I ($||P|| = 8000$), **(a)** points with $DT(P)$, **(b)** 6 clusters, (c) cluster boundaries, **(d)** cluster regions

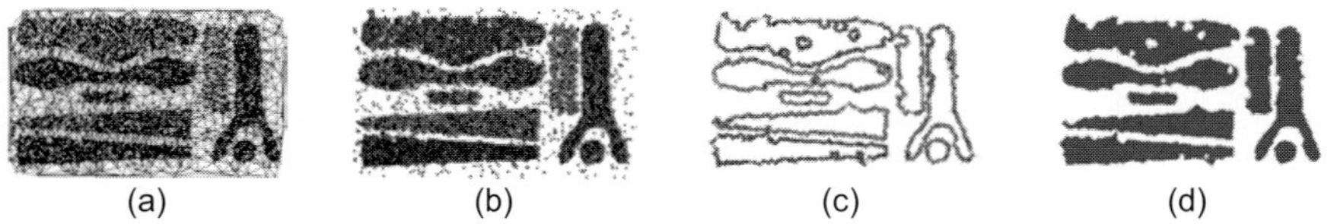

Fig. 6. Synthetic dataset II ($||P|| = 8000$), **(a)** points with $DT(P)$, **(b)** 8 clusters, **(c)** cluster boundaries, **(d)** cluster regions

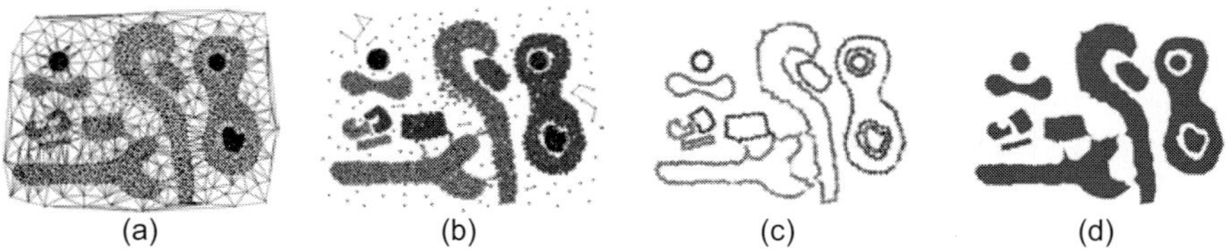

Fig. 7. Synthetic dataset III ($||P|| = 8000$), **(a)** points with $DT(P)$, **(b)** 12 clusters, **(c)** cluster boundaries, **(d)** cluster regions

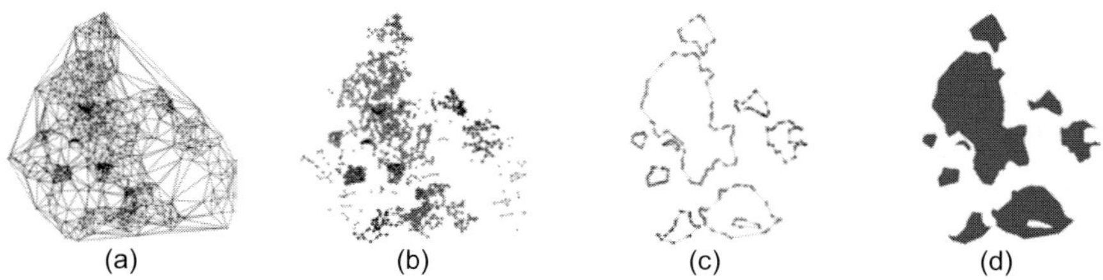

Fig. 8. Real Dataset representing sexual offences ($||P|| = 1695$), **(a)** points with $DT(P)$, **(b)** 9 clusters, **(c)** cluster boundaries, **(d)** cluster regions

Fig. 5 and Fig. 6 illustrate our approach applied to two datasets of the CHAMELEON's benchmark suite (Karypis *et al.*, 1999). Cluster boundaries are successfully derived. Fig. 5(c) depicts boundaries for the first dataset and Fig. 6(c) for the second set, respectively. Boundaries reveal shapes of clusters more easily than human inspection of the dataset or of the clustered data. Cluster regions depicted in Fig. 5(d) and Fig. 6(d) reveal holes within clusters. Note that, visual inspection is insufficient to find holes in the regions of clusters, when presented simply as clusters (Fig. 5(b) and Fig. 6(b)). However, the holes become visually apparent with polygonization (Fig. 5(d) and Fig. 6(d)). Fig. 7 depicts a dataset containing many heterogeneous clusters: small and large clusters, non-convex clusters, clusters with heterogeneous densities, clusters inside clusters and clusters linked by multiple bridges. Cluster boundaries and regions shown in Fig. 7(c) and Fig. 7(d) illustrate the robustness of our approach. Real datasets representing geographical phenomena are more complex than synthetic datasets, thus it is more difficult to extract cluster boundaries. A real dataset shown in Fig. 8 indicates locations of sexual offences that occurred in 1997 around urban areas of Queensland,Australia. A large cluster is spread around Brisbane, the capital city of Queensland and several crime concentrations are discovered around the suburbs of Brisbane. Although shapes of clusters are heterogeneous, irregular and complex, our approach provides cluster boundaries for crime cluster regions. Altogether, these experimental results demonstrate that our approach effectively approximates shapes of clusters.

4 Applications

4.1 Choropleth Mapping with Cluster Boundaries

One of the difficulties of presenting point data is to manage privacy. Another problem with point data is that theoretically, the probability of an event at a point is zero. To resolve the privacy problem or to make inferences on regions rather than locations, point data are often aggregated based on polygonal maps. For example, one way of presenting discrete point data is to use choropleth maps. These maps display point data related to a specific topic with respect to a boundary map such as political or administrative area map. The display of a choropleth map fills polygons of the reference areas with colours or gray tones according to densities (the number of points per unit of area). One key aspect about choropleth mapping is that its visual presentation is dependent on the base map used for plotting. Dent (1999) warns that distributions of continuous geographical phenomena are not governed by political or administrative subdivisions. Thus, choropleth mapping of points with political or administrative layers is less informative. Using the set of cluster regions of *P* as a basic area map overcomes the problem of traditional choropleth maps, since boundaries of cluster regions are derived from the distribution of *P*. Thus, it minimises artificial constraints on choropleth mappings, which is important in exploratory spatial analysis.

(a) (b) (c) (d)

Fig. 9. Choropleth maps with cluster regions, **(a)** synthetic dataset I, **(b)** synthetic dataset II, **(c)** Synthetic Dataset III, **(d)** real dataset

Fig. 9 displays choropleth maps for the datasets discussed in Section 3. Here, the densest cluster is in black (RGB(0,0,0)) and the sparest cluster is in light gray (RGB(200,200,200)). Intermediate clusters are shaded with gray tones in proportion to their respective densities. Visual inspection of the original datasets (see for example Fig. 5) is insufficient to report relatively dense or sparse clusters. In Fig. 5, one can hardly see any difference in density. The datasets seem too large for such visual inspection. However, the choropleth map in Fig. 9(a) clearly indicates that two clusters on the left-hand side are relatively dense. A pattern that shows density decreasing from left to right is now clearly visible. By contrast, Fig. 9(b) indicates that density among clusters does not have a global pattern. The denser clusters of synthetic dataset II are randomly mixed with other clusters. Since this spread is not easily recognised from Fig. 6, the corresponding choropleth map is more informative. Fig. 9(c) shows that clusters are either high density or low density. Thus, we can easily find three high concentrations. Fig. 9(d) reveals that most densities of clusters are about the same although there are substantial discrepancies in the cluster sizes.

4.2 Cluster Associations

Clustering for data mining is to summarise the distribution of *P* in an effort to suggest a manageable number of patterns of concentrations for further explorations. Thus, clusters are representatives of the phenomena recorded with locations in *P* and suggesting interesting groups. Several approaches (Knorr *et al.*, 1997; Estivill-Castro and Murray, 1998) have been proposed to detect associations among geographical layers using clusters in data-rich environments. Boundaries of clusters and representatives of clusters (medoids or means) are the most popular candidates for reasoning with clusters. However, these candidates are summaries. They constitute only partial information about clusters, thus we need special care when we use these information when mining for associations.

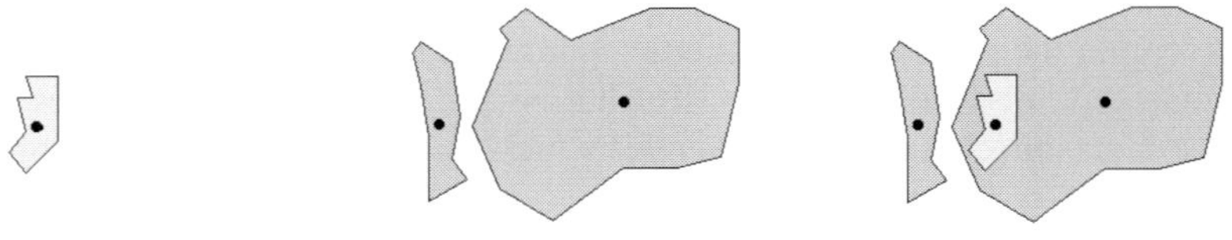

(a) (b) (c)

Fig. 10. Cluster associations, **(a)** layer 1, **(b)** layer 2, **(c)** overlay of layer 1 and layer 2

Fig. 10 illustrates the problem of using such partial information. Consider two layers shown in Fig. 10(a) and Fig. 10(b). Layer 1 displayed in Fig. 10(a) has a small cluster while the Layer 2 depicted in Fig. 10(b) has two clusters small and large. Black dots represent medoids of clusters. Although the small cluster in Layer 1 has a high association with the large cluster in Layer 2 (since the small cluster lies within the large cluster), traditional approaches fail to detect this association. Association analysis using boundary matching (Knorr *et al.*, 1997) does not succeed in detecting this correlation, since boundaries of the small cluster in Layer 1 do not match with those of the large cluster in Layer 2. This aspect is shown in Fig. 10(c). Similarly, association analysis using medoids (Estivill-Castro and Murray, 1998) is unable to report this association, since the medoid of the small cluster in Layer 1 is closer to the medoid of the small cluster in Layer 2 than that of the large cluster in Layer 2. We can detect this relationship by computing intersection area of cluster regions. The intuition of this approach is that two clusters exhibit a high association if most phenomena in Layer 1 take place where concentrations of phenomena in Layer 2 occur. The small cluster in Layer 1 intersects with the large cluster in Layer 2, while it does not intersect with the small cluster in Layer 2. Thus, it is more associated with the large cluster in Layer 2 rather than the small one.

5 Final Remarks

Clustering methods are becoming more sophisticated and effective. Post-clustering processes that seek to identify possible lures (positive associations) are now in demand. Detection of cluster boundaries is a natural choice for reasoning about clusters such as matching boundaries with various feature data, polygonizing clusters and mining for associations. We propose an automatic cluster boundary extraction method that is well-suited for data-rich environments. In this approach, shapes of clusters are governed by data, not by users. The automatic approach derives the shape of a cluster not only from the distribution of points in the cluster, but also from that of points in different clusters. Thus, this approach is more informative and less biased since it is data-driven, which is very important in exploratory spatial data analysis (Openshaw, 1987). The cluster boundary extraction has been extended to cluster polygonization that has potential as a tool for spatial analysis and mining. Choropleth mapping with cluster regions as a base map exhibits unbiased visualization. Further, intersecting polygonized cluster regions enables us to find positive associations among layers with ease. As a results a supporting application has been developed. The application is implemented in the C++ programming language using LEDA (Library of Efficient Data types and Algorithms) version 4.2. The application supports visualization of cluster boundaries, polygonized cluster regions and choropleth maps.

References

Dent BD (1999) Cartography: Thematic Map Design. McGraw-Hill, Boston.

Edelsbrunner H, Kirkpatrick D, Seidel R (1983) On the Shape of a Set of Points in the Plane. IEEE Transactions on Information Theory 29(4):551-559

Ester M, Kriegel H-P, Sander J, Xu X (1996) A Density-Based Algorithm for Discovering Clusters in Large Spatial Databases with Noise. In: Simoudis E, Han J, Fayyad UM (eds) Proc. of the 2nd Int. Conf. on Knowledge Discovery and Data Mining. AAAI Press, pp 226-231

Estivill-Castro V, Lee I (2000) AUTOCLUST: Automatic Clustering via Boundary Extraction for Mining Massive Point-Data Sets. In: Abrahart J, Carlisle BH (eds) Proc. of the 5th Int. Conf. on Geocomputation.

Estivill-Castro V, Murray AT (1998) Discovering Associations in Spatial Data - An Efficient Medoid based Approach. In: Wu X, Ramamohanarao K, Korb KB (eds) Proc. of the 2nd Pacific-Asia Conf. on Knowledge Discovery and Data Mining. LNAI 1394, Springer, pp 110-121

Karypis G, Han E, Kumar V (1999) CHAMELEON: A Hierarchical Clustering Algorithm Using Dynamic Modeling. IEEE Computer: Special Issue on Data Analysis and Mining 32(8):68-75

Knorr EM, Ng RT, Shilvock DL (1997) Finding Boundary Shape Matching Relationships in Spatial Data. In: Scholl M, Voisard A (eds) Proc. of the 5th Int. Symposium on Spatial Databases. LNCS 1262, Springer, pp 29-46

Miller HJ, Han J (2001) Geographic Data Mining and Knowledge Discovery: An Overview. Taylor & Francis, New York

Ng RT, Han J (1994) Efficient and Effective Clustering Method for Spatial Data Mining. In: Bocca JB, Jarke M, Zaniolo C (eds) Proc. of the 20th Int. Conf. on Very Large Data Bases. Morgan Kaufmann, pp 144-155

Okabe A, Boots BN, Sugihara K, Chiu SN (1999) Spatial Tessellations: Concepts and Applications of Voronoi Diagrams. John Wiley & Sons, West Sussex

Openshaw S (1987) A Mark 1 Geographical Analysis Machine for the automated analysis of point data sets. Int. Journal of GIS 1(4):335-358

The Impact of Filtering on Spatial Continuous Queries

Thomas Brinkhoff

Institute for Applied Photogrammetry and Geoinformatics (IAPG), FH Oldenburg/Ostfriesland/Wilhelmshaven (University of Applied Sciences), Ofener Str. 16/19, D-26121 Oldenburg, Germany, tbrinkhoff@acm.org

Abstract

Spatiotemporal database systems (STDBS) are primarily oriented to applications that track and present details about moving objects. Such applications must be kept informed about new, relocated, or removed objects that fulfil a given query condition. Consequently, a STDBS must inform its clients about these updates. Such queries are called *continuous queries*. The volume and frequency of transmissions is influenced by technical restrictions like the computing power of a client, the spatial distances a client is able to distinguish, and the maximum speed and throughput of the network connection. In this paper, filtering algorithms are presented that reduce the number of transmitted update operations. Two contradictory optimisation goals can be observed: First, to reduce the memory requirements of the STDBS for buffering these operations and, second, to reduce the volume and frequency of transmissions. Delaying or even not transmitting updates to a client may, however, decrease the quality of the query result. The impact of these algorithms is presented through discussion of a series of experiments.
Keywords: spatiotemporal database systems, moving objects, continuous queries

1 Introduction

Spatiotemporal database systems (STDBS) are an enabling technology for applications such as Geographic Information Systems (GIS), environmental information systems, and multimedia. In particular, the storage and retrieval of *moving objects* are central tasks of a STDBS. The investigation of spatiotemporal database systems is especially oriented to applications which are required to track and visualise moving objects. Many of these applications originate from the field of *traffic telematics*. This is a field in which techniques from the areas of telecommunication and computer science are combined and used in establishing traffic information and assistance services. Such applications require the management of moving objects, such as vehicles of a fleet (Wolfson *et al.*, 1999) (Brinkhoff, 1999).

An important issue is the support of *mobile and location-based applications*. Mobile applications refer to locations and require the transmission of spatial or spatiotemporal data. The appearance of mobile applications has also had an impact on devices used for presenting data: Personal digital assistants (PDAs) or mobile telephones are used as clients. The computing power of such devices, however, is restricted compared to traditional computers. In addition, speed and throughput of wireless networks are subject to large variations.

The work presented in this paper is motivated by the two trends mentioned above. Applications tracking, and presenting moving objects require current and appropriate information about any new, relocated, or removed objects. Consequently, a STDBS must inform its clients about such update operations. The query causing this process is called *continuous query* (Terry *et al.*, 1992). The 'result set' of a *spatial continuous query* is influenced by the update operations occurring in the database and by a given query condition. The query condition includes two aspects and consists of spatial predicates (e.g., a window query) and optionally of non-spatial predicates defining further selections (e.g., a selection of vehicles that are cars or motorbikes). For mobile applications, the processing capabilities of the client must also be taken into account.

Typical technical restrictions concern the computing power of the client, the spatial distances the client is able to (or wants to) distinguish, and the maximum speed and throughput of the connection between the STDBS and the client. Therefore, it is not advisable to transmit the complete result of a continuous query. Instead, a reasonable *filtering* must be performed. Two contradicting optimisation goals can be observed: First, to reduce the memory requirements of the STDBS for buffering the operations and, second, to reduce the volume and frequency of transmissions to the client. \Delaying or even not transmitting update operations to a client may, however, decrease the quality of the query result. Therefore, algorithms for filtering the result of a spatial continuous query are required to maintain a sufficient quality of the query result.

This paper starts with a short definition of the model used for describing moving objects and presents the main properties of spatial continuous queries. The second section introduces a first algorithm for processing continuous queries. More sophisticated algorithms are presented and analysed in the third section. They limit the memory requirements of the STDBS and reduce the volume and frequency of the transmissions. The quality of the query results and other properties of these algorithms are then experimentally investigated. Finally, the paper concludes with a summary and an outlook on future work.

2 Continuous Queries

2.1 Definitions

The following discussion assumes a STDBS, which stores the positions and other attributes of moving objects. In a temporal database, *valid time* and *transaction time* is distinguished. The valid time describes the time when a record is valid in the modelled reality. The transaction time is the time when a record is committed in the database system.

In the following, it is assumed that a moving object *obj* has an identifier *obj.id*. The object is described by a sequence of records obj_i ($i \in N$). Each record consists of a spatial location $obj_i.loc$ (short: loc_i), of a time stamp $obj_i.time$ (short: $time_i$) giving the beginning of the valid time, of a time stamp $obj_i.trTime$ (short: $trTime_i$) giving the transaction time, and of non-spatiotemporal attributes $obj_i.attr$ (short: $attr_i$). A *time stamp t* is represented by a natural number ($t \in N$). If the records obj_i and obj_{i+1} exist, the valid time of record obj_i corresponds to the interval $[time_i , time_{i+1})$. If no record obj_{i+1} exists for a record obj_i, the record obj_i is the current state of the corresponding moving object from the point of view of the STDBS. Furthermore, a final record obj_i may exist with $i > j$ for all records obj_j of this moving object. It indicates the end of the lifetime of the object. In order to simplify the discussion, we assume that $time_i \leq trTime_i$ holds.

We distinguish three *basic types of updates* concerning a moving object: 1, the insertion of a new object, 22, the modification of an existing object, and 33, the deletion of an existing object. With respect to the query condition of a distinct client, the type of an (modifying) update operation may change (Brinkhoff and Weitkämper, 2001). For example, the position of an object representing a vehicle has been modified in the database. If this vehicle leaves the query window of a client, this modification must be reclassified to a deletion for this particular client. Table 1 gives a summary of such reclassifications. The vehicle leaving the window is represented by the (yes, no) row and the modification column. Updates that do not need to be reclassified are shown as I1 and D1. While I2 and D2 denote reclassified updates.

Table 1. Reclassification of update operations

fulfils query condition?		original type of operation:		
previous record	current record	Insertion	deletion	modification
obj_{i-1}	obj_i	reclassified type of operation:		
no	no	-	-	-
no	yes	insertion (I1)	./.	insertion (I2)
yes	no	./.	deletion (D1)	deletion (D2)
yes	yes	./.	./.	modification (M)

The reclassified type of an update operation may also determine *the interest of a client* in this operation: Deletions are typically of *high interest*. The same holds for insertions. For modifications, the situation may be different. In general, the number of modifications considerably exceeds the number of other operations. Therefore, it may be acceptable to skip some modifications, especially if the distance to the last reported position is small or the topology has not changed (e.g., the car is still on the motorway or is still in the same county). Then, the result set received by a client is not identical to the complete result set of a continuous query. In contrast to the assumptions done in queuing theory, not only a delaying but also a skipping of operations is acceptable. Another restriction concerns the database: the STDBS cannot store a reflection of all update operations each client has received as a result of the query condition – the cost would be prohibitive.

2.2 A First Algorithm

If a moving object is changed in the database, the STDBS will determine the affected clients and will reclassify the update operation type accordingly. If the update is of interest for a distinct client, the STDBS will call the procedure `collectUpdates` (see Fig. 1). In general, the new update operation is added to the set *client.ops*, which collects the operations intended for the client. *newOp* consists of the identifier of the corresponding object *(objId)*, of the current object representation *(curr)*, and the reclassified type of the operation (*opType)*. If an element concerning the same object already exists in the set *client.ops*, this element will be updated. As a result, and depending on the current type of operation, the element will be deleted or modified. `collectUpdates` guarantees that for each object at most one operation exists in the set *client.ops*.

The function `computeTransmission` determines the set of operations to be sent to a client. The STDBS calls this function before the updates are transmitted to a client. As a result, this first solution returns *client.ops* and then empties the set..

```
void collectUpdates (Client client, Operation newOp) {
// Adds an update operation newOp to a collection of a client.
 // case 1: the operation concerns no object referenced in the set
 if (newOp.objId ∉ {op.objId | op ∈ client.ops})
     client.ops = client.ops ∪ {newOp};
 // case 2: the operation concerns an object referenced in the set
 else {
     // determine the stored operation
     Operation oldOp = op∈client.ops with op.objId == newOp.objId;
     // if necessary, delete the operation from the set
     if ((oldOp.opType∈{I1,I2}) && (newOp.opType∈{D1,D2}))
         client.op = client.op \ {oldOp};
     // or update the type of operation and the description
     else if ((oldOp.opType∈{D1,D2}) && (newOp.opType∈{I1,I2})) {
         oldOp.opType = M;        // delete plus insert becomes modification
         oldOp.curr = newOp.curr;
     }
     // or update only the description
     else oldOp.curr = newOp.curr;
    }}

Set computeTransmission (Client client) {
// Determines the update operations to be sent to a client.
 Set sendOps = client.ops;
 client.ops = ∅;
 return sendOps;
}
```

Fig. 1. First version of the filtering algorithm

An aspect to consider is how best does one determine the point at which a set of updates should be transmitted to a client. One solution is to send a transmission as soon as the transaction time has exceeded a given period Δt. In this case, the size of set *client.ops* is only limited by the number of updates that a STDBS is able to process in the given period Δt. This number can be quite large. Assuming many clients use a STDBS in parallel, it may result in excessive memory requirements and poor scalability. Further, the performance of the client or the network connection to the client may restrict the number of operations that can be processed during a given period.

By reducing the period Δt, the first disadvantage may be reduced. However, the sum of transferred operations would increase. The reason for this increase is that the probability of replacing operations in the set *client.ops* decreases with shorter periods Δt. Time restrictions, that require a minimum period between two data transmissions, are another rationale against reducing Δt. The same argumentation will hold if the transmission is triggered by the size of the set *client.ops*. Only in the case of Δt=1, the algorithm computes the complete result of a continuous query. Otherwise, the transmission of operations may be delayed. By replacing outdated entries, the result set may be smaller than the complete result.

We can observe two contradictory optimisation goals: First, to reduce the memory requirements of the STDBS for buffering update operations and, second, to reduce the volume and frequency of transmissions to the client. In the following, we try to balance between these two objectives by modifying the initial algorithm.

3 Improving the Algorithm

According to (Brinkhoff and Weitkämper, 2001), table 2 summarises the parameters and functions, which can be used for describing the restrictions of a client. These parameters are used by the algorithms presented below.

Table 2. Parameters and functions used for describing the restrictions of a client

parameter	Description
maxOps	The maximum number of update operations that can be sent to a client by one transmission.
minOps	The minimum number of operations reasonable to be sent to a client by one transmission; it holds: minOps ≤ maxOps.
minPeriod	The minimum period between two transmissions to a client.
thr	A threshold for the measure of interest (see section 3.1).
function	**Description**
intr(ob_{prev},obj_{curr})	The measure of interest for operations that are not of high interest.
isRelevant(ob_{prev},obj_{curr})	Boolean function determining whether an update operation is relevant for a client or not.

3.1 Algorithm Observing the Restrictions of a Client

An algorithm, which determines the next update operations to be sent to a client for performing the continuous query, should observe the restrictions and measures described above. Like in section 2.2, the algorithm presented in Fig. 2 consists of the operations `collectUpdates` and `computeTransmission`.

The procedure `collectUpdates` is similar to the first version. A previous object description *(prev)* and an attribute *time* have been added to the elements of the set *client.ops*. The parameter *newOp* also includes an attribute *prev* representing the previous object representation in the database. A STDBS should be able to determine *newOp.prev* efficiently. For a new element in the set *client.ops*, the attribute *time* is generally set to the valid time of *newOp.prev*. An exception from this rule is the insert operation I1. Then, *time* is set to the valid time of the new object. If an element concerning the same object already exists in the set *client.ops*, this element will be updated. Note that the attribute *time* is not changed in this case. It still represents the time when the operation was inserted into *client.ops*.

```
void collectUpdates (Client client, Operation newOp) {
// Adds an update operation newOp to a collection of a client.

  // case 1: the operation concerns no object referenced in the set
  if (newOp.objId ∉ {op.objId | op ∈ client.ops}) {
     newOp.time = (newOp.opType == I1) ? newOp.curr.time : newOp.prev.time;
     client.ops = client.ops ∪ {newOp};
  }
  // case 2: the operation concerns an object referenced in the set,
  //         the attribute oldOp.time remains unchanged!
  else {
     // determine the operation
     Operation oldOp = op∈client.ops with op.objId == newOp.objId;
     // if necessary delete the operation from the set
     if ((oldOp.opType∈{I1,I2}) && (newOp.opType∈{D1,D2}))
        client.op = client.op \ {oldOp};
     // or update the type of operation and the description
     else if ((oldOp.opType∈{D1,D2}) && (newOp.opType∈{I1,I2})) {
        oldOp.opType = M;        // delete plus insert becomes modification
        oldOp.curr = newOp.curr;
     }
     // or update only the description
     else oldOp.curr = newOp.curr;
}  }

Set computeTransmission (Client client, Time currTime) {
// Determines the updates to be sent to a client. currTime: the current time

  // initialize the set of operations to be sent
  Set sendOps = ∅;
  // if the period is too short: return nothing
  if (currTime-client.timePrev) < client.minPeriod)
     return sendOps;
  // determine the operations of high interest
  Set o1 = {op∈client.ops | (op.opType)∈{I1,I2,D1,D2}) ∨
     (intr(op.prev,op.curr)≥client.thr)) ∧ isRelevant(op.prev,op.curr) };
  if ( |o1| > client.maxOps )
     sendOps = {op∈o1| client.maxOps elements having the oldest time stamps};
  else
     sendOps = o1;
  // determine further operations of interest
  if (|sendOps| < client.minOps) {
     Set o2 ={op∈client.ops∧op∉sendOps|isRelevant(op.prev,op.curr)};
     sendOps = sendOps ∪ {op∈o2| client.minOps-|o1| elements
                                  having the highest intr(op.prev,op.curr) };
  }
  // final actions
  if (sendOps ≠ ∅)
     client.prevTime = currTime;
  client.ops = client.ops \ sendOps;
  return sendOps;
  }
```

Fig. 2. Filtering algorithm observing the restrictions of a client

The function `computeTransmission` has been completely modified: it determines the set of operations to be sent to the client observing the parameters and functions shown in table 2. First, the algorithm tests whether the time interval between the time, when update operations were sent to the client last, and the current time is sufficient. Then, a set of operations is determined. This set contains all operations of high interest and the operations whose measure of interest exceeds a given threshold *thr*. The elements are ranked according to the attribute *time*, which was determined by the operation `collectUpdates`. If necessary, this sequence will be cut by *maxOps*. If reasonable, further operations are added. The selected elements form the result of the function `computeTransmission`. These elements are removed from the set *client.ops*.

Note that non-relevant update operations are not removed from the set *ops*. There are two reasons for this. The first reason is to accumulate the movement. The STDBS cannot derive the last object representation obj_{last} sent to a client without explicitly storing previous information; the attribute *prev* of a new update operation *newOp* is often not identical to obj_{last}. Furthermore, keeping these operations in the set allows the value of the attribute *time to be preserved.* Otherwise, a sequence of non-relevant movements would repeatedly change the value of *time*. As a result, the ranking of this operation within other operations would stay on a low level. By keeping the first value of time, the ranking of the operation is improved over the duration of time. Fig. 3 illustrates these effects.

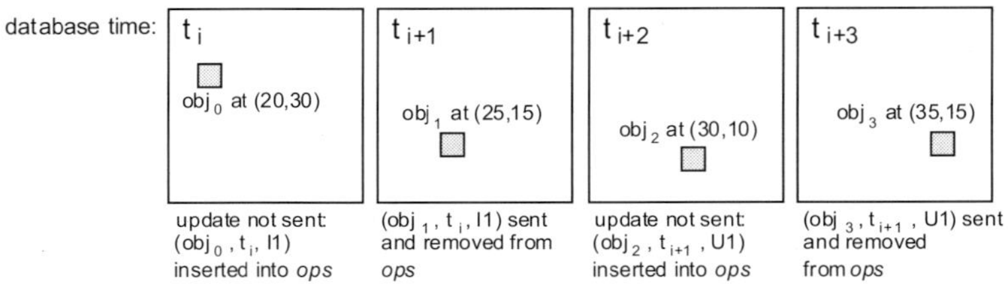

Fig. 3. Illustration of setting and keeping the attribute *time*

3.1.1 Discussion of the Time Complexity

The filtering algorithm depicted in Fig. 2 observes the parameters of table 2. However, its design has not considered any optimisations for reducing the time or the space complexity of the algorithm. Let us first discuss time complexity.

We can distinguish two different rankings using the attribute *op.time* ($op \in client.ops$) and the function *intr* computing the measure of interest. Therefore, we consider the two subsets ops_1 and ops_2 separately. ops_1 consists of all operations of high interest plus the relevant operations of lower interest. ops_2 consists of the other operations of lower interest. The assignment of an operation to one of these two subsets will only change, if the object description *op.curr* is changed. This is not problematic because in this case the algorithm `collectUpdates` is called, which can handle this case. The ordering of ops_1 is trouble-free because the attribute *op.time* will not be changed and the property *isRelevant* is static as long as the corresponding object *op.curr* is not changed. The same holds for the ordering of ops_2 because the function *intr* result does not change without changing *op.curr*.

The operations performed on the sets consists (a) of the insertion of elements, (b) of the search for an existing element (and its deletion) and (c) of the retrieval (and deletion) of the first *k* elements according to the ordering of the elements of

ops_1 and ops_2, respectively. To support these operations, the following options exist (*n* denotes the number of elements in *client.ops*):

- Operation (b) determines the existence of an operation in the set by using the identifier of the corresponding object. For an efficient search, we must know the value of the attribute, which defines the ordering: the valid time of the object representation originally inserted into *client.ops* and the object representation inserted before into *client.ops*. However, the original valid time is unknown for the calling STDBS. Therefore, the effort for performing operation (a) is *O(log(n))* and for operation (c) *O(maxOps*log(n))* if a balanced search tree according to the orders of ops_1 and ops_2 is used. However, the worst-case search time of (b) will be of *O(n)* in this case.
- Organising the sets by two redundant balanced search trees allows performing the operations (a) and (b) in *O(log(n))* and operation (c) in *O(maxOps*log(n))*. Each update operation, however, must be performed twice and as a result maintaining the two search trees increases the space requirements.

3.1.2 Discussion of the Space Complexity

The number of all current valid moving objects fulfilling the spatial and other non-spatial query conditions only limits the number of elements in the set client.ops. This number is denoted by *N*; it holds: $n \leq N$. In the worst case for each client performing the continuous query, the status of all moving objects fulfilling the query condition at the beginning or throughout the duration must be recorded in the set *client.ops*. Then, *n* will be quite large. Assuming a STDBS used by many clients in parallel, that means a huge memory and maintenance overhead. Therefore (and because of the time complexity), it is necessary to reduce this overhead by limiting the size of the set *client.ops*. This is the topic of the next section.

3.2 Restricting the Space Complexity

In an effort to restrict the memory demands the algorithm is modified by using a parameter, *maxSize*. The minimum value of *maxSize* is determined by the parameter *maxOps*: $maxSize \geq maxOps$. However, a higher value of *maxSize* would improve the results of the continuous query. If we restrict the size of the set, we will need a *replacement strategy*, which is required for a new important operation in the case that the set is full. The obvious replacement strategy is to remove the least important element from the set *client.ops* if the set has the size *maxSize*. One exception, however, must be observed. If we removed delete operations, this would have drastic impacts for the client: the client in most cases would never remove the corresponding object. Therefore, we must not remove such operations from the set *client.ops*. Instead, we remove another operation having an older time stamp *op.time* from the set. In the case where only delete operations exist, however, this approach does not work. A solution to neglect the space restrictions or to disregard the time restriction *minPeriod*, should be considered. This implies that the data would be sent earlier and therefore more data would be sent to the client than originally expected.

Removing elements from the set of operations has a further impact: A client may receive update or delete operations concerning objects unknown for the client because it has not received any insert operation for this object before. Consequently, such an update must be executed as an insertion and the client can ignore such a deletion.

4 Experimental Investigation

The experiments presented in the following investigate the applicability of the algorithms. Especially, the impact of the technical restrictions and of limiting the memory on the quality of the query results should be examined.

4.1 Test Data and Queries

For generating suitable test data, the generator for spatiotemporal data presented in (Brinkhoff, 2000) was used. This generator allows moving objects to be computed and observes several rules effective in simulating typical traffic situations. In our case, a street network consisting of 6,065 edges was used, and can be downloaded from the web site referenced in (Brinkhoff, 2000).Six object classes were defined. The maximum distance done by an object was between 1/250 and 1/8000 of the sum of the x-extension and the y-extension of the data space. The probability of a move per time stamp was 25%. The query condition used for the queries in the following tests is quite simple: it selects all objects lying in a query window having a size of 10% of the data space.

4.2 The Tests and Their Results

The following experiments were performed through an an implementation of the continuous query programmed in Java using Oracle 8i. The continuous queries were started at time stamp 640 and stopped at time stamp 1280. At time stamp 640, 281 moving objects were within the query window and at the end 387 objects. During the query, the complete number of operations to be transmitted to a client was 56,712. The measure of interest *intr* was computed as follows:

(1) $intr(obj_{prev}, obj_{curr}) := (time_{curr} - time_{prev}) + w_{loc} * dist\ (time_{curr}, time_{prev})$

The factor w_{loc} scales the Euclidean distance such that the influence of time and space is equalised. The threshold *thr* is set on a value that would be exceeded for $dist = 0$, if the period between the two operations was larger than 4**minPeriod.*

The first test series investigate the results of the continuous query for different minimum periods *minPeriod.* Fig. 4 gives an overview of the results. The results of a *minPeriod* of 1 correspond to the results computed by the algorithm presented in section 2.2; the other results are computed using the algorithm of section 3.1. The *omitting degree* describes the quality of the query result according to definition of (Brinkhoff and Weitkämper, 2001). The smaller the omitting degree, the better the quality of the query results. The omitting degree consists of two components: *timeOD* describes the temporal quality and *distOD* the spatial quality. For the sake of brevity, the definition of the omitting degree has not been presented here.

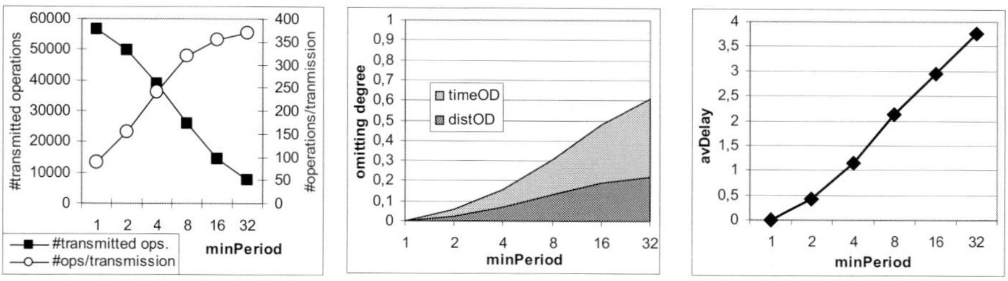

Fig. 4. Results depending on the minimum period *minPeriod*

The number of transmitted operations declines from 56,712 for a *minPeriod* of 1 to 7,769 for a *minPeriod* of 32, i.e. by a factor of 7.3. The number of operations per transmission increases by a factor of about 4.2. With increasing values of *minPeriod* the probability increases that an operation is updated by a new operation before it is transmitted to the client. As a result, the quality of the response decreases considerably; for a *minPeriod* of 32, we observe an omitting degree of about 0.61. The average delay is barely affected because it is only measured for transmitted operations and not for operations being replaced before sending them to the client.

In the next test series, the number of operations transmitted to the client *(maxOps)* was limited. The value of *minOps* was always set to *maxOps*/2. Fig. 5 shows the main results for a *minPeriod* of 4 and 8. In the unlimited case (unl), up to 292 and 362 operations per transmission occur.

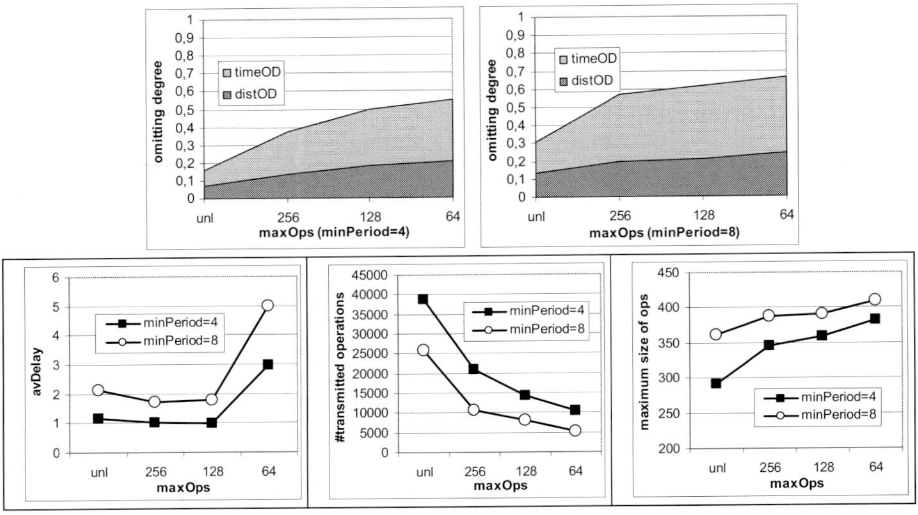

Fig. 5. Results depending on the number of operations per transmission

By limiting the number of operations, the total number of transmission operations also decreases: we observe factors of 3.8 and 4.9 between the unlimited case and a maximum number of 64 operations for a *minPeriod* of 4 and 8, respectively. Consequently, the measure *omitDeg* increases. However, the increase is relatively moderate, especially for the distance measure *distOD*. This observation demonstrates that the heuristics used by the algorithm for selecting the transmitted operations have success and compensate some of the loss of quality. The graphs depicting the average delay are quite interesting. They show that up to a certain point, the effect of limiting *maxOps* is that older operations in the set *client.ops* are replaced by newer operations. In this case, no impact on *avDelay* can be observed. Beyond this point, the transmission of operations is really delayed and *avDelay* increases. Another observation concerns the size of *client.ops*: the smaller *maxOps*, the larger the maximum size of this set. Therefore, the maximum size of *client.ops (maxSize)* is limited in the test series of Fig. 6. That means we investigate a version of the algorithm of section 3.1, which observes the space restrictions presented in section 3.2.

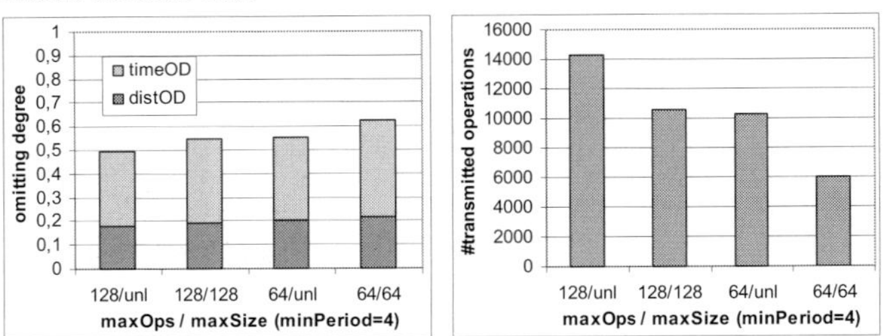

Fig. 6. Results depending on *maxOps* and *maxSize* (unl = unlimited)

Limiting *maxSize* leads to a further decline of the number of transmitted operations. Again, the impact on the quality of the query result is rather moderate. Especially, the distance portion of the omitting degree is almost unaffected.

We can summarise that the experiments have shown that the proposed algorithm allows limiting the number of transmitted operations as well as the number of operations buffered by the STDBS without a huge loss of quality.

5 Conclusions

In this paper, filtering algorithms for processing spatial continuous queries in a spatiotemporal database system (STDBS) have been presented. After presenting a first algorithm, we have observed two contradicting optimisation goals: First, to reduce the memory requirements of the STDBS and second, to reduce the volume and frequency of transmissions to the clients. To balance between these two objectives, an algorithm has been presented that observes different parameters modelling technical restrictions as well as the interest of a client in a distinct update operation. A restriction of the memory requirements of the algorithm has been achieved by using an adapted replacement strategy.

Delaying and not transmitting update operations to a client, however, decreases the quality of the query result. In an experimental investigation, the proposed algorithms have been examined measuring the quality of the query results. These tests have shown that the algorithm, which was finally proposed, allows limiting the number of transmitted operations as well as of the number of the operations buffered by the STDBS without a huge loss of quality. In particular, the impact on the distance between the locations of the transmitted object descriptions is rather moderate.

The definition of continuous queries in this paper was based on a simple model of moving objects. Therefore, future work should cover a definition using a more expressive data model. The same holds for the application. More complex is, for example, the detection of collisions for 3D moving objects (Mirtich, 2000). The experimental investigations presented in this paper have been based on a standard database system. A major drawback of using such a database system is the considerable effort necessary for determining the previous object description of an updated object. This disadvantage must be eliminated by extending the database system by a suitable buffering technique or by using (prototypes of) spatiotemporal database systems. More detailed performance investigations could then include the measurement of the processing time for performing continuous queries.. Another aspect to consider is the behaviour of the restricting parameters. In this paper, it is assumed that they do not change over time with respect to a client. However, the resolution of a client may be changed by performing a zoom operation. The parameters *minOps, maxOps* and *minPeriod* may be affected by the traffic of other network users or by a changed capacity of the connection (e.g., using the new mobile telephone standard UMTS, the maximum speed of a connection will depend on the distance of the mobile telephone to the next base station.) Therefore, efficient filter algorithms are required, which observe varying restrictions.

References

Brinkhoff T (1999) Requirements of Traffic Telematics to Spatial Databases. In: Proceedings 6th International Symposium on Large Spatial Databases. Hong Kong, China. Lecture Notes in Computer Science, Vol.1651, Springer, pp 365-369

Brinkhoff T (2000) Generating Network-Based Moving Objects. In: Proceedings 12th International Conference on Scientific and Statistical Database Management, Berlin, Germany, pp 253-255 (extended version accepted for GeoInformatica, Kluver)

Brinkhoff T, Weitkämper J (2001) Continuous Queries within an Architecture for Querying XML-Represented Moving Objects. In: Proceedings 7th International Symposium on Spatial and Temporal Databases, Redondo Beach, CA. Lecture Notes in Computer Science, Vol.2121, Springer, pp 136-154

Mirtich B (2000) Timewarp Rigid Body Simulation. In: Proceedings ACM SIGGRAPH 27th International Conference on Computer Graphics and Interactive Techniques, New Orleans, LA, pp 193-200

Terry D, Goldberg D, Nichols D, Oki B (1992) Continuous Queries over Append-Only Databases. In: Proceedings ACM SIGMOD International Conference on Management of Data. San Diego, CA, pp 321-330

Wolfson O, Sistla AP, Chamberlain S, Yesha Y (1999) Updating and Querying Databases that Track Mobile Units. Distributed and Parallel Databases 7(3):257-287

An Object Model for Geologic Map Information

Boyan Brodaric[1] and Jordan Hastings[2]

GeoVISTA Center, Pennsylvania State University *and* Geological Survey of Canada, 234B-615 Booth St., Ottawa, ON K1A 0E9, brodaric@NRCan.gc.ca[1]
Department of Geography, University of California Santa Barbara, Ellison Hall 3611, Santa Barbara, CA 93106, jordan@geog.ucsb.edu [2]

Abstract

National geologic map databases are presently being constructed in the U.S. and Canada, as well as in several other countries. Here, we describe an object-based model for geologic map information, specifically designed to represent digital geologic maps and related geoscientific information. Although oriented to geoscience, several fundamental issues in representing geospatial information are explored in this design, including the philosophic and cognitive basis of mapping in general, and the overall framework in which map-related information can be represented. Thus, we take an ontologic approach to geospatial representation, supplemented by an epistemic view of the scientific process, which culminates in a very general model—a meta-model—for map information. Practical as well as theoretical considerations motivate this approach. Primarily, however, we describe the theoretical foundations of our meta-model, specifically semiotics, category theory, and ontology in geospatial information. Finally, we report briefly on a specific prototype data model derived from the meta-model and implemented in a commercial object-oriented GIS.
Keywords: geologic map information, data models, ontology, classification.

1 Introduction

The public agencies that provide geologic maps and related information in the U.S. and Canada are just now beginning to actively develop on-line systems for their work. These systems simultaneously address two basic needs: improving the efficiency of routine information handling within an agency, and promoting the non-traditional usage of geologic information within and outside the agency. Though conflicts in information policy do exist among agencies, the need to cater to on-line science is cautiously overriding such concerns as several multi-agency systems are emerging. In North America, for example, three major systems exemplify this approach: (1) the Canadian Geoscience Knowledge Network (CGKN; http://cgkn.net/), a co-operative initiative to link the public geoscience data providers in Canada; (2) the U.S. National Geologic Map Database project (NGMDB; http://ncgmp.usgs.gov/ngmdbproject/), a Congressionally-mandated repository for map information from many state agencies as well as the USGS; and (3) GeoInformatics, a proposed network of U.S. academic geoscience databases (GEON; http://www.geoinformaticsnetwork.org/).

In this paper we discuss a project undertaken for NGMDB in particular[1]: an object-based geologic map information meta-model derived from the North American Digital Geologic Map Data Model standards effort (NADM; Johnson *et al.*, 1999; http://geology.usgs.gov/dm/). The design of our meta-model rests on two ambitious objectives regarding the underlying geospatial[2] map information:

1. To supplement *feature-based* geospatial information with *ontologic* context, explicitly represented; and further
2. To supplement the ontologic aspects of geospatial information with *epistemic* considerations, drawn from geoscience directly.

The first objective expresses our view that scientific meta-modelling, i.e. abstracting the many types of models within geoscience, is critical for the effective representation of geoscientific information (Bennet, 1997; Gahegan & Brodaric, 2002; Raper & Livingstone, 1995; Langley, 2000). Specifically, in our design we designate and encode four fundamental types of models, symbolic, conceptual, occurrence, and descriptive, that underpin the structure of geoscientific information and its visual presentation.

The second objective reinforces the point that our meta-modelling emphasis proceeds from epistemology, the fundamental knowledge acquisition and evaluation task (Clancey, 1993). In particular, geologists' evolution of environmental concepts from their interpretation of observed occurrences is often incremental, complex, uncertain and dynamic, and frequently results in multiple valid models for a geographic region (Brodaric & Gahegan, 2001). This behaviour contrasts with many non-scientific domains in which concept development is a minor component, and where indeed, concepts are often known *a priori* rather than dynamically obtained and revised, and where feature occurrences are distinct.

In general, ontologically-driven geospatial information frameworks are well suited to relatively fixed domains, for example in land cadastre, and urban utility networks. In these contexts, they serve to partition geospatial concepts into perspectives and to instantiate relatively certain and unambiguously described occurrences within them (e.g. Benslimane, 2000; Fonseca *et al.*, 2000). However, geospatial information frameworks for dynamic and exploratory work, in science particularly, also need to be epistemologically-driven to account for the creation

[1] Variants of this architecture are also being implemented within CGKN

[2] We use "geospatial" to subsume the terms "geographic", "geologic", and "geoscientific" uniformly

and evolution of concepts and occurrences, and to engage the uncertainties and ambiguities consequently introduced.

In this paper, we begin tackling the central representational issues inherent in geoscience by exploring the complex of relationships between concepts and occurrences, and by modelling them. This requires a general structure for modelling concepts and their relation to phenomena, which is a level of abstraction higher than traditional knowledge engineering approaches that directly model domain concepts rather than the structure organising them (Clancey, 1993). Thus, we arrive at a general object-based structure, a meta-model for geoscientific information that embeds very broad ontological and epistemological factors and can be refined into domain-specific data models. This design complements related work on geological maps that considers relational data model design (Baines & Giles, 1997; Johnson at al., 1999), web-based and multi-media applications (Brodaric *et al.*, 1999; Heyn, *et al.*, 2000), geometric representations (Balovnev, *et al.*, 1997), reasoning (Voisard, 1998; 1999), and cognition (Brodaric & Gahegan, 2001). It is novel in terms of (1) integrating aspects of cartographic presentation with concept development and information ontology into (2) an object-based meta-model for geologic map information.

The paper is organised as follows: Section 2 describes the meta-model developed; Section 3 describes a prototype implementation; Section 4 sketches future work; and Section 5 concludes with a short summary.

2 Theory and Design

Geologic maps depict geologic occurrences within a specific geographic region and spanning particular periods of geologic time. Typical maps are 2D, representing the intersection of 3D occurrences with a surface, usually the Earth's. These maps often represent a causal-process history that explains the evolution in time of the occurrences and their interactions. The history is typically located in accompanying texts and diagrams, and is supported by the occurrences themselves. Not all occurrences, or their parts, can be directly observed or measured, as some are historical and others inaccessible mainly due to sampling constraints, causing them to be inferred. Even observables can be variously described and identified according to the specific expertise of different investigators. Reasoning is thus affected by the amount and type of evidence, by the known theory, by the physical situation, and by the mapper's cognitive orientation. Geologic map representations are thus *meta*-models of geologic space-time-process models, including observed and inferred occurrences, their causal history, contexts and other factors impacting reasoning (Voisard, 1998; 1999).

In this section we describe our meta-model and indicate how individual geologic map reference models can be derived from it. To attain generality we ground the meta-model in semiotics, category theory, and geospatial information ontology.

2.1 Semiotics

Semiotics, the study of signs (Noth, 1990), provides a useful initial framework for representing a scientific view of a geologic map. In cartographic semiotics the meaning of a map symbol derives from the relationship held between the symbol, the concept being symbolised per some interpreting agent, and the occurrence being referred to (MacEachren, 1995). We implement these semiotic primitives as foundational objects in our meta-model (Fig. 1):

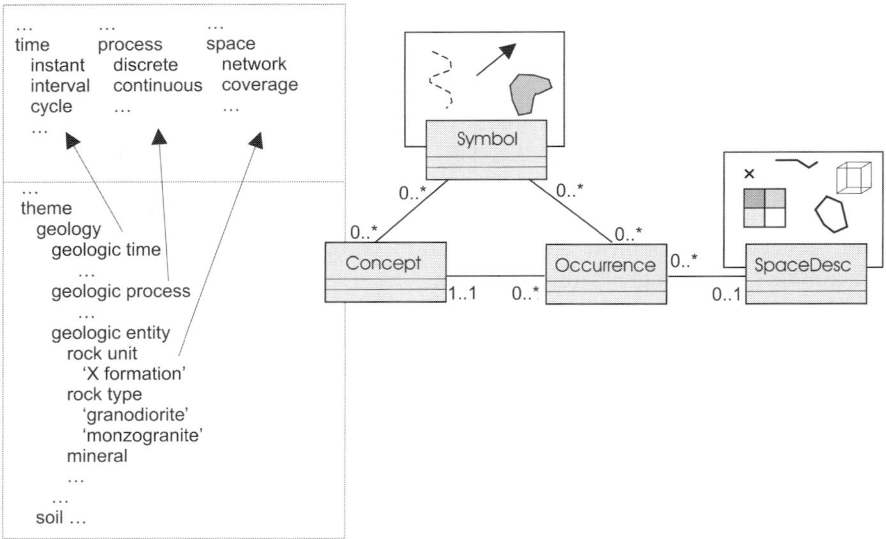

Fig. 1. The semiotic triangle is shown using UML notation (Rumbaugh *et al.*, 1999); some relations for concept, symbol and occurrence primitives are not depicted. Insets show example instances of spatial descriptions (geometry; right), symbols (top) and concepts (left); indents and arrows denote concept specialisation.

1. Concept: refers to the *abstract* objects that constitute geoscientific vocabulary; these can instantiate specific occurrences (below) and relations, and serve as values for attribute domains: e.g. "X formation", "granodiorite", "fault", "intrudes", etc. Concepts may tier from very general and perhaps universal, such as Aristotle's "substance" (Sowa, 2000, 57), through generic, such as "space", "time", "process", among other aspects, to particular (Guarino, 1998), such as those for a specific discipline, theory or even a geographic region.
2. Occurrence: refers to the *tangible* object observable in the field and/or on maps (geospatial features), and their relations in space, time, and otherwise. Note that an occurrence is an instance of exactly one concept and optionally possesses one spatial description, its geometry. Thus, a geospatial feature possessing constant identity might be represented by several distinct occurrences distinguished by conceptual, spatial, temporal or other variation. Though this

notion of geospatial feature is implicit and not explicitly modelled, the meta-model does allow multiple concepts to be applied to the same spatial object.

3. Symbol: refers to the *(carto)graphic* objects in the visual display, presented as areas, line and point markers, charts, tables, and so on. Independent treatment of symbols enables cartographic behaviour related to scale dependencies, symbol overlap, etc., to be associated with both concepts and occurrences. This permits the establishment of cartographic norms via concept symbolisation, and allows for cartographic exceptions in specific occurrences; e.g. though the default designation for all occurrences of the 'X formation' concept is blue, a particular pluton occurrence might be showcased in red.

2.2 Category Theory

The human interpretative process involves categorisation where concepts and occurrences (Fig. 1) are obtained in parallel and mutually affective. The meaning of concepts is thought to become increasingly contextual in lower conceptual tiers, affected by epistemological factors such as methodologies, actions, physical situations, and other factors related to how humans categorise, construct/recognise concepts (Smith & Mark, 2001; Frank, 2001). This insight would seem to hold for geologic mapping, suggesting more complex relations between concepts and occurrences than typically modelled in databases (Baker, 1999; Brodaric & Gahegan, 2001; Dehn *et al.*, 2001).

Specifically, in the mapping process, occurrences interact with concepts in two ways: as *instances* of established concepts, such as various map polygons labelled according to a geoscientific concept, e.g. 'X formation'; and also as *evidence* leading to the development of these concepts, such as the field observations employed in determining 'X formation' (Fig. 2). In information science these interactions are typically referred to as instantiation and classification, respectively. Instances are unique identifications, whereas classifications are prototypical statements based on evidence of varying typicality in which concept meaning is closely related to either (1) an average summary of evidence, or (2) an ideal description (Lynch *et al.*, 2000), viz. a 'type locality' in geologic parlance.

Thus, occurrences are distinguished by their role as instance or evidence in relation to concepts. Identifying and modelling these roles (Fig. 2) then permits them to be described so that, for example, human or machine classification methods can attribute evidence, increasing the context explicitly modelled in the database.

A consequence of varying typicality of evidence is that evidence descriptions, consisting both of structure, such as attributes and behaviour signatures, and of content, such as attribute values and behaviour functions, will be variously similar to concept descriptions. In particular, concept descriptions will vary dramatically with descriptions for outlying evidence, but concepts will share descriptions with ideal evidence. Descriptions may also be shared between occurrences, as understandings converge within and between individual mappers in a specific mapped area (Brodaric & Gahegan, 2001). The meta-model therefore possesses a free-standing data store for descriptions (Fig. 2), enabling them to be shared among and between concepts and occurrences. For example, in Fig. 2 the concept 'X formation' shares a 'granodiorite' rock type description with ideal evidence at a site; however, a single polygonal instance of the 'X formation' concept is described by 'monzogranite' instead.

Thus, descriptions are the central information repository in the meta-model, containing the bulk of traditional data and information. Descriptions may be subclassed and also inter-related among themselves, as needed (not depicted in Fig. 2). The remainder of the meta-model can be seen as a knowledge superstructure adding conceptual, cartographic and occurrence-driven context to the descriptions. Loading specific concepts and description types into the meta-model can generate reference models.

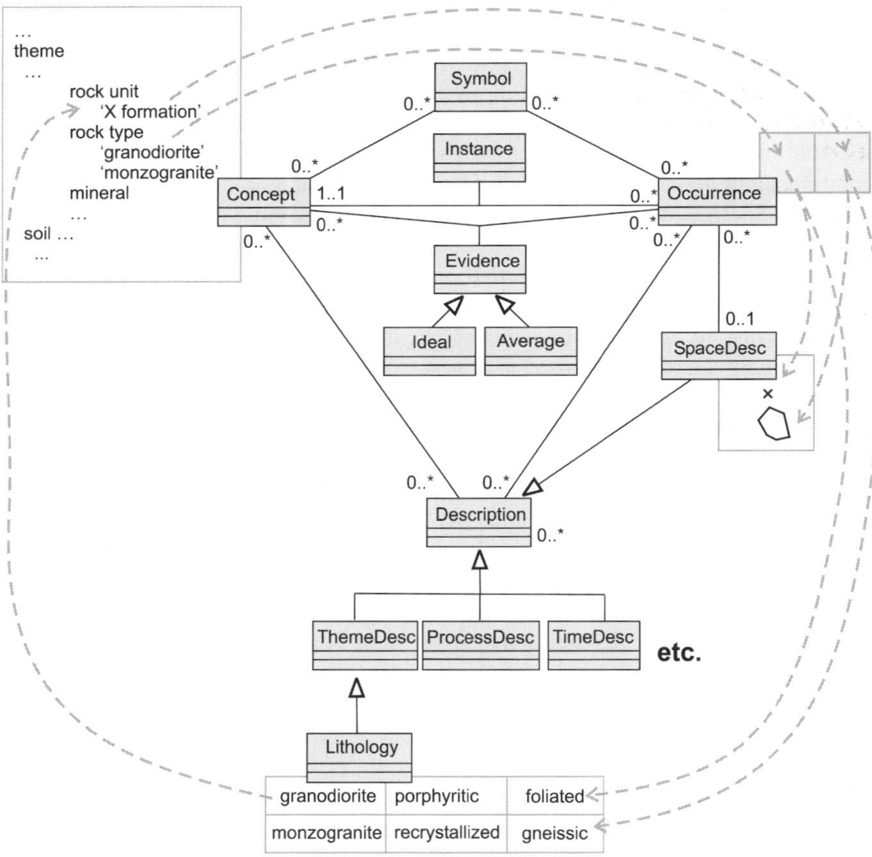

Fig. 2. Insets show example instances of concepts (left) and occurrences (right), with spatial (right) and thematic (bottom) descriptions. The dashed lines illustrate concepts evolving from evidence. See text for details.

2.3 Information Ontology

The knowledge structure described above emphasises the relationships among and between symbols, concepts, occurrences, and descriptions. Specific knowledge representations can be achieved by grouping these primitives into arrangements, called models. We designate four types of models, conceptual, occurrence, cartographic and descriptive, and denote legends and maps as combinations of these basic types.

In information science, conceptual models are formally known as *ontologies* (Guarino, 1998).In the practice of geoscience these are commonly expressed as vocabularies, taxonomies, or classification schemes, such as those for geologic time (e.g. "Precambrian"), rock units (e.g. "X formation "), rock types (e.g. "granodiorite"), or as more complex models such as those for petrography, stratigraphy, genesis, among others. (Heyn, *et al.*, 2000). By extension, we consider occurrence models as part of *epistemologies*, formalising how we know/evaluate geologic realities (Raper, 1999). Epistemologies and ontologies are fundamentally linked: an ontology provides a set of concepts and logic for how occurrences might be arranged in space and time, such as typically described in a map legend and in accompanying notes. Epistemology, however, provides a specific arrangement of geologic occurrences and their causal explanations emplaced in space and time, thus demonstrating on the map the validity of the concepts and related logic. A map in which the conceptual and occurrence models are inter-consistent is deemed to 'work', expressing this connection.

Ontologies (conceptual models) are represented in the meta-model as logically consistent collections of concepts, their relations, descriptions, and evidence (Fig. 3). Similarly, epistemologies (occurrence models) bind together logically consistent occurrences, their relations, descriptions and instances.

Conceptual models enable concepts to be bound in different arrangements, thereby providing various conceptual perspectives on, say, a single occurrence model (Fonseca, *et al*, 2000). Likewise, multiple occurrence models can derive from a single conceptual model, recognising that, for example, a common taxonomy may lead to different maps for the same area.

The meta-model also accommodates *cartographic* models for symbols and *descriptive* models for descriptions. For example, a palette is a model of symbols from a specific symbol library or agency-approved cartographic standard. A geospatial model formalises the spatial descriptions and relations (geometry and topology) that denotes a dataset/layer in a GIS.

A map legend may then be thought of as a symbolised conceptual model devoid of occurrences. Applying a legend to a valid set of occurrences enables their instantiation, classification and visualisation; furthermore, applying an alternate legend to the same set of occurrences effectively generates a re-conceptualised (derivative) visualisation for the area.

In conclusion, we define a geologic map view as a complex model consisting of a legend (symbolised perspective) manifesting a conceptual model (ontology) that in turn exemplifies an occurrence model (epistemology) for the region of interest.

This map view draws objects from one or more geospatial models (GIS layers), and/or utilises descriptions from some aspatial models (attribute databases).

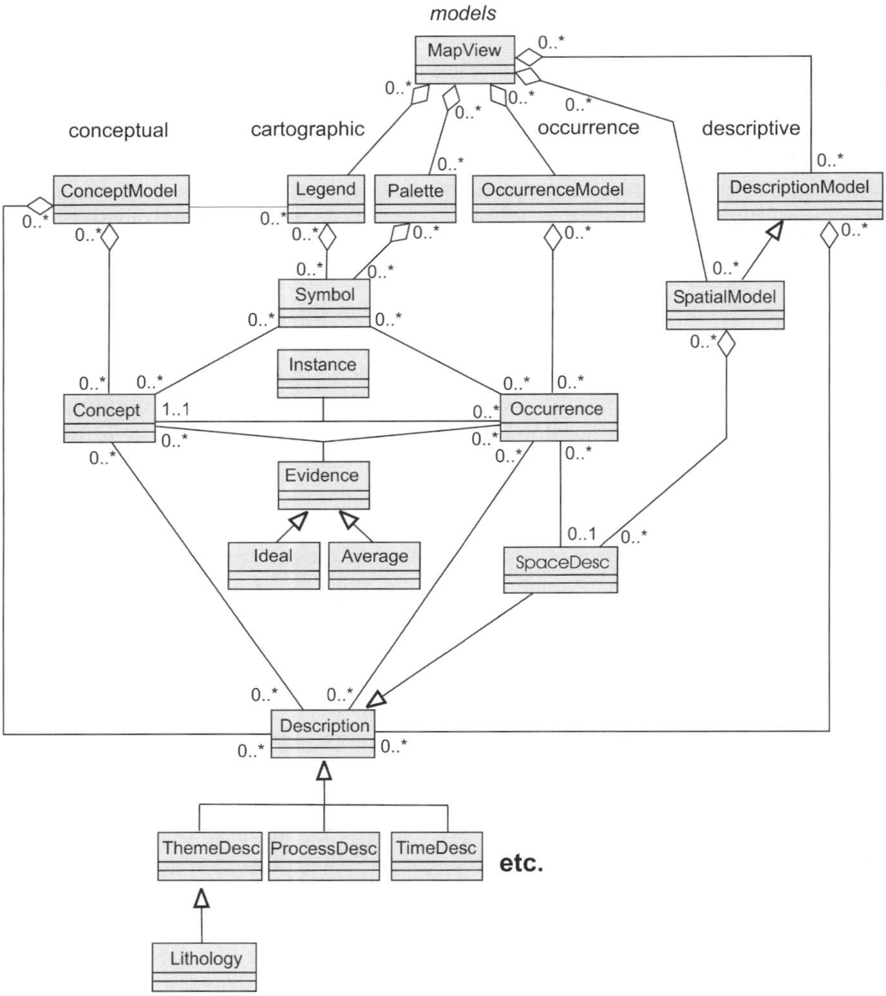

Fig. 3. The definition of a map in the meta-model. See text for details. Note, for reasons of clarity and space some details are omitted

3 Implementation

Testing of the meta-model is proceeding on several fronts: within NGMDB (Hastings & Brodaric, 2001), in a web-based geologic map database project (Davenport *et al.*, 2001) and in a prototype digital library for sustainable development (Journeay, *et al.*, 2000). Testing within NGMDB has thus far concentrated on evaluating (1) the suitability of the meta-model to geological data and (2) the ease of execution of some common operations, putting aside for the moment issues of efficiency and scalability.

The NGMDB test data consisted of four 1:24K geologic maps provided by the Kentucky Geological Survey (KGS), who collaborated with the USGS and GE Smallworld ™ (GESW) for the evaluation. As a preparatory step, the meta-model was configured to suit the maps: concepts and description types were specified for the geologic formations, faulted structures, and coal beds, found in the region. Once configured, the meta-model was implemented virtually 'as is' inside GESW's case tool, requiring only minor custom programming to implement specialisation relations such as those between the description object and its subclasses (Fig. 3). Custom methods were also added to some meta-model objects to achieve the desired mapping functionality (see below). Testing involved loading data and performing some typical visualisation and analysis operations.

3.1 Loading Data

The Kentucky data were supplied in ESRI "shapefile" and MS Excel spreadsheet formats. Populating the data-model with map information involved disaggregating this "legacy" feature-based structure—thematically distinct layers of geospatial features each with spatial, symbol, and descriptive attributes, as well as external database links—and re-organising it into conceptual, occurrence, symbol and descriptive models. It also involved unifying related information scattered amongst disparate data sources, such as externally held stratigraphic lexicons. Concepts embedded in feature attribute values, theme labels and external sources were collected and organised into conceptual models for rock units (Table 1), rock types, and minerals. The symbol attributes of features were collected and organised into a single cartographic palette. The spatial attributes of features from all maps became distinct GESW native spatial objects that comprised a single spatial model. The descriptive attributes became descriptions in a single descriptive model; and the external database links, which inherited GESW's external data access methods, became descriptions in a description model for each external source. This transformation resulted in a dataset that was on the one hand more normalised, as concept, symbol, and attribute descriptions were not replicated for features, and on the other hand more integrated, as some scattered data were unified and previously external data links were directly embedded.

Table 1. Developing conceptual models. Some rock unit concepts and relations intended for input to the database are listed in this table

...	...	...
Stearns coal zone	Contains	Beaver Creek coal horizon
Stearns coal zone	Contains	Stearns No. 2 coal horizon
Breathitt Group	Contains	Princess Formation
Breathitt Group	Contains	Grundy Formation
Breathitt Group	Contains	Alvy Creek Formation
Grundy Formation	Contains	Corbin Sandstone Member
Alvy Creek Form.	Contains	Livingston Conglomerate
...	...	...

3.2 Visualization

The map view as defined by the meta-model above provides sufficient structure for performing map-based visualisation from a geospatial database. We tested this structure under three visualisation scenarios: (1) direct display of a stored map; (2) re-display based on reclassification of existing spatial objects; and (3) dynamic symbolisation of concepts and occurrences at multiple scales.

Display: a **mapview→display** method was added for displaying a map's occurrence model from the database (Fig. 4, left). This method also activates various model information, so that, for example, querying a spatial object returns only those concepts, descriptions, and symbols belonging to the map view.

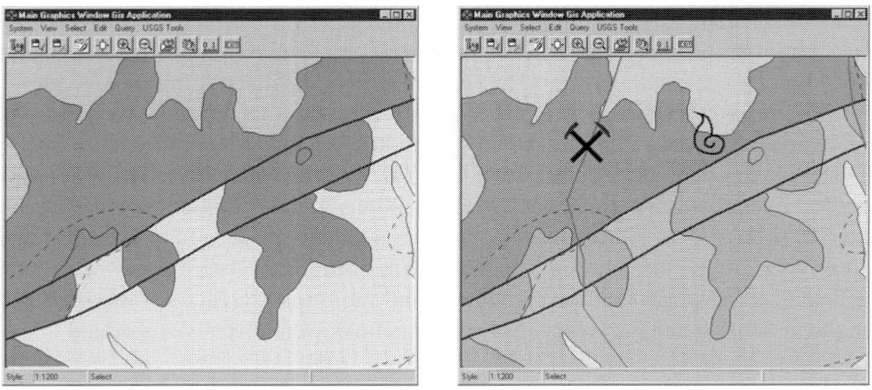

Fig. 4. Displaying and generating maps using the meta-model and test data; rock unit occurrences (left) are dynamically reclassified into their dominant rock types (right)

Reclassification: a **mapview→reclass** method was added for re-conceptualising a map, i.e. deriving a new map view from the active one. In this method, concepts and spatial objects from the active map view are retained, whereas additional occurrences and symbols are created or activated; models for all these are then built for the new map view. Fig. 4 shows rock unit occurrences (left) and a derived map view displaying their dominant rock types (right) dynamically derived from associated rock unit concept descriptions.

Symbolization: an **occurrence→display** method was added for dynamically displaying an occurrence. Display happens when a scale-appropriate symbol is found, first for the occurrence, or second for its associated concept. This allows for default and unique occurrence appearance, and it enables both appearance and presence to vary dynamically by scale, such as when zooming.

3.3 Analysis

A simple network analysis was performed in which dominant rock type occurrences derived from the input geologic maps (see above) were evaluated in terms of their downstream proximity to pollutant discharge and their resultant material susceptibility to contamination. Stream and digital elevation model information, organised as simple features with attributes, was used in conjunction with the model-based information held in the meta-model. The resulting map high-lighted several potential contamination sources.

Beyond its scientific application, this analysis exercise has practical importance. It demonstrates the meta-model can interact with common data sources in typical GIS environments where objects are associated primarily in terms of spatial relations, with little regard for the knowledge and highly semantic relations inherent in geoscience and maintained by the meta-model.

4 Future Work

Results from the Kentucky prototype and other implementations indicate the meta-model provides a normalised and integrative environment for organising and manipulating geoscientific map information. Several aspects require further testing and development; these remain to be explored in future work:

(1) scalability and efficiency of the design with respect to increased data volume;

(2) multi-map, multi-resolution, multi-ontology environments;

(3) modelling geoscientific explanations (Voisard, 1998; 1999);

(4) modelling and implementing process and simulation models;

(5) using the meta-model as a mediator for interoperability;

(6) mapping to existing geospatial standards;

(7) guidelines for incorporating geological phenomena; and

(8) developing an ontology for geologic mapping.

5 Conclusions

Developing broadly applicable standards for geologic map information is critical to improving the overall usability of such information within the geosciences and beyond. We believe that attaining such agreement depends on: (1) embracing a very general meta-model; and (2) deriving specific, but adaptable, data models from this meta-model. Here we have described one such meta-model design and briefly demonstrated a model implementation from it.

Acknowledgements

We thank R. Laudati, R. Wahl, J. Weisenfluh, for their assistance with design and implementation of the NGMDB Kentucky prototype; the GSC, USGS, GeoVISTA Center at Pennsylvania State University, and the Department of Geography at the University of California Santa Barbara for their collegial support.

References

Bain KA, Giles JRA (1997) A standard model for storage of geologic map data. Computers and Geosciences 23(6):613-620

Baker V (1999) Geosemiosis. Geological Society of America Bulletin 111(5):633-645

Balovnev O, Breunig M, Cremers AB (1997) From GeoStore to GeoToolKit: The second step. In: Scholl M, Voisard A (eds) Advances in Spatial Databases, 5th International Symposium, SSD'97. July, 1997, Berlin. LNCS 1262, Springer, Berlin

Bennet DA (1997) A framework for the integration of geographical information systems and modelbase management. International Journal of Geographic Information Systems 11(4):337-357

Benslimane D, Leclercq E, Savonnet M, Terrasse MN, Yetongnon K (2000) On the definition of generic multi-layered ontologies for urban applications. Computers, Environment, and Urban Systems 24:191-214

Brodaric B, Gahegan M (2001) Learning geoscience categories in-situ: implications for geographic knowledge representation. Proceedings of the Ninth ACM International Symposium on Advances in GIS. Atlanta, GA, Nov. 9-10, 2001. ACM Press, New York, pp 130-135

Brodaric B, Journeay M, Talwar S (1999) Using the proposed U.S. national digital geologic map data model as the basis for a web-based geoscience library prototype [online]. In: Soller D (ed) Digital Mapping Techniques '99 - Workshop proceedings, USGS Open File Report 99-386. pp 83-91 Available from: http://pubs.usgs.gov/openfile/of99-386/brodaric1.html

Clancey WJ (1993) The knowledge level reinterpreted: modeling socio-technical systems. International Journal of Intelligent Systems 8:33-49

Davenport P (2001) The bedrock geology map database for the Canadian Geoscience Knowledge Network. In: Proceedings, International Association of Mathematical Geology Annual Meeting. Cancun, Mexico, Sept. 9-12

Dehn M, Gartner H, Dikau R (2001) Principles of semantic modeling of landform structures. Computers & Geosciences 27:1005-1010

Frank AU (2001) Tiers of ontology and consistency constraints in geographical information systems. International Journal of Geographic Information Science 15(7):667-678

Fonseca FT, Egenhofer MJ, Clodoveu AD Jr., Borges KAV (2000) Ontologies and knowledge sharing in urban GIS. Computers, Environment, and Urban Systems 24(3):251-272

Gahegan M, Brodaric B (2002) Computational and visual support for geographical knowledge construction: Filling in the gaps between exploration and explanation. In: Proceedings, Spatial Data Handling 2002. July 9-12, Ottawa, Canada (in press)

Guarino N (1998) Formal ontology in information systems. In: Guarino N (ed) Formal Ontology in Information Systems, Proceedings of FOIS'98. 6-8 June, 1998, Trento, Italy, IOS Press, Amsterdam, pp 3-15

Hastings J, Brodaric B (2001) Evolution of an object-oriented, NADM-based data model prototype for the USGS national geologic map database project. In: Proceedings, International Association of Mathematical Geology Annual Meeting. 9-12 September, Cancun, Mexico

Heyn G, Kuebler S, Richter B, Skala W, Voisard A (2000) The geohyp project: representing knowledge in geologic hypermaps. GEO-Informationssysteme 13(13):24-28

Johnson BR, Brodaric B, Raines GL, Hastings JT, Wahl R (1999) Digital geological map data model v4.3 [online]. Available from: http://geology.usgs.gov/dm/model/Model43a.pdf

Journeay M, Robinson J, Talwar S, Walsh M, Biggs D, McNaney K, Kay B, Brodaric B, Harrap R (2000) The georgia basin digital library: Infrastructure for a sustainable future. In: Proceedings, GeoCanada 2000. May -- June 1, Calgary

Langley P (2000) The computational support of scientific discovery. International Journal of Human-Computer Studies 53:393-410

Lynch EB, Coley JD, Medin DL (2000) Tall is typical: central tendency, ideal dimensions, and graded category structure among tree experts and novices. Memory & Cognition 28(1):41-50

MacEachren AM (1995) How Maps Work: Representation, Visualization and Design. Guilford Press, New York

Noth W (1990) Handbook of Semiotics. Indiana University Press, Bloomington

Raper J (1999) Spatial representation: the scientist's perspective. In: Longely PA, Goodchild MF, Maquire DJ, Rhind DW (eds) Geographical Information Systems: Principles and Technical Issues, 2nd edn. Wiley, New York, pp 61-70

Raper JF, Livingston D (1995) Development of a geomorphological spatial model using object-oriented design. International Journal of Geographical Information Systems 9(4): 359-384

Rumbaugh J, Jacobson I, Booch G (1999) The unified modeling language reference manual. Addison-Wesley, Reading, MA

Smith B, Mark D (2001) Geographical categories: an ontological investigation. International Journal of Geographic Information Science 15(7):591-612

Sowa JF (2000) Knowledge Representation: Logical, Philosophical, and Computational Foundations. Brooks/Cole, New York

Voisard A (1998) Geologic hypermaps are more than clickable maps. In: Laurini R, Makki K, Pissinou N (eds) ACM-GIS '98, Proceedings of the 6th international symposium on Advances in Geographic Information Systems. 6-7 November, 1998, Washington, DC

Voisard A (1999) Abduction and deduction in geologic hypermaps. In: Guting RH, Papadias D, Lochovsky F (eds) SSD'99, LNCS 1651. Springer-Verlag, Berlin pp 311-329

Indexing and Browsing Digital Maps with Intelligent Thumbnails

Christoph Schlieder and Thomas Vögele

Centre for Computing Technologies (TZI), University of Bremen, PO Box 33 04 40, 28334 Bremen, Germany, {cs|vogele@tzi.de}

Abstract

With the increasing number of digital maps and other geo-referenced data that are available over the Internet, there is a growing need for access to techniques that allow us to preview the content and to evaluate it relative to the requirements for complex spatial and thematic queries. . Analogous to the digital indices for full-text searches on text documents we introduce highly condensed, machine-readable indices of digital maps. The purpose of these "intelligent thumbnails" is to support sophisticated queries of the type concept@location. An intelligent thumbnail is based on a projection of the thematic content of a digital map onto a standard reference tessellation. To make the thumbnail exchangeable in a distributed and heterogeneous environment, the underlying Standard Reference Tessellation (SRT) is qualitatively abstracted from a polygonal representation to a connection graph. In combination with a hierarchical place name structure and concept ontology, it is used to evaluate the spatial and thematic relevance of the indexed data sources with respect to spatio-thematic queries.
Keywords: digital maps, metadata, spatial queries, data reduction, spatial reasoning

1 Introduction

Digital maps and other geospatial data are the valuable assets of many private and public organisations. For that reason, more and more geospatial data are made available through online (meta)data catalogues and Internet portals. To a potential user or buyer, metadata catalogues offer browse services of available data sources, and allow an individual to select datasets best suited for a specific task. The selection process should be based on queries of the type concept@location, i.e. an evaluation of which data sets contain information about a specific thematic concept at a specific geographic location.

Typically, metadata catalogues provide tools to select and rank the available data sets based on their semantic relevance with respect to certain thematic concepts. Most state-of-the-art systems still use simple keyword matches. Considerable research effort has been made, however, for the development of more sophisticated terminological queries, for example with the help of formalised domain ontologies and terminological reasoning (Wache et al., 2001).

On the other hand, the mechanisms available to solve spatial queries and to evaluate the spatial relevance of a data source are still fairly simplistic. Typically, they are based on a match between a bounding box representation of the area covered by a data source, and in most cases, a rectangular search area. All data sets for which the corresponding bounding box intersects with the search area are assumed spatially relevant with respect to the query. Even if the general shortcomings of bounding box reasoning are neglected, the fact that a digital map covers an area delimited by its bounding box does not necessarily imply that the map actually contains information about the concept in question at every location within this area.

For example, the digital map of *Federal Land Features of the United States* published by the U.S. Geological Survey (USGS) (USGS, 2000) covers the contiguous and non-contiguous United States. The data set contains the polygon features of all federally administered lands larger than 640 acres, their name identifiers, and information about the federal agencies in charge.

For a metadata description of the data set, the USGS uses the Content Standard for Digital Geospatial Metadata published by the Federal Geographic Data Committee (FGDC) (FGDC, 1994), and expresses the spatial coverage of the map in terms of a bounding box. As a result, somebody interested in, for example, the outlines of *National Parks* in *Contra Costa County, California*, will be pointed to the *Federal Land Features* map because a) it contains information about national parks, and b) it covers the whole contiguous United States, which *Contra Costa County* is a part of. After a time-consuming download (the data set exceeds 50 MB), a detailed analysis of the data with the help of a geographic information system (GIS) would reveal that there are in fact no *National Parks* in *Contra Costa County*, making the data set unsuitable for the intended purpose.

The example above shows, that metadata for geospatial data often do not provide enough information to make educated decisions about the usability of data sets for specific tasks. Obviously, there is a need for tools that support more sophisticated screening-level queries, which do not require the download of large data files and the application of complex GISs. Some GISs, like for example ESRIs ArcGIS 8.x, therefore introduced data management tools like ArcCatalog (ESRI, 2001). In ArcCatalog, small raster images of digital maps, the so-called "thumbnails", can be created to provide a visual preview of the data.

We argue, that these purely visual indices do not suffice to support sophisticated spatial queries. Firstly, they can only represent a single thematic concept at a given location, and secondly they have to be analysed and evaluated "manually" by a human, thus providing no means for effective automatic searches. We therefore propose to create "intelligent thumbnails", i.e. small, machine-readable indices of the thematic and spatial content of digital maps, analogous to the thematic indices used for full text searches in digital text documents.

2 Intelligent Thumbnails

2.1 Components of an Intelligent Thumbnail

An intelligent thumbnail is a machine-readable and highly condensed index of the thematic and spatial information-contents of a digital map. It is designed to support reasoning about whether the map is relevant with respect to a specific query of the type *concept@location*. The level of relevance assigned to a map depends on how closely the content of the map matches the query. This takes into account direct matches of the specified *concept* and *location*, but also near matches, which result from terminological specialisation or generalisation of the concept, as well as spatial generalisations of the location.
The index references the thematic and spatial contents of a data source (i.e. a digital map) to both thematic domain ontology and a spatial reference model; see Fig. 1. The spatial reference model consists of a standard reference tessellation (SRT) and a place name structure.

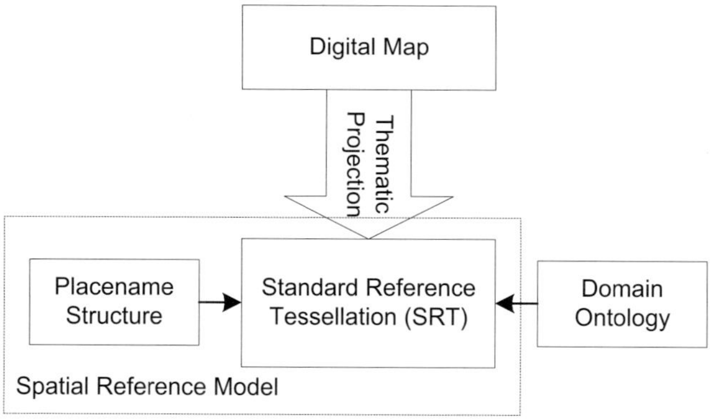

Fig. 1. Components of an Intelligent Thumbnail

2.1.1 Thematic Projection onto a Reference Tessellation

An intelligent thumbnail links concepts to locations. In this context, the term "concept" refers to the thematic attributes of the features represented in the digital map and the term "location" to geographic entities represented in a spatial reference model. The thumbnail is created through a projection of the thematic layers of a digital map onto a spatial reference model (Fig. 1).

The spatial reference model is based on a polygonal tessellation of spatial entities with a coverage equal to or larger than the extent of the digital map to be indexed. To make the indices of different digital maps comparable, the spatial reference model has to be standardised; i.e. it has to use a well-defined set of spatial entities and descriptors. In addition, to allow for complex spatial queries as described above, the spatial reference model has to support reasoning about the topological and partonomic relations between spatial entities.

One option to build a standardised spatial reference model is to use a uniform grid. This approach is applied by some of the more advanced gazetteer services such as the gazetteer integrated in the German Environmental Information Network (GEIN) (Riekert, 1999), (Bilo et al., 2000). Gazetteers are place name-lists that link the names of geographic features to geographic footprint representations (Hill, 2000). The projection of the polygonal footprint of a geographic feature onto a regular, homogeneous grid is depicted in Fig 2. Projecting geographic footprints onto reference grids allows us to apply spatial reasoning in topologically related ways. For example, if the grid cells occupied by object A are also occupied by object B, it can be inferred that object A is contained in object B. Other topological relations like disjunction, connection, and overlap, can also be evaluated using this approach.

Instead of uniform grids, we propose to use polygonal tessellations as the basis for spatial reference models. Maps of postal code areas, administrative units, and census districts are examples of such Standard Reference Tessellations (SRTs). Polygonal SRTs have a number of advantages over uniform grids:

- Many polygonal SRTs represent well-known and officially named spatial entities, which a user can relate to more easily than to arbitrarily created and cryptically named grid cell rasters. For example, it is much easier to refer to the place name *Contra Costa County* than to a grid cell descriptor like *CA1089*.
- From a user perspective, polygonal SRTs are a conceptually more logical way in which to organise spatially distributed data. Many companies, for example, arrange their marketing areas along the lines of postal code areas or other popular reference tessellations. As a result, polygonal SRTs are available in many organisations in digital form, including GIS data formats.
- Administrative units and other SRTs are typically associated with a hierarchical partonomic structure. Each state in the U.S., for example, consists of a number of counties which consists of a number of communities. As we will see in section 2.2.1, the evaluation of such hierarchical partonomies can be part of a metric to compute spatial closeness in an attempt to evaluate the spatial and thematic relevance of a data set.

As we pointed out above, an intelligent thumbnail is created by mapping the thematic layers of a digital map onto the a polygonal SRT; (see Fig. 2) This can be achieved effectively within a GIS, using standard GIS functionality. Good results were achieved with a prototypical extension for ESRI's ArcView desktop GIS. The extension uses a GIS-specific polygonal representation of the SRT (i.e. ESRI shape files) for the mapping task. The result of the process is an XML-encoded list of thematic concepts, assigned to the name descriptors of the SRT polygons.

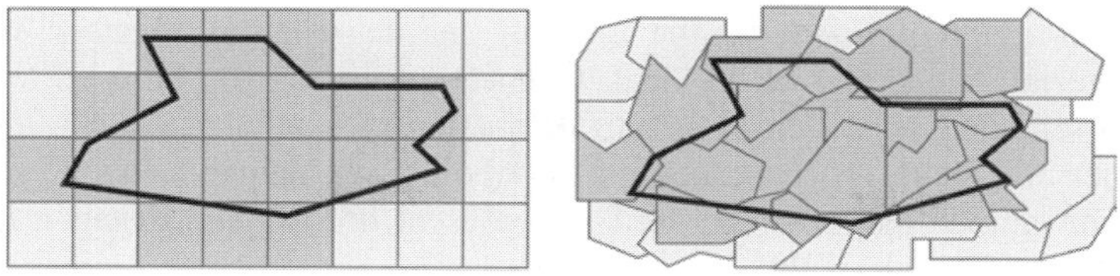

Fig. 2(a) Thematic Mapping onto a Regular Grid, and **(b)** a Polygonal Standard Reference Tessellation (SRT)

2.1.2 Qualitative Spatial Models

In the previous section we argued, that polygonal tessellations representing well-known spatial units, like administrative boundaries and postal code areas, provide better and more intuitive spatial reference models for intelligent thumbnails than regular grids. Polygonal SRTs do, however, have some disadvantages as well. One disadvantage is that different polygonal representations of the same SRT may not be identical in terms of an exact match of vertex co-ordinates. Such mismatches can have a number of reasons, like different levels of accuracy and spatial resolution, or different coordinate systems and projections. As a result, each data provider should or must use the same, "officially approved" polygonal SRT to create a meaningful intelligent thumbnail.

The second disadvantage is, that by using polygonal SRTs in the form of GIS data sets as a basis for the search through intelligent thumbnails, the spatial reasoning needed to resolve spatial queries has to rely heavily on GIS functionality. This requires the handling of potentially large volumes of GIS data and the availability of complex GIS software even for a browsing-level information retrieval. Especially in highly heterogeneous and distributed data-exchange infrastructures, this may hamper the flow of information and limit the number of participants (Vögele et al., 2002).

We argue, that for the spatial reasoning tasks needed in the context of information retrieval, most of the spatial information-content of polygonal GIS data is superfluous. In order to make the information retrieval independent of potential GIS-related inconsistencies and complexities, and to create light-weight, exchangeable intelligent thumbnails, we propose to use qualitative spatial reference models. Such models are based on a combination of qualitative abstractions of polygonal spatial reference tessellations and take the form of connection graphs, and hierarchical place name structures.

2.1.3 Connection Graphs

We use the concept of connection graphs as it was defined in (Schlieder et al., 2001). Connection graphs are an extension of the well-known neighbourhood graphs (Molenaar, 1998) such that the topological neighbourhood relations of polygons in an homogeneous tessellation are encoded, with their ordering, and, if applicable with the connection to an external area. Technically speaking, the connection graph consists of the dual of the tessellation together with the combinatorial embedding of the dual. Connection graphs can be used to encode standard reference tessellations as part of qualitative spatial models. Although in this paper we focus on the encoding of topological (neighbourhood) relations, connection graphs may also be used to represent ordinal and distance relations (Stuckenschmidt et al., 1999).

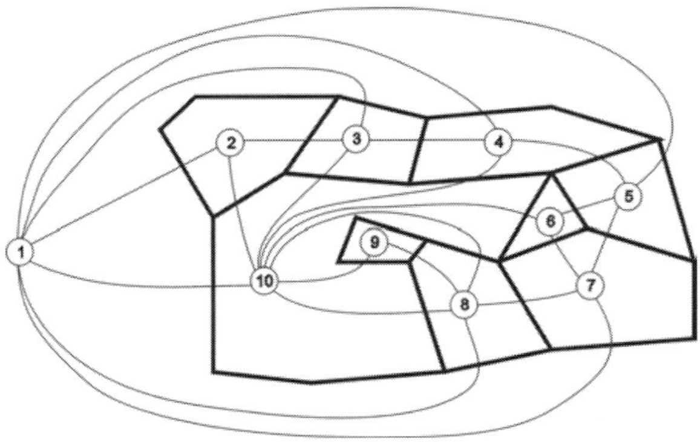

Fig. 3. Connection Graph G_c of Polygonal Tessellation D

Fig. 3 shows the connection graph G_c obtained by homogeneous decomposition of a polygonal tessellation D. Each polygon from D is represented by a vertex from G_c. In addition, there is the node 1 representing the external polygonal region. The edges from G_c, which are incident with a vertex, are easily obtained together with their circular ordering by scanning the contour of the corresponding polygon. For polygon 10 the following circular sequence of neighbours is obtained: 1, 2, 3, 4, 6, 8, 9, 8. Note that polygon 8 appears twice in this list because it shares with 10 two disconnected polygon edges. On the other hand, polygon 9, which shares three edges with 10, appears only once because the three edges are connected. As the example shows, the connection graph is a multi-graph in which several edges can join the same pair of vertices.

2.1 4 Place Name Structures

Place name structures link the entities of standard reference tessellations (SRTs) to place names, which are the basis for intuitive and user-friendly spatial queries. A place name structure can be seen as an hierarchical tree, where the nodes of the tree represent well-known name descriptors for geographic features, and the edges reflect the binary part-of relations between these features. In a qualitative spatial model, the leaves of the tree coincide with the nodes of the connection graph representing the SRT (**Fig. 4**. Hierarchical Tree of a Place name Structure).

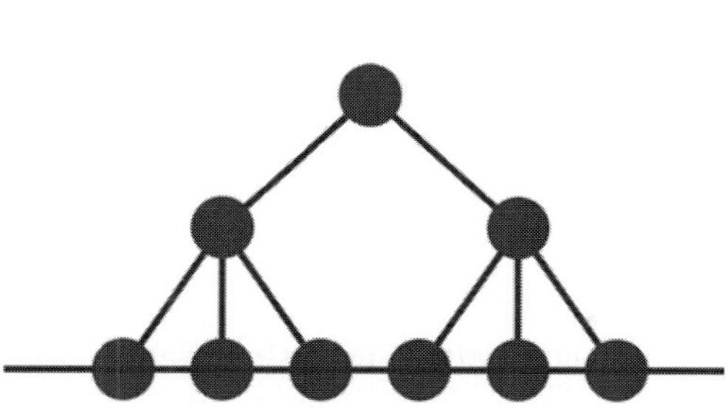

Fig. 4. Hierarchical Tree of a Place name Structure

For many SRTs, place name structures are readily available: Administrative units, for example, are typically represented by hierarchical trees that describe the part-of relations between entities like countries, states, counties, communities etc. Used as a framework for qualitative spatial reference models, such standard place name structures provide a common vocabulary for spatial references. They ensure that all spatial references and spatial queries in an intelligent thumbnail can ultimately be mapped onto the same, unambiguous and consistent set of place names.

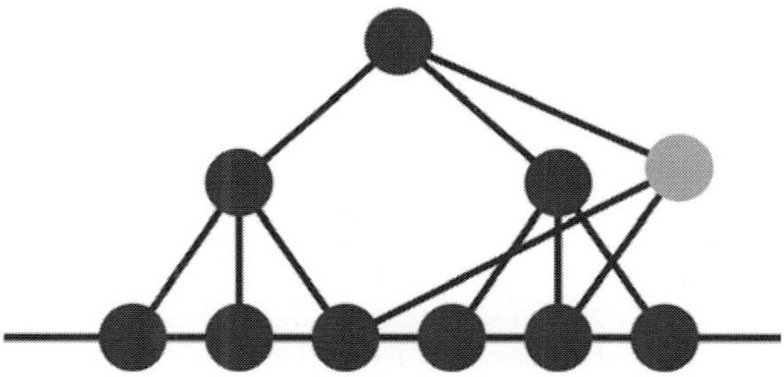

Fig. 5. Poly-Hierarchy of an Extended Place name Structure

However, spatial reference models and standard place name structures have to be extensible in order to allow for the incorporation of "colloquial" place names, i.e. commonly used descriptors for geographic features, as well as user-specific terms. For example, the place name *East-Bay* is a popular term to describe a number of counties situated on the eastern side of the *San Francisco Bay*. To solve a typical user-driven query like "Are there any lands managed by the National Park Service in the *East-Bay*?", the system has to have access to a place name structure that formalises the spatial semantics of the term *East-Bay*.

A qualitative spatial model of the administrative units of *California* can be extended to "understand" the meaning of the term *East-Bay* by establishing the respective part-of relations based on an existing place name structure. The result is a poly-hierarchic, Directed Acyclic Graph (DAG) representing an extended, user-specific place name structure (Fig. 5). Because qualitative spatial models do not rely on GIS functionality and complex binary data formats, it should be easy to provide simple tools that support the manipulation of place names in a user-friendly way. This is the basis for highly distributed data exchange infrastructures, where users may modify standard place name structures to match their individual needs and use them to specify highly personalised spatial queries (Vögele et al., 2002).

2.1.5 Domain Ontologies

A domain ontology is the basis for the evaluation of the semantic connection between thematic concepts specified in the query and concepts indexed in the intelligent thumbnail. For example, in a query like *NPS-lands @ Contra Costa County*, we could use a domain ontology representing the organisational structure of US federal agencies and federal lands to find out which types of federal lands are managed by the *National Park Service (NPS)* (Fig. 6).

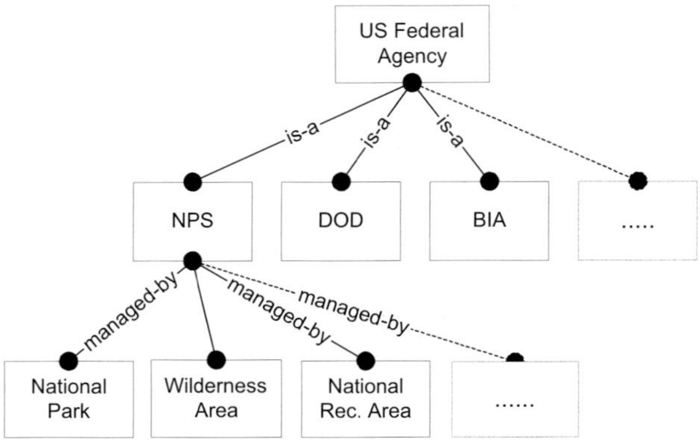

Fig. 6. Ontology of US federal agencies and federal lands (schematic)

This paper is focused on the solution of the *location*-part of a query of type *concept@location* using spatial reasoning. For brevity, however, we will not explore the representation of domain ontologies and terminological reasoning. There are a number of approaches and systems available for this task (Wache et al., 2001). For the prototypical implementation of the intelligent thumbnail, we used description-logics based on the knowledge representation language OIL (Ontology Interchange Language) (Fensel et al., 2000; Stuckenschmidt, 2000), in conjunction with the RACER theorem prover (Haarslev et al., 2001).

2.2 Reasoning about Spatial and Thematic Relevance

With the help of concepts described above, queries of type *concept@location*, have been specified and resolved. In addition, the data sets retrieved have been ranked based on their spatial and thematic relevance with respect to the query. The basis for such evaluations are metrics that express the degree of spatial and terminological distance between locations and concepts represented in the thumbnail, and locations and concepts specified in the query.

2.2.1 Metrics for Spatial Distance

In an intelligent thumbnail, metrics for spatial distance can be computed based on the connection graph representation of the SRT and the DAG representation of the place name structures, respectively. In the simplest case, spatial closeness can be expressed using a metric based on graph-theoretical node distances.

In the connection graph, the Euclidian distance ν between two nodes N1 and N2 is a measure for the actual spatial proximity of the two areas represented by these nodes. In the DAG, the distance δ between the same nodes N1 and N2 indicates their degree of separation with respect to a hierarchical partonomy. The total distance, or spatial relevance measure, D(N1,N2) between N1 and N2 is obtained by a linear combination of ν and δ:

$$D(N1,N2) = \alpha\, \nu(N1,N2) + (1-\alpha)\, \delta(N1,N2),$$

By manipulating the weighting factor α, a spatial query may be fine-tuned to favour either locations that are spatially close ($\alpha = 1$) to the place name of interest, or locations that belong to the same part of a hierarchical partonomy ($\alpha = 0$). The value of α depends mostly on the intent of the query and the semantics of the concepts in question. In the case of federal lands mapped on administrative units, for example, hierarchical closeness is less important than spatial proximity. This is the case, since the indexed concepts (i.e. the federal agencies managing the lands) do not change between the administrative partonomy units. In the case of lands managed by state agencies, however, hierarchical distance could be more important because the same concepts (i.e. state parks) have different properties (i.e. are managed by different agencies) in different units of the partonomy (i.e. different states).

In the prototypical implementation of the intelligent thumbnail, α is set to a default value of 0.5, resulting in an equal weighting of spatial and hierarchical proximity. It is up to the user to change α and fine-tune the query. However, concept-specific default values for α could also be included in the terminological domain ontology, allowing for an automatic fine-tuning of queries.

2.2.2 Metrics for Thematic Distance

To evaluate the thematic or semantic distance between two concepts, both concepts have to be part of a formalised concept hierarchy, or ontology. As described in section 2.1.5, we use the knowledge representation language OIL to encode concept ontologies, and a theorem prover based on description logics to reason about the semantic distance σ between two concepts C_q and C_t. In the simplest case, a binary metric based on subsumption could be defined: C_t is either a sub-concept of C_q, or it is not, making σ be either 1 or 0.

Of course, such a simplistic approach does not suffice and in the long run has to be replaced by a more powerful approach. Jones (Jones et al., 2001), for example, proposes a weighted shortest path procedure to evaluate the semantic distance between two concepts C_q and C_t in a semantic net. However, as we focus on spatial queries in this paper, we will not discuss metrics for thematic distance in more detail.

2.2.3 Combined Metric for Spatial and Thematic Distance

The result of a query on a set of data sources indexed by an intelligent thumbnail, is the ranking of these data sources based on their spatial and thematic relevance with respect to the query. As a first approach, we use a spatial relevance metric R_S, which is a linear combination of the total spatial distance D and the semantic distance σ:

$$R_S = \beta D + (1-\beta)\sigma$$

Again, the weighting factor β can be used to bias the query towards spatial relevance (β=1), or thematic relevance (β=0).

2.3 Data Reduction

The main objective for creating intelligent thumbnails is to be able to index and preview the spatial and thematic contents of digital maps without having to access the data set in full. Especially in distributed and heterogeneous data exchange infrastructures, it is very important to keep the thumbnail size and underlying qualitative spatial reference model as small as possible. Further, open, non-proprietary data formats should be used for their representation. At the same time, enough information has to be retained to support complex queries of the type *concept@location*, and to rank digital maps based on the thematic and spatial relevance metrics described above. To achieve this objective, a number of data reduction and generalisation measures must be applied:

1. The first generalisation step is the selection of a subset of all the thematic layers in a digital map. This selection reflects the data provider's choice of which thematic contents of the data source is important, and should therefore be included in the intelligent thumbnail.
2. In a second generalisation step, thematic concepts are abstracted from objects and lumped into a number of feature types. A specific site (i.e. *John Muir National Historic Site*) is lumped together with other sites and represented as one feature type (i.e. *National Historic Site*). Depending on the level of thematic detail wanted, the index can be condensed even further by using concepts that are higher up in the concept hierarchy. The feature type *National Historic Site* , for example, can be abstracted as *land managed by the NPS.*
3. Maybe the largest generalisation and data reduction takes place during the thematic projection. All selected feature types in a digital map are projected

onto a standard reference tessellations, but if two identical concepts happen to map onto the same spatial entity, only one reference is maintained. The concept *National Historic Site*, for example, would be assigned only once to *Contra Costa County*, even if several such sites exist in the same spatial unit. Applied to the *USGS Federal Land Features* map, the initial 53 MB of map data (ESRI shape format) can be reduced to an index of about 500 KB (XML ASCII format). This amounts to a data reduction factor of approximately 1: 100.

4. Considerable data reduction can also be achieved by using a qualitative, graph-based representation of polygonal standard reference tessellations. For example, the 9.8 MB ESRI shape file holding the polygonal tessellation of all counties within the contiguous United States could be reduced by a factor of 6 to a connection graph of 1.6 MB (XML ASCII format).
5. In addition to data reduction, generalising polygonal standard reference tessellations as connection graphs help to avoid the typical problems of heterogeneous GIS data, like incompatible discretizations, proprietary data formats, and different coordinate-systems and map projections.

3 Summary and Discussion

In this paper, we describe intelligent thumbnails and qualitative spatial models as the basic components to create machine-readable indices of the spatial and thematic contents of digital maps. Such indices are necessary to support queries of the type concept@location, i.e. to select from a large number of data sources the ones that are relevant with respect to the thematic and spatial criteria specified in a query.

Although it is the ultimate goal of our work to combine the evaluation of thematic and spatial relevance, this paper was focused mainly on the representation and reasoning about spatial concepts. The reason for this bias is the fact, that while considerable research was concerned with thematic and terminological knowledge representation and reasoning in the last few years, comparatively little attention was given to the development of qualitative spatial representations and reasoning within the geographical data application domain.

Jones (Jones et al., 2001) recently proposed the use of "parsimonious" spatial models for geographical information retrieval. Similar to the intelligent thumbnail, his system uses metrics for hierarchical, spatial (euclidean), and semantic distance to come up with an overall score for geo-referenced data objects in a database. In addition, in the area of place name structures, Hill (Hill, 2000) is working on a data standard for gazetteer services.

Many aspects of the digital thumbnails presented in this paper are still based on simple assumptions and methods. One example is the metrics for spatial and thematic distance evaluations. Our future work will be concerned with refining these assumptions and incorporating more sophisticated methods into the intelligent thumbnail.

References

Bilo M, Streuff H (2000) Das Umweltinformationsnetz Deutschland - GEIN2000 - Fachliche Anforderungen an ein Forschungs- und Entwicklungsvorhaben. 3rd workshop Hypermedia im Umweltschutz, Ulm

ESRI (2001) System Design Strategies. Environmental Systems Research Institute, Inc.

Fensel D, Horrocks I, Harmelen F V, Decker S, Erdmann M, Klein M (2000) OIL in a Nutshell. In: 12th International Conference on Knowledge Engineering and Knowledge Management EKAW 2000. Juan-les-Pins, France.

FGDC (1994) Content Standards for Digital Geospatial Metadata [online]. Washington, D.C., US Government, Federal Geographic Data Committee. Available from: ftp://fgdc.er.usgs.gov.

Haarslev V, Möller R (2001) RACER System Description. University of Hamburg, Hamburg, Germany

Hill LL (2000) Core elements of digital gazetteers: place names, categories, and footprints [online]. In: ECDL 2000, Lisbon, Portugal. Available from: http://www.alexandria.ucsb.edu/~lhill/paper_drafts/ECDL2000_paperdraft7.pdf.

Jones C, Alani H, Tudlope D (2001) Geographical Information Retrieval with Ontologies of Place. In: COSIT 2001. Morro Bay, California

Molenaar M (1998) An Introduction to the Theory of Spatial Object Modelling. Taylor & Francis, London Bristol

Riekert W-F (1999) Erschließung von Fachinformationen im Internet mit Hilfe von Thesauri und Gazetteers. In: Management von Umweltinformationen in vernetzten Umgebungen, 2nd workshop HMI. Nürnberg

Schlieder C, Vögele T, Visser U (2001) Qualitative Spatial Reasoning for Information Retrieval by Gazetteers. In: COSIT'01. Morro Bay.

Stuckenschmidt H (2000) Using OIL for Intelligent Information Integration. In: Workshop on Applications of Ontologies and Problem-Solving Methods at the European Conference on Artificial Intelligence ECAI 2000. Berlin

Stuckenschmidt H, Visser U, Schuster G, Vögele T (1999) Ontologies for Geographic Information Integration. In: Workshop Intelligent Methods in Environmental Protection: Special Aspects of Processing in Space and Time, 13. International Symposium of Computer Science for Environmental Protection (CSEP '99). University of Bremen

USGS (2000) Federal Land Features of the United States [online]. Reston, VA, U.S. Geological Survey. Available from: http://nationalatlas.gov/atlasftp.html.

Vögele T and Schlieder C (2002) The Use of Spatial Metadata for Information Retrieval in Peer-to-Peer Networks. Proceedings of AGILE2002, Palma de Mallorca, Spain (in press)

Wache H, Vögele T, Visser U, Stuckenschmidt H, Schuster G, Neumann H, Hübner S (2001) Ontology-Based Integration of Information - A Survey of Existing Approaches. In: IJCAI 2001.

Modelling and Manipulating Multiple Representations of Spatial Data

Christelle Vangenot[1], Christine Parent[2], Stefano Spaccapietra[1]

Swiss Federal Institute of Technology, Database Laboratory, Lausanne, Switzerland, christelle.vangenot@epfl.ch, stefano.spaccapietra@epfl.ch[1]
University of Lausanne, INFORGE, Lausanne, Switzerland, christine.parent@hec.unil.ch[2]

Abstract

One of the requirements which is poorly supported by spatial data models is a consistent management of different representations of the same spatial phenomena from different viewpoints or at different resolutions. This need is well known by users and designers. Modelling of applications, where users share the same database for different contexts and cartographic applications are examples of environments where such a need arises. This paper proposes a conceptual data model providing full support for multiple representations of the same real world data. The model addresses two complementary aspects: the integrated approach, that leads to the definition of customised database items, and the inter-relationship approach, where the representations are linked through inter-representation links. Both approaches use the stamping technique to differentiate among multiple representations of a given phenomenon and to access a particular representation. Finally, we focus on consequences of multiple coexisting representations on data manipulation. This proposal has been tested and validated with users, and implemented as a front-end to existing DBMS[1].
Keywords: multi-representation, multi-resolution, spatial data modelling, databases.

---

[1] We gratefully acknowledge support for this research by the European Economic Community (MurMur Project - "Multiple Representation – Multiple Resolution" - IST 10723), by the Swiss OFES (grant 99999), by EEC Interreg II Program (Cobalt Project), and by the Swiss National Research Fund (grants 8765-2000 & 112-2002).

1 Introduction

Interoperability supports communication and access among information repositories, giving users an opportunity to locate complementary information about the same or related facts from various sources that have been independently developed. Unfortunately, semantic interoperability is not easy to achieve, as related knowledge is most likely to be described in different terms, using different assumptions and different data structures. Reconciling this heterogeneity to build a fully integrated database is a complex problem and largely unresolved.. A simpler step is the identification of related knowledge and the provision of a mechanism that in some way establishes the relationships among different representations of the same elements, i.e. that supports multiple representations of data. One well known example of applications where multiple representations is a crucial need and are not supported by current technology is cartographic applications. Map producers often need to build maps of the same geographic region at different levels of detail. Cartographic generalisation, the process of deriving a less detailed map from a more detailed one, is a complex process that usually remains partially interactive (Müller *et al.* 1995). Often, map producers have to maintain one database per scale, and do not directly maintain interrelationship between multiple scale databases. As a result, there is generally no update propagation and inter-database consistency is lost. A multiple representation GIS would allow storing all representations in a single database, and support the consistency of the different representations through appropriate and automatic update propagation. Beyond cartography, multiple representations of geographic data are needed to be able to serve multi-disciplinary user communities, as the same piece of land may support analysis, planning, and forecast activities by city administrations, environmentalists, sociologists, botanists, zoologists, among others.

Research on multi-representation has often been addressed in specific contexts such as multi-scale databases, views, and multi-instantiation, each of which are separate research areas with specific requirements. A significant part of the research in multi-scale databases was inspired by the largely hierarchical nature of transitions between scales (Weibel *et al.* 1999). (Timpf 1998) proposed to keep the representations of maps at different scales in hierarchical data structures, where levels correspond to increasing detail and the representations of the same object are linked together. In (Kilpelaïnen 1998), these multiple representation links are used to propagate updates. (Martel 1999) proposes patterns to deal with multi-representation problems. His solution, implemented in VUEL (Bernier *et al.* 2001) allows the association of several thematic, graphic and semantic characteristics to the same geographic object. (Stell *et al.* 1998) see the database organised as a stratified map space with transformation functions, where each map gathers objects that share the same semantic and spatial granularity. (Devogele *et al.* 1998) built a federated database on top of mono-scale databases. Geodyssey (Jones *et al.* 1996), a deductive knowledge-based system allowing multi-scale and temporal representations, incorporates reasoning processes to maintain consistency and propagate updates.

This paper presents a set of modelling concepts allowing each information fact to be described through multiple, consistent, and possibly irreducible representations. The next section defines which aspects of multiple representations we deal with. Section 3 presents our suggested framework and data structures that adequately support multi-representation. Sections 4 and 5 detail the concepts and design principles for multiple representation modelling. Section 6 suggests some extensions to algebraic operators to be able to manipulate our new modelling concepts. Section 7 presents the implementation of the model. The conclusion points at additional work that complements the results reported in this paper.

2 What is Multiple Representation?

Databases are intended to keep an integrated and consistent set of data that provides the information needed to support application requirements from one or several user communities. These data represent real-word phenomena that are of interest to users of the database. While the real world is assumed unique, the way it is represented depends on the intended use. Thus, different applications that share interest in the same real-word phenomena may have different perceptions and therefore require different representations; i.e. different sets of objects, links, properties, and/or different values. Their needs differ because they don't share the same viewpoint, they don't need data acquired at the same instant or with the same cartographic resolution. Multiple resolutions may be used in each of these cases as well as in others. In this paper, we explicitly deal with two of them, but our solution is generic and can be applied to other aspects. We selected the viewpoint aspect, which is fundamental for data sharing, and we address the aspect concerning the level of detail (or resolution) of the geographic data.

A **viewpoint** is the expression by a group of users about their specific interests in data management. It acts as a complex abstraction process where objects and links are filtered, leaving out whatever in the real world is not of interest for the particular viewpoint. This delimits the so-called universe of discourse. Then each phenomenon in the universe of discourse is described according to the viewpoint. This induces multiple design decisions, such as the choice of a representation concept (the same phenomenon may be represented as an object, an attribute or a relationship), the elaboration of its type (among all possible properties of the phenomenon, only a subset is of interest for the viewpoint), and for each property the choice of its representation.

Every representation conveys a simplified description of the reality. This materialises into a level of detail that characterises both spatial and thematic characteristics of information (Molenaar 1998). In the spatial dimension, the level of detail, called **spatial resolution**, determines the geometric aspects of the phenomena. Spatial resolution is defined by thresholds (Peng 1997). For instance, spatial resolution could be defined as a process of filtering out all 2D area objects whose surface is less than 100 square meters and all 1D linear objects whose total length is less than 150 meters. Beyond filtering, spatial resolution also has a smoothing effect: given a detailed geometry, a less-detailed representation will retain a simplified geometry that leaves out all irregularities whose size is less than the given threshold. It also has a merging effect: distinct geometries whose distance is less than the threshold, collapse into a unique geometry. A set of buildings, for instance, may collapse to form a single built-up area when the details on individual buildings are no longer of interest. The concept of multiple levels of detail also applies to thematic data, and is called **semantic resolution.** Semantic resolution allows filtering out objects/relationships/attributes that are not relevant while working at a specific level of detail. Semantic resolution may be specified by the number of classes, the membership rules of classes, the number of attributes of classes, the granularity of hierarchical value domains (Rigaux *et al.*1995), the depth of is-a and aggregation hierarchies. In the following sections, we use the term resolution to cover both spatial and semantic resolution.

3 The Multi-Representation Framework

3.1 MADS Data Model

Our objective is to define a set of concepts necessary for describing datasets with multi-representation. Our prototype was built as an extension to an already existing data model, the MADS data model (Parent *et al.* 1998). MADS is an object+relationship, spatio-temporal, conceptual data model. In this model, we assume that the real world of interest that is to be represented in the database is composed of objects, their links, and their static and dynamic properties (attributes and methods). Spatial and temporal aspects may be associated at the different structural levels: object, attribute and relationship. The spatial characteristics of an object convey information about its location and its extent, while its temporal criteria describe its lifecycle. Attributes may have spatial (e.g. point or area) or temporal (e.g. instant or time interval) domains of values. They also may be space- or time-varying. Specific relationships describe topological constraints between spatial and temporal objects. In MADS icons express information in an unambiguous and visual way. An example of a schema, drawn with the MADS schema editor, is presented in Fig. 7.

3.2 Proposed Framework

The two facets of multi-representation, viewpoint and resolution are introduced in the data model through the definition of representation stamps. A stamp is a (viewpoint-value, resolution-value) pair, for instance ("cartography", 1 meter). Stamps characterise the different representations of the real world objects. They are like meta-data specifying the context of elaboration (here the viewpoint and the resolution) of each representation. Any schema or database element, i.e. object type, relationship type, attributes, instance, value, holds one or more stamps. An

element bears several stamps when it is shared by several contexts. For instance, Fig. 1. Stamping attributes and their values
in which the object type Road holds two stamps, e1 and e2 (for ("cartography", 1) and ("hydrography", 5)), which means that this object type is shared by the contexts defined by the stamps e1 and e2. Beyond modelling, stamps are also used during data manipulation for filtering or directing the access to data. In our model, users' transactions also bear stamps (see section 6) that specify which data the transaction may access.

We propose two complementary approaches to define multiple representations for the same real world phenomenon. We denote the first strategy as the **integrated approach**. The idea is to have a unique name and container for various representations while identifying the specifics of each representation using the stamping technique. Each construct of the data model can be turned into a multi-representation construct by stamping it with a representation stamp. A multi-representation construct (object, attribute, link) embeds within an integrated definition the different representations, each representation being characterised by one or several stamps. Section 4 below discusses multi-representation object types, attributes, and relationship types.

When representational needs for the same phenomena are so diverse that they can hardly be integrated into a common definitional framework, it may be more convenient to independently define the needed representations, and to interrelate them through dedicated links. The latter allow an expression of the multi-representation semantics. We call this the **inter-relationship approach**. These links, named correspondence relationship types in section 5, relate two representations of the same real-world entity described by different instances and pertaining to different viewpoints and/or resolutions.

The two approaches are complementary and both essential to describe multiple representations. A model offering only the integrated approach does not allow the description of correspondences between groups of objects. This may prevent the model to be applicable in situations where the multi-representation database has to be built on top of (or derived from) multiple existing data sets. Conversely, a data model with only correspondence links does not provide 'orthogonal modelling' tasks, as links can only be defined between objects but not between attributes or values. This means that differences in viewpoint or resolution would invoke the modelling of the related representations as objects but not as attributes or values. The designer's freedom in choosing a modelling construct would be restricted.

4 Multi-Representation Database Elements

4.1 Stamping Object Types, Attributes and Their Values

Let us use an example to illustrate the definition of multi-representation database elements. Consider two representations of roads, identified by stamps e1 and e2. Roads are spatial objects whose spatial extent, depending on the representation, is represented as an area (most precise representation, stamp e1) or a line (less precise one, stamp e2). Moreover the road number, the name of the road, its administrative classification, its type, and the name of its main manager are interests of representation e1. Representation e2 also needs the road number, its name (but with another naming scheme), its type (but with another classification), the department in charge of it, and its set of managers. Fig. 1 shows the stamped object type Road and its attributes in which both representations (e1 and e2) are embedded. The stamps of the object type are shown in the rectangle between the rectangles containing the object type name and the list of its properties. Stamps are also written next to each attribute definition, thus specifying the stamps for which the definition is pertinent. The Road object type is stamped with two stamps, e1 and e2; that means that this object type is shared by the applications identified by the stamps e1 and e2. It can be accessed by the transactions having the stamp e1, or the stamp e2, or both stamps. Its attributes are also stamped, i.e. the structure (in term of sets of attributes) of the object type changes according to the stamps. It behaves as if it has two definitions, one for stamp e1 and one for stamp e2.

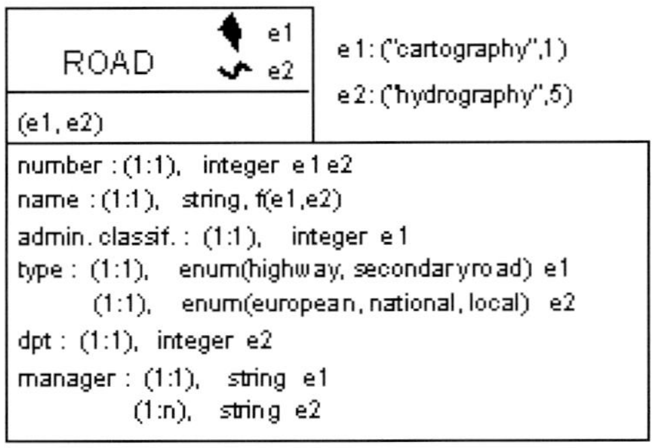

Fig. 1. Stamping attributes and their values

Furthermore, if we consider its attributes:

- The attribute *number* has the same stamps as its object type Road: This attribute is shared by both representations. It is accessible by the transactions having the stamp e1, or the stamp e2, or both.
- The annotation 'f'(e1, e2) after the attribute *name* means that its value is varying according to stamps. It is thus possible to store several values for this attribute, one per specified stamp (or less if several stamps share a value). It is thus possible to store two values for name, like for instance, "RN7" for stamp e1 and "Route Napoléon" for stamp e2.
- The attribute *administrative classification* (respectively *dpt*) belongs to the representation e1 (resp. e2). It is only accessible by the transactions having stamp e1 (respectively e2) or both stamps.
- The attributes *geometry*, *type* and *manager*, have two definitions and therefore two values, one for the stamp e1 and one for the stamp e2. The stored values of the attributes geometry and type pertain to different value domains. It is thus possible to store the geometry of an area for stamp e1 and a line for stamp e2.

The values of the attribute manager belong to the same domain but the cardinality of this attribute is different. A single value is stored for transactions of stamp e1 and several values are stored for transactions of stamp e2.

Instances of object types are also stamped (so are instances of relationship types). They can have either the same set of stamps as their type or only a subset of it. A stamped instance has values for the attributes that are defined for its stamps. An instance of Road with the stamps (e1, e2) has values for all the attributes. Transactions of stamp e1, or e2, or (e1, e2) can see this instance: transactions of stamp e1 with the attributes defined for e1 or (e1, e2); transactions having both stamps with all attributes. An instance of Road with the stamp e1 has values for the attributes belonging to the definition at stamp e1. Transactions of stamp e1 or (e1, e2) can view this particular instance.

All the database items are stamped, however, as they are not independent, they can not be stamped irrespective of each other. We have defined rules specifying stamping consistency. The first rule says that the stamps of database elements must be included in the set of stamps of the element they are part of: Stamps of object/relationship types have to be included in the set of stamps of the schema they belong to, stamps of attributes in the stamps of the type they are defined in, stamps of attribute values in the stamps of the attribute, and stamps of instances in the stamps of their type. For instance, the attributes of the road object type may be stamped with (e1, e2) link number, or just with e1, or just with e2, link dpt.

4.2 Stamping linked elements

Accessing a relationship type is different from accessing an object type. Indeed, relationship types cannot be considered regardless of the object types they link: access to a relationship type instance must always enclose access to the object type instances linked. Hence, we rule that a transaction may only access a relationship type if its stamps also give access to all the linked object types. By default relationship types are not stamped. With respect to object types, it is possible to define multi-representation relationship types. Stamping a relationship type allows one to characterise the properties and instances of the type and to add access conditions. For example, stamping the instances of the topological relationship type 'near' shown in Fig. 2 allows one to describe cartographic representations at different scales, where the same house is adjacent to a road in the less detailed representation (e.g. stamp e1) and no longer adjacent in the most detailed representation (e.g. stamp e3) (Jen 1995). To describe this situation, the instance of the near relationship linking this house and this road has to be stamped with e1 only.

Fig. 2. Stamping a topological relationship type

If the relationship type is explicitly stamped, the requesting transaction must have at least one of the stamps of the relationship type, in addition to stamps giving access to the linked object types.

Topological and synchronisation relationship types are a particular kind of relationships because they invoke spatial/temporal constraints between the linked objects. Thus, transactions have to make sure they have the stamps of the geometry/ lifecycle of the linked object types.

5 Correspondence Relationship Types

Correspondence relationships allow designers to relate different representations of the same real-world entity when described by different instances. This approach allows users to deal with more situations of multiple representations. Indeed the correspondence relationship types can be of different kinds and link one or more instances. In this approach, we use the following correspondence links:

The *identity* relationship □ type is used to link instances that are alternative representations of the same real world entity, like for instance an object that is represented as a building by some users and as an historical monument by others (see Fig. 3). The two linked objects have the same object identifier. This relationship shares its semantics with the is-a link but it is more generic as it can have properties describing the link (e.g. the attribute matching-date) or useful for the propagation of the updates of one object to the other one.

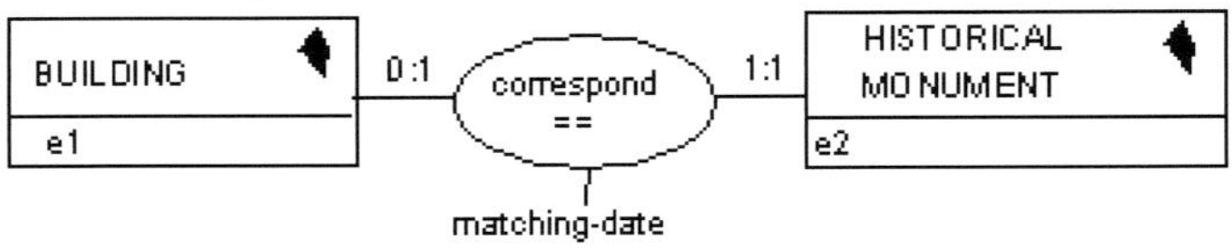

Fig. 3. Identity relationship type between two object types

The *aggregation* relationship type may be used as a correspondence relationship. The aggregation describes the situation where a real world phenomenon is decomposed in a representation. For instance in Fig. 4, a university is seen as a simple object University from one viewpoint and from another viewpoint as a set of component objects: the buildings and playgrounds of the university.

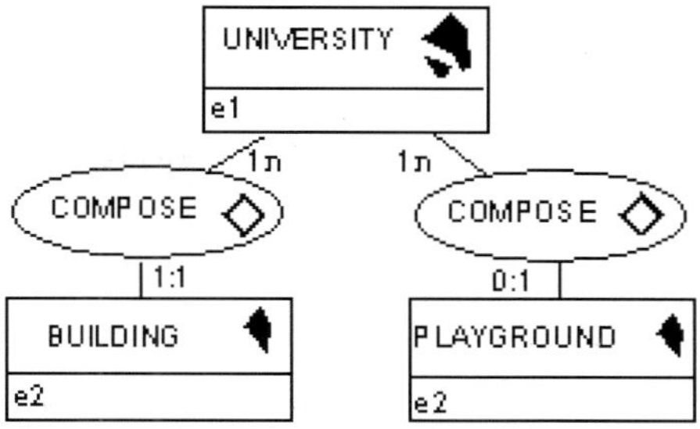

Fig. 4. Example of aggregations

The *SetToSet* relationship type is a new kind of relationship type that conveys the semantics that a group of objects corresponds to another group of objects. This correspondence cannot be captured by traditional relationships that only allow users to link one instance of each participating object type. The "set to set" correspondence is usual when describing data coming from cartographic generalisation. Indeed objects on a map do not necessarily match 1:1 with the real objects. For instance a set of aligned buildings in a built up area may be represented in a map by a set of building icons whose spatial configuration, shape, and spacing looks similar to their real world presentation.

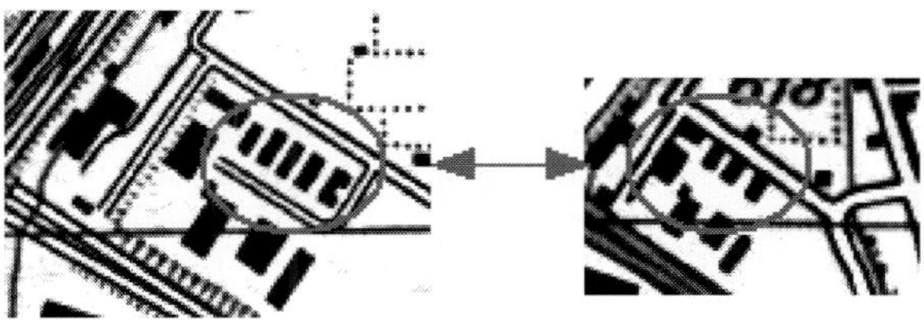

Fig. 5. An example of a SetToSet correspondence.

Fig. 5 shows an example for the same area, in which the more detailed map on the left has five buildings, while the less detailed map on the right includes only three buildings. Fig. 6 is a possible schema of the example of Fig. 5. Note that as each instance of a SetToSet relationship may link several instances of each participating object type; each role has an additional cardinality specification defining how many object instances the role may link to one relationship instance.

Fig. 6. An example of the SetToSet relationship

As for relationship types, in order to get access to a correspondence relationship a query must have at least one stamp of each of the two linked objects.

6 Data Manipulation

This section discusses the requirements that characterise the manipulation of data structures with multi-representation. To avoid ambiguities in the manipulation of multi-representation structures, queries have to convey the specification about which representations are to be used, i.e. to use stamps. The combination of the stamps specified by the query and the stamps that characterise data and metadata in the database defines which data is visible.

The first step when manipulating our multi-representation database is to specify the subset of the database that will be used. This is done with the OPEN SCHEMA command in which a set of stamps is specified. This command allows transactions to access all object types, relationship types, and attributes whose set of stamps intersect the OPEN SCHEMA set of stamps. Access to a relationship requires also that the stamps held by the transaction give access to all the linked object types. For instance, OPEN SCHEMA (My Schema,{e1}) selects the sub-schema composed of object types and attributes that have the stamp e1 and relationship types and link object types with (at least) stamp e1. Access to the object type Road of Fig.1 would be limited to the following attributes: number, name with the value of stamp e1, admin. classif., type, manager, and geometry with the e1 definition.

The schema selected with the OPEN SCHEMA command is not necessarily a mono- representation schema. Thus, queries are likely to manipulate stamped representations. To unambiguously manipulate stamped data, they also have to contain stamps. Those stamps make up additional selection criteria when accessing data. Queries within a transaction by default inherit stamps from the transaction. Consider, for instance, a query on the schema of Fig. 1 to retrieve Road instances where number=7. The query expressed with the MADS algebra is:

SELECTION [number = 7] Road

The query is automatically complemented with the stamps associated to the transaction. There is no stamp specification for the attribute number as there is no ambiguity accessing it. If the user wants to limit his/her query to the roads of stamp e1, he/she needs to complete it as:

SELECTION [number = 7] (Rstamp=e1) Road

This query gives access to the definition and instances of the Road type that holds for e1. The schema of the result is composed of attributes from Road whose stamps include e1. The selected instances are those whose number is 7 and that bear at least the stamp e1. As presented above, an attribute may hold several values and/or definitions. To access such attributes, the query also has to specify unambiguously which value or definition is required. For instance, the following query:

SELECTION [(R-stamp=e1) type = "highway"] Road

returns the roads whose attribute type has the value "highway" for its definition at stamp e1.

SELECTION [name = (R-stamp=e1) "Beaulieu"] Road

returns the roads whose attribute name has the value "Beaulieu" at stamp e1.

Due to multi-representation types, two operations are needed to insert instances in the database. The first one inserts a new instance in a type. Users have to provide a value for all the mandatory attributes corresponding to the format at the given stamp. For instance, the following expression adds a new instance of Road that is stamped with e1:

```
INSERT INTO ROAD VALUES ( stamp = {e1} ,
    geometry = area {(x1,y1), ...(xn,yn)} , number = 7, name = "RN7" ,
    admin.classif. = 5, type = “secondary road”, manager = “Dupont”)
```

Another operation is needed to add a new representation to an instance that was previously defined. For instance, to add a new representation to the RN7 Road instance the following expression reads:

```
ADDREP TO ROAD WHERE number = 7 VALUES ( stamp = {e2} ,
    name = “Route Napoléon” , type = “national”, dpt = 21,
    manager = {“Dupont”, “Durant”, “Rochat”})
```

7 Implementation and Tests

The results presented in this paper have been implemented and tested as part of the work done in the European project MurMur [MUR00]. The goal of the project is to develop, test and validate a software layer implemented on top of the Oracle 8i DBMS. This layer should allow users to:

- Define a spatio-temporal database schema using a multiple representations data model,
- Store their data in the underlying database (Oracle 8i) in a totally transparent way (translations to the model of Oracle are done by the Murmur software),
- Formulate and run interactive queries on the database.

The schema of Fig. 7 shows a screen copy of a multi-representation schema defined with the MurMur schema editor. Stamps are visualised by different colours (in black and gray in the figure). A test version of the prototype can be downloaded at http://lbd.epfl.ch.

The proposed data model and tools have been tested and validated on two real case studies. The first one is a cartographic application from IGN, the French national mapping agency, and involves three existing databases describing the French territory at different scales. The second one is a risk management application from the Cemagref research centre, and involves integrating different thematic databases with temporal data.

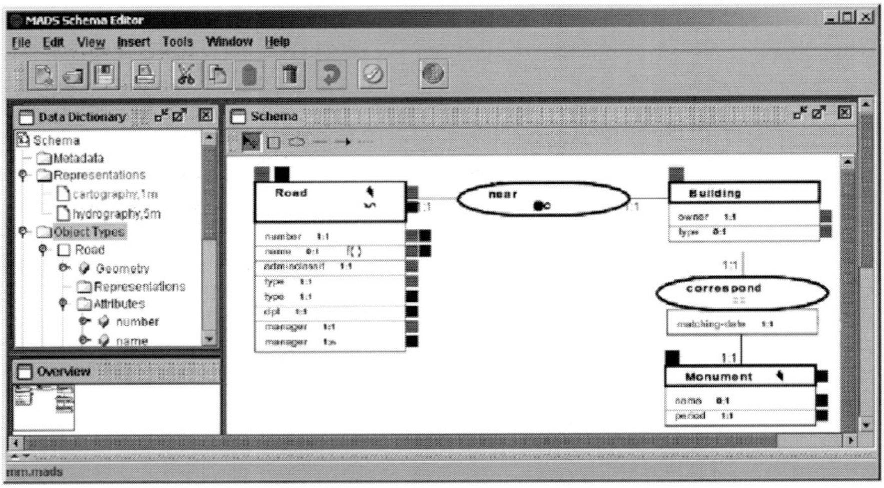

Fig. 7. Screen copy of a session of the MurMur schema editor

8 Conclusion

In this paper we propose a set of concepts as an extension of the MADS data model to support multiple representation of data. Representations may vary according to different criteria, the viewpoint, that is the materialisation of user's needs, and the resolution, that specifies the level of detail of a representation. We propose two approaches to describe multiple representations of real world objects: the integrated approach where the stamping technique is used to define customised data structures and the inter-relationship approach where representations are linked through correspondence links. MADS operators are then extended to be able to manipulate stamped data and metadata. More information about modelling and manipulation facilities is available in (Vangenot 2001). Work is now underway to address problems of consistency between representations. Our goal is to deal with consistency between data of different resolution or viewpoint in the database system (Egenhofer *et al.* 1994), and to find generic rules valid for whatever the application.
Further, our objective is to propose a language and mechanisms allowing designers to define application specific rules.

References

Bernier E, Bédard Y, Devillers R (2001) Automatic generalization and multiple representation for Spatial OLAP. In: Proceedings Geo Information Fusion and Revision. Laval University, Quebec, Canada

Devogele T, Parent C, Spaccapietra S (1998) On Spatial Database Integration. International Journal of Geographical Information Systems 12(4): 335-352

Egenhofer MJ, Clementini E, Di Felice P (1994) Evaluating inconsistencies among multiple representations. In: Proceedings of the Sixth International Symposium on Spatial Data Handling. Edinburgh, Scotland, pp 901-920

Jen TY (1995) Evolution of spatial relationships for interoperability of heterogeneous spatial databases. In: COSIT'95 Doctoral Consortium, Geoinfo-Series. Department of Geoinformation of Technical University, Vienna, pp 58-66

Jones CB, Kidner DB, Luo LQ, Bundy GL, Ware JM (1996) Database design for a multi-scale spatial information system. International Journal of Geographical Information Systems 10(8): 901-920

Kilpelaïnen T (1998) Maintenance of topographic data by multiple representations. In: Proceedings for the Annual Conference and Exposition of GIS/LIS '98. Forth Worth, Texas

Martel C (1999) Développement d'un cadre théorique pour la gestion des représentations multiples dans les bases de données spatiales. M.Sc. thesis, Laval University, Canada

Molenaar M (1998) Composite objects and multiscale approaches. In: An introduction to the theory of spatial object modeling for GIS. Taylor & Francis, pp 161-191

Müller JC, Lagrange JP, Weibel R, Salgé F (1995) Generalization: State of the art and issues. In: Müller JC, Lagrange JP, and Weibel R (eds) GIS and Generalization: Methodology and Practice. pp 3-17

MurMur Consortium (2000) Supporting Multiple Representations in Spatio-Temporal databases [online]. In: Proceedings of the 6th EC-GI & GIS Workshop. Lyon, France. Available from: http://lbd.epfl.ch/e/MurMur

Parent C, Spaccapietra S, Zimanyi E, Donini P, Plazanet C, Vangenot C (1998) Modeling Spatial Data in the MADS Conceptual Model. In: Proceedings of 8th International Symposium on Spatial Data Handling. Vancouver, Canada

Peng W (1997) Automated generalization in GIS. Ph.D. thesis, Wageningen Agricultural University and International Institute for Aerospace Survey and Earth Science (ITC), Enschede, Hollande

Rigaux P, Scholl M (1995) Multi-scale partitions: Applications to spatial and statistical databases. In: Proceedings of the 4th International Symposium on Advances in Spatial Databases, SSD'95, LNCS 951. Springer-Verlag, pp 170-183

Stell S, Worboys M (1998) Stratified Map Spaces: A formal basis for multi-resolution spatial databases. In: Proceedings of the 8th International Symposium on Spatial Data Handling. pp 180-189

Timpf S (1998) Hierarchical structures in map series. Ph.D. thesis, Technical University Vienna

Vangenot C (2001) La multi-représentation dans les bases de données géographiques (in French). Ph.D. thesis, no.2430, EPFL

Weibel R, Dutton G (1999) Generalizing spatial data and dealing with multiple representations. In: Longley P, Goodchild MF, Maguire DJ, Rhind DW, Geographical Information Systems: Principles, Techniques, Management and Applications, (1), 2nd Edition. Geoinformation International

Terrain Modelling Based on Contours and Slopes

Christopher Gold and Maciej Dakowicz

Department of Land Surveying and Geo-Informatics, Hong Kong Polytechnic University, Hung Hom, Kowloon, Hong Kong, Tel: (852)2766-5955; Fax: (852)2330-2994, christophergold@voronoi.com; maciej.dakowicz@dtm.prv.pl

Abstract

Good quality terrain models are becoming more and more important, as applications such as runoff modelling are being developed that demand better surface orientation information than is available from traditional interpolation techniques. A consequence is that poor-quality elevation grids must be massaged before they provide useable runoff models. This paper describes improved methods for extracting good quality terrain models from topographic contour maps, which despite modern techniques are still the most available form of elevation information. Recent work on the automatic reconstruction of curves from point samples, and the generation of medial axis transforms (skeletons) has greatly helped in the visualisation of the relationships between sets of boundaries, and families of curves. The insertion of skeleton points guarantees the elimination of all flat triangles. Additional assumptions about the local uniformity of slopes give us enough information to assign elevation values to these skeleton points. Various interpolation techniques were compared using visualisation of the enriched contour data. Examination of the quality and consistency of the resulting maps indicates the required properties of the interpolation method in order to produce terrain models with valid slopes. The result provides us with a surprisingly realistic model of the surface - that is, one that conforms well to our subjective interpretation of what a real landscape should look like.
Keywords: terrain modelling, contours, skeleton, slopes, Voronoi diagram

1 Introduction

This paper concerns the generation of interpolated surfaces from contours. While this topic has been studied by many people (including the first author) for over 20 years, this work is interesting for a variety of reasons. Firstly, contour data remains the most readily available data source. Secondly, valid theorems for the sampling density along the contour lines have only just been discovered (Amenta et al. 1998). Thirdly, the same publications provide simple methods for generating the medial axis transform, or skeleton, which definitively solves the "flat triangle" problem, which often occurs when triangulating contour data, by inserting additional points from this skeleton. Fourthly, the problem of assigning elevation values to these additional ridge or valley points can be resolved, using the geometric properties of this skeleton, in ways that may be associated with the geomorphological form of the landscape. In addition, comparisons of the methods used in a variety of weighted-average techniques throw a lot of light on the key components of a good interpolation method, using three-dimensional visualisation tools to identify what should be "good" results – with particular emphasis being placed on reasonable slope values, and slope continuity. This last is often of more importance than the elevation itself, as many issues of runoff, slope stability and vegetation are dependent on slope and aspect – but unfortunately most interpolation methods cannot claim satisfactory results for these parameters.

2 Geometric Preliminaries

The methods discussed here depend on a few fundamental geometrical constructs that are well known – the Voronoi diagram and its dual, the Delaunay triangulation (Fig. 1). The first is often used to partition a map into regions closest to each generating point; the second is usually used as the basis for triangulating a set of data points, as it is guaranteed to be locally stable. It may easily be constructed using its "empty circumcircle" property – this circle is centered at the Voronoi node associated with each triangle, and goes through the three vertices.

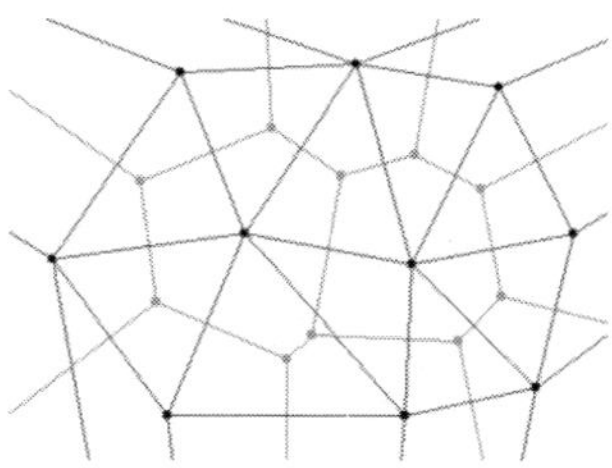

Fig. 1. Delaunay triangulation and Voronoi diagram

The Voronoi diagram and Delaunay triangulation are associated with other geometric structures, shown in Fig 2a, known as the crust and the skeleton (or "medial axis transform"), using algorithms introduced by Amenta et al. (1998). They examined the case where a set of points sampled from a curve or polygon boundary were triangulated, and then attempted to reconstruct the curve. They showed that this "crust" was formed from the triangle edges that did not cross the skeleton, and that if the sampling of the curve was less than 0.25 of the distance to the skeleton formed by the remaining Voronoi edges the crust was guaranteed to be correct. Gold (1999) and Gold and Snoeyink (2001) simplified Amenta's algorithm for the extraction of the crust by showing that, in every

Delaunay/Voronoi edge pair, either the Delaunay edge could be assigned to the crust or else the dual Voronoi edge could be assigned to the skeleton. The Delaunay edge belongs to the crust when there exists a circle through its two vertices that contain neither of its associated Voronoi vertices; otherwise, the corresponding Voronoi edge belongs to the skeleton. A simple InCircle test applied to each Delaunay/Voronoi edge pair (Fig. 2b) distinguishes these cases.

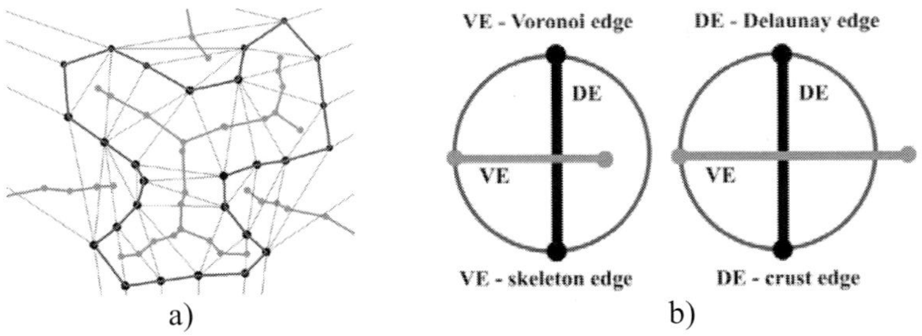

Fig. 2. Crust and skeleton, **(a)** crust (thick black lines) and skeleton (thick gray lines), **(b)** crust/skeleton test

3 Generation of Ridge and Valley Lines

In our particular case, the data is in the form of contour lines that we assume are sufficiently well sampled – perhaps derived from scanned maps. Despite modern satellite imaging, much of the world's data is still in this form. An additional property is not sufficiently appreciated – they are subjective, the result of human judgement at the time they were drawn. Thus they are clearly intended to convey information about the perceived form of the surface – and it would be desirable to preserve this, as derived ridges and valleys.

Fig. 3a shows our raw data set (which is completely imaginary), and Fig. 3b shows the resulting crust, which reconstructs the contour lines and the skeleton. Fig. 3c shows the crust and only those skeleton points that provide unique information – ridge and valley lines that separate points on the same contour ("flat

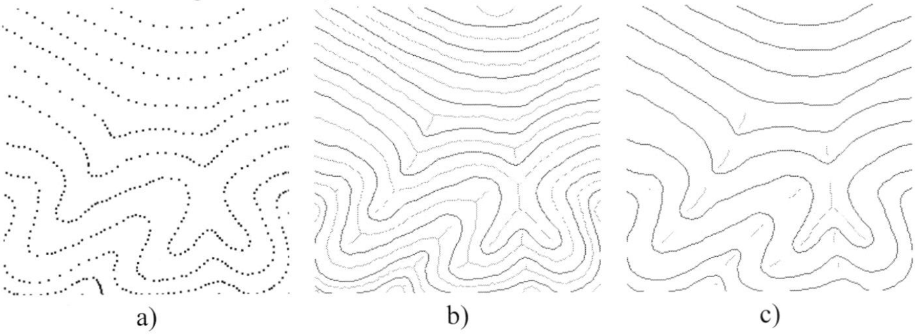

Fig. 3. Contours, **(a)** data points, **(b)** crust and skeleton, **(c)** crust and skeleton branches

triangles"), rather than merely those points that separate adjacent contours. Aumann et al. (1991) produced somewhat similar results by raster processing.

Fig. 4 shows a close-up of the test data set with shaded flat triangles having all vertices at the same elevation. In a Delaunay triangulation all circumcircles must be empty, and the insertion of a Voronoi vertex (circumcentre or skeleton point) will force the deletion of its forming triangle. Thus, the insertion of the skeleton point of a flat triangle guarantees that it is replaced by new triangles with the skeleton point as a vertex.

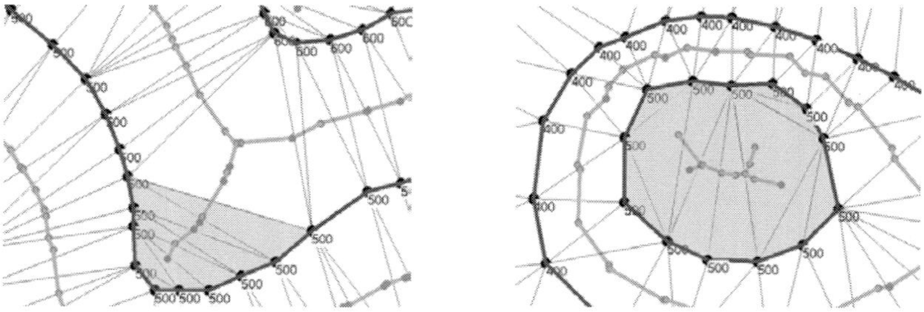

a) b)

Fig. 4. Skeleton and "flat triangles": a) ridge; b) summit

Two techniques for estimating heights at skeleton points have been developed, each with its own physical interpretation. The first, following Thibault and Gold (2000), uses Blum's (1967) concept of height as a function of distance from the curve or polygon boundary, with the highest elevations forming the crest at the skeleton line. This is illustrated in Figs. 5a and 5b, where points on a simple closed curve are used to generate the crust and skeleton. In Fig. 5a the circumcentres of the skeleton points are given an elevation equal to the circumradius. The resulting interpolated model is shown in Fig. 5b. This model is based on the idea that all slopes are identical, and thus the radius is proportional to the height of the skeleton point. Of course, in the case of a real summit as in Fig. 4b, the slope would initially be unknown, and would be estimated from the circumradii of the next contour level down.

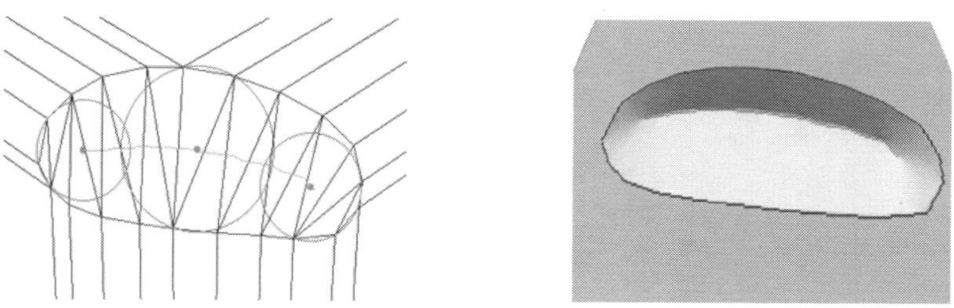

a) b)

Fig. 5. Triangulation of a summit, **(a)** skeleton and circumcentres, **(b)** elevation model after adding skeleton vertices with assigned height values

In the case of a ridge or valley, the circumradius may also be used, as in Fig. 6a, to estimate skeleton heights based on the hypothesis of equal slopes. The larger circle, at the junction of the skeleton branches, has a known elevation – half way between the contours – and may be used to generate the local slope. The elevation of the center of the smaller circle is thus based on the ratio of the two radii; see Thibault and Gold (2000) for details.

While this method is always available, it is not always the preferred solution where constant slope down the drainage valley, rather than constant valley-side slope, is more appropriate. In a second approach, illustrated in Fig. 6b, the line of the valley is determined by searching along the skeleton, and heights are assigned based on their relative distance along this line. This may be complicated where there are several valley branches – in which case the longest branch is used as the reference line. This involves careful programming of the search routines, although the concept is simple. In practice, an automated procedure has been developed, which uses the valley length approach where possible, and the side-slope method when no valley head can be detected, such as at summits and passes. This refines and amplifies the work of Thibault and Gold (2000). In particular, instead of enriching the model with all skeleton vertices, only the skeleton vertices of flat triangles are used, and new interpolation techniques are introduced to estimate skeleton point elevations.

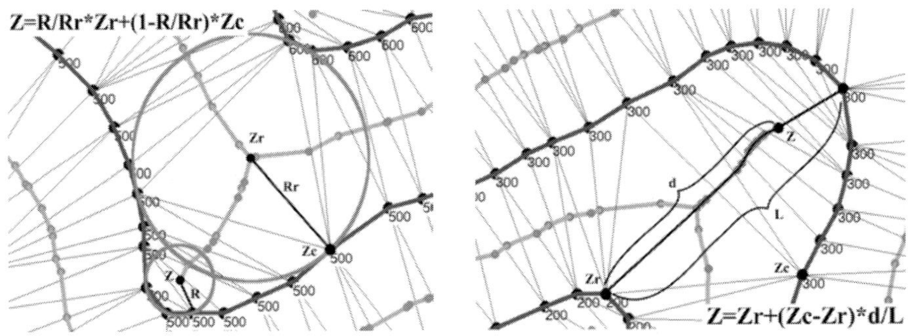

a) b)

Fig. 6. Estimating skeleton heights, **(a)** from circumradii, **(b)** from valley length

4 Components of an Interpolation Model

On the basis of a sufficient set of data points, we then wanted to generate a terrain model with satisfactory elevations and slopes, as the basis of a valid rainfall runoff model. Our approach was to interpolate a height grid over the test area, and to view this with an appropriate terrain visualisation tool. To obtain perspective views we used Genesis II, available from www.geomantics.com. Vertical views were generated using version 5 of the Manifold GIS, available from www.Manifold.net. We feel that 3D visualisation has been under-utilised as a tool for testing terrain modelling algorithms, and the results are often more useful than a purely mathematical, or even statistical, approach.

We have restricted ourselves to an evaluation of several weighted-average methods, as there are a variety of techniques in common that can be compared. All of the methods were programmed by ourselves – which left out the very popular Kriging approach, as too complicated. Nevertheless, many aspects of this study apply to Kriging as well, since it is a weighted-average method with the same problems of neighbour selection and the inclusion of slope values at data points as the methods we attempted.

In general, we may ask about three components of a weighted-average interpolation method. Firstly: what is the weighting process used? Secondly: which set of neighbours is used to obtain the average? Thirdly: is it the data point elevation only that is being averaged? (Often it is the data point elevation alone, but sometimes it is a plane through the data point incorporating slope information.)

One of the simplest weighted-average models is triangle-based interpolation in which a linear interpolation is performed within each triangle. Fig. 7 shows the result, including the skeleton draped over the flat triangles. Fig. 8 shows the improvement when estimated skeleton points are added, and all flat triangles are automatically removed.

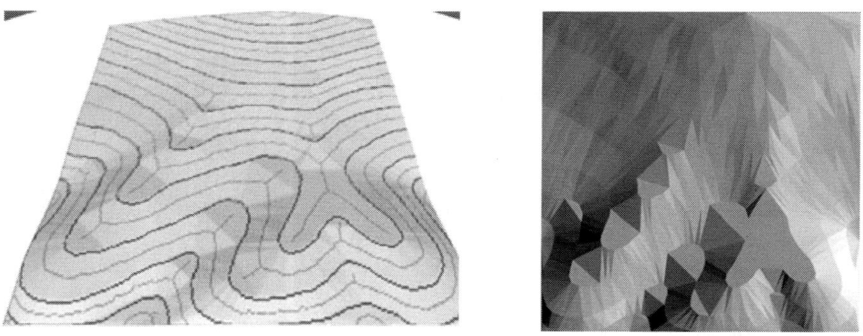

a) b)

Fig. 7. Triangle-based interpolation, **(a)** perspective view, **(b)** vertical view

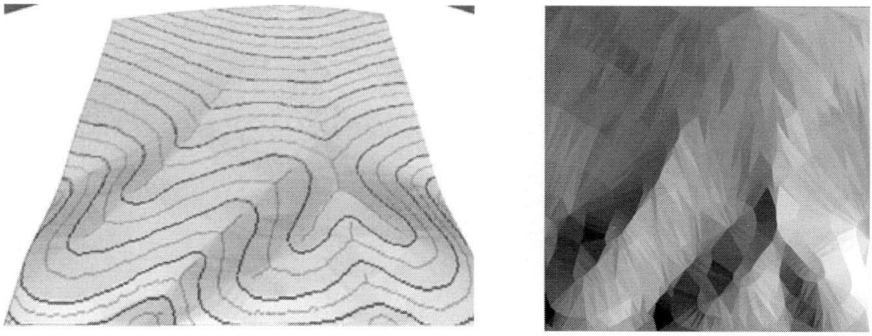

a) b)

Fig. 8. Adding skeleton points to Fig. 7, **(a)** perspective view, **(b)** vertical view

The other weighted average models that were tested were the traditional gravity model, and the more recent "area-stealing" or "natural neighbour" or perhaps more properly "Sibson" interpolation methods (Sibson, 1980; Watson and Philip, 1987; Gold, 1989). Here the number of neighbours used may vary.

In the case of the gravity model, the weighting of each data point used is inversely proportional to the square of the distance from the data point to the grid node being estimated, although other exponents have been used. There is no obvious set of data points to use, so one of a variety of forms of "counting circle" is used. Fig. 9 shows the resulting surface for a radius of about a quarter of the map. Data points form bumps or hollows.

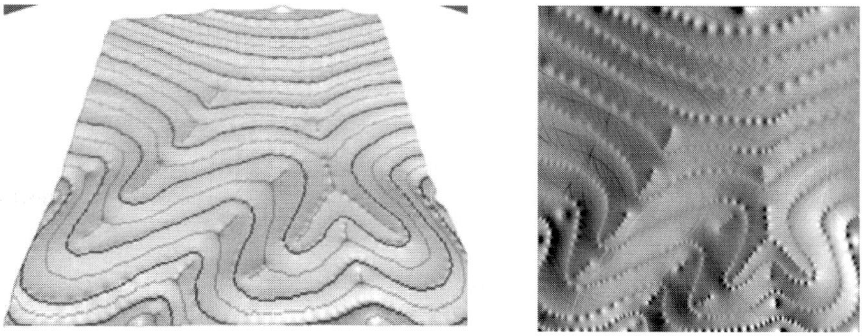

a) b)

Fig. 9. Interpolation using the gravity model with medium radius, **(a)** perspective view, **(b)** vertical view

If the radius is reduced there may be holes in the surface where no data is found within the circle (Fig. 10a). If the radius is increased the surface becomes somewhat flattened, but the bumps remain (Fig. 10b). The result depends on the radius, and the other selection properties being used. Clearly estimates of slope would be very poor, and very variable.

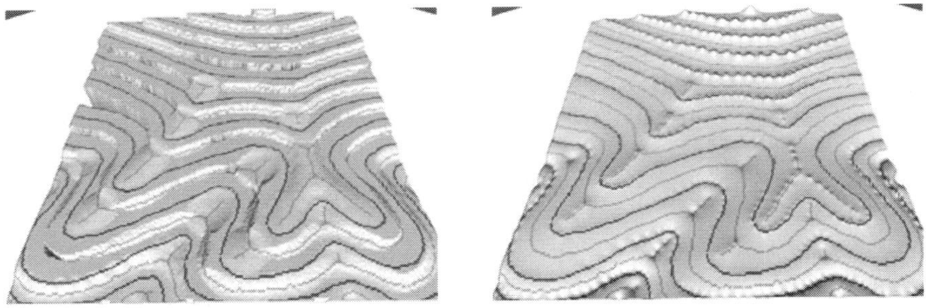

a) b)

Fig. 10. Gravity model interpolation, **(a)** with small radius, **(b)** with large radius

The Sibson method, illustrated in Fig. 11a, is based on the idea of inserting each grid point temporarily into the Voronoi diagram of the data points, and measuring the area stolen from each of a well-defined set of neighbours. These stolen areas are the weights used for the weighted average.

Figs. 11b and 11c show a sample data set and the neighbour selection for the same point in both the gravity and Sibson method. In the Sibson method natural neighbour selection results in a reasonable set of neighbours, but the circle used in the gravity method may not select a sufficient number of neighbours to produce a valid elevation value for the interpolated point. The Sibson method is particularly appropriate for poor data distributions as the number of neighbours used is well defined. In the gravity model, when the data distribution is highly anisotropic, there is considerable difficulty in finding a valid counting circle radius.

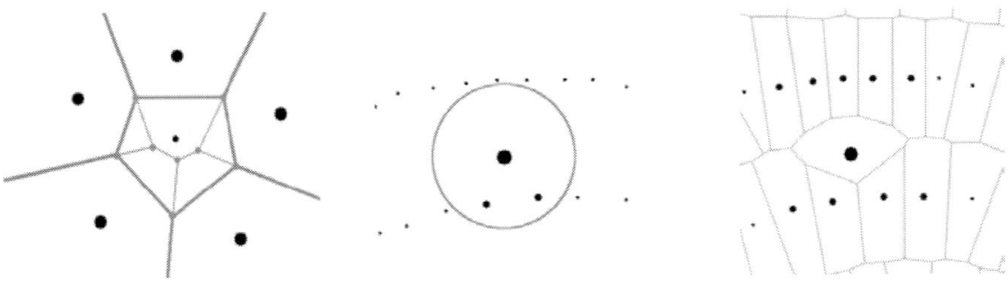

a) b) c)

Fig. 11(a) Sibson interpolation, **(b)** neighbour selection using a counting circle, **(c)** neighbour selection using Voronoi neighbours

Fig. 12 shows the results of Sibson interpolation. The surface behaves well, but is angular at ridges and valleys. Indeed, slopes are discontinuous at all data points (Sibson, 1980). One solution is to re-weight the weights, so that the contribution of any one data point not only becomes zero as the grid point approaches it, but the slope of the weighting function approaches zero also (Gold, 1989). Fig. 13 shows the effect of adding this smoothing function. While the surface is smooth, the surface contains undesirable "waves" – indeed, applying this function gives a surface with zero slope at each data point.

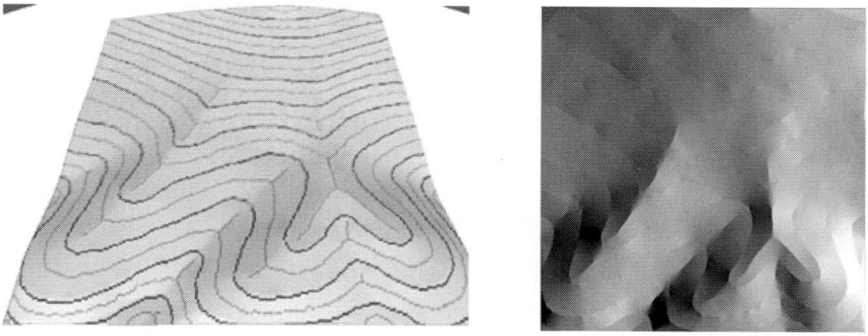

a) b)

Fig. 12. Sibson interpolation, **(a)** perspective view, **(b)** vertical view

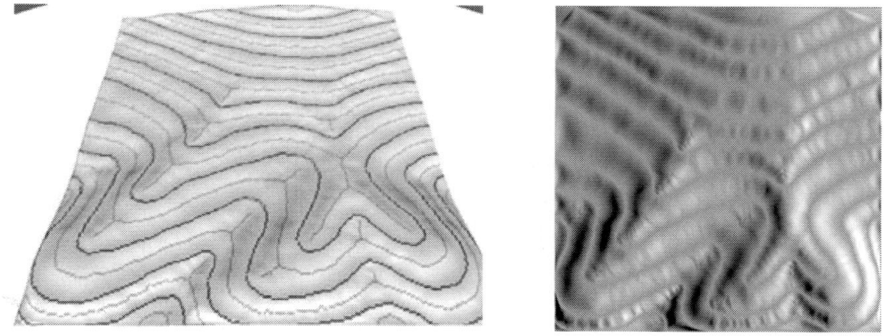

a) b)

Fig. 13. Adding smoothing to Fig. 12, **(a)** perspective view, **(b)** vertical view

5 Slopes – The Ignored Factor

This brings us to a subject often ignored in selecting a method for terrain modelling – the slope of the generated surface. In real applications, however, accuracy of slope is often more important than accuracy of elevation – for example in runoff modelling, erosion and insolation. Clearly, an assumption of zero slope, as above, is inappropriate. However, in our weighted average operation we can replace the height of a neighbouring data point by the value of a function defined at that data point – probably a planar function involving the data point height and local slopes. This function may be estimated by a variety of methods. Thus at any grid node location we find the neighbouring points and evaluate their planar functions for the (x, y) of the grid node. These z estimates are then weighted and averaged as before (Gold, 1989).

Fig. 14 shows the result of using Sibson interpolation with data point slopes. The form is good, but slight breaks in slope can be seen at contour lines. When using smoothing and slope information together, the surface is smooth, but has unwanted oscillations (Fig. 15). Clearly an improved smoothing function is desirable to eliminate these side effects.

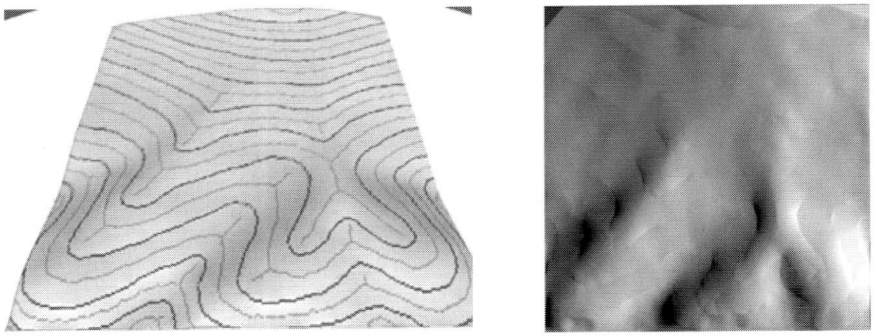

a) b)

Fig. 14. Sibson interpolation using slopes at data points, **(a)** perspective view, **(b)** vertical view

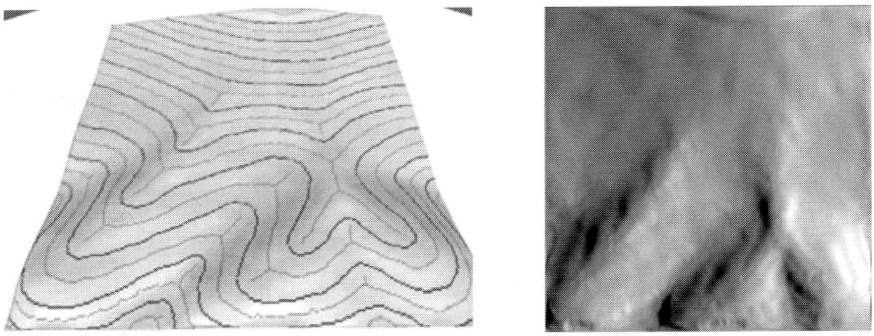

a) b)

Fig. 15. Sibson interpolation using slopes and smoothing function at data points, **(a)** perspective view, **(b)** vertical view

While it is impossible to show the results of all our experiments in this paper, we also used the method of Burrough and Mcdonnell, (1998) to calculate slopes and profile curvature of grids created from various combinations of our available weighted-average methods. Sibson interpolation with slopes and without smoothing gives more consistent regions of coherent slopes. Adding slopes to the simple TIN model (i.e. using the position in the triangle to provide the weights, as in Fig.16a) produced results that were almost as good as the Sibson method when the sample points were closely spaced along the contours. However, the Sibson method is much superior for sparser data, or where the points do not form contour lines. The gravity model does not provide particularly good slope estimates, but even here, including the data point slope function produces a significant improvement, as in Fig. 16b.

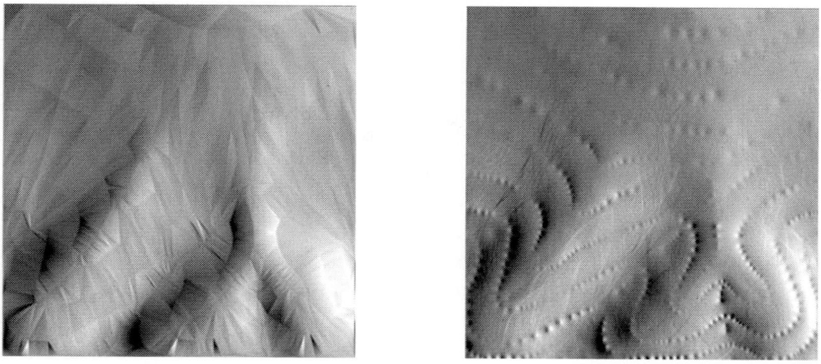

a) b)

Fig. 16. Adding slopes at data points, **(a)** triangle-based interpolation, **(b)** gravity interpolation

6 Summary and Conclusions

For the common problem of deriving surfaces from contours, we propose a general approach:

1. Generate skeleton points by the method of Aumann et al. (1991) or of Thibault and Gold (2000). Ignore skeletons between contours.
2. Assign elevations to these skeleton points by the methods described here based on either longitudinal or lateral slope consistency, or other suitable techniques.
3. Eliminate flat triangles by the insertion of these skeleton points into the original TIN.
4. Estimate slope information at each data point by any appropriate technique.
5. Perform weighted-average interpolation using the previously estimated slope information. Avoid methods such as the gravity model, which require user specified parameters. Sibson interpolation appears to be the best choice.

From our work, several broad generalisations may be made. To produce good surface models with reasonable slopes from contour maps, the single most valuable contribution is the addition of skeleton points with estimated elevations, in order to eliminate flat triangles. The second most important contribution is the addition of slope information at the data points, and its use in the weighted average interpolation process - even poor interpolation methods are significantly improved. Also important is the selection of a meaningful set of neighbours around the estimated point. Of lesser importance is the particular interpolation method used, although this statement is highly dependent on the data distribution and density. Gravity models in general should be avoided if possible. Surprisingly, mathematically guaranteed slope continuity is not usually critical, although we are continuing to work on an improved smoothing function that guarantees both slope continuity and minimum curvature – probably based on the work of Anton et al. (1998). Nevertheless, the moral is clear: both for finding adjacent points and for skeleton extraction, a consistent definition of neighbourhood is essential for effective algorithm development.

We conclude with another imaginary example. Fig. 17a shows four small hills defined by their contours, modelled by a simple triangulation. Fig. 17b shows the result using Sibson interpolation, slopes and skeletons. Skeleton heights were obtained using circumcircle ratios, as no valley-heads were detected. While our evaluation was deliberately subjective, we consider that our results in this case, as with the previous imaginary valley, closely follow the perceptual model of the original interpretation. Thus, for the reconstruction of surfaces from contours, or generation of DEMs, we believe that our methods are a significant improvement on previous work.

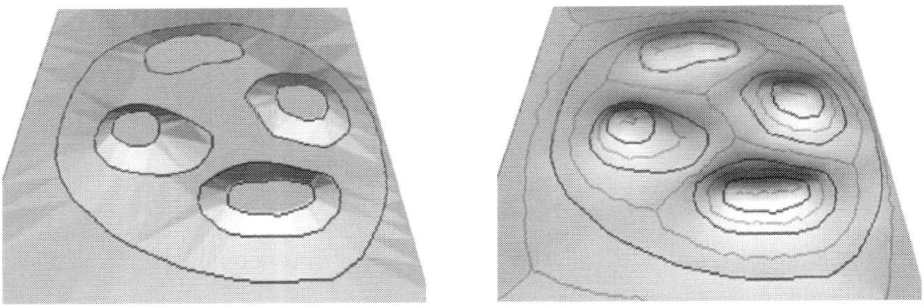

a) b)

Fig. 17. Triangulation of several small hills, **(a)** triangle-based interpolation, **(b)** Sibson interpolation with slopes

Acknowledgements

The work described in this paper was substantially supported by a grant from the Hong Kong Polytechnic University (Project No. PolyU A-PB79).

References

Amenta N, Bern M, Eppstein D (1998) The crust and the beta-skeleton: combinatorial curve reconstruction. Graphical Models and Image Processing 60:125-135

Anton F, Gold CM, Mioc D (1998) Local coordinates and interpolation in a Voronoi diagram for a set of points and line segments. In: Proceedings 2nd Voronoi Conference on Analytic Number Theory and Space Tillings. 1998, Kiev, Ukraine, pp 9-12

Aumann G, Ebner H, Tang L (1991) Automatic derivation of skeleton lines from digitized contours. ISPRS Journal of Photogrammetry and Remote Sensing 46:259-268

Blum H, (1967) A transformation for extracting new descriptors of shape. In: Whaten Dunn W (eds) Models for the Perception of Speech and Visual Form. MIT Press, pp 153-171

Burrough P, Mcdonnell RA (1998) Principles of Geographical Information Systems, 2nd ed. Oxford University Press

Gold CM (1989) Chapter 3 - Surface interpolation, spatial adjacency and GIS. In: Raper J (ed) Three Dimensional Applications in Geographic Information Systems. Taylor and Francis, Ltd., London, pp 21-35

Gold CM (1999) Crust and anti-crust: a one-step boundary and skeleton extraction algorithm. In: Proceedings of the ACM Conference on Computational Geometry. Miami, Florida, pp 189-196

Gold CM, Snoeyink J (2001) A one-step crust and skeleton extraction algorithm. Algorithmica 30:144-163

Sibson R (1980) A Vector Identity for the Dirichlet Tessellation. Math. Proc. Cambridge Philos. Soc. 87:151-155

Thibault D, Gold CM (2000) Terrain Reconstruction from Contours by Skeleton Construction. GeoInformatica 4:349-373

Watson DF, Philip GM (1987) Neighborhood-based interpolation. Geobyte 2:12-16

Siting Observers on Terrain

Wm Randolph Franklin

Rensselaer Polytechnic Institute, Troy, New York 12180-3590 wrf@ecse.rpi.edu, http://www.ecse.rpi.edu/Homepages/wrf

Abstract

This paper presents an experimental study of a new algorithm that synthesises separate programs, for fast viewshed, and for fast approximate visibility index determination. It has been implemented in a working testbed for siting multiple observers jointly to cover terrain from a full level-1 DEM, and to do it so quickly that multiple experiments are easily possible. Both the observer and target may be at a given fixed distance above the terrain. The process operates as follows. (1) An approximate visibility index is calculated for each point in the cell under consideration. (2) A set of tentative observers is selected from the highly visible points. (3) The actual observers are selected from them, so as to cover as much of the cell as possible, using a greedy algorithm. Various experiments with varying parameters were performed on the Lake Champlain West cell (raster data set), with observations such as the following. (1) Forcing tentative observers to be well spaced was more important than using the most visible tentative observers. (2) Most of the new observers added (because they covered the most unseen points) were already visible to an existing observer. (3) Randomly deleting many tentative observers before final selection did not reduce the final area covered.
Keywords: terrain visibility, viewshed, line of sight, multiple observers

1 Introduction

Consider a terrain elevation database, and an observer, O. Define the *viewshed* as the terrain visible from O within some distance R of O. The observer might be situated at a certain height above ground level, and might also be looking for targets at a certain height above the local ground. Also, define the *visibility index* of O as the fraction of the points within distance R of O that are visible from O. This paper combines an earlier fast viewshed algorithm with an earlier approximate visibility index algorithm, to site multiple observers so as to jointly cover as much terrain as possible.

This paper extends the earlier visibility work in Franklin (2000) and Franklin and Ray (1994), which also survey the terrain visibility literature. Notable pioneer work on visibility includes De Floriani et al. (1993); De Floriani and Magillo (1994); Fisher (1993); Lee (1992); Shannon and Ignizio (1971). Shapira (1990) studied visibility, and provided the Lake Champlain W data used in this paper. Ray (1994) presented new algorithms and implementations of the visibility index, and devised the efficient viewshed algorithms that are used here.

This multiple observers case is particularly interesting and complex, and has many applications. A cell phone provider wishes to install multiple towers so that at least one tower is visible (in a radio sense) from every place a customer's cellphone might be. Here, the identities of the observers of highest visibility index are of more interest than their exact visibility indices, or than the visibility indices of all observers. One novel future application of siting radio transmitters will occur when the moon is settled. The moon has no ionosphere to reflect signals, and no stable satellite orbits. The choices for long-range communication would seem to include either a lot of fiber optic cable or many relay towers. That solution is the multiple observer visibility problem.

As another example, a military planner needs to put observers so that there is nowhere to hide that is not visible from at least one. This leads to a corollary application, where the other side's planner may want to analyse the first side's observers to find places to hide. In this case, the problem is to optimise the targets' locations, instead of the observers'.

Again, a planner for a scenic area may consider each place where a tourist might be to be an observer, and then want to locate ugly infrastructure, such as work yards, at relatively hidden sites. We may wish a forest clearcut site to be invisible to observers driving on a highway sited to give a good view. Finally, an architect may be trying to site a new house while following the planning board's instruction that, "You can have a view, but you cannot be the view."

Speed of execution on large datasets is of more importance than may be apparent. Many prototype implementations, demonstrated on small datasets, do not scale up well. That may happen either because of the size and complexity of the data structures used, or because of the asymptotic time behaviour. For instance, even an execution time proportional to $N*log(N)$, where N is the size of the input, is problematic for $N=1,000,000$. In that case, the $log(N)$ increases the time by a factor of 20. Some preliminary published algorithms may even be exponential if performing a naive search. Therefore, we strive for the best time possible.

In addition, large datasets may contain cases, which did not occur in the small test sets, which require tedious special programming by the designer. In a perfect software development process, all such cases would have been theoretically analysed *a priori*, and treated. However, in the real world, testing on the largest available datasets increases some confidence in the program's correctness.

Next, a large enough quantitative increase in execution speed leads to a qualitative increase in what we can do. Only if visibility can be computed efficiently, can it be used in a subroutine that is called many times, perhaps as part of a search, to optimise the number of observers. This becomes more important when a more realistic function is being optimised, such as the total cost. E.g., for radio towers, there may be a trade-off between a few tall and expensive towers, and many short and cheap ones. Alternatively, certain tower locations may be more expensive because of the need to build a road. We may even wish to add redundancy so that every possible target is visible from at least two observers. In all these cases, where a massive search of the solution space is required, success depends on each query being as fast as possible.

Finally, although the size of available data is growing quickly, it is not necessarily true that available computing power is keeping pace. There is a military need to offload computations to small portable devices, such as a Personal Digital Assistant (PDA). A PDA's computation power is limited by its battery, since, (approximately), for a given silicon technology, each elemental computation consumes a fixed amount of energy. Batteries are not getting better very quickly; increasing the processor's cycle speed just runs down the battery faster.

There is also a compounding effect between efficient time and efficient space. Smaller data structures fit into cache better, and so page less, which reduces time. The point of all this is that efficient software is at least as important now as ever.

The terrain data structure used here is a 1201 by 1201 matrix of elevations, such as from a USGS level-1 Digital Elevation Model cell (raster data set). The relative advantages and disadvantages of this data structure versus a triangulation are well known, and still debated; the competition improves both alternatives. This paper utilises the simplicity of the elevation matrix, which leads to greater speed and small size, which allows larger data sets to be processed.

For distances much smaller than the earth's radius, the terrain elevation array can be corrected for the earth's curvature, as follows. For each target at a distance *D* from the observer, subtract D*D/(2E) from its elevation, where *E* is the earth's radius. (The relative error of this approximation is (D/(2E))**2.) It is sufficient to process any cell once, with an observer in the center. The correction need not change for different observers in the cell, unless a neighbouring cell is being adjoined. Therefore, since this can be easily corrected in a pre-processing step, our visibility determination programs ignore the earth's curvature.

The radius of interest, *R*, out to which we calculate visibility, is not related to the distance to the horizon, but is determined by the technology used by the observer. E.g., if the observer is a radio communications transmitter, doubling *R* causes the required transmitter power to quadruple. If the observer is a searchlight, then its required power is proportional to *R**4*.

In order to simplify the problem under study enough to make some progress, this work also ignores factors such as vegetation that need to be handled in the real world. The assumption is that it is possible, and a better strategy, to incorporate them only later. Franklin (2002) is an expanded version of this paper, containing equations and colour figures.

2 Detailed Process

Our system of selecting a set of observers to cover a terrain cell consists of four core programs coordinated with a shell script. Assorted auxiliary programs, to format output for viewing, etc, were also written.

VIX. This calculates approximate visibility indices of every point in a cell. It takes a radius of interest, the observer and target height, and *T*, a sample size. *VIX* reads an elevation cell. For each point in the cell in turn, *VIX* considers that point as an observer, picks *T* random targets uniformly and independently distributed within the radius of the point, and computes what fractions are visible. That is this point's estimated visibility index.

FINDMAX. This selects a manageable subset of the most visible tentative observers from *VIX*'s output. This is somewhat subtle since there may be a region containing all points of very high visibility. A lake surrounded by mountains would be such a case. Since multiple close observers are redundant, we force the tentative observers to be spread out as follows.

1. Divide the cell into smaller blocks of points. If necessary, first perturb the given block size so that all the blocks are the same size, within one.
2. In each block, find the *K* points of highest approximate visibility index, for some reasonable *K*. If there were more than *K* points with equally high visibility index, then select *K* at random, to prevent a bias towards selecting points all on one side of the block. (This was implemented by using a multiplicative hash function of the point's coordinates as a secondary sort key.) If a block has fewer than *K* points, then return all its points.

VIEWSHED. This finds the viewshed of a given observer at a given height out to a given radius, *R*. The procedure, which is an improvement over Franklin and Ray (1994), goes as follows.

1. Define a square of side *2R* centred on the observer.
2. Consider each point around the perimeter of the square to be a target in turn.
3. Run a sight line out from the observer to each target calculating which points adjacent to the line, along its length, are visible, while remembering that both the observer and target are probably above ground level.
4. If the target is outside the cell, because *R* is large or the observer is close to the edge, then stop processing the sight line at the edge of the cell.

One obvious "improvement", when the target is outside the cell, would be to move the target into the edge of the cell before running the sight line. However, this would cause the computed viewshed to depend slightly on *R*, which looks poor. The above procedure, due to Ray (1994), is an approximation, but so is representing the data as an elevation grid, and this method probably extracts most of the information inherent in the data. There are combinatorial concepts, such as Davenport-Schintzel sequences, (Agarwal and Sharir, 1999), i.e., which present asymptotic worst-case theoretical methods.

SITE. This takes a list of viewsheds and finds a quasi-minimal set that covers the terrain cell (data set or domain) as thoroughly as possible. The method is a simple greedy algorithm. At each step, the new tentative observer whose viewshed will increase the cumulative viewshed by the largest area is included, as follows.

1. Let *C* be the cumulative viewshed, or set of points visible by at least one selected observer. Initially, *C* is empty.
2. Calculate the viewshed, *Vi*, of each tentative observer *Oi*.
3. Repeat the following until it is not possible to increase *area(C)*, either because all the tentative observers have been included, or (more likely) because none of the unused tentative observers would increase *area(C)*.
   (a) For each *Oi*, calculate *area(union(C,Vi))*.
   (b) Select the tentative observer that increases the cumulative area the most, and update *C*. Not all the tentative observers need to be tested every time, since a tentative observer cannot add more area this time than it would have added last time, had it been selected. Indeed, suppose that the best new observer found so far in this step would add new area *A*. However we have not checked all the tentative new observers yet in this loop, so we continue. For each further tentative observer in this execution of the loop, if it would have added less than *A* last time, then do not even try it this time.

3 Lake Champlain W Cell Experiments

The above process was tested on the level-1 USGS DEM Lake Champlain West, Fig. 1, containing 1201 by 1201 points. (The intensities of all the figures in this paper have been nonlinearly transformed to enhance visibility.) This cell has a variety of terrain, ranging from 1591 meter high Mt Marcy in the Adirondack Mountains of New York State, to a large flat lowland and Lake Champlain. A viewshed radius of 100 points was used. The observers and targets are both 100 m above the terrain surface.

In *VIX*, the approximate visibility indices of the points were determined with 10 random targets per observer. The resulting visibility map is shown in Fig. 2.

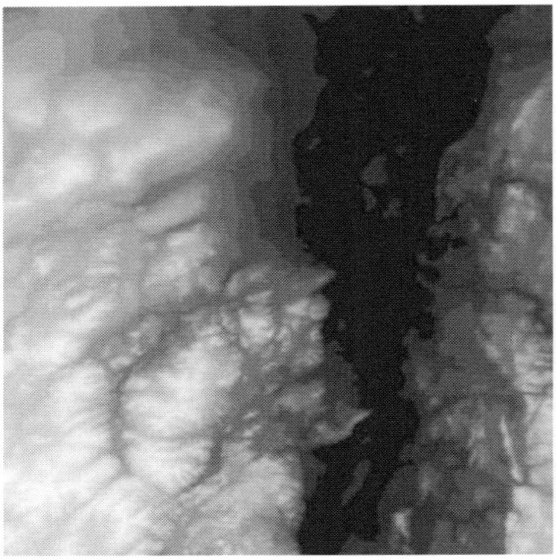

Fig. 1. Lake Champlain W Cell (light=high elevation, dark=low elevation)

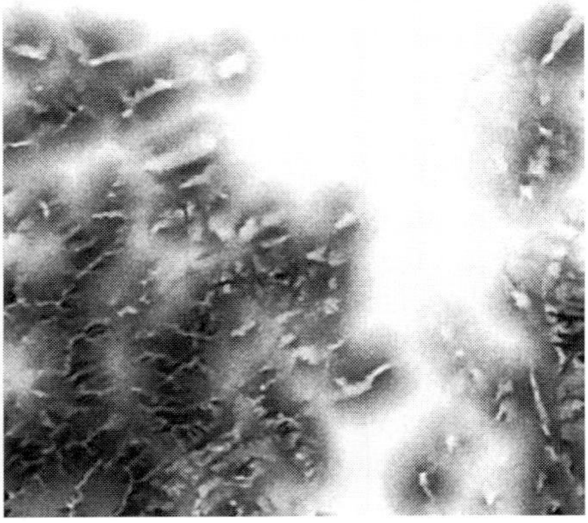

Fig. 2. Visibility Indices for Radius = 100, Observer and Target Ht=100 (light=high VI, dark=low VI)

In *FINDMAX*, the cell was divided into blocks of size 101 by 101 points, and 7 tentative observers were selected in each block, for a total of 1008 tentative observers. Fig. 3 shows some sample viewsheds returned by *FINDMAX*.

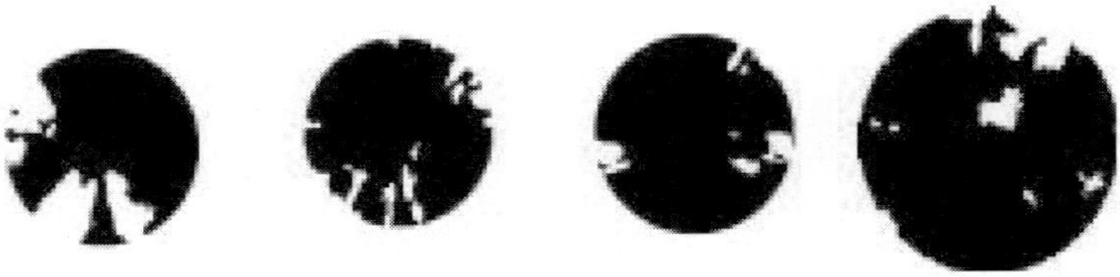

Fig. 3. Sample high visibility viewsheds

Several observations, only some obvious, were made from this experiment.

1. Initially the cumulative area grows quickly, but then that slows since most of the so-far unused viewsheds cover very few so-far unseen points. Fig. 4 shows this. One set of points shows the cumulative visible area growing with the number of included observers. The other shows the decreasing amount of new visible area added by each new observer. The latter set is scaled up by a factor of 30 so that both datasets could be meaningfully shown on one plot.
2. The order that the tentative observers were included was not strongly related to their visibility indices. There was an exception at the start, when few of the viewsheds overlapped, because the viewsheds' radius, 100, was so much smaller than the cell size, 1201.
3. Starting with the 50th observer to be added, almost every new observer was already visible to an observer that had been added earlier. This suggests that it

would be a bad idea to select new observers only from terrain that was not yet covered.
4. After 180 tentative observers were chosen, the process stopped. None of the remaining 828 tentative observers' viewsheds covered a single new point that was not already visible. This tells us that covering the whole cell would require some observers of a lower visibility index.
5. When the process stopped with the 180 observers, 99.91% points were visible from some observer. Indeed after only 90 observers, Fig. 4 shows that already 98% of the points are visible.

We recognise that running these experiments on other data sets may produce varying conclusions, and are currently testing that aspect.

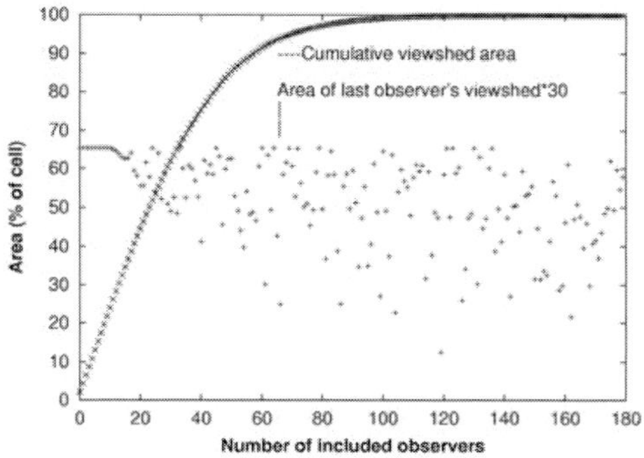

Fig. 4. Area of last inserted viewshed *30, and cumulative visible area, as observers are inserted, for Lake Champlain W Cell

3.1 Execution Times

The worth of the above system lies, not in its theoretical depth, but in its speed. Great care was taken, first, in algorithm and data structure selection, and then in balancing the system and in micro-optimisation. Observations from preliminary experiments, not described here, were used to refine the system. Remember that quantitative speed translate into a qualitative change in the science, since faster SW allows more and larger experiments to be run, hypotheses to be tested, and observations to be made.

Here are some sample times, on a 600MHz Pentium for the preceding experiment. These are real, elapsed, times. Those are larger than CPU times, because of disk I/O, and the time needed to uncompress the input before feeding it to the program, and then to compress the output. Therefore, elapsed times more accurately reflect a program's performance.

VIX took 114 seconds to test 10 random targets against each of the 1442401 observers. *FINDMAX* took 7 seconds to pick 1008 tentative observers from the 1442401 points. *VIEWSHED* took 188 seconds to calculate 1008 viewsheds. *SITE* took 149 seconds to select 180 observers from the 1008 tentatives. The total elapsed time was only 458 seconds to find 180 observers to cover 99.9% of this level-1 DEM cell. If necessary, the time could be reduced with further fine tuning.

Various tradeoffs are possible among the above components. For a fixed cell size, the time for *VIX* depends linearly on the number of random targets per observer. More targets mean that the estimated visibility indices are more accurate. However, *SITE* often seems to do about as well w/o using exactly the most visible tentative observers. *VIEWSHED*'s time depends linearly on the radius multiplied by the number of tentative observers. *SITE*'s time depends roughly quadratically on the number of tentative observers. In addition, storing all those viewsheds requires memory, 190MB in the above experiment. Further, this memory should be real memory, not virtual; otherwise, the paging would slow the program's real elapsed time by a factor of several.

3.2 Not Forcing Tentative Observers to be Spaced

Because subdividing the cell into blocks seemed inappropriate, we selected extra tentative observers per block, combined them, and returned the top 1000 tentative observers, which now had slightly higher visibility indices on average, but were not as well spaced out as before. The results from *SITE* became noticeably worse. It stopped after including 181 observers, because no new potential observer saw even one unseen point. At this point, only 92% of the cell was visible from some observer. Therefore, forcing the tentative observers to be spaced out is good.

3.3 Requiring New Observers to be Hidden from Existing Ones

Selecting new tentative observers from, as yet, uncovered terrain is an obvious heuristic, Franklin (2000), thus, *SITE* was modified to address this aspect.. In an earlier experiment, this had led to much worse performance. In the specific experiment here, the result was that the cumulative area grew slightly more slowly. The process stopped after 90 observers, covering 96% of the cell, had been selected.

3.4 Arbitrarily Culling Tentative Observers

Since there were so many highly visible tentative observers, a certain redundancy was suspected. Therefore the previous experiment was modified as follows. Of the 1008 tentative observers written by *FINDMAX*, every second observer was deleted, and *SITE was then rerun*. Surprisingly, there was essentially no deterioration in the results. *SITE* found 180 observers to cover 1439522 points, in only 26 elapsed seconds.

More experimentation is needed to determine a reasonable minimal number of tentative observers, as a function of the input parameters.

3.5 Lower Height

One might object that the relatively high observer and target height (100m) in the above experiments, which led to many highly visible points, made the process easier. Therefore, those heights were reduced to 30m and the process was rerun. In this case, *SITE* needed 324 observers to cover 99% of the cell. The elapsed time for the whole process was only 527 seconds, with the increase due to *SITE* needing to use more observers. The issue is not just that the average visibility of even the top observers is a little lower, but also that the viewsheds have more holes that require other viewsheds to fill.

We next stressed the system by repeating the process with the observer and target height at 10m. This time, *SITE* needed 515 observers to cover 98% of the cell, using 517 elapsed seconds. One reason for needing more observers was that, in some blocks, even the best observers had only 80% visibility index. Fig. 5 shows samples of some of the best viewsheds.

Fig. 5. Some of the best viewsheds for H=10

Fig. 6 shows the cumulative viewshed after 10, 20, 30, ..., 90 observers have been included.

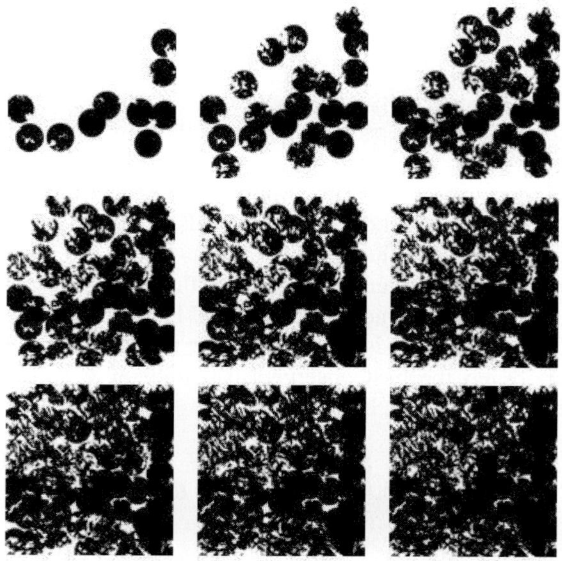

Fig. 6. Cumulative Viewshed for R=100, H=10, After 10, 20, ... 90 Observers Added, for Lake Champlain W

3.6 Larger Radius

We then stressed the system by increasing the viewshed radius from 100 points to 1000 points, while keeping the height at 10. By the time *SITE* had added 53 observers, 99% of the cell had been covered, although *SITE* was able to place 217 different observers before no more would cover even a single new point. This process took 3800 elapsed seconds, mostly in *VIEWSHED*, since its time depends on the radius.

3.7 Hailey E Cell

Finally, the USGS Hailey East level 1 DEM cell was tried, with observer and target height of 10, and radius 100. This cell is more mountainous, so the average visibility index would be expected to be less. For this test, *SITE* used 991 of the 1008 tentative observers before giving up with only 94% of the area covered. However, the first 240 observers covered a full 90% of the area. This suggests that trying to cover the last few points may be counterproductive, unless more tentative observers are added to cover the points that are still hidden.

Fig. 8 shows the cumulative viewshed, and the area of the latest added viewshed, as a function of the number of viewsheds added. As before, there is little relation between a tentative observer's visibility index and how soon it is added (after the first observers with a high visibility have been added).

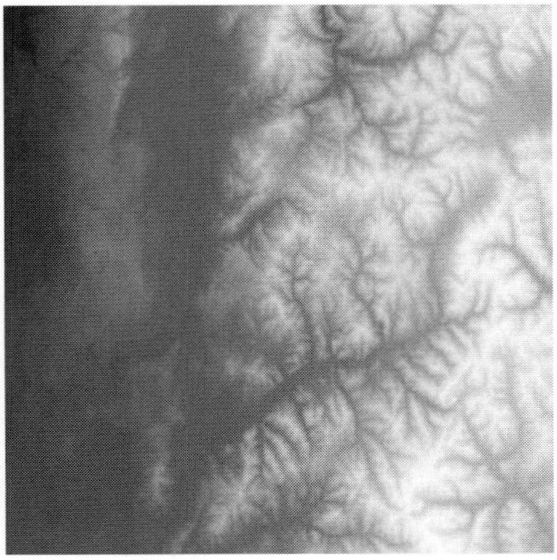

Fig. 7. Hailey East Cell (again, light=high elevation, dark=low elevation)

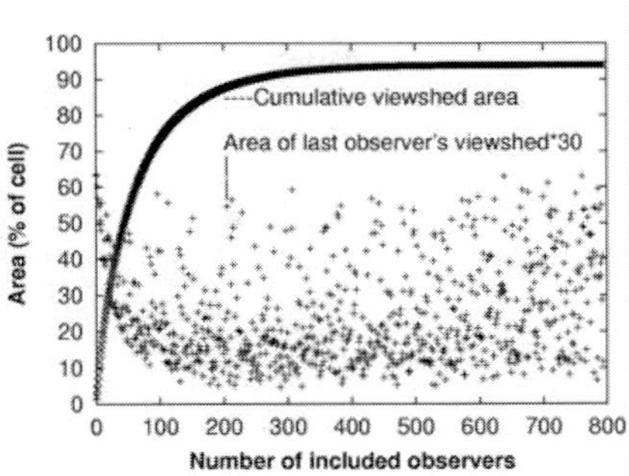

Fig. 8. Area of Last Inserted Viewshed *30, and Cumulative Visible Area as Observers are Inserted, for Hailey-E Cell

4 The Future

The various tradeoffs mentioned above, and the above experiments illuminate a great opportunity. They tell us that shortcuts are possible in siting observers, which will produce just as good results in much less time.

Another area for investigation is the connectivity of either the viewshed, or its complement. Indeed, it may be sufficient for us to divide the cell into many separated small hidden regions, which could be identified using the fast connected component program described in Nagy et al. (2001).

There is also the perennial question of how much information content there is in the output, since the input dataset is imprecise, and is sampled only at certain points. A most useful, but quite difficult, problem is to determine what, if anything, we know with certainty about the viewsheds and observers for some cell. For example, given a set of observers, are there some regions in the cell that we know are definitely visible, or definitely hidden?

This problem of inadequate data is also told by soldiers undergoing training in the field. Someone working with only maps of the training site will lose to someone with actual experience on the ground there.

Finally, the proper theoretical approach to this problem would start with a formal model of random terrain. Questions could then be asked about the number of observers theoretically needed, as a function of the parameters. Until that happens, continued experiments will be needed.

Acknowledgements

This paper was supported by the National Science Foundation while the author was a Program Director there, in CISE/C-CR. The SW environment, which greatly aided our productivity, consisted of GNU C++, zsh, SuSE linux, xemacs, LaTeX, gnuplot, and xv. The HW was a 600MHz Pentium III with 192MB of memory.

References

Agarwal P, Sharir M (1999) Davenport-Schinzel sequences and their geometric applications. In: Sack JR, Urrutia J (eds) Handbook of Computational Geometry. Elsevier Science Publishers B.V., North-Holland Amsterdam

De Floriani L, Jeanne P, Nagy G (1993) Visibility-related image features. Pattern Recognition Letters 13:463-470

De Floriani L, Magillo P (1994) Visibility algorithms on DTMs. Int. J. Geographic Information Systems 8(1):13-41

Fisher PF (1993) Algorithm and implementation uncertainty in viewshed analysis. Int. J. Geographical Information Systems **7**:331-347

Franklin WR (2000) Applications of analytical cartography. Cartography and Geographic Information Systems 27(3):225-237

Franklin WR (2002) Siting observers on terrain - color version [online]. Available from: http://www.ecse.rpi.edu/Homepages/wrf/research/site.pdf

Franklin WR, Ray C (1994) Higher isn't necessarily better: Visibility algorithms and experiments. In: Waugh TC, Healey RG, (eds) Advances in GIS Research: Sixth International Symposium on Spatial Data Handling. Taylor & Francis, Edinburgh, pp 751-770

Lee J (1992) Visibility dominance and topographic features on digital elevation models. In: Bresnahan P, Corwin E, Cowen D (eds) Proceedings 5th International Symposium on Spatial Data Handling, Vol. 2. International Geographical Union, Commission on GIS, Humanities and Social Sciences Computing Lab, U. South Carolina, Columbia, South Carolina, USA, pp 622-631

Nagy G, Zhang T, Franklin W, Landis E, Nagy E, Keane D (2001) Volume and surface area distributions of cracks in concrete. In: 4th International Workshop on Visual Form. Capri, Italy

Ray CK (1994) Representing Visibility for Siting Problems. Ph.D. thesis, Rensselaer Polytechnic Institute

Shannon RE, Ignizio JP (1971) Minimum altitude visibility diagram - MAVD. Simulation pp 256-260

Shapira A (1990) Visibility and terrain labeling. Master's thesis, Rensselaer Polytechnic Institute

An Agent-Based Framework for Polygonal Subdivision Generalisation

Martin Galanda and Robert Weibel

Department of Geography, University of Zurich, Winterthurerstr. 190, 8057 Zurich, Switzerland, {mgalanda,weibel}@geo.unizh.ch

Abstract

Comprehensive techniques for the generalisation of polygonal subdivisions are still lacking in research as well as in commercial GIS, but required for the derivation of arbitrary scales from a master database. This paper outlines a framework for the automated generalisation of polygon mosaics (polygonal subdivisions) based on a multi agent system by extending previous research carried out by the AGENT consortium. After listing generic properties of agents spatial levels of polygon generalisation agents are identified, namely *map*, *group*, *polygon* and *line*. Each of them is linked to a specific agent type. Both the process of polygon generalisation based on a multi agent system as well as the evolution of an agent during the generalisation process is discussed theoretically. Then, a worked example clarifies and illustrates the concepts and methods presented before. Finally, foreseeable challenges of the framework's implementation based on preliminarily experiments are pointed out.
Keywords: cartographic generalisation, polygonal subdivision, multi agent system

1 Introduction

One of the goals of current research in map generalisation is to enable the automated derivation of arbitrary scales from a master database (Bengtson 2001, Ruas 2001). Although progress in research is reported continuously, some key generalisation tasks are not yet solved satisfactorily. Amongst other aspects, a framework that includes both methods and concepts for the generalisation of categorical data in the vector data model (polygon mosaics or polygonal subdivisions) are missing.

A categorical data set is collectively space exhaustive over the entire map, i.e. all its points must belong to only one category and polygon. Thus, whenever one polygon object is generalized, that is, its geometry is changed all polygons sharing a geometric primitive with that object[1] are affected, too. In practice, however line-based algorithms are commonly used to generalise polygonal data. Since the specific structure and topology of polygons is ignored, unsatisfying results are not surprising. Existing approaches dedicated to the generalisation of categorical data concern either very specific kinds of data (e.g. Jaakkola 1998, Mueller and Wang 1992) or the isolated implementation of single geometric algorithms (e.g. Bader and Weibel 1997, Galanda 2001). Comprehensive techniques for polygon generalisation have neither been addressed in the research community nor through commercial GIS vendors.

This paper outlines a framework for the automated generalisation of polygonal subdivisions (polygon generalisation) based on a multi agent system (MAS). The research and results presented extend previous work carried out by the AGENT consortium (Lamy et al. 1999, Ruas 1999, Barrault et al. 2001, Regnauld 2001). Section 2 gives a short introduction to MAS and discusses the applications in map generalisation. Section 3 presents the specifications for spatial levels of polygon generalisation. In section 4, an outline of the process of polygon generalisation based on agents is presented, while section 5, illustrates a first worked example. Section 6 discusses the selected implementation issues, and finally, some conclusions are drawn and an outlook on the next steps of the research work is presented in section 7.

2 MAS in Cartographic Generalisation

Borrowing from a definition given by (Luck 1997) an agent for cartographic generalisation denotes a geographic entity capable of controlling its own decision making and generalisation guided by a set of cartographic constraints[2]. Thus, an agent (Barrault et al. 2001, Duchêne et al. 2001, Ruas 1999)

- is linked to a set of constraints;
- possesses a method to determine its so-called happiness, i.e., it evaluates and summarises the satisfaction of all its constraints;
- aims to reach a perfect state (= perfect happiness; all constraints are satisfied);
- proposes plans (cartographic operations) according to the violated constraints in order to improve its happiness;
- triggers plans autonomously starting with the one supposed to be best;
- is able to compare and store different states;

[1] According to the characteristics of a polygon mosaic at least one other polygon is concerned.

[2] "A constraint is a condition similar to the predicate in a production rule. The distinction is that a constraint is not bound to a particular action."(Beard 1991, p.122) The concept of a constraint-based approach to cartographic generalisation has been emphasised by (Weibel 1996) and (Ruas 1999), among others.

- can return (backtrack) to any previous state;
- may instantiate, trigger and coordinate other agents[3].

A MAS designates several independent agents cooperating to solve problems at both a local (related to a single agent) and a global (related to a group of agents) level (Luck 1997, p. 1227). Cartographic generalisation is an holistic and subjective task, that is, each conflict should be solved in its spatial context, and several 'correct' solutions are conceivable. A solution often constitutes a compromise between several local and global constraints. Modelling the generalisation process by a MAS means that "a sub-optimal but acceptable solution can often be reached" (Lamy et al. 1999). Compared to other technologies used in map generalisation like expert systems or neural networks a MAS supports an holistic approach and allows a dynamic adaptation to a changing environment. As a result of these properties, the MAS technology seems to provide an adequate framework for map generalisation (Regnauld 2001).

In the context of the project AGENT funded by the European Commission, a consortium of experts in multi agent systems, experts in automated map generalisation and GIS specialists have succeeded in setting up a MAS for generalisation tasks of topographic mapping (Lamy et al. 1999, Barrault et al. 2001)[4]. The AGENT package (data schema and the generic agent engine) is implemented in a commercial GIS and has already proven useful in map production (Bengtson 2001). This paper extends methods and concepts - developed with respect to the generalisation of road networks and urban settlements during the AGENT project - for use with polygonal subdivisions.

3 Spatial Levels of Polygon Generalisation

The more a cartographic conflict is narrowed in on a spatial context, and the more precisely the situation is characterised, the more specific transformations are applied and the faster an adequate result (i.e., a compromise) is achieved (Ruas 1999, Barrault et al. 2001). Thus, map space is organised in so-called spatial levels of polygon generalisation. So each constraint, cartographic operation etc. is delegated to a specific spatial level[5], that is, it holds a specific scope. For instance, the constraint of minimal size refers to a single polygon object while the minimal distance between objects is related to a group of polygons. Related to polygon generalisation, 4 different spatial levels, namely

- map,

---

[3] In the following an agent supervising other agents is generally called a *parent agent* while an agent triggered and controlled by a parent agent is termed a *child agent*.

[4] More information on this project can be found at http://agent.ign.fr/.

[5] The spatial level of constraints can be found in the complete list of constraints to polygon generalisation at http://www.geo.unizh.ch/~mgalanda/poly-www/constraints.html. The spatial level of generalisation operators is listed in section 4.2.

- group,
- polygon and
- line,

seem to be reasonable. Agent prototypes for these different levels are available in the framework. The generic properties and behaviours listed in the previous section are independent of an agent's spatial level.

Map Agent. Every polygon mosaic holds only one map agent (cf. Fig. 1). It is responsible for constraints and cartographic operations concerning the whole polygon map (e.g. reclassification) as well as the instantiation of the group agents.

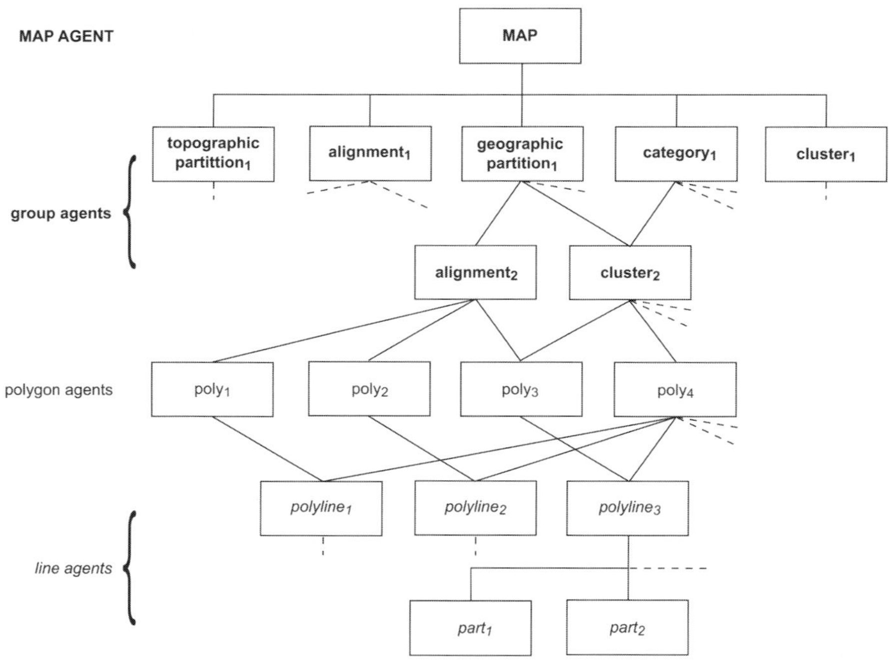

Fig. 1. The organisation of agents at the four different spatial levels of polygon generalisation. See also Fig. 2 below for some graphic examples

Group Agents. Group agents handle contextual generalisation, i.e. conflicts between polygon objects. They are composed of several polygon objects sharing a common geometric or semantic relation. Thus, reasonable group agents may be attached to a cluster, an alignment, a category, a topologic partition (e.g. neighbours of 1st and 2nd order) or a geographic partition (e.g. a collection of polygons bounded by rivers or roads). Group agents can recursively subdivide themselves if needed, that is, a group agent can spawn off other group agents besides the polygon agents. For instance, a geographic partition agent may supervise a cluster agent - compare *geographic partition*$_1$ in Fig. 1.

Polygon Agents. A polygon agent coordinates the generalisation of an area object. Constraints and operations acting on an individual polygon are evaluated and performed without considering the agent's spatial context (e.g. enlargement). However, as already mentioned above the geometric transformation of one polygon induces always at least the modification of one other polygon. So every change must be propagated to those objects sharing a common geometric primitive with the transformed object.

Line Agents. Line agents are delegated to polylines bordering a polygon object and their generalisation (e.g. simplification). Exactly two polygon agents can supervise - assuming a clean topology - a line agent of the first generation[6]. As group agents they are enabled to recursively subdivide themselves in order to

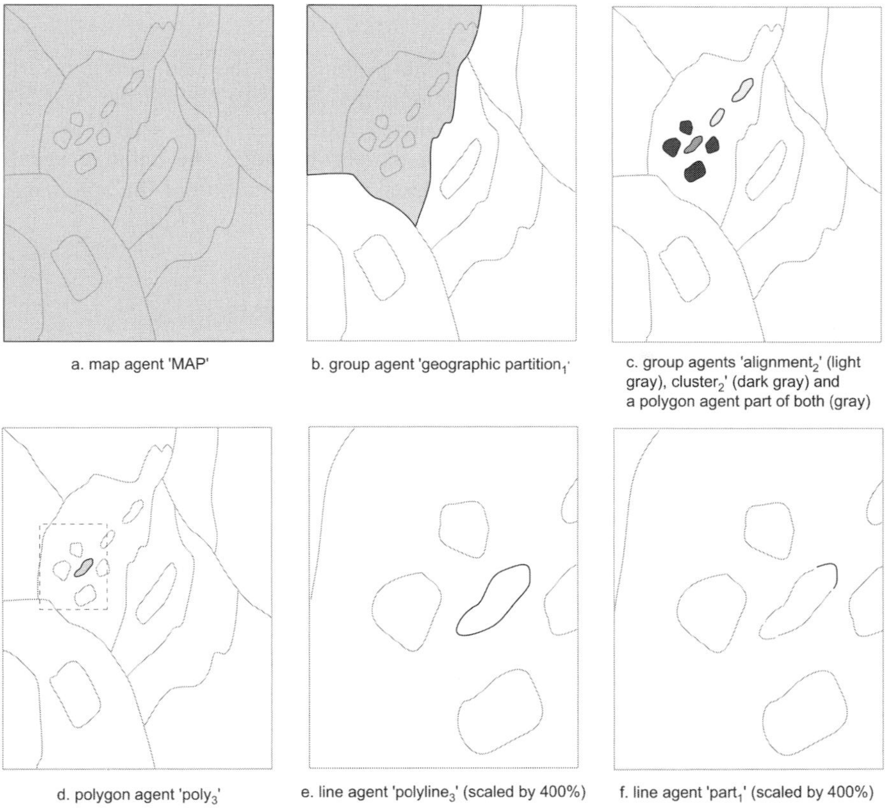

Fig. 2. Some examples of different spatial levels of polygon generalisation in a polygon mosaic according to the organization of agents shown in Fig. 1

[6] A child agent of the first generation always has parents of a superior agent type - e.g. a line agent of the first generation is exclusively supervised by polygon agents. A child agent of the second generation always has parents of the same agent type - e.g. a group agent of the second generation always has another group agent as a parent.

If necessary parent agents build their child agents at run time. Parent agents are enabled to specify the child agents' constraints according to an analysis of all their child agents or the failure of a previous plan, e.g. a group agent can tell a polygon perform generalisation on homogenous line parts (Duchêne et al. 2001), see *polyline*$_3$ in Fig. 1 and 2(e,f).

agent to not enlarge itself when it knows about a lack of free map space. Different parent agents can supervise one child agent in turn. For instance, *poly*$_3$ in Fig. 1 and Fig. 2c is part of the group agents *alingment*$_2$ and *cluster*$_2$. Thus, the agents' generalisation must be done in a sequential process.

4 Agent Life Cycle in Polygon Generalisation

The generalisation process for polygonal data presented in the next section reverts to experiences gained and concepts developed by (Ruas 1999) and the AGENT project (Barrault et al 2001). It is organised into three main stages. A pre-processing stage prepares the database for the second stage, the iterative agent-based generalisation. Finally, the user undertakes an evaluation of the automated generalisation results during the third stage.

4.1 Pre-Processing: Data Specification and Analysis

The first step in the pre-processing stage is that all constraints are specified according to the basic conditions of the generalisation task. In addition, supplementary information (auxiliary data) is calculated on the database objects.

4.2 Generalisation: Life Cycles of Agents

At the very beginning of the generalisation process the map agent is activated. It has to care about its own constraints but also to identify and trigger group agents on the fly. Procedural knowledge of polygon generalisation is needed to decide either to try first to satisfy a parent agent's constraint and then work on the child agent's constraint or to first initialise the child agent's generalisation and then the parent's constraints. In practice, a mixture of these approaches may be applied. Again, procedural knowledge can help in choosing which child agent is to be triggered first amongst all the child agents.

The AGENT package uses a constraint-based approach to generalisation (Beard 1991, Weibel 1996, Ruas 1999), that is, constraints initialise and control the entire generalisation process. Constraints to polygon generalisation are the subject of (Weibel 1996), (Peter and Weibel 1999) and (Edwardes and Mackaness 2000). Details of the agent engine to be used in our project (i.e. the one developed for the AGENT project) are described in (Regnauld 2001). In a MAS for map generalisation constraints are linked to agents that aim to satisfy the *offensive constraints* without violating one of the *defensive constraints*[7]. To every constraint a measure[8] is linked to evaluate its satisfaction (= to determine the severity of violation). A list of possible plans is attached, too, that propose cartographic operations (geometric and semantic transformations) for improving the agent's happiness – Fig. 3. The final ordered list of plans results from the weighted sum of all plans suggested by any violated constraint of the agent. The current best plan is then triggered.

constraint	The distance between two polygon objects should not be less than the minimal distance (e.g. minimum visual separability distance).
agent	Group
measure	proximity measure (e.g. Hausdorff distance)
plan	a. Displacement - algorithm based on 'snakes' (Galanda 2001) b. Displacement - vector displacement c. Exaggeration - algorithm based on 'snakes' (Galanda 2001) d. Aggregation - convex hull e. Typification

Fig. 3. Example of an offensive metric constraint at the group agent level in polygon generalisation, the attached measure and list of possible plans.

The process of improving an agent's happiness starts when its parent agent changes its state to *active*. The following sequence of constraints evaluation [evaluate constr.][9], proposing plans [propose plans], triggering the best plan [trigger best plan] and re-evaluation [re-evaluate] is the same generic behaviour of all agents and called an agent's life cycle. A modified 'hill-climbing' algorithm is applied within this life cycle - see Fig. 4 - to find the best compromise to all constraints of an agent (Regnauld 2001). The life cycle ends when either all constraints are satisfied completely (perfect state) or there is no plan left to try. Nevertheless, the database is updated by the best state ever reached in the life cycle [update database]. The agent is set to *passive* again, that is, the control of the generalisation process is returned to the parent agent.

---

[7] An *offensive constraint* is an indicator for the need of generalisation, e.g. a minimum size constraint, while a *defensive constraint* controls the preservation of a certain property of an agent such as its topologic relation.

[8] For details on measures in polygon generalisation refer to (Peter 2001).

[9] Terms in brackets relate to life cycle steps displayed in Fig. 4.

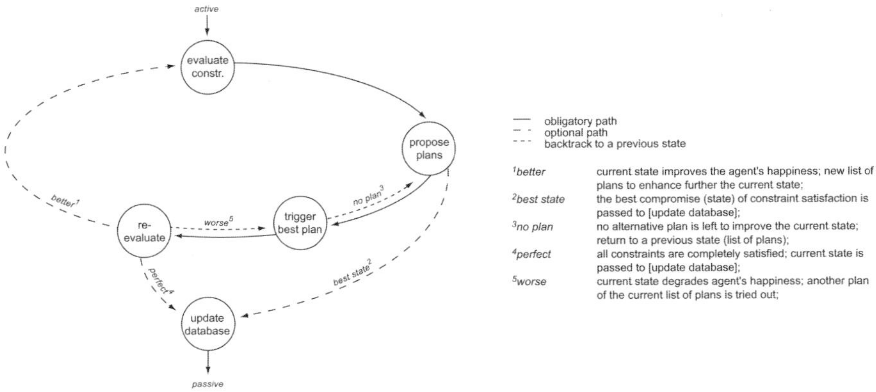

Fig. 4. An agent's life cycle in polygon generalisation with the incorporated 'hill-climbing' algorithm.

One constraint attached to every parent agent assesses the happiness of its child agents. A violation of that constraint requires either plan *a* or *b*:

a. the instantiation of the child agent's life cycles; the autonomous generalisation of the child agents, that is, the control of the generalisation process is passed to the individual child agents in turn; each of them tries to independently improve its happiness.
b. the state of the conflicting child agents is set to *reactive* in turn; each of them executes an order (a method and parameters) given by the parent agent[10].

In both cases, the child agent reports modified properties (geometry and semantics) to its parent agent that then continues its life cycle at the re-evaluation step.

The example of a constraint to polygon generalisation shown in Fig. 3 presents a set of possible plans (cartographic operations) to enhance the agent's happiness if required. In polygon generalisation we distinguish the following cartographic operators (spatial agent type given in parentheses):

For semantic transformation

- *Reclassification* (map) changes the category an object belongs to.

For geometric transformation

- *Aggregation* (group) combines a polygon with other polygons of the same or a similar class to a new polygon object.
- *Typification* (group) reduces the complexity of a group of polygon objects by removing, displacing, aggregating and enlarging single objects. The readability is ensured and the significant alignment pattern is preserved.

---

[10] In general a reactive order can be given to an agent by any other agent.

- *Displacement* (polygon or group) denotes the movement of a whole polygon object - no shape distortion.
- *Collapse* (polygon) reduces the polygon geometry either to a line or to a point geometry.
- *Elimination* (polygon) removes an object from the polygon mosaic; the freed space is assigned to other categories.
- *Enlargement* (polygon) denotes a global increase (decrease) of a polygon's geometry.
- *Exaggeration* (polygon or group) defines a local increase (decrease) of a polygon's geometry.
- *Simplification* (line) reduces the granularity of a linear geometry by removing vertices.
- *Smoothing* (line) improves the visual appearance of a line object.

Algorithms implement the concepts described by cartographic operators. Both a review of existing algorithms and some newly developed algorithms for polygon generalisation based on optimisation techniques can be found in (Galanda 2001).

4.3 Post Processing: Final Evaluation

After the map agent's life cycle is completed the user evaluates the final result with the help of a detailed report (severity of remaining conflicts, happiness per agent or category etc.) output automatically by the system (Ruas 2001). If necessary some interactive (re)generalisation is performed.

5 A Worked Example

In this section, we present a worked example for the agent-based framework for polygon generalisation. The example is intended to both clarify and illustrate the ideas and concepts discussed theoretically above. It demonstrates some excerpts of the agent-based generalisation of one cluster agent (Fig. 6). It consists of five polygons belonging to two categories (light and dark gray) and is embedded in a polygon environment - compare *state0* in Fig. 6. The different states of the cluster agent and the cartographic operations they result from (shown in Fig. 5) as well as the evolution of the constraint satisfaction (diagrams displayed in Fig. 6) summarise how the 'hill-climbing' algorithm finds the best compromise between several competing constraints.

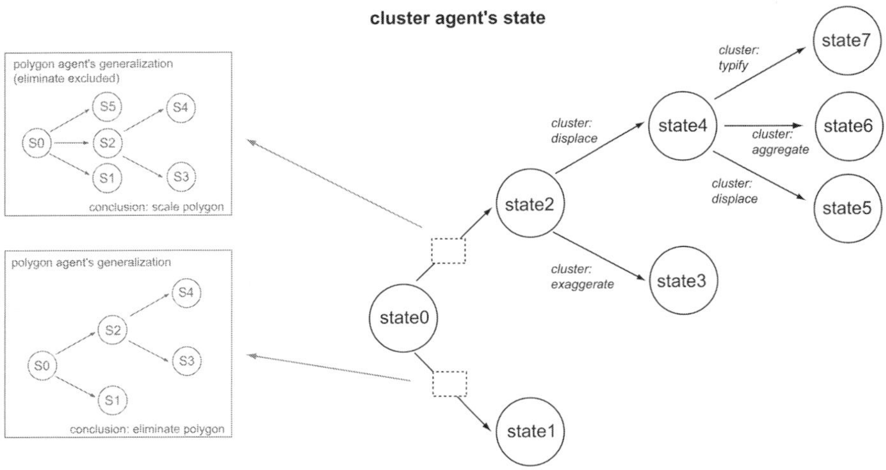

Fig. 5. The different states of the cluster agent (state0 to state7) in the generalisation process and the cartographic operations they result from. The states (s0 to s5) of the triggered polygon agent are indicated in the two small insets to the left.

In the example the following three constraints[11] related to cluster agents are considered:

- \`minimal distance' (constraint A)
  The distance between two polygon objects should not be less than the minimal distance.
- \`preserve alignment' (constraint B)
  The alignment of polygons of different classes (characteristic patterns of alignment) should be preserved.
- \`child agents' happiness' (constraint C)
  The constraints of a parent's child agents must reach a defined level of satisfaction.

state0 The *state0* in Fig. 6 shows the agent's state when activated. The life cycle (cf. Fig. 4) starts with an evaluation of the constraints in order to determine if there is a need for generalisation or not. Those constraints directly related to a property of the example agent (constraint *\`minimal distance'* and *\`preserve alignment'*) are satisfied. However, the third constraint observing the happiness of the supervised child agents is not fulfilled, because the area property of one polygon agent (the polygon in the centre of the cluster) falls below the threshold defined in a minimum area constraint. The polygon agent's generalisation

[11] In the given example the satisfaction of the individual constraints is rated on a continuous scale ranging between perfect satisfaction and total violation (see the diagrams in Fig. 6.)

performed in a separate life cycle (*s0* to *s5* in Fig. 5) results in the elimination of that polygon.

state1 The modified geometry is reported to the cluster agent that subsequently re-evaluates the changed situation (*state1*).

state0 As *state1* does not conform to the *'preserve alignment'* constraint a backtrack to the previous state (*state0*) is made, the only plan provided for such a conflict. Next, the polygon agents are again activated. A restriction is then set with respect to the conflicting polygon, namely the elimination operator is removed from the list of possible plans.

state2 The cluster agent's *state2* results from an enlargement of that polygon. Due to the enlargement of this polygon the distance to the other polygons falls below the threshold defined in the *`minimal distance'*-constraint. From an agent point of view, an improvement of the overall constraint satisfaction (see Fig. 6) has been achieved, that is, *state2* is considered to be better than *state0*. Possible plans to solve that conflict are proposed (compare also the list of plans in Fig. 3).

state3 The application of an exaggeration operation (*state3*) decreases the cluster agent's happiness because the *`preserve alignment'*-constraint is violated even further as a result of the shape distortion at the polygon level.

state2 Subsequently, a backtrack to *state2* is performed and the next plan of the list is applied.

state4 The displacement of polygons resulting in *state4* represents the best compromise so far with respect to the different constraint satisfaction. As this state is considered better than its previous state (*state2*) a new list of plans is proposed and tried out.

state5-7 No further improvement of the agent's happiness occurs; each time a backtrack to *state2* is performed.

state2 Thus, the life cycle of the cluster agent continues at *state2* by searching for alternative plans to improve its happiness. In this example it is assumed that at *state2* no further plans are left (typification and aggregation are not considered)

state0 No plans to be tried out are left at *state0*, too. Consequently the best solution, the one stored in *state4*, is used to update the agent's geometry. The cluster agent is set to *passive* again.

6 Implementation Issues

Although the MAS-based approach is not new and advances have been made, extensive additional research efforts are required, in particular for subdivision and generalization of polygons. For instance, a comprehensive set of automated, robust and efficient algorithms specifically designed for use with polygonal data along with measures to determine the satisfaction of all constraints attached to one spatial level of polygon generalization need to be explored.

Applying the existing AGENT package, experiments were undertaken with the landscape model, land use data (VECTOR25 and VECTOR200) of the Suisse Federal Office of Topography and geological data of the Swiss National Park at a scale of 1:25,000. These real data tests identified amongst other things the following major challenges in the implementation of the proposed framework:

- the retrieval of procedural knowledge of polygon generalisation (choice of the `best plan'; which child agent first? etc.);
- the interplay and interaction of different agents - both of the same and of different spatial levels - acting on the same subset of polygons or the same polygon object;
- the improvement (implementation) of the communication between agents, both of the same and of different spatial levels;
- the determination of reasonable plans and ordered lists of possible plans related to every constraint, that are independent of both the kind of data and the scale range;
- the automated identification of group agents belonging to a map agent or a superior group agent by spatial and/or semantic analysis.

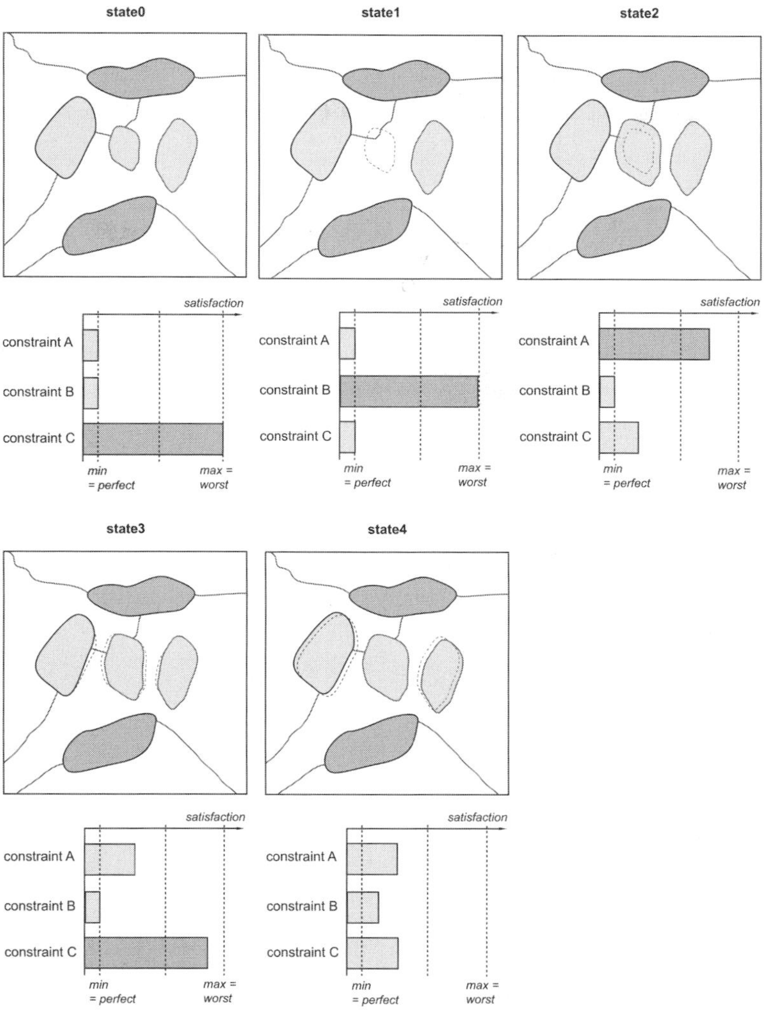

Fig. 6. The different states of the example agent and the corresponding constraint satisfaction

7 Conclusions and Future Work

The main goal of this research project is the evaluation of the suitability of MAS for automated polygon generalization. This paper outlined a framework for the generalization of polygonal subdivisions based on a MAS. Concepts and methods developed by the AGENT consortium (Lamy et al. 1999, Barrault et al. 2001) for the generalisation of topographic maps (road networks and urban settlements) were adapted and extended. The key points of the presented research concerned:

- the definition of the agents' organisation in a polygonal subdivision (sect. 3);
- a first acquisition of constraints related to the different agent types including measures and potential plans - for an example cf. Fig. 3;
- the interplay of different agents and agent types within the so-called agent's life cycle (sect. 4).

While a worked example (sect. 5) clarified the theoretical discussion of the framework and demonstrated the underlying ideas and concepts section 6 pointed out relevant implementation issues.

As indicated in the previous section our work concentrates currently on the extension and fine-tuning of the existing AGENT package (Lamy et al. 1999, Barrault et al. 2001) in order to put the presented framework into practice. First results of real world data experiments will be forthcoming in July 2002.

Acknowledgements

This research is part of project 'GENDEM: Map Generalisation for Thematic and On-Demand Mapping in GIS' supported by the Swiss National Science Foundation under contract 20-52759.97. Many thanks go to Mathieu Barrault for valuable assistance and comments.

References

Bader M, Weibel R (1997) Detecting and Resolving Size and Proximity Conflicts in the Generalisation of Polygon Maps. In: Proceedings of the 18th International Cartographic Conference. Stockholm, pp 1525–1532

Barrault M, Regnauld N, Duchêne C, Haire K, Baeijs C, Demazeau Y, Hardy P, Mackaness W, Ruas A, Weibel R (2001) Integrating multi-agent, object-oriented and algorithmic techniques for improved automated map generalization. In: Proceedings of the 20th International Cartographic Conference. Beijing, pp 2110-2116

Beard M (1991) Constraints on Rule Formation. In: Buttenfield B, McMaster R (eds) Map Generalization: Making Rules for Knowledge Representation. Longman, London, pp 121–135

Bengtson M (2001) Design and implementing of automatic generalisation in a new production environment for datasets in scale 1:50000 (and 1:100000). In: Proceedings of the 20th International Cartographic Conference. Beijing, pp 1076–1083

Duchêne C, Barrault M, Haire K (2001) Road Network Generalization: A Multi Agent Approach. In: Proceedings of the 20th International Cartographic Conference. Beijing, pp 2166–2177

Edwardes A, Mackaness W (2000) Modelling knowledge for automated generalization of categorical maps – a constraint based approach. In: Atkinson P, Martin D (eds) GIS and Geocomputation (Innovations in GIS 7). Taylor&Francis, London, pp 161-173

Galanda M (2001) Optimization techniques for polygon generalization [online]. In: Proceedings of the 4th Workshop on Progress in Automated Map Generalization.

Beijing, Available from: http://www.geo.unizh.ch/ICA/docs/beijing2001/galanda_v0.pdf

Jaakkola O (1998) Multi-Scale Categorical Databases with Automatic Generalization Transformation Based on Map Algebra. Cartography and Geographic Information Systems 25(4):195–207

Lamy S, Ruas A, Demazeau Y, Jackson M, Mackaness W, Weibel R (1999) The Application of Agents in Automated Map Generalisation. In: Proceedings of the 19th International Cartographic Conference. Ottawa, pp 1225–1234

Luck M (1997) Foundation of Multi Agent Systems: Issues and Directions. Knowledge Engineering Review 12(3): 307–318

Mueller J, Wang Z (1992) Area-Patch Generalization: A Competitive Approach. The Cartographic Journal 29(2): 137–144

Peter B (2001) Measures for the Generalization of Polygonal Maps with Categorical Data [online]. In: Proceedings of the 4th Workshop on Progress in Automated Map Generalization. Beijing, Available from: http://www.geo.unizh.ch/ICA/docs/beijing2001/Peter_v2.pdf

Peter B, Weibel R (1999) Using Vector and Raster-Based Techniques in Categorical Map Generalization. In: Proceedings of the 19th International Cartographic Conference. Ottawa, pp 1135–1145

Regnauld N (2001) Constraint based mechanism to achieve automatic generalization using agent modeling. In: Proceedings of the GIS Research UK 9th Annual Conference. Glamorgan, pp 329–332

Ruas A (1999) Modèle de généralization de données géographiques à base de contraints et d'autonomie. Ph.D. thesis, Université de Marne-la-Vallée

Ruas A (2001) Automating the generalisation of geographical data: the age of maturity?. In: Proceedings of the 20th International Cartographic Conference. Beijing, pp 1943–1953

Weibel R (1996) A Typology of Constraints of Line Simplification. In: Proceedings of the 7th International Symposium on Spatial Data Handling. Delft, pp 9A.1–9A.14

Template Matching in Support of Generalisation of Rural Buildings

Desmond Rainsford and William Mackaness

Geography Department, The University of Edinburgh, Drummond St, Edinburgh EH8 9XP, 0131 650 8163, wam@geo.ed.ac.uk

Abstract

A range of methodologies has been proposed to derive generalised forms of buildings at coarser scales. In this paper, we explore the use of simple pattern matching algorithms in order to select from a set of templates, a building outline that best characterises a more detailed form. This template matching process is used to simplify the form of rural buildings (farmsteads) in Danish mapping, and is currently a manual task within the National Mapping Agency of Denmark (KMS). This research has explored the feasibility of automating this approach, and reports on its implementation, and provides an evaluation. The challenge in pattern matching is to minimise the misidentification of patterns (type I error) and failures to find any match (type II error) in the assignment of templates. The initial set of results was very encouraging.
Keywords: building generalisation, pattern recognition, automated cartography

1 Building Generalisation

Levels of generalisation are controlled by the map scale, theme and the importance attached to map objects. While manual generalisation was entirely appropriate for the creation of paper products at fixed scales in a pre-digital age, it is an impediment to the rapid production of electronically derived maps at arbitrary scales. It is usually still necessary to maintain separate databases of different levels of detail in order to produce maps of the same area at different scales or for different themes. Such a constraint is hugely inefficient in terms of data maintenance and storage. With good reason, efforts at automated generalisation have focussed almost entirely on emulating the skills of human cartographers. Curiously, little attempt has been made to use automated techniques to change and perhaps improve upon the cartographic conventions that have evolved with paper maps over the centuries. It could be argued that the subject of this project is a case where a better solution could be not to try to mimic a traditional map making technique, but to adopt a more flexible approach offered by automated simplification methods.

Litchner (1979) was one of the earliest workers to successfully demonstrate an automated approach specific to building generalisation. Using seven "elementary" procedures (simplification, enlargement, displacement, combination, selection or elimination, classification or exaggeration) he was able to produce a satisfactory generalisation of city buildings. However, his approach did not take into account context. Rule-based approaches to building generalisation have since become more prevalent. Edwardes et al. (1998) demonstrates the analysis of cartometrics measures (squareness, compactness etc.) to create a set of rules that are used to select procedures to effect generalisation. Glover and Mackaness (1999), on the other hand, devised a rule set based on display scale, map type and object class). This paper explores an alternative approach based on template matching – inspired by manual methods used by the Danish Mapping Agency KMS for the generalisation of rural buildings. A broad range of techniques exist for both pattern matching (Veltkamp 2001a; Veltkamp 2001b; Loncarid 1998), and automatic feature extraction from remotely sensed images (for example the work by Stassopolou et al. (2000) on building detection). This paper builds on ideas of pattern matching which show favourable comparison more conventional approaches to map generalisation.

2 Shape Recognition

Farm buildings in Denmark are displayed on small scale maps, by representing them using a series of simple alphabetic templates with similar shapes. The use of templates in effect creates caricatures of the farm building outlines. This approach has been adopted by the KMS to provide visual consistency in their small scale maps such that meaning of these features can be understood from their shapes. At such scales (1:50,000), it can be argued that the implied meaning of a small feature is more important to the user than its representational accuracy. Thus, it is common to see the generalisation of buildings such as churches represented by a uniform symbol. Templates, on the other hand, preserve more of the character of the physical object. Fig. 1 shows a small sample from a KMS product showing the original building in grey and the generalised form (a fitted template) in outline.

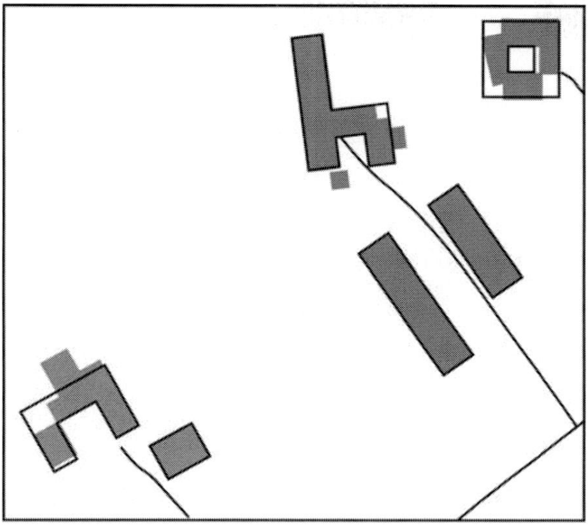

Fig. 1. Manual template matching

The aim of this project was to automate a process which is presently carried out manually by the Danish Mapping Agency. In order to perform this task, nine polygon templates have been created (identified by the alphabetical characters they resemble), based on a specification document provided by KMS. A template matching process requires the recognition of polygon shapes as well as the application of simplification procedures. Shape recognition has not been a central theme in map generalisation, although the use of "descriptive trees" and "minimum spanning trees", for this purpose, are described by Weibel and Dutton (1999). Most of the literature on this subject exists outside of the geographic realm. A review of shape characterising procedures by Ehler *et al.* (1996) cites single values measures (e.g. circularity and form), fourier measures (for sinuous lines), image content queries (IBM development) and the use of binary shape matrices. The first of these requires the judicious use of several measures but is relatively easy to implement. The fourier method is not well suited for polygonal shapes. Image content queries we will discuss later. Binary shape matrices are used to fit shapes in the raster domain by the application of a roving spinning filter. In principal, this approach could be applied to the current task, but would require the rasterization of the building shapes and would be complicated to apply.

Perhaps the richest vein of pattern recognition expertise is related to image analysis. One technique, skeletonisation, provides a synthetic and thin representation of objects that are useful for the description of shape (Attali and Montanvert, 1997). While skeletonisation is a powerful technique for extracting topology from shapes, its use is beyond the scope of this exercise. Also of interest is the use of Hausdorff-based image comparison techniques using the generalised Hausdorff measure. This technique has been shown to be effective for recognising partly obscured targets in photographic images and is obviously well suited for surveillance applications. Doubtless, this technique could be used for shape selection, but it would be an overly complex for the scope of problem addressed here.

3 Methodology

Because we are dealing with single building polygons only, we do not have to address the problems of aggregation and displacement and we can concentrate on the steps required for simplification. The national mapping agency of Denmark (KMS) has a set of defined templates that are used in the manual process of characterising farmsteads. The dimensions of each template are altered to fit those of the farm buildings that it represents. In practice, this means the template shapes are most often stretched or flattened to achieve satisfactory results. For the purpose of this research, a subset of templates was modelled (illustrated in Fig. 2).

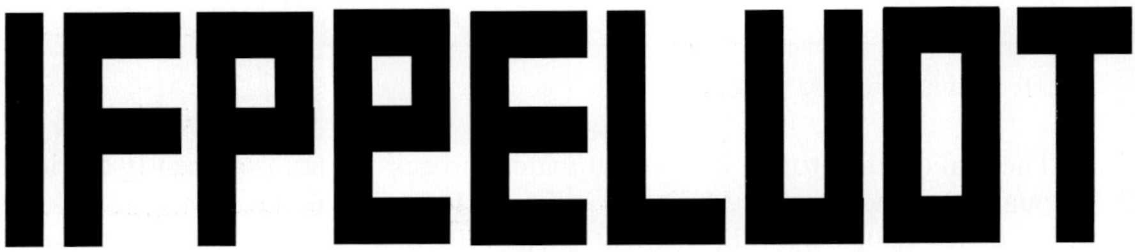

Fig. 2. The templates: I F P G E L U O T

The template fitting scheme uses a two step process of simplification followed by a template selection procedure. This is figuratively shown in Fig. 3. The latter operation progressively narrows the choice of templates, based on the number of holes in the object, the number of vertices and the sequence of internal angles, until a unique selection is arrived at.

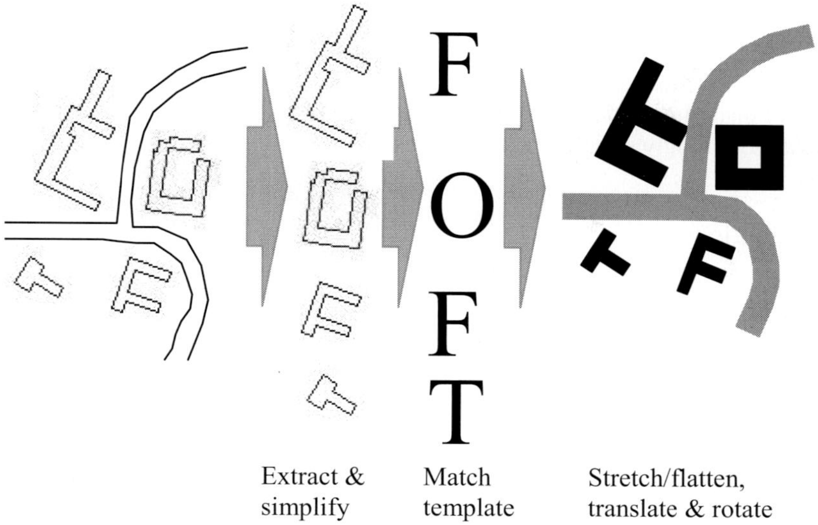

Fig. 3. The complete process

3.1 Simplification

The simplification part of the algorithm begins by checking that at least one dimension is 25 metres or greater. If not, the building is ignored on the basis that, at 1:50,000 scale, it would be less than 0.5 mm and so not worth plotting. The shapes are then squared and buffered followed by reduction to close any open loops. Minor appendages ("nubs") are removed and this is followed by fitting a Minimum Bounding Rectangle (MBR) which is used to determine the parameter required for the second round of simplification.

A study was made of template geometries in order to establish rules that would specify when to delete "limbs" on a building shape. Conceptually, we wish to eliminate limbs that are less than half the length of those on the templates. Because the templates are not fixed in size we needed to examine the ratio of limb lengths to template dimensions. It was found that the most reliable measure was the ratio of limb length to template width (MBR width). It appeared that the limb cut-off threshold should lie in the range 0.18 to 0.36 of the MBR width.

Initially the simplification algorithm used a fixed length of 7.5 metres, which translates to 0.15 mm at 1:50,000 scale. It was subsequently found that a range of values could be used, the figure becoming simpler in form as the simplification tolerance was increased. In the evaluation phase, we noted that, in several instances, a larger factor would sometimes result in a fit being made where none had been obtained before but that the match was erroneous. (i.e. the larger the tolerance, the less likely that the match would be correct). This simplification tolerance provided a simple index of how likely that the match was correct. Starting by simplifying with a small factor (0.15 * MBR width) an attempt is made to see if a template could be fitted. If the shape is unmatched it is further simplified by applying a factor of 0.20 * MBR width and attempting again to match it. A final attempt is made with the simplification factor increased to 0.25 * MBR width. If no match is found at this attempt, "NO MATCH" is returned and no additional simplification is carried out lest the shape is oversimplified and a template is wrongly selected.

3.2 Selection

Some building shapes have central courtyards or 'holes' in them. It is a simple process to count the number of boundaries or 'rings' that a feature has. Knowing the number of rings, we select either template groups I, L, U, T, F, E (1 ring) or O, P (2 rings).

3.3 Closing rings

Some features look like closed courtyards, but were frequently not closed structures (such as Fig. 4a). In order for these building shapes to be matched by one of the 2 ring templates it was necessary to close them. An algorithm that counted the number of rings of an object, coupled with a 'shrink wrap' convex hull algorithm, enabled us to assess how close a building was to being a closed courtyard and thus enable us to categorise a feature like the one in Fig. 4a as a '2 ring' feature (one ring describing the perimeter and an additional ring shown as the dark outline generated using a shrink wrap convex hull function.

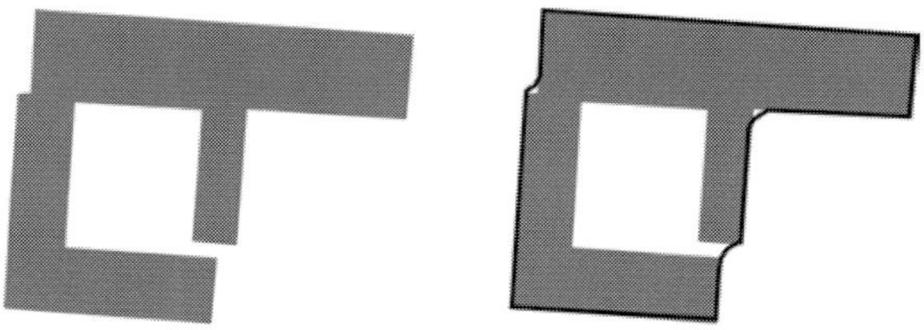

(a) (b)

Fig. 4. A farm building with a central courtyard and an open loop **(a)** is processed using a 'shrink wrap' convex hull function. Comparing the two footprints enables the courtyard to be identified.

Using the number of vertices, the program selects within one of these groups and returns either a single candidate or more. The next stage is to examine the angle sequence that characterises the shape in order to differentiate within the '1' or '2' ring group.

3.4 Angle Sequencing

Internal angles of the simplified and squared polygons were measured and recorded in order to characterise their shapes. Within tolerance, the angle should either be 90^0 or 270^0. We deem 90^0 to be a "left" turn and store this information as an "L" in a string. In the same way 270^0 is deemed to be a "right" turn and an "R" is stored. Thus, we end up with a word string describing the angle sequence of the shape (e.g. **RLRRLLRRR**). This string is then passed to our **match_sequence** function.

3.5 Sequence Matching

The purpose of the **match_sequence** operation is to match the measured angle sequence with those of a selected template. As we can make no assumptions as to where in the shape we started recording the angles, we need to cycle the angle sequence and attempt a match with the template sequences until either a match has been found or the sequence is completely cycled. The process is illustrated in Fig. 5.

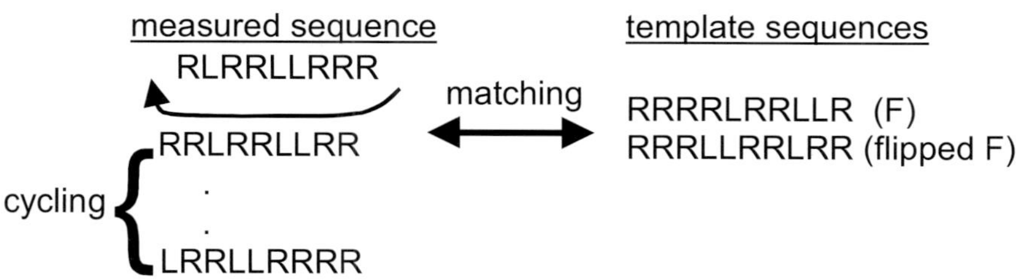

Fig. 5. Schematic representation of the operation of the match_sequence function

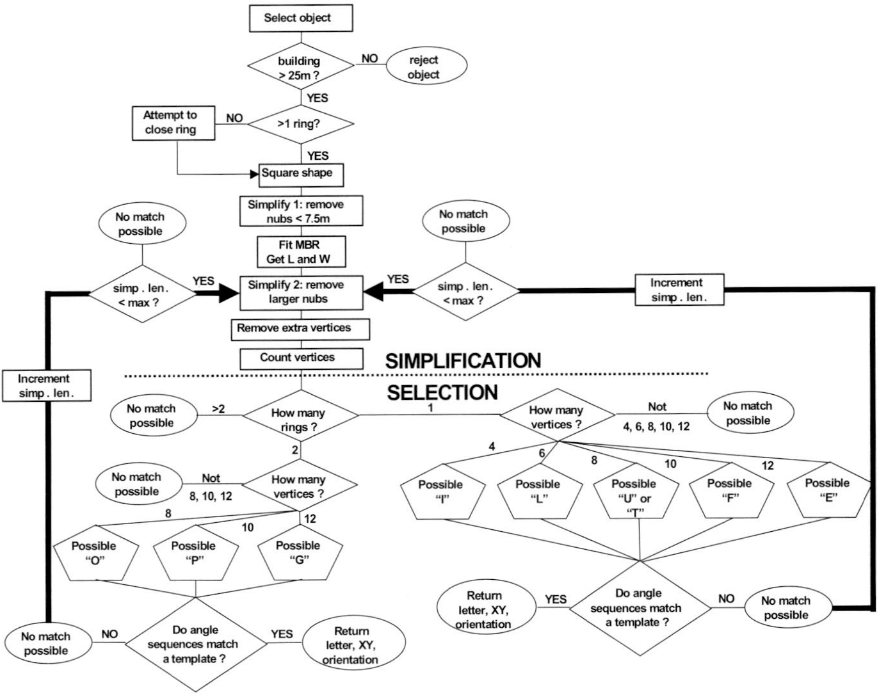

Fig. 6. Workflow of the template matching algorithm

As we cannot assume the facing direction of the template, we include angle sequences for the templates facing forward and backwards. We are then able to return the facing direction of the shape as well as the template ID. This information is, of course, only useful for non-symmetric templates (e.g. "F" and "G"). This final test reduces the candidates to a single template (if there is more than one) and eliminates spurious shapes which happen to have the same number of vertices as the candidate template(s). The entire process is summarised in the flow diagram illustrated in Fig. 6. Fig. 7 shows the transformation stages and sequencing process for an example building.

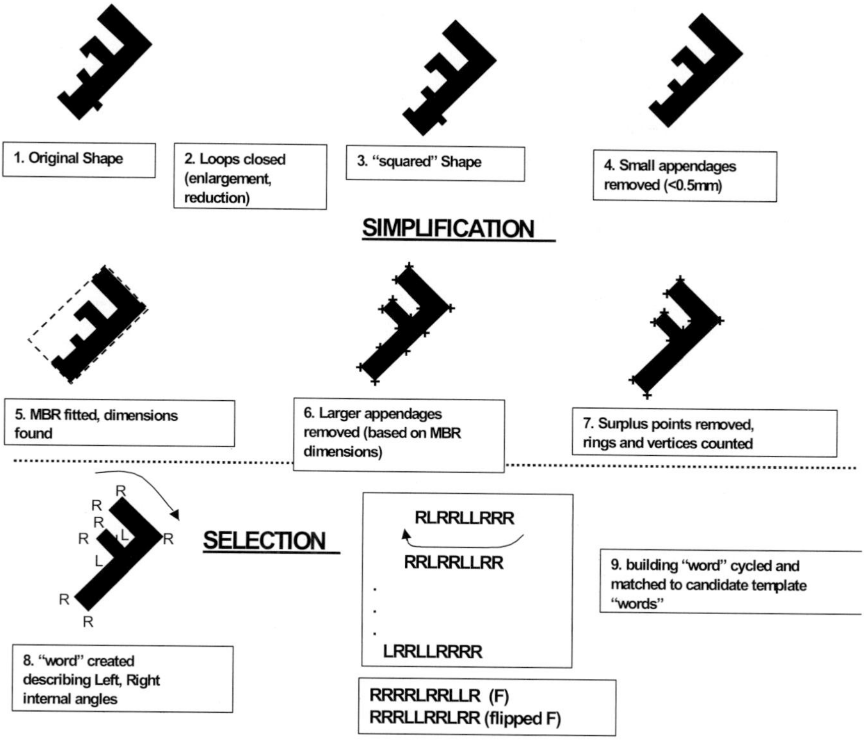

Fig. 7. The various processing stages in identifying an 'F'

Fig. 8 shows diagnostics from the algorithm and reveals how the amount of simplification is iteratively increased in the search for the best template match. In this example, the first iteration attempts to match the shape unsuccessfully to an "F". Additional simplification in the second iteration results in the successful match of the shape to a "U". In this case, because a "U" is symmetrical, its facing direction is meaningless. The results are displayed in the last two lines of the printout. Fig. 9 shows the transformation that took place in the simplification process as part of the diagnostics detailed in Fig. 8.

```
number of outer vertices = 10
number of holes = 0
Attempting to close ring...
First round of building simplification..
building was NOT simplified
Fitting bounding box...
Second round of building simplification..
Iteration 1:
calculated simplification length = 8.650238:
final number of outer vertices (geom_id)= 10
Getting angle sequence..
i: 1  angle sequence: L
i: 2  angle sequence: LR
i: 3  angle sequence: LRL
i: 4  angle sequence: LRLL
i: 5  angle sequence: LRLLL
i: 6  angle sequence: LRLLLL
i: 7  angle sequence: LRLLLLR
i: 8  angle sequence: LRLLLLRR
i: 9  angle sequence: LRLLLLRRL
i: 10  angle sequence: LRLLLLRRLL
Begining template selection..
First pass selection of candidate templates is: 'I', 'L', 'U', 'T', 'F', or 'E'
Second selection of candidate templates is: 'F'
Matching angle sequence to 'F'
Iteration 2:
calculated simplification length = 11.533651:
building was simplified
final number of outer vertices (geom_id)= 8
Getting angle sequence..
i: 1  angle sequence: L
i: 2  angle sequence: LL
i: 3  angle sequence: LLR
i: 4  angle sequence: LLRR
i: 5  angle sequence: LLRRL
i: 6  angle sequence: LLRRLL
i: 7  angle sequence: LLRRLLL
i: 8  angle sequence: LLRRLLLL
Beginning template selection..
First pass selection of candidate templates is: 'I', 'L', 'U', 'T', 'F', or 'E'
Second selection of candidate templates is: 'U' or 'T'
Matching angle sequence to 'U'
match found!!!
The selected template letter is: U, Facing: forward
Centroid (X, Y) = 570390.606220, 6246590.148621   Orientation (radians) = 1.75035
```

Fig. 8. Diagnostics from the algorithm in the search for a best fit

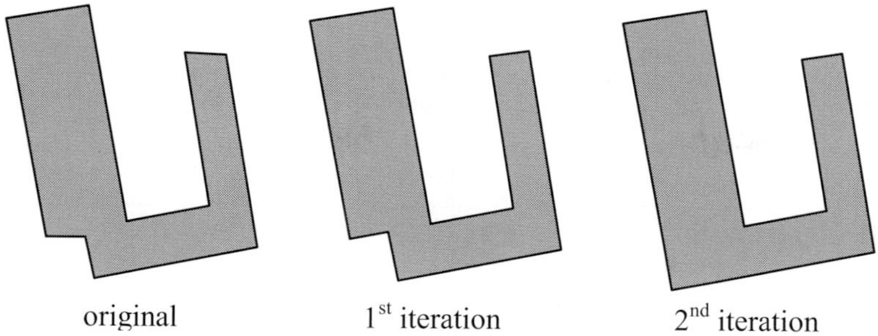

Fig. 9. Simplification of a building outline leading to the selection of a "U" template. The sequence illustrates the processes detailed in Fig. 8

4 Results

In the evaluation phase, we assessed each building shape and subjectively assigned it a template letter. Because this was a subjective process, we recognised that some of the assessments were questionable and that in some cases it was appropriate to suggest more than one answer because of ambiguity (Fig. 10).

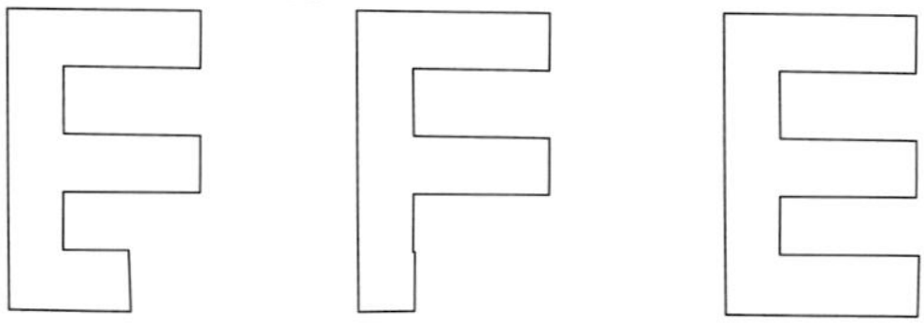

Fig. 10. Which is it: an E or an F?

Fig. 11 shows the results of the matching. In each box is an outline of the farmstead building, and two sets of letters. The lower letters are what we subjectively deem correct solutions. The upper letter is the template proposed by the algorithm. The program was 'right' if it produced a result that was the same as our subjective assignment (black building), 'wrong' if an incorrect template was identified (white building) and 'unclassified' if no template was returned if a reasonable match existed (shaded building). 'No' means no template was found (upper letter) or no template was expected to be found (lower letter). We define the 'wrong' and 'unclassified' results as type I and type II errors respectively. A summary of the testing is displayed in Fig. 12. By opting to iterate through three levels of simplification we were able to improve the success rate of template matching about 3% to 78% with 15% wrong answers (the remainder being unclassified). We also kept a record of the facing directions returned for the asymmetric templates. The results were completely accurate for the "F" template, random for the "P" template and the absence of suitable shapes did not allow the "G" template to be tested. The inability of the program to detect the proper facing direction of "P" shapes was due to the fact that the angle sequencing took into account only the outer ring. The angles of the outer ring of a "P" are the same as that of an "L" shape, which is symmetrical and cannot therefore be faced.

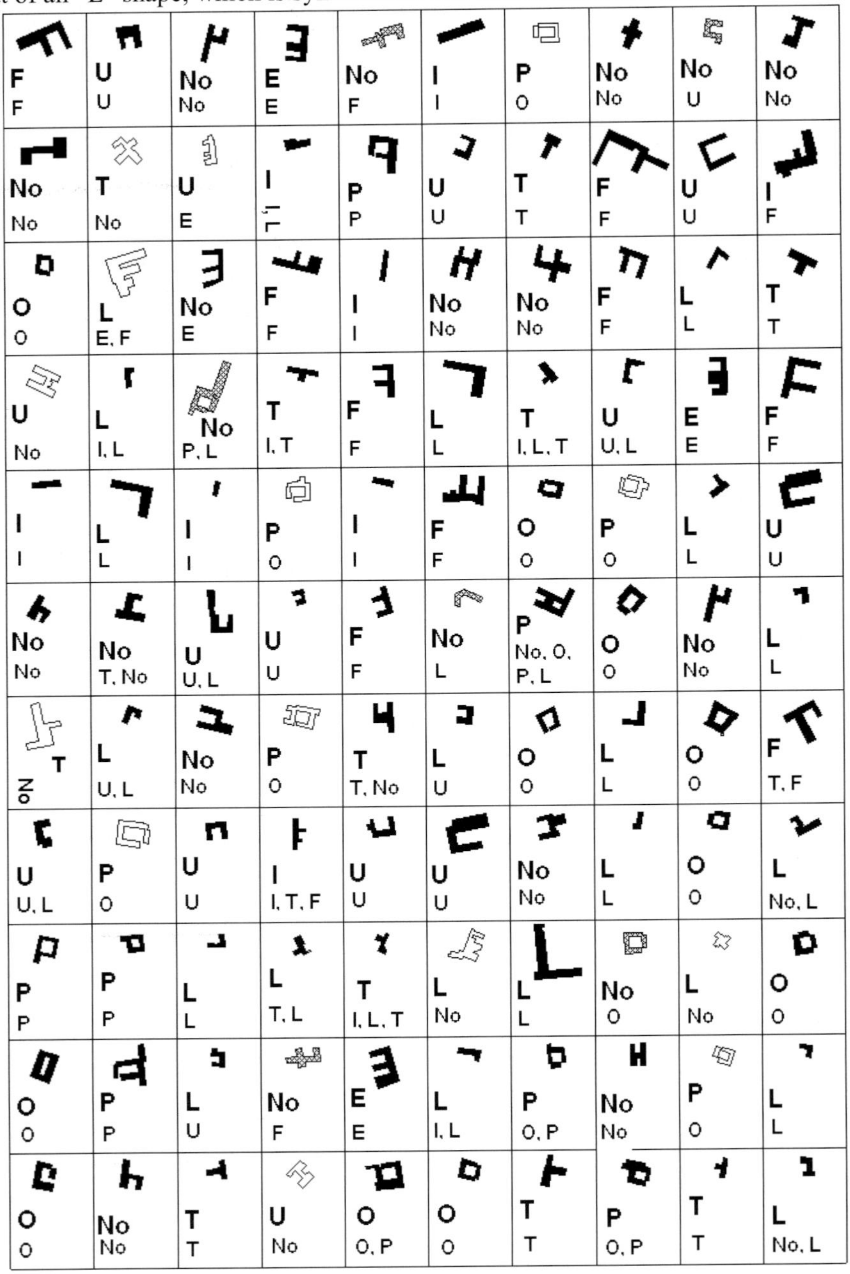

Fig. 11. Testing of a subset of farm buildings

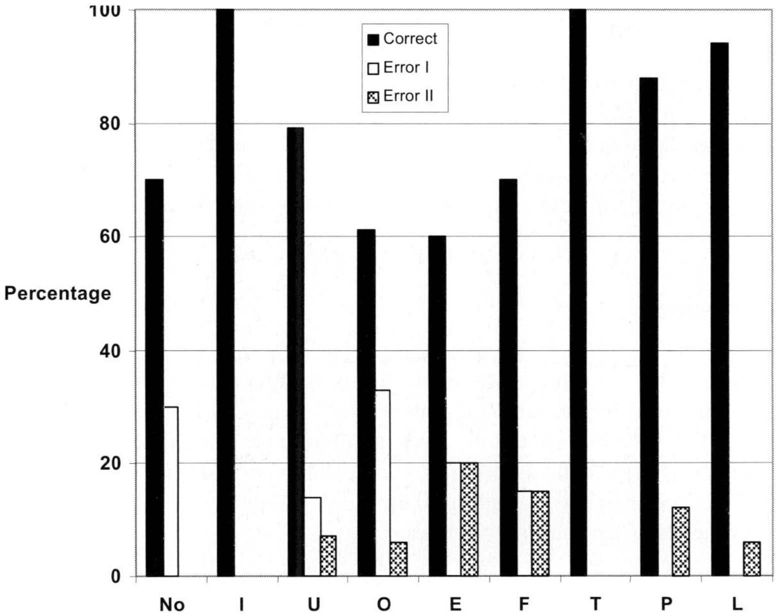

Fig. 12. Degrees of success for each of the eight templates modelled

5 Evaluation

When we examine the results in detail and in particular the failures, we noted some deficiencies in the algorithm and saw how some improvements might be made. These are examined below.

5.1 "O" Shapes that are Identified as "P's"

Several of the failures were as a result of irregular "O" shapes that were insufficiently simplified and were identified as "P's" instead. We attempted to rectify this by increasing the simplification parameter, but this resulted in other shapes being oversimplified and mis-identified as a result. Instead, a better alternative might have been to apply a higher level of simplification specifically to those shapes that were found to contain a courtyard.

5.2 Buildings with More than One Hole

The algorithm assumes that, if the building has one inner ring, no further rings need to be closed. This may not always be the case, as a farm could conceivably have two or more courtyards. Ideally, all shapes should be tested regardless of whether they already have a closed courtyard.

5.3 Lack of Templates

Owing to the fact that only eight templates were considered, the algorithm would often try in vain to find an appropriate template. With each iteration and associated higher level of generalisation, the risk of oversimplification increases. If more templates were introduced, matches could be made earlier in the process and the risk of misidentification through oversimplification reduced.

6 Conclusion

The template matching scheme, outlined in this document, has the advantage that it is fairly simple to implement and uses a small number of measures to identify the shapes. The scheme should be able to unambiguously detect confounding shapes and reject these. The solution is also scalable and additional templates can easily be incorporated by including their angle sequences, in the appropriate program branches, after counting vertices.

As far as limitations are concerned, the scheme is highly dependent on a satisfactory simplification being obtained before vertices, loops and angles are measured. Although rules have been devised for the simplification process, further testing will be required to see if these are adequate. While a squaring method is invoked, it is expected that the program will be less effective in matching shapes with angles markedly different from right angles.

Without too much additional effort the templates could be displayed in the place of the farm building. Our algorithm already returns all the information needed to accomplish this. Knowing the identity of the template, a copy could be made, flipped (if required), rotated and translated to the shape centroid. In addition, using the dimensions of the MBR, that we had obtained in the course of the simplification process, we could scale the template in X and Y so its size and aspect approximates that of the farm building. This step is required so that simple transformations in x and y can be used stretch and flatten accordingly. Further refinement and extension of the range of templates is required before the system can be commercialised. Reducing the occurrence of mis-matches (type I errors) to a very low level would be highly desirable even if it were to require modifications that reduced the number of successful matches as well. In such an automated environment, where user intervention is required, it is important to minimise the effort of the operator and to direct them to buildings for which a template has not been found. The algorithm provides a simple measure of template matching quality, through the incremental increase in the simplification tolerance. From the operator perspective, templates could be colour coded according to how quickly a solution was reached.

Acknowledgments

The researchers are grateful to Laser Scan for the use of their software, and to the National Mapping Agency of Denmark for the use of their data, and the template specification document on which this work was based.

References

Attali D, Montanvert A (1997) Computing & Simplifying 2D & 3D Continuous Skeletons. Computer Vision & Image Understanding 67:261-273

Edwardes A, Regnauld N, Mackaness W (1998) A rule based approach to algorithm selection in building generalisation. Internal AGENT report EDIN-001-10-05-1998

Ehler GB, Cowen DJ, Mackey HE Jr. (1996) Development of a shape fitting tool for site evaluation. In: Advances in GIS Research II: Proceedings 7th International Symposium on Spatial Data Handling. Delft, I:4A pp 1-12.

Glover E, Mackaness WA (1999) Dynamic generalisation from single detailed database to support web based interaction. 19th International Cartographic Conference, vol 2. pp 1175-1183

Litchner W (1979) Computer Assisted Processes of Cartographic Generalization in Topographic Maps. Geo-Processing 1:183-199

Loncaric S (1998) A survey of shape analysis techniques. Pattern Recognition 31(8):983-1001

Stassopolou A, Caelli T, Ramirez R (2000) Building Detection using Bayesian Networks. International Journal of Pattern Recognition and Artificial Intelligence 14(6):715-734

Veltkamp RC (2001a) Shape Matching: Similarity Measure and Algorithms. In: Proceedings Shape Modelling International. 7-11 May, 2001, Genova, Italy, IEEE Press, pp 188-197

Veltkamp RC, Hagedoorn M (2001b) State-of-the-art in shape matching. In: Lew M (ed) Principles of Visual Information Retrieval. Springer Verlag, pp 87-119

Weibel R, Dutton G (1999) Generalising spatial data and dealing with multiple representations. In: Longley P, Goodchild, MF, Maguire DJ, Rhind, DW (eds) Geographical Information Systems. John Wiley, New York, pp 125-156

Minimising Positional Errors in Line Simplification Using Adaptive Tolerance Values

Nadia Shahriari[1] and Vincent Tao[2]

The University of Calgary, Department of Geomatics Engineering,
2500 University Dr. NW, Calgary, AB, Canada T2N 1N4, nshahria@ucalgary.ca[1]
York University, Department of Earth and Space Science & Engineering, 4700 Keele Street, Toronto, ON, Canada M3J 1P3, Tel: (416) 736-5221, Fax: (416) 736-5817, tao@yorku.ca[2]

Abstract

Line simplification is an important function in GIS and cartography and is widely used in commercial GIS software packages. Most line simplification algorithms require the user to supply a tolerance value, which is used to determine the extent to which simplification is to be applied.. All simplification algorithms induce positional errors in the data set, because they produce a discrepancy between the original line and its simplified version. The amount of this error depends on both the tolerance value and the shape of the line. This is the reason that many researchers have focused on measuring the geometric characteristics (or complexity) of lines. Using one tolerance value for all lines in the data set results in different positional errors for different lines. What is usually important for the user, is to maintain a specific level of quality, and not the tolerance value itself. The question is, 'how does one specify a tolerance value for each line based on the user specified level of accuracy'? This paper presents a solution to solve this problem. In this approach, the user supplies the target level for desired accuracy and the simplification tolerance value is calculated accordingly.
Keywords: line simplification, complexity of line, adaptive tolerance value, positional error

1 Introduction

Cartographic generalisation involves selecting the features to be maintained at the targeted scale, simplifying non-relevant characteristics, enhancing significant shapes, displacing without defacing global and local shapes, and finally harmonising the final presentation (Plazanet, 1997). Until about a decade ago, trained cartographers usually practised map generalisation, by using rules, examples and intuition. Traditionally, map-making agencies applied generalisation using manual techniques. Many mapping organisations, especially the larger national, state or provincial establishments had their own guidelines, standards and procedures for manual map generalisation. Only recently has this situation begun to change, but it has done so dramatically and definitively, as maps – along with every other form of human communication – have become digital (Dutton, 1999).

Recent reviews of digital generalisation procedures have identified a number of distinct processes. Most of the techniques focus on either the manipulation of vector- or raster-mode data. Much of the work on raster-mode generalisation of images has come from the field of remote sensing and includes such techniques as high- and low-frequency filtering (McMaster, 1989). High-frequency filters emphasise fine details and edges by passing the higher frequencies while low-frequency filters emphasise the more generalised trends. Fourier analysis, two-dimensional convolution, linear edge detection, non-linear edge enhancement are some of the common generalisation techniques for raster imagery. Since the mid-1960s, however, most work in this area has addressed generalisation in vector-mode. Specifically, researchers have thoroughly addressed the generalisation of digital lines, while leaving other aspects of generalisation as yet unresolved. For example, many algorithms have been developed for the simplification, smoothing, enhancement, displacement and merging of linear features (McMaster, 1989).

Line simplification is an important function in cartography and GIS. In line simplification, source data are transformed to reduce data volume, to merge databases with different scales, or to maintain cartographic quality when scale is reduced. Line simplification involves the selective elimination of vertices along a cartographic line to remove unwanted information. Line simplification is only one component of cartographic generalisation. It is, however, the most commonly applied generalisation operation and is widely used in commercial GIS software packages (Veregin *et al*, 1999).

Several approaches for the simplification of digital lines have been presented in the cartographic, computer science, remote sensing and mathematics literatures. While some of these algorithms are extremely simplistic in nature, the more sophisticated ones consider geometric properties of the line. McMaster's (1987) classification of simplification algorithms includes: independent point routines, localised processing routines, constrained extended local processing routines, unconstrained local processing routines and global routines. Global routines focus on the careful selection of the critical points, or the salient geometric characteristics of the line (McMaster, 1989). Simplification is considered successful if the simplified caricature of a line closely resembles the original version. Most simplification algorithms need a tolerance value. All simplification algorithms introduce positional error. The amount of this error depends on both the tolerance value used for simplification and the shape of the line. The problem arises when the tolerance value stays the same for all lines in the data set, regardless of the variations in their shape.. The result is that simplification introduces different amounts of positional errors on different lines and the user usually does not have a direct understanding of these effects on data quality. In other words, the user specifies a tolerance value for simplification, without knowing how much the positional accuracy of each line is affected by this tolerance value. Therefore, adaptive tolerance values should be used for line simplification of a data set.

This paper presents a solution to this problem. First some measures of line complexity will be discussed and applied to a data set to illustrate the differences in geometry of different linear features. Then, the paper describes a new approach by which the user specifies the maximum positional error instead of a tolerance value. The tolerance value will be specified for each individual line based on this maximum valid positional error. In fact, the tolerance value will be adapted to each line. Although applying one tolerance value for each line is not logically correct, if the line is not geometrically homogenous, however, the assumption here is that the lines are homogenous. This new approach may be extended to line sections if the lines are not homogenous and the result will be different tolerance values for different parts of a line.

2 Complexity Measures of Lines

Researchers have introduced a variety of mathematical measures to evaluate the complexity of a line. Some of these measures may be applied to a single line while the others are applied to compare the geometry of a line before and after simplification. McMaster (1986) called these categories of measures as "single attribute measurements" and "displacement or comparative measurements" respectively. Most of the research in the field of complexity measures focuses on four categories of measures: *length*, *density*, *angularity* and *curvilinearity*. These categories are discussed in the following subsections.

2.1 Length Measures

The length of a line is probably the simplest geometric attribute of it. The length, however, cannot singularly represent the complexity of the line. Nevertheless, it can be used to represent the effects of the simplification.. Because much of the sinuosity is eliminated, a line becomes shorter, as it undergoes the simplification process. McMaster (1986) introduced "the percentage change in line length" as a length measure. This measure has been expressed as the total length of the simplified line divided by the length of the original line and presented as a percentage. This ratio is expected to decrease as the line is more simplified.

Jasinski (1990) introduced another measure, "error variance", which is an average perpendicular displacement of every point in a line to the anchor line. Error variance measures the deviation of the line from its straight-line approximation, its anchor line. Buttenfield (1984) defined the anchor line as a straight segment linking the first and the last point of the coordinate string. Error variance is expected to increase as the line is more simplified. As the mean value of a distribution does not say anything about dispersion of values in the data set, Jasinski (1990) introduced another measure called "coefficient of variation of error variance" which gives one the relative variability of a distribution as a ratio of the standard deviation to the mean.

2.2 Density Measures

Density of a line represents the frequency of detail that exists in it. Number of points of a line, as a measure of density, solely cannot represent the complexity of the line. It can be used, however, to represent the effect of simplification on a line by comparing the density of the line before and after simplification. All simplification algorithms, principally, reduce the number of points along a line. Therefore, density of the line is expected to decrease as the line is more simplified.

McMaster (1986) introduced three coordinate measures to compare the density of the simplified line with the density of the original line. These measures are "ratio of change in the number of coordinates", "difference in average number of coordinates per inch" and "ratio of change in the standard deviation of the number of coordinates per inch".

Jasinski (1990) introduced another density measure called "average segment length" which is the average length of all segments between the points of the line. The value of this measure is expected to increase as the line is more simplified because fewer points are retained and segments automatically get longer. Jasinski (1990) also introduced another density measure called "coefficient of variation of average segment length" because the average segment length does not give much information whether most of the segments are similar or whether the line consists of some very short and some very long ones.

2.3 Angularity Measures

Angularity of a line is one of its primary geometric characteristics. Angularity measures evaluate specifically the individual angular changes along a line. McMaster (1986) defined nine measures for angularity of a line to compare its angularity before and after simplification. The "percentage change in angularity" is one of these measures, which was expressed as the sum of the angles between consecutive vectors on the simplified line divided by this sum on the original line. He defined the "absolute angle of change between each pair of consecutive vectors" as a basic angularity measure. Jasinski (1990) defined another formula to calculate the "average angularity", which ranges from 0 (straight line) to 1 (the line backs on itself) and another measure, which is called "Coefficient of variation of average angularity". The latter measure is expected to decrease as the line is more simplified.

2.4 Curvilinearity Measures

Curvilinearity of a line is defined by the number of inflection points in it. Inflection points divide the line into curvilinear segments, which are the portions of a line in which all angles are in the same direction, either positive or negative. Curvilinearity is a measure of direction of angular changes while angularity is the measure of magnitude of angular changes. McMaster (1986) defined four curvilinearity measures. These measures compare the curvilinearity of a line before and after simplification. McMaster (1986) defined the "ratio of the change in the number of curvilinear segments" as the basic curvilinearity measure, which is the number of curvilinear segments of the simplified line divided by the number of curvilinear segments of the original line. Jasinski (1990) defined the ratio of total curvilinearity to the number of all turns as the "curvilinearity ratio".

3 Discussion of Data and Measurements

A set of four lines was selected for the project. Each of these lines is a section of a river in the USA and provided by ESRI data which has come with ArcView software. For simplicity, the rivers have been named river1, river2, river3 and river4 which are parts of the rivers "Platte", "Arkansas", "Brazos" and "Red River of the North" respectively. In order for the set of lines to be representative of a wide range of shapes, it was decided to select the rivers with different irregularities. River1, which is less irregular, has 459.46 km length and includes 19 points. The length of river2 is 714.99 km and includes 40 points. River3 is 962.17 km and has 87 points. River4 which is 443.72 km, has almost the same length as river1, but has more irregularities and includes 137 points. Fig. 1 illustrates the selected rivers.

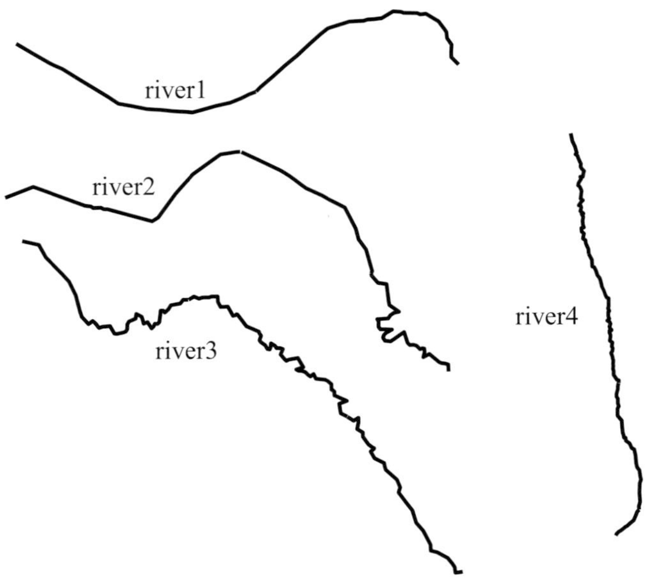

Fig. 1. Data set used for complexity measures

The Douglas-Peucker's (1973) line simplification algorithm has been selected in this project. There are several reasons for this selection. This algorithm is the most accurate at selecting critical points (White, 1983). Additionally, the algorithm's selection of these critical points generates simplifications that mimic those generated by manual generalisation, and retains details critical for map-reader recognition (Buttenfield, 1991).

Visually, it is obvious that these four lines have different shapes. The difference can be shown mathematically using the complexity measures. For each aspect of line complexity, namely, *length*, *density*, *angularity* and *curvilinearity*, different measures have been applied on the test data set, however, only some of them will be discussed in this paper. The results have been plotted as graphs, with the measurement value against the tolerance value set for simplification.

"Error variance" (Jasinski, 1990) has been applied on the data set as a length measure. Fig. 2 illustrates the results. As could be expected, the error variance increases as the lines are more simplified. This is logical, since the progressive simplification routine by Douglas-Peucker retains characteristic points which tend to have a high error variance (Jasinski, 1990).

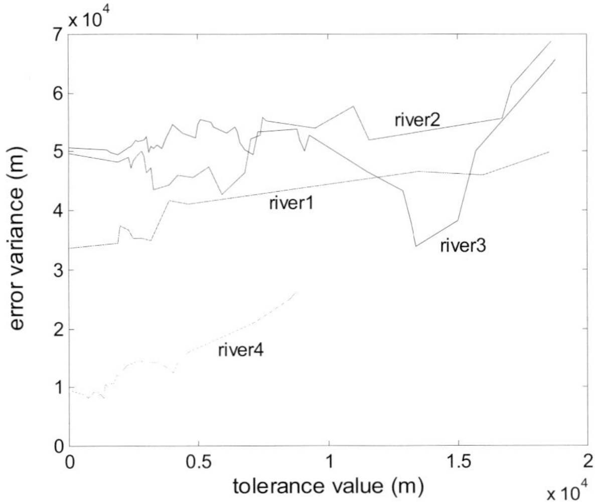

Fig. 2. Error variance

"Percentage change in number of coordinates" (McMaster, 1986) has been used to illustrate the density differences of four lines. Fig. 3 illustrates the result of this measure. As could be expected, the percentage change in number of points is decreasing as the lines are further simplified. Despite it's greater irregularity, the decrease is faster in the case of river4, because its points are close to its anchor line. Therefore even with a small tolerance value, many points are removed by the simplification.

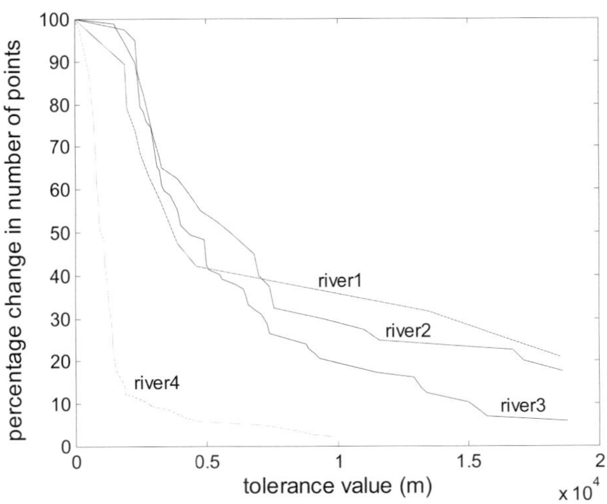

Fig. 3. Percentage change in number of points

The third measure applied to the data set was the "percentage change in angularity" (McMaster, 1986). Fig. 4 illustrates the results of this measure. As expected, the percentage change in angularity is decreasing as the lines are further simplified. Again, this decreasing is faster in the case of river4, because the line is mostly simplified with low tolerance values. Therefore, it loses its angularity much faster than the other three lines.

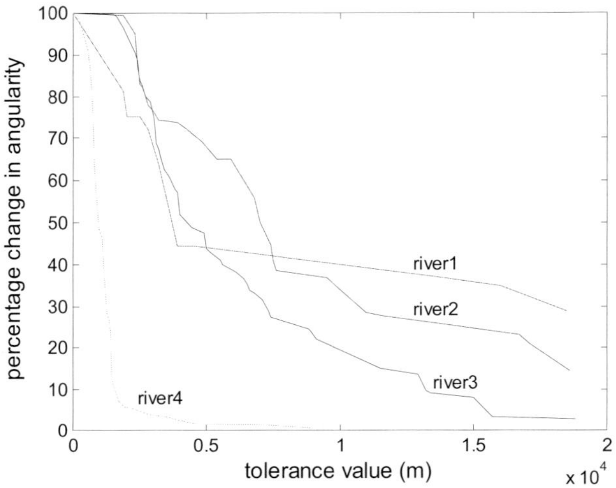

Fig. 4. Percentage change in angularity

Fig. 5 illustrates the results of the last measure, "ratio of the change in the number of curvilinear segments" (McMaster, 1986).

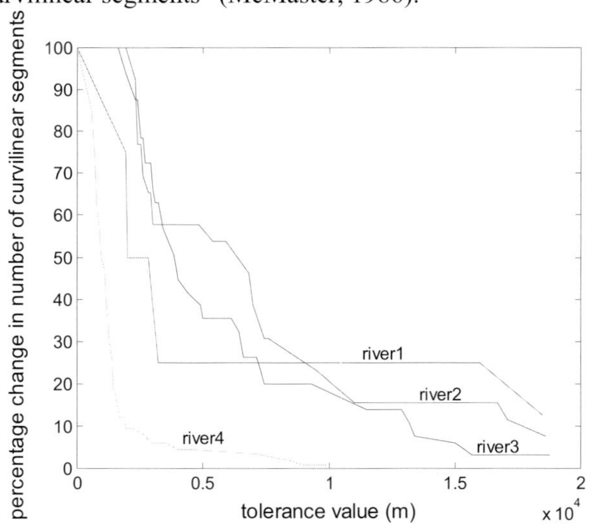

Fig. 5. Percentage change in curvilinearity

The "ratio of the change in the number of curvilinear segments" is a curvilinearity measure. As expected, the percentage change in curvilinearity is decreasing with more simplification. Fig. 5 shows how the simplification affects the curvilinearity of lines with different shapes. The straight lines in Fig. 5 means that the associated line is more simplified, however, the number of curvilinear segments does not change.

4 Data Quality

Line simplification algorithms simplify the lines by selective elimination of vertices (shape points) along the line. This process results in positional error because it produces a discrepancy between the location of the original line and that of the simplified line (Veregin *et al*, 1999). McMaster (1986) defined three groups of displacement or comparative measurements, which are used to evaluate differences between the base line and the simplified line. These measurements are "vector displacement", "polygonal displacement" and "perimeter displacement" measures. The vector displacement measure has been used in this project to investigate the effect of simplification in inducing positional errors in the data set. As a result of coordinate elimination, a vector difference is produced each time part of the simplified line is displaced. This distance may be measured as the perpendicular distance from the eliminated coordinates on the base line to the new vector on the simplified line. Sum of the length of all vectors between the two lines divided by the length of the original line is a measure of vector displacement (McMaster, 1986). This measure is expected to increase when the line is more simplified. This is logical, because the discrepancy between the location of the original line and that of the simplified line increases as the line is more simplified. Fig. 6 illustrates the results of applying vector displacement measure on the four lines of the data set.

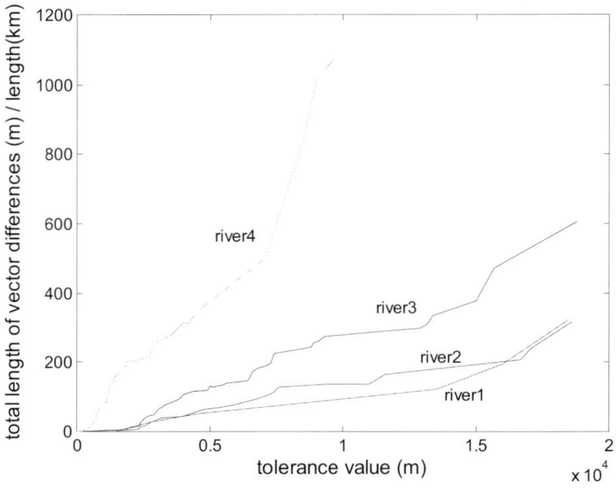

Fig. 6. Vector displacement

Figure 6 shows that the amount of vector displacement, which represents the induced positional error, is different for each line when one tolerance value is used for all four lines. In most parts of the graph for each specific tolerance value, there are four different values for vector displacement. For example, if the user specifies 7000m as a tolerance value, the vector displacements will be 53.05, 101.25, 181.25 and 350.59 m per km of line length for river1, river2, river3 and river4 respectively. This results due to the difference in the complexity of the lines, as discussed in the previous section. When using one tolerance value for all lines in the data set, simplification induces more positional errors in river4 then the other rivers in the data set. This occurs because river4 has more density, angularity and curvilinearity than the others.

5 Adaptive Tolerance Value

Line simplification, as discussed above, can have significant effects on the data quality; however, the user does not have any control on these effects. When using a simplification algorithm, the user generally specifies a tolerance value to control the degree of simplification but is unable to specify a target level of quality that must be attained. What is missing is a way for users to select a weed tolerance value that ensures a certain level of positional accuracy is maintained (Veregin *et al*, 1999). Using one tolerance value for all lines in the data set induces different degrees of positional errors for different lines. Therefore, in order to keep a certain level of positional accuracy for all features in the data set, the different tolerance values should be used for different features. The question to explore is how should one specify a tolerance value for each feature based on the user-specified level of accuracy.

This paper presents a new way to solve this problem. In the approach presented, the user supplies a certain level of positional accuracy (either vector displacement or polygonal displacement or perimeter displacement or any other measure of positional accuracy depending on the user's application) instead of supplying a tolerance value. As a result, the tolerance value will be specified automatically for each line in the data set. In order to obtain the tolerance value for each line, the data set has to be pre-analysed. In this pre-analysis, vector displacement (or any other positional displacement) measure is applied on all lines in the data set using a set of tolerance values for each line, which can start from 0 to maximum tolerance value for the line. Then the proper tolerance value for each line is specified using the set of tolerance values and associated vector displacement from the pre-analysis phase. The proper tolerance value for each line is a value by which maximum simplification is achieved while maintaining the user specified positional accuracy. An example of this process is shown in Fig. 7. In this example the user specifies 150 m per km as the maximum vector displacement and the values of 14438.86 m, 11271.90 m, 6436.25 m and 1398.41 m is calculated as tolerance values for riverl, river2, river3 and river4 respectively. Therefore, four rivers will be simplified using four different tolerance values and maximum vector displacement will remain under 150m per km for all four features.

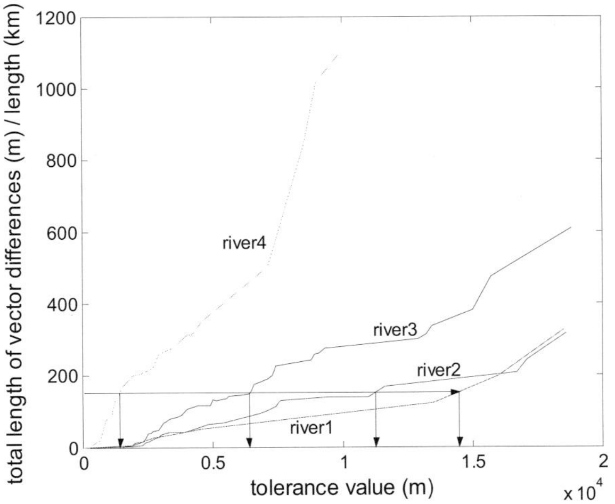

Fig. 7. Adaptive tolerance values.

This algorithm has also been tested on a larger data set. The test data set is an Arcview shape file including Canadian rivers, located in UTM zone12 (NAD83) at the scale of 1:7.5 million (National Atlas of Canada, 1997). The size of the shape file is 467KB and includes 1353 lines. The first step is the pre-analysis in which the vector displacements are calculated for each line using different tolerance values. In order to specify a set of tolerance values for each line, first maximum possible tolerance value (T_{max}) for that line (by which the line will be simplified to a straight line connecting the start point and end point of the line) was calculated. Next a set of tolerance values was selected between zero and T_{max}. After assigning tolerance values for each line, vector displacement of the line is calculated for each tolerance value in the set of assigned tolerance values for that line. All these tolerance values and corresponding vector displacements were automatically recorded in a text file, along with feature identifiers indicating corresponding lines. The next step was simplification using adaptive tolerance values. For this step, the value of 100m/km was specified as maximum valid vector displacement for the lines of the map (as user specified level of positional accuracy). Then the proper tolerance value for each line was automatically selected using the set of tolerance values and vector displacement values stored in the text file in the previous step. Next, each line of the map was simplified using the selected tolerance value. Fig. 8 shows a section of the data set (top), together with the results of using constant tolerance value (1500 m) and the result of using adaptive tolerance value (100m/km). As was expected, the amount of vector displacement induced by line simplification is completely different for different lines when using a constant tolerance value (Fig. 8, left). The vector displacement, however, has been kept under 100m/km for all lines in the simplification test using adaptive tolerance values (Fig. 8, right). This means the user has control on the positional errors induced by line simplification. The processing time for the case of a constant tolerance value was 0.601 s while for the adaptive case was 0.911 s plus 13.309 s for pre-analysis phase. The reason that more processing time is needed for the adaptive approach is partly because of the need to search the text file to find the proper tolerance value for a line. As the order of the lines in the text file is the same as their order in the map file, the program does not have to search the whole text file to find the proper tolerance value for one line but only the section of the file that starts with the identifier of that line. Although the time difference is not that much, however, some methods can be investigated to improve the efficiency such as a binary search tree to find the appropriate tolerance value.

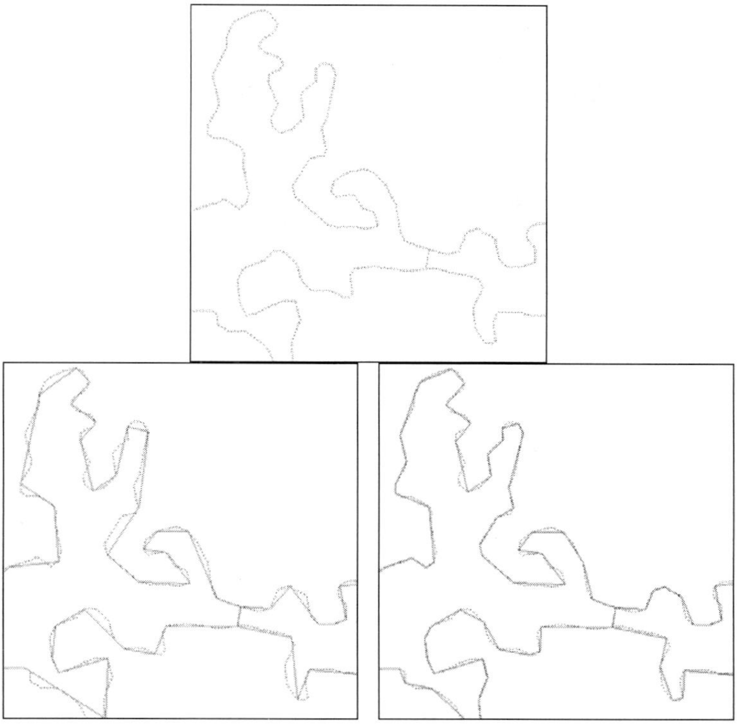

Fig. 8. A section of the test data set (top), the results of using constant tolerance value (left) and the result of using adaptive tolerance value (right)

6 Conclusion

The focus of this research has been the development of a method to calculate the adaptive tolerance values. In the proposed approach, the user supplies the target level of positional accuracy instead of a simplification tolerance value. The tolerance value is then calculated based on the given level of accuracy. The result is that each line in the data set is simplified using the calculated tolerance value specific for that line. In this paper, vector displacement has been used to define a target level of positional accuracy. However, based on a user's application, areal displacement, perimeter displacement or any other measure of discrepancy between the original line and the simplified line can be used to define the target level of accuracy. It is assumed, in this paper, that the lines are homogenous. However, the reality is that there are many linear features that are composed of non-homogenous sections. Further work may focus on extending and evaluating the implementation of the proposed approach to segments of lines in the data set.

References

Buttenfield B (1991) A rule for describing line feature geometry. In: Buttenfield B, McMaster RB (eds) Map generalization. Longman Scientific & Technical, London, pp 150-171

Buttenfield B (1984) Line structure in graphic and geographic space. Unpublished Ph.D. dissertation, University of Washington

Douglas DH, Peucker TK (1973) Algorithms for the reduction of the number of points required to represent a digitized line or its caricature. The Canadian Cartographer 10:112-122

Dutton GH (1999) Introduction and problem statement. In: A hierarchical coordinate system for geoprocessing and cartography. Lecture notes in Earth Science 79. Springer-Verlag, Berlin

Jasinski MJ (1990) The Comparison of Complexity Measures for Cartographic Lines. NCGIA Technical report 90-1. National Center for Geographic Information & Analysis, Department of Geography, State University of New York at Buffalo, Buffalo, New York 14260

McMaster RB (1989) The integration of simplification and smoothing algorithms in line generalization. Cartographica 26:101-121

McMaster RB (1987) Automatic Line generalization. Cartographica 24(2):74-111

McMaster RB (1986) A statistical analysis of mathematical measures of linear simplification. The American Cartographer 13(2):103-116

National Atlas of Canada (1997) Canadian Rivers [online]. GeoGratis website. Available from: geogratis.cgdi.gc.ca

Plazanet C (1997) Modeling Geometry for Linear Feature Generalization. In: ESF/NSF 1995 Summer Institute in Geographic Information Research: Bridging the Atlantic, in Portland Maine (USA) Part 3. Taylor & Francis, pp 264-279

Veregin H, Dai X (1999) Minimizing positional error induced by line simplification. Proceedings of the International symposium on spatial data quality 1999. The HongKong polytechnic University, HongKong

White ER (1983) Perceptual Evaluation of line generalization algorithm. Unpublished Master's thesis, University of Oklahoma.

A New Merging Process for Data Integration Based on the Discrete Fréchet Distance

Thomas Devogele

Naval academy Research Institute (IRENav), Lanvéoc, BP 600, F-29 240 Brest Naval, FRANCE, devogele@ecole-navale.fr

Abstract

The overlay process is currently one of the main computational solutions used to integrate several data layers from different sources. Unfortunately, it is problematic when trying to overlay many layers. This leads to several geometric problems such as the management of sliver polygons. This paper proposes a new merging process to complement the vector overlay for data integration of several layers. This process, based on measures derived from the Fréchet distance, matches common points (either lines or polygons). It also merges an ordered set of pairs of matching points (vertices) into a single geometry.
Keywords: data integration, overlay, merging, data matching, Fréchet distance

1 Introduction

Overlaying geospatial data is a frequently required and computationally complex procedure in a Geographic Information System (GIS). The overlay process was one of the first ways through which users combined map information to produce new maps.

Most GISs use map layers to structure geographical objects and each layer describes some particular aspect of the real world. Organisation of data into several map layers is a well-known technique, which often strives to achieve efficiencies in data storage and manipulation. However, these structures can be computationally costly when data from different layers, or when several layers of data must be combined in efforts to resolve a spatial query. Often the data layers can be distributed among remote databases and must be identified and accessed using solutions such as those proposed by the NSDI (USGS 1998).

Currently, four kinds of constraints limit a successful application of the overlay process:

- **Projections and coordinate systems** can be different. To combine two data layers, the projection and coordinate system of one of these data sets must often be converted. This conversion process can be a source of error.
- **Accuracy and scale** are generally difficult aspects with which to deal. Although each layer may be reasonably accurate within the scale limitations, differences in input errors between the two layers will most likely cause mismatches between the two geospatial references. Moreover, the scales and levels of generalisation of these data sets can be incongruent. Some other conflicts (e.g. semantics, resolution) between the object representations complicate the overlay process (Parent *et al.* 1996)
- **Limited arithmetic precision** of computers.
- **Homologous objects**. Some objects can be represented in both data sets. If the two layers are combined, the objects may be redundant in the resulting data set. However, integration of these redundant objects is necessary to control the resulting layer accuracy (Chrisman 2001) and to integrate complementary a-spatial descriptions of some phenomena (Devogele et al. 1998). By extension, homologous geometries are defined as geometries of homologous objects. In the same way, homologous points or vertices are defined as points that represent the same part of an object. (e.g. points that represent the same turn of a road).

Therefore, to ensure an efficient overlay of layers, some semantic and computational constraints must be defined. First, layers must have the same scale and projection. Secondly, the object geometry should not be replicated in two different layers. As data can come from different sources, however, these conditions cannot always be guaranteed. Therefore, several manual pre-processes and post-processes must be implemented to adjust the overlay result. These processes are unwieldy, time consuming and error prone.. Therefore, identification of a generic and automated overlaying process is, as yet, an unresolved objective for future GIS research.

This paper proposes a new merging operation that can be defined as an automatic pre-process that improves the overlay results of either lines or polygons. It is based on a data matching process and measures derived from the Fréchet distance. The reminder of this paper is organised as follows. Section 2 briefly outlines the basics of the overlay process, fuzzy tolerances and merge process. Section 3 describes data matching between homologous geometries (either lines or polygons) and the discrete Fréchet distance used by the merging process. Section 4 introduces the new merging process based on the matching process and a weighting function. Finally Section 5 concludes our paper and outlines some further work.

2 Overlay Process

The overlay process can handle different types of data layers (e.g. tiles, network, Digital Terrain Model). So far many overlay processes have been defined (e.g. CAD-type overlay, Boolean overlay, rules overlay) (cf. (Chrisman 2001) for a survey). Overlay processes depend on the type of geometry (e.g. polygon, line) and the type of expected result. Whatever the kind of overlay process, the input is made of two or more layers issued from the same or different sources, and the output is a new layer in which the new geometries (e.g. points, line) are defined as a function of the input geometry.

2.1 Errors and Corrections

Constraints that limit a successful application of the overlay process create **sliver polygons** when a basic overlay process is applied (see Fig. 1c). These polygons are small artefacts created during an overlay and result from slight differences in the geometric presentation of boundaries that should have been the same.

To resolve these errors, a fuzzy tolerance (Fig. 1d) can be associated with the overlay process (Dougenik 1980). The fuzzy tolerance is a distance where intersections and points are treated as coincident. In other words, it is the smallest distance between two geometries. If the distance between two points (vertices or segment point) is less than the fuzzy tolerance, these two points are merged. The fuzzy tolerance also resolves dangling lines and dangling nodes that should *logically* be connected at two ends. Dangling lines are dead-end lines that are connected to other lines at one end only. The other end is a dangling node.

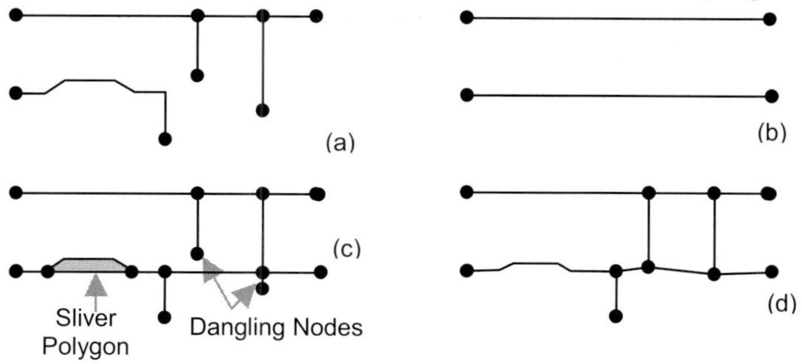

Fig. 1. Overlay examples; **(a)** and **(b)** input layers, **(c)** output without fuzzy tolerance, **(d)** output with fuzzy tolerance

One challenge is to distinguish between small geometric differences of homologous objects from different objects that are close or nearby. Either the distance is set too small and many errors are not removed, or it is set too large and some distinct points are merged (for example, a real dead-end close to a road). Moreover, the overlay with fuzzy tolerance leads to new inaccuracies. For example, new positional errors can "creep" into a GIS (Pullar 1993). (Harvey and Vauglin 1996) improves this process by introducing multiple tolerances. The setting of the tolerance is fixed according to the application purpose, scale and on statistical analysis of line form and error.

Fuzzy tolerance is a method used to merge geometry locally. Unfortunately, this method does not take into account the 'global' geometry. Indeed, different objects, which are close, cannot be distinguished from homologous objects, due to measures between their geometries. Similarly, the fuzzy tolerance process merges closest points (e.g. using Euclidean distance (d_E)). However, the two points that may be merged are not necessarily the homologous points (Alt and Godau 1995).

2.2 Merging Process for Homologous Geometry

For homologous geometry, the data integration process must use a better merging process than the one of fuzzy tolerance. The input of this process is two geometries and the output is a single geometry. Few merging processes are defined due to the fact that is very difficult to distinguish homologous geometries from closed geometries and to match the points of a common geometry.

Some processes have been proposed like a pre-process of Adjust (TCI 1999). Adjust is a rubber sheeting process. Rubber sheeting transformation is non-linearly distorted to force-fit some geographic elements (e.g. points, lines) to specific positions. As implemented in Adjust, a semi-automatic process assembles AutoCAD layers. A user selects two sets of points (e.g. crossroads in network layers) and then correlates them using From-To relationships. It also takes some lines, which should logically coincide. The process automatically interpolates any number of matching points (so-called calibration pairs). These matching points are intermediate points on these lines. Two modes are available to automate the setting of calibration pairs: "vertex to vertex" or "divide distance" (see Fig. 2).

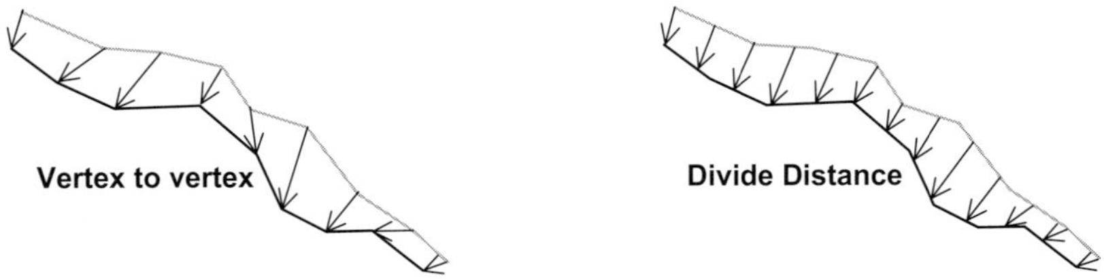

Fig. 2. Two modes of automate matching points of Adjust

These matching points are merged when the rubber sheeting are used. The "From" point and the "To" point are merged into the "To" point.

We can remark that:

- The "From-To" is a one-to-one relationship. A point cannot relate to many points. Section 3 will show that these relationships can be one-to-many or many-to-one.

- The output point is the "To" point. This solution is adequate if the layer of the point "To" is much more accurate than the "From" one. If the **accuracy** (relationship between a measurement and the reality represented) of the layers are equivalent, a "middle" point should be a better solution.
- Calibration pairs take into account the order of the vertices or points of lines. In other words, for the vertex to vertex mode, for two oriented line L_1 and L_2, the calibration pair following ($L_{1.i}$, $L_{2.j}$) are ($L_{1.i+1}$, $L_{2.j+1}$) with $L_{1.i}$ $L_{1.i+1}$ are two consecutive vertices of L_1 and $L_{2.j}$ $L_{2.j+1}$ are two consecutive vertices of L_2. Similarly for the divide distance mode, the calibration pair following ($L_{1.i'}$, $L_{2.j'}$) are ($L_{1.i'+1}$, $L_{2.j'+1}$). $L_{1.i'}$ $L_{1.i'+1}$ are two consecutive points derived from the divide distance of L_1 and, $L_{2.j'}$ $L_{2.j'+1}$ are two consecutive points derived from the divide distance of L_2.

3 Data Matching Between Points of Homologous Geometries

To obtain a high-quality merging process, several pairs of matching points are required using **data matching** processes. These allow one to identify groups of homologous objects that represent the same part of the real world from two sets of geographic data. To obtain an accurate data matching between homologous objects, three different types of data matching must be combined (Devogele *et al.* 1996):

- **Semantic** matching puts objects in correspondence according to their semantic attributes, which are discriminated (Cohen 2000) (Spéry *et al.* 2001). For example, the values of the attributes "identifier of a road" can be used to match roads from two sets of data.
- **Topologic** matching uses composition or topologic relationships between the different objects to match a given object. If two relationships correspond, then this correspondence can be used to find homologous objects linked by this relationship.
- **Geometric** matching employs location of data. Commonly, some measures of distance between objects are computed. Other geometric characteristics, such as the direction (Gabay and Doytsher 1994) have been proposed to match these data. This kind of matching process is the most popular.

In this article, the data matching objective is to identify homologous points (vertex or intermediate point) for geometry of homologous objects. These homologous objects must be found beforehand. Therefore, we focus only on the geometric data matching between points of homologous geometries (points of lines or points of borderlines of polygons.

3.1 Discrete Fréchet Distance

The Fréchet distance is the better maximal linear distance to match a set of ordered points as linear geometries of homologous objects.. For details see for example (Alt and Godau 1995) for a discussion.

The Fréchet distance is the maximal distance between two oriented lines. Each oriented line is equivalent to a continuous function f: [a, a']→V (g: [b, b']→V) where a, a', (b, b') ∈ $\Re$, a < a' (b<b') and (V, d) is a metric space. Then d_F denotes their Fréchet distance defined as:

$$d_F(f,g) = \inf_{\substack{\alpha:[0,1]\to[a,a'] \\ \beta:[0,1]\to[b,b']}} \max_{t\in[0,1]} d(f(\alpha(t)), g(\beta(t)))$$

An illustration of the Fréchet distance follows: a man is walking with a dog on a leash. He is walking on the one curve, the dog on the other one. Both may vary their speed, but backtracking is not allowed. Then the Fréchet distance of the curves is the minimal length of a leash that is necessary. The Fréchet method has the advantage of computing distances only on a limited number of homologous points.

Eiter and Mannila (Eiter and Mannila 1994) give a good approximation: the discrete Fréchet distance (d_{dF}) that computes in time O(n m). L_1 and L_2 are interpreted as two oriented sets of vertices: $<L_{1.1}\ldots L_{1.n}>$ and $<L_{2.1}\ldots L_{2.m}>$. While d_{dF} is the minimal length of leash i.e. away from the pair of beginning vertices $(L_{1.1},L_{2.1})$ to the pair of ending vertices $(L_{1.n},L_{2.m})$.. This gives an ordered set of $(L_{1.i},L_{2.j})$ such as the following pair of $(L_{1.i},L_{2.j})$ which is one of these three pairs: - $(L_{1.i+1},L_{2.j+1})$ man and dog are walking, - $(L_{1.i+1},L_{2.j})$ only the man is.

To identify a set of homologous points, we focus on vertices for three reasons. The number of homologous point must be limited, the vertices are more accurate than intermediate points (Veregin 1999) and semantically more important. So, we can interpret L_1 and L_2 as two order sets of vertices: $<L_{1.1}\ldots L_{1.n}>$ and $<L_{2.1}\ldots L_{2.m}>$. The discrete Fréchet between L_1 and L_2 is computed recursively as the followings:

$$d_{Fd}(L_1,L_2) = \max\begin{pmatrix} d_E(L_{1.n}, L_{2.m}) \\ \min\begin{pmatrix} d_{Fd}(<L_{1.1}\ldots L_{1.n-1}>,<L_{2.1}\ldots L_{2.m}>)\forall n\neq 1 \\ d_{Fd}(<L_{1.1}\ldots L_{1.n}>,<L_{2.1}\ldots L_{2.m-1}>)\forall m\neq 1 \\ d_{Fd}(<L_{1.1}\ldots L_{1.n-1}>,<L_{2.1}\ldots L_{2.m-1}>)\forall n\neq 1, m\neq 1 \end{pmatrix}\end{pmatrix}$$

$<L_{1.1}\ldots L_{1.n-1}>$ and $<L_{2.1}\ldots L_{2.m-1}>$ represent lines. Hence, it is possible to recursively apply this d_{dF} process with parameters: $<L_{1.1}\ldots L_{1.n-1}>$, $<L_{2.1}\ldots L_{2.m-1}>$… This process is finished when both lines are reduced to both points ($<L_{1.1}>$, $<L_{2.2}>$) and d_{dF} ($<L_{1.1}>$, $<L_{2.2}>$) = $d_E(L_{1.1}, L_{2.2})$.

For the example in Fig. 3, the matrix of d_E between points $L_{1.i}$ and $L_{2.j}$ is given in table 1 to compute and visualise $d_{dF}(L_1, L_2)$. Note that d_{dF} is equal to 1.90.

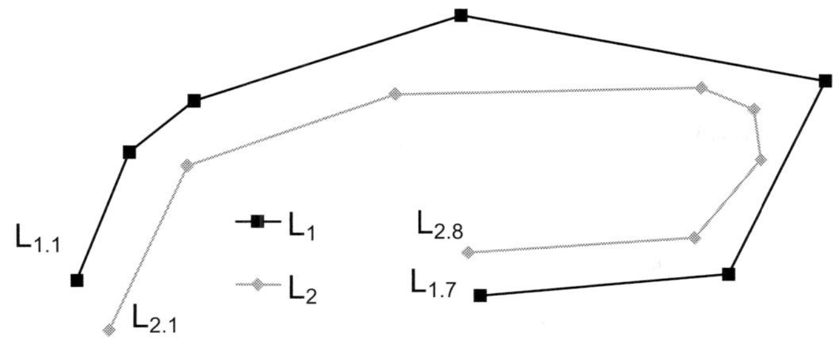

Fig. 3. Example of homologous lines

Table 1. Matrix of d_E between ($L_{1.i}$, $L_{2.j}$) distance and ways

	L2	L2.j.x	0.5	1.7	4.9	9.6	10.4	10.5	9.5	6
L1		L2.j.y	0	2.3	3.3	3.4	3.1	2.4	1.3	1.1
L1.i.x	**L1.i.y**	**i. j**	**1**	**2**	**3**	**4**	**5**	**6**	**7**	**8**
0	0.7	**1**	**0.86**	2.33	5.55	9.97	10.67	10.64	9.52	6.01
0.8	2.5	**2**	2.52	**0.92**	4.18	8.85	9.62	9.70	8.78	5.39
1.8	3.2	**3**	3.45	**0.91**	3.10	7.80	8.60	8.74	7.93	4.70
5.9	4.4	**4**	6.97	4.70	**1.49**	3.83	4.68	5.02	4.75	3.30
11.5	3.5	**5**	11.54	9.87	6.60	**1.90**	**1.17**	**1.49**	2.97	6.00
10	0.8	**6**	9.53	8.43	5.68	2.63	2.33	1.68	**0.71**	4.01
6.2	0.5	**7**	5.72	4.85	3.09	4.47	4.94	4.70	3.40	**0.63**

d_{dF} can be used because the length of the longer segment (LengthMaxSeg) (Eiter and Mannila 1994) limits the generated approximation:

$$d_F(L_1, L_2) \leq d_{dF}(L_1, L_2) \leq d_F(L_1, L_2) + \text{LengthMaxSeg}$$

A sampling can be applied to both lines to limit this approximation to ε,. New intermediary vertices can be added such as the length of each segment is inferior to ε (cf. Fig. 5 for an example of sampling). In our case, samplings are needed when the length of segments is important and when the resolutions of the data sets are different.

3.2 Data Matching of Line's Points

d_{dF} are computed to measure the maximal distance between two lines. Hence, d_{dF} can be combined with other processes to match homologous lines. We propose to re-use d_{dF} to define a data matching process between vertices from homologous lines. Indeed, One of d_{dF} ways can be chosen to match the vertices. After using a maximal criteria (d_{dF}), an average criteria is employed. The chosen way, so-called the minimal way (W_m), is defined by the case in which the average distances between its pair ($L_{1.i}$, $L_{2.j}$) is minimal. In other words, between all ways, the one, which has the less taut leash, is chosen.

For the homologous lines of Fig. 3, three ways (the grey cells of table 1 give the pairs of these ways) are possible:

- **W_1**: $(L_{1.1},L_{2.1})$ $(L_{1.2},L_{2.2})$ $(L_{1.3},L_{2.2})$ $(L_{1.4},L_{2.3})$ $(L_{1.5},L_{2.4})$ $(L_{1.5},L_{2.5})$ $(L_{1.5},L_{2.6})$ $(L_{1.6},L_{2.7})$ $(L_{1.7},L_{2.8})$ average of d_E between $(L_{1.i},L_{2.j}) = 1.12$
- **W_2**: $(L_{1.1},L_{2.1})$ $(L_{1.2},L_{2.2})$ $(L_{1.3},L_{2.2})$ $(L_{1.4},L_{2.3})$ $(L_{1.5},L_{2.4})$ $(L_{1.5},L_{2.5})$ $(L_{1.6},L_{2.6})$ $(L_{1.6},L_{2.7})$ $(L_{1.7},L_{2.8})$ average of d_E between $(L_{1.i},L_{2.j}) = 1.14$
- **W_3**: $(L_{1.1},L_{2.1})$ $(L_{1.2},L_{2.2})$ $(L_{1.3},L_{2.2})$ $(L_{1.4},L_{2.3})$ $(L_{1.5},L_{2.4})$ $(L_{1.5},L_{2.5})$ $(L_{1.5},L_{2.6})$ $(L_{1.6},L_{2.6})$ $(L_{1.6},L_{2.7})$ $(L_{1.7},L_{2.8})$ average of d_E between $(L_{1.i},L_{2.j}) = 1.18$

Intuitively, the man and the dog can walk only on this grey cell. For example, if the man is on $L_{1.5}$ and the dog is on $L_{2.5}$, two displacements are possible: - the dog is walking to $L_{2.6}$ and the man is standing in $L_{1.5}$ - the man is walking to $L_{1.6}$ and at the same time the dog is walking to $L_{2.6}$. Moreover, the average distances between pairs of W_m can be computed, 1.12 is inferior as 1.14 and 1.18. So W_1 is the W_m (in bold typeface in table 1). Fig. 4 shows pairs of $(L_{1.i},L_{2.j})$ associated with this minimal way.

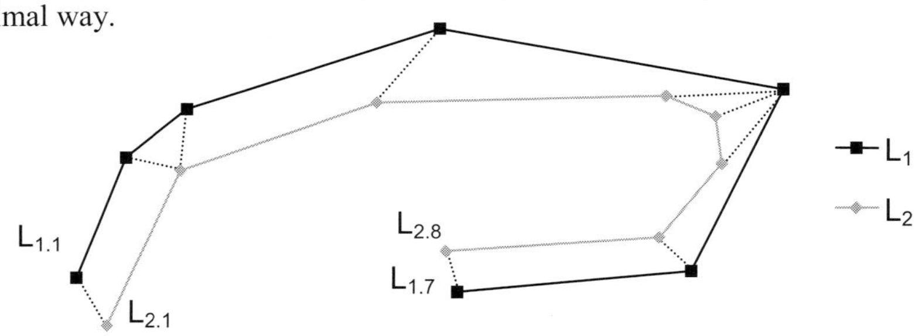

Fig. 4. Pairs of $(L_{1.i},L_{2.j})$ of the minimal way represented by dot lines

We can remark that:

- This matching can be a:
  - one-to-one (between $L_{1.1}$ and $L_{2.1}$ for example)
  - many-to-one (between $L_{1.5}$ and $L_{2.4}$, $L_{2.5}$, $L_{2.6}$ for example)
  - one-to-many (between $L_{1.2}$, $L_{1.3}$ and $L_{2.2}$ for example).

  Generally, a matching of vertices L_1 and L_2 would imply a one-to-one mapping. In our case, many-to-one, or one-to-many matching are not errors, it is only a "turn" with more detail in one data set.
- The average distance method forgives the many-to-many matching.
- Pairs take into account the order of vertex of lines. Graphically, Pairs are a collection of non-crossing dotted lines

This shows that the discrete Fréchet distance can be used to match the points of homologous lines.

3.3 Partial Data Matching of Line's Points

Some pairs of lines, such as the pair shown in Fig.5 can only be partially matched. More precisely, only parts of the lines are matched. For example, still using the lines of Fig. 5, we can visually show that L_1 can be matched to the partial line <$L_{2.5}$…$L_{2.14}$>. To identify this kind of data matching, d_{dF} cannot be employed. Some other parts of line (<$L_{2.1}$…$L_{2.5}$> and <$L_{2.14}$…$L_{2.17}$>) cannot be used to compute d_{dF}. Therefore, a new measure, so-called the partial discrete Fréchet distance (d_{pdF}), is introduced:

- To detect the partial homologous line < $L_{2.begin}$…$L_{2.end}$ >.
- To compute d_{pdF}. d_{pdF} is equal to d_{dF}(L1, < $L_{2.begin}$…$L_{2.end}$ >) $L_{2.begin}$ and $L_{2.end}$ are chosen such as begin < end and d_{dF}(L1, < $L_{2.begin}$…$L_{2.end}$ >) are smaller. Phase 1 and 2 are simultaneous.
- To chose the minimal way W_m. This phase is similar to the one of data matching between homologous lines.

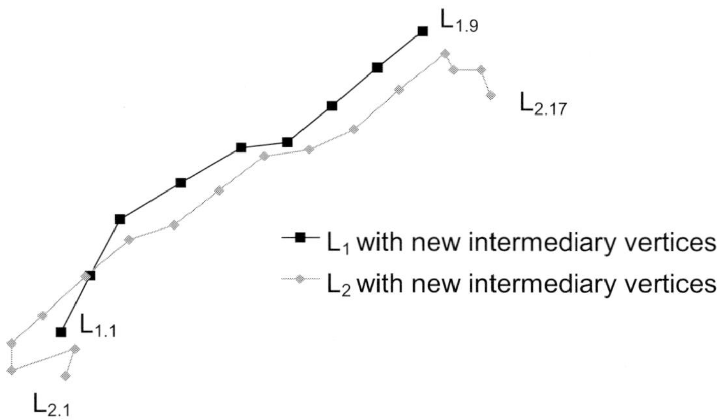

Fig. 5. Example of partial homologous lines with new intermediary vertices

A non-optimal algorithm to compute this measure is:

$B = \{L_{2.1}, L_{2.2}, \ldots, L_{2.m-1}\}$; $E = \{L_{2.m}, L_{2.m-1}, \ldots, L_{2.2}\}$;
$d_{Fdp} = +\infty$;
For $L_{2.j}$ in B
    If $d_E(L_{1.1}, L_{2.j}) < d_{Fdp}$ then
        For $L_{2.jj}$ in E
            If $j \le jj$ and $d_E(L_{1.m}, L_{2.jj}) < d_{Fdp}$ then
                If d_{Fd}(<$L_{1.1}$…$L_{1.n}$>,<$L_{2.j}$… $L_{2.jj}$>) < d_{Fdp} then
                    {d_{Fdp} = d_{Fd}(<$L_{1.1}$…$L_{1.n}$>,<$L_{2.j}$… $L_{2.jj}$>);
                    $L_{2.begin} = L_{2.j}$;
                    $L_{2.end} = L_{2.jj}$;}

For the lines of Fig. 5, the matrix of d_E between points of $L_{1.i}$ and $L_{2.j}$ is given in table 2 to illustrate the result. Note that d_{pdF} is equal to 1.22. Only one way (grey cells in the matrix) is possible for this example. So the W_m is ($L_{1.1}$,$L_{2.5}$) ($L_{1.2}$,$L_{2.6}$) ($L_{1.3}$,$L_{2.7}$) ($L_{1.3}$,$L_{2.8}$) ($L_{1.4}$,$L_{2.9}$) ($L_{1.5}$,$L_{2.10}$) ($L_{1.6}$,$L_{2.11}$) ($L_{1.7}$,$L_{2.12}$) ($L_{1.8}$,$L_{2.13}$) ($L_{1.9}$,$L_{2.14}$).

Table 2. Matrix of d_E between the partial homologous lines of Fig. 5

L_1	L_2	**L2.j.x**	4.6	4.8	3.4	3.4	4.1	5.05	6	7	8	9	10	11	12	13	13.2	13.8	14
		L2.j.y	1.6	2.6	1.8	2.8	3.8	5.15	6.5	7	8.25	9.5	9.75	10.5	11.9	13.2	12.6	12.6	11.7
L1.i.x	L1.i.y	**i.j**	**1**	**2**	**3**	**4**	**5**	**6**	**7**	**8**	**9**	**10**	**11**	**12**	**13**	**14**	**15**	**16**	**17**
4.5	3.2	**1**	1.60	0.67	1.78	1.17	**0.72**	2.03	3.62	4.55	6.14	7.74	8.55	9.77	11.4	13.1	12.8	13.2	12.7
5.15	5.2	**2**	3.64	2.62	3.82	2.97	1.75	**0.11**	1.55	2.58	4.17	5.77	6.65	7.89	9.58	11.2	10.9	11.3	10.9
5.8	7.2	**3**	5.73	4.71	5.91	5.01	3.80	2.18	**0.73**	**1.22**	2.44	3.94	4.91	6.16	7.78	9.37	9.16	9.65	9.35
7.15	8.5	**4**	7.36	6.35	7.68	6.82	5.60	3.95	2.31	1.51	**0.89**	2.10	3.11	4.34	5.92	7.50	7.31	7.81	7.56
8.5	9.8	**5**	9.08	8.10	9.49	8.66	7.44	5.79	4.14	3.18	1.63	**0.58**	1.50	2.60	4.08	5.64	5.47	5.99	5.82
9.5	10	**6**	9.72	8.77	10.2	9.44	8.22	6.58	4.95	3.91	2.30	0.71	**0.56**	1.58	3.14	4.74	4.52	5.02	4.81
10.5	11.3	**7**	11.3	10.4	11.8	11.0	9.86	8.22	6.58	5.54	3.94	2.34	1.63	**0.94**	1.62	3.14	3.00	3.55	3.52
11.5	12.7	**8**	13.0	12.1	13.5	12.7	11.5	9.93	8.29	7.26	5.66	4.06	3.31	2.26	**0.94**	1.58	1.70	2.30	2.69
12.5	14	**9**	14.7	13.7	15.2	14.4	13.2	11.5	9.92	8.90	7.30	5.70	4.93	3.81	2.16	**0.94**	1.57	1.91	2.75

Partial data matching is required to match one line to a part of another line. To match a part of a given line to another line, detection of homologous parts of lines is a more complex process. It is always possible to reduce the d_{dF} between parts, by part reduction. This more complex data matching process is not treated in this paper.

3.4 Data Matching of Polygon's Points

This process can also be applied to match vertices of homologous polygon borderlines. However, for oriented lines, the beginning pair of points and the end pair of points are known. Unfortunately, for polygon borderlines, these pairs are not predetermined. Thus, the process must define a function T to translate polygons borderlines P_1 and P_2 into lines L_1 and L_2 such as the d_{Fd} between L_1 and L_2 is minimal. Subscripts of L_1 and P_1 are identical. On the other hand, the L_2 subscripts correspond to P_2 subscripts only by a circular translation (if $L_{2.j'} = P_{2.m}$ then $L_{2.j+1} = P_{2.1}$ else $L_{2.j'+1} = P_{2.j+1}$).

A new method is defined as follows:

- To find j' such as <$P_{2.j'}$, $P_{2.j'+1}$… $P_{2.m}$, $P_{2.1}$…$P_{2.j'-1}$> is the ordered set of vertex of L_2.
- To compute $d_{dF}(L_1, L_2)$, j' is chosen such as $d_{dF}(L_1, L_2)$ is minimal. So, phase 1 and 2 are simultaneous.
- To chose the minimal way (W_m). This phase is similar to the data matching between homologous lines.
- To translate this W_m from L_1, L_2 into P_1,P_2 using the inverse circular translation

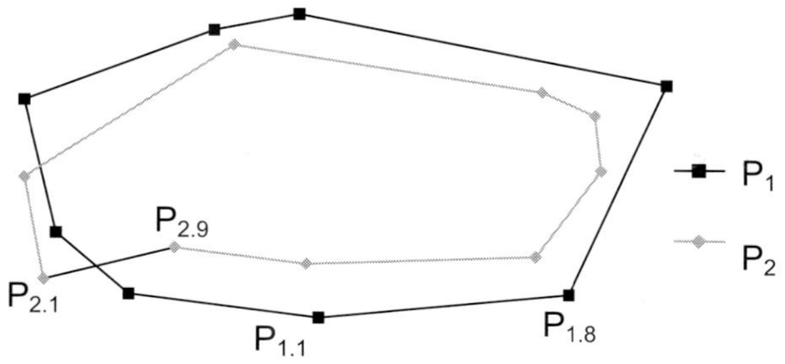

Fig. 6. Example of homologous polygons

Table 3. Matrix of d_E between homologous polygons of Fig. 6

	L2	L2.j.x	2	1.7	4.9	9.6	10.4	10.5	9.5	6	4
L1		L2.j.y	1	2.3	4	3.4	3.1	2.4	1.3	1.2	1.4
L1.i.x	L1.i.y		1	2	3	4	5	6	7	8	9
6.2	0.5	1	4.23	4.85	3.73	4.47	4.94	4.70	3.40	**0.73**	2.38
3.3	0.8	2	1.32	2.19	3.58	6.82	7.46	7.38	6.22	2.73	**0.92**
2.2	1.6	3	**0.63**	0.86	3.61	7.62	8.34	8.34	7.31	3.82	1.81
1.7	3.3	4	2.32	**1.00**	3.28	7.90	8.70	8.85	8.05	4.79	2.98
4.6	4.2	5	4.12	3.47	**0.36**	5.06	5.90	6.17	5.69	3.31	2.86
5.9	4.4	6	5.17	4.70	**1.08**	3.83	4.68	5.02	4.75	3.20	3.55
11.5	3.5	7	9.82	9.87	6.62	**1.90**	**1.17**	**1.49**	2.97	5.96	7.79
10	0.8	8	8.00	8.43	6.02	2.63	2.33	1.68	**0.71**	4.02	6.03

The matrix of d_E between points $P_{1.i}$ and $P_{2.j}$ in Fig. 6 is given in table 3 to visualise the d_{dF} and ways. In this example, j' is equal to 8. So $d_{dF}(P_1, P_2)$ is equal to 1.90 and four ways are possible. Pairs of points are the grey cells of table 1. The value W_m is given in bold typeface.

4 Weighting Function of the Merging Process

The new merging process proposed in this paper, requires the minimal way (W_m) and the weighting function (f_w). The kth vertex of the resulting line (L_M) is defined thanks to the kth pair ($L_{1.i.}$, $L_{2.j}$) of the W_m and f_w, it is defined as follows:

$$L_{F.k} = f_w\left(L_{1.i}, L_{2.j}\right) = \frac{\alpha_{1.k} \times L_{1.i} + \alpha_{2.k} \times L_{2.j}}{\alpha_{1.k} + \alpha_{2.k}}$$

For each pair, the weight ($\alpha_{1.k}$, $\alpha_{2.k}$) is proportional to:

- The accuracy of the input layers,

- The kind of points considered. The weight of a vertex is more important than the weight of a new intermediary vertex. According to (Vergin 1999), vertices are more accurate than the other points of the line.
- The cardinality of the pairs of W_m where this point is included. The weight of a point included in a single pair is more important than the weight of a point included in several pairs. Indeed, a point included in several pairs, represents a curve more detailed in the other layer. The resulting curve must be closer to the more detailed curve.

In the case of the two lines of Fig. 3, the merging line can be computed thanks to their W_m and f_w (cf. table 4). For the example, we suppose L_1 less accurate than L_2. A default weight of 0.8 is associated with the points of L_1 and a default weight of 1 is associated with the points of L_2. For each point, the default weights are divided by the number of pairs in the line.. Weights: $\alpha_{1.5}$, $\alpha_{1.6}$, $\alpha_{1.7}$ associated with $L_{1.5}$ are divided by 3 and weights $\alpha_{2.2}$, $\alpha_{2.3}$ are divided by 2. The merging line is given in Fig. 7.

This example visually shows that this merging process gives a suitable result. In order to define the merging line, this process uses the accuracy of lines and the "local" accuracy. This last one is taken into account by the cardinality of pairs. In the case of Fig. 7, the merging line is:

- Globally closer to L_2 (the more accurate line),
- Locally closer to more accurate than part of line ($L_{2.2}$, $L_{2.3}$ and $L_{1.4}$, $L_{1.5}$, $L_{1.6}$).

Table 4. Merging process of lines of Fig. 3, thanks to the pairs of W_m and f_w

Wm						weight		Merging line	
$L_{1.i}$	$L_{2.j}$	$L_{1.i.x}$	$L_{1.i.y}$	$L_{2.j.x}$	$L_{2.j.y}$	$\alpha_{1.k}$	$\alpha_{2.k}$	$L_{mk.x}$	$L_{mk.y}$
$L_{1.1}$	$L_{2.1}$	0	0,7	0,5	0	0,80	1,00	0,28	0,31
$L_{1.2}$	$L_{2.2}$	0,8	2,5	1,7	2,3	0,80	0,50	1,15	2,42
$L_{1.3}$	$L_{2.2}$	1,8	3,2	1,7	2,3	0,80	0,50	1,76	2,85
$L_{1.4}$	$L_{2.3}$	5,9	4,4	4,9	3,3	0,80	1,00	5,34	3,79
$L_{1.5}$	$L_{2.4}$	11,5	3,5	9,6	3,4	0,27	1,00	10,00	3,42
$L_{1.5}$	$L_{2.5}$	11,5	3,5	10,4	3,1	0,27	1,00	10,63	3,18
$L_{1.5}$	$L_{2.6}$	11,5	3,5	10,5	2,4	0,27	1,00	10,71	2,63
$L_{1.6}$	$L_{2.7}$	10	0,8	9,5	1,3	0,80	1,00	9,72	1,08
$L_{1.7}$	$L_{2.8}$	6,2	0,5	6	1,1	0,80	1,00	6,09	0,83

For this example, all points are vertices, thus, new intermediary vertices are not considered. Empirically, to consider the lower accuracy of the additional point, the weight of these points should be multiplied by 0.8.

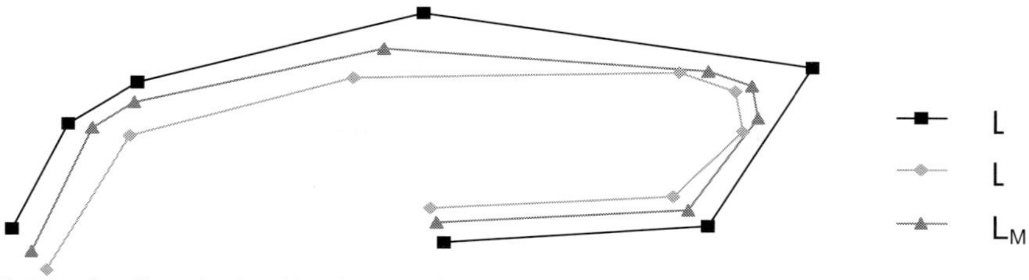

Fig. 7. Merging line obtained by the merging process

To merge partial homologous lines or polygons, the kth vertex of L_M or P_M is also defined thanks to the kth pair and a similar weighting function. Weights are proportional to the three same arguments (accuracy of the input layers, kind of points and cardinality of pairs). Fig. 8 shows the resulting line L_M (see (a)) and the polygon P_M (see (b)) computed for our examples.

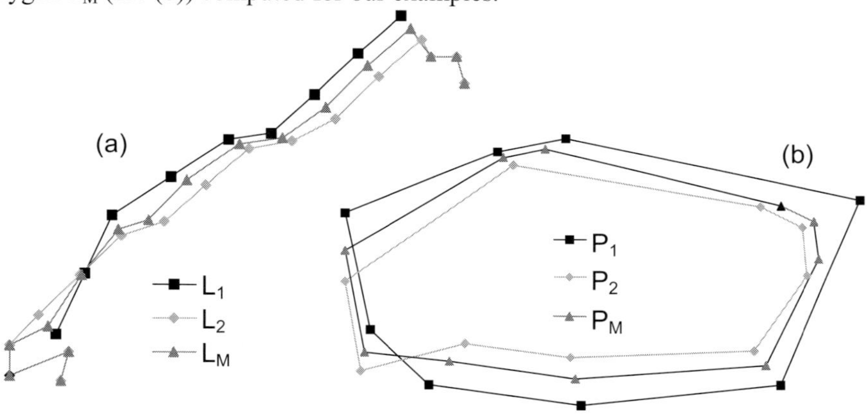

Fig. 8. Merging line obtained by the partial merging process and merging polygon obtained by the merging process

5 Joining Process

When the merging process transforms a point ($L_{i.j}$) into several points ($L_{M.j'}$, $L_{M.j''}$…) there is a join problem at the line (L_k) which is connected to $L_{i.j}$. To preserve the topology, L_k must be joined with one point of L_M.

Two situations are distinguished:

- $L_{i.j}$ is transformed into an odd number of points. The average point is used to join L_k (see Fig. 9a)
- $L_{i.j}$ is transformed into an even number of points. A new point is defined at the halfway of the middle edge (see Fig. 9b).

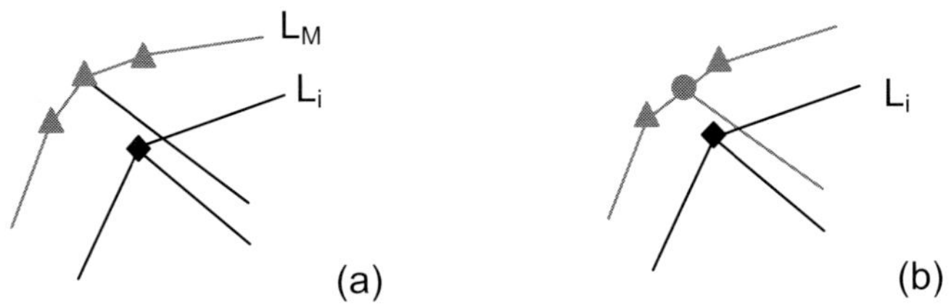

Fig. 9. Examples of junction between the merging line and another line

6 Conclusion

This paper introduces a new merging process that can be an alternative to fuzzy tolerance methods for a better data integration of homologous objects. It makes the distinction between different and homologous objects, thanks to data matching and measures derived from the Fréchet distance. Thus, homologous objects are merged and different objects are overlaid.

Moreover, it uses homologous points and not closest points to compute merging lines. Another advantage of this merging process relies on the fact that it can be automatic and generic. In others words, it can be applied to homologous lines, homologous polygons and partial homologous lines. Additionally, this process takes into account connections of merging lines to other lines.

This approach must be combined, however, with other data integration processes like rubber sheeting (to smooth displacements induced by the merging and joining processes). Any data integration based on this novel merging (e.g. overlaying, merging, joining and rubber sheeting) must be organised in terms of data and process schedule. The proposed merging technique is defined to integrate data from different layers in the same geographic area, it can also be extended to integrate data from adjacent coverages.

References

Alt H, and Godau M (1995) Computing the Fréchet distance between two polygonal curves. International Journal of Computational Geometry & Applications 5(1-2):75-91

Chrisman N (2001) Exploring Geographic Information Systems, 2nd Edition. John Wiley & Sons, Inc.

Cohen W (2000) Data Integration using similarity joins and a Word-Based Information representation language. ACM Transactions on Information Systems 18(3):288-321

Devogele T, Trevisan J, Raynal L (1996) Building a Multi-Scale Database with Scale-Transition Relationships. In: Kraak et Molenaar (eds) Spatial Data Handling. pp 337-351

Devogele T, Parent C, Spaccapietra S (1998) On spatial database integration. International Journal of Geographical Information Science 12(4):335-352

Dougenik JA (1980) WHIRLPOOL: A geometric processor for polygon coverage data. In: 4th Auto-Carto. pp 304-311

Harvey F, Vauglin F (1996) Geometric Match-processing: Applying Multiple Tolerances. In: Kraak and Molenaar (eds) Spatial Data Handling. pp 155-171

Gabay Y, Doytsher Y (1994) Automatic adjustment of line maps. In : GIS/LIS, ASPRS. pp 233-241

Parent C, Spaccapietra S, Devogele T (1996) Conflicts in Spatial Database integration. In: Parallel and Distributed Computing Systems. pp 772-778

Pullar D (1993) Consequences of using a tolerance paradigm in spatial overlay. In: 11th Auto Carto. pp 288-296

Spéry L, Claramunt C, Libourel T (2001) A Spatio-Temporal Model for the Manipulation of Lineage Metadata. GeoInformatica 5(1):51-70

TCI software (1999) Adjust : True Rubber Sheeting inside AutoCAD [online]. Baker City, Oregon. Available from: http://www.tcicorp.com/tci_adjust.html

USGS (1998) About the USGS Geospatial Data Clearinghouse [online]. Available from: http://nsdi.usgs.gov/

Veregin H (1999) Data quality parameters. In : Longley, Goodchild, Maguire, Rhind (eds) Geographical Information Systems. John Willey & Sons, pp 177-189

Above-Ground Biomass Estimation of Successional and Mature Forests Using TM Images in the Amazon Basin

Dengsheng Lu[1], Paul Mausel[2], Eduardo Brondizio[3] and Emilio Moran[3]

Center for the Study of Institutions, Population, and Environmental Change (CIPEC), Indiana University, Bloomington, Indiana 47408, USA, ph. 812-856-5767, fax 812-855-2634, e-mail: dlu@indiana.edu[1]
Department of Geography, Geology, and Anthropology, Indiana State University, Terre Haute, Indiana 47809, USA, ph. 812-237-2254, fax 812-237-8029, gemause@scifac.indstate.edu[2]
Anthropological Center for Training and Research on Global Environmental Change, Indiana University, Bloomington, Indiana 47405, USA, ph. 812-855-6182, fax 812-855-3000, ebrondiz@indiana.edu and moran@indiana.edu[3]

Abstract

Above-ground biomass estimation of successional and mature forests in moist tropical regions is attracting increasing attention. Because of complex stand structure and abundant vegetation species, rarely has remote-sensing research been successfully conducted in biomass estimation for moist tropical areas. In this paper, two study areas in the Brazilian Amazon basin—Altamira and Bragantina— with different biophysical characteristics were selected. Atmospherically corrected Thematic Mapper (TM) images and field vegetation inventory data were used in the analysis, and different vegetation indices and texture measures were explored. Multiple regression models were developed through integration of image data (including TM bands, different vegetation indices, and texture measures) and vegetation inventory data. These models were used for biomass estimation in both selected study areas. This study concludes that neither TM spectral bands nor vegetation indices alone are sufficient to establish an effective model for biomass estimation, but multiple regression models that consist of spectral and textural signatures improve biomass estimation performance. The models developed are especially suitable for above-ground biomass estimation of dense vegetation areas.
Keywords: biomass estimation, successional and mature forests, Amazon basin, Thematic Mapper, vegetation index

1 Introduction

Biomass estimation for tropical forests has received much attention in recent years because the change of biomass regionally is associated with important components of climate change. Biomass determines potential carbon emission that could be released to the atmosphere due to deforestation or conversion to non-forest land use. Therefore, accurate biomass estimation is necessary for better understanding deforestation impacts on global warming and environmental degradation.

Computer-processed remotely sensed data are possibly the best data for economical biomass estimation in the tropical region over large areas. Although various methods for biomass estimation have been tested (Gillespie et al. 1992, Overman et al. 1994, Roy and Ravan 1996), rarely has research successfully been conducted in a large area in moist tropical regions due to the complex tropical forest stand structure and abundant tree species. Previous research mainly used spectral signatures to establish biomass estimation models (Nelson et al. 1988, Franklin and Hiernaux 1991, Leblon et al. 1993, Nelson et al. 2000, Steininger 2000), but ignored spatial information largely due to a lack of understanding about the relationships between biomass and spatial signatures. However, spatial information could be an important factor in improving biomass estimation accuracy even though many uncertainties exist. For example, what types of texture measures are appropriate to extract biomass information? What size of moving window for selected texture measures can most effectively extract biomass information? Which models can be used to estimate biomass using remote-sensing technology? How do different biophysical environments influence biomass estimation? In order to explore biomass estimation using remote-sensing data, two study areas, each having different vegetation growth rates, soil conditions, and land-use history, are selected. Different types of vegetation indices and texture measures are tested. This paper develops biomass estimation models for successional and mature forest biomass estimation in the Amazon basin.

2 Study Areas

The Altamira study area is located along the Transamazon Highway in the Brazilian state of Para (Fig. 1). The city of Altamira and the Xingu River anchor the eastern edge of the study area. In the 1950s an effort was made to attract colonists from all over Brazil, who came and settled along streams as far as 20 km from the city centre. With the construction of the Transamazon Highway in 1970, this population and older caboclo settlers from earlier rubber eras claimed land along the new highway and legalised their land claims. Geopolitical goals and political economic policies that transferred production of staples like rice, corn, and beans from the most southern Brazilian states to more northern regions drove Early settlement. The region has had a gradual shift to a more diverse set of land uses: pasture, cocoa, sugar cane, black pepper, and staple crops. Mahogany is beginning to be planted in cocoa groves as a diversification strategy and can be expected to benefit landowners who have the best soils in the area (Moran et al. in press). The dominant native types of vegetation are mature moist forest and liana forests. Nutrient-rich alfisols, as well as nutrient-poor ultisols and oxisols were found in the Altamira study area. Altamira has experienced high rates of deforestation and secondary succession associated with implementation of agro-pastoral projects. Annual rainfall in Altamira is approximately 2,000 mm and is concentrated from late October through early June. Average temperature is about 26°C.

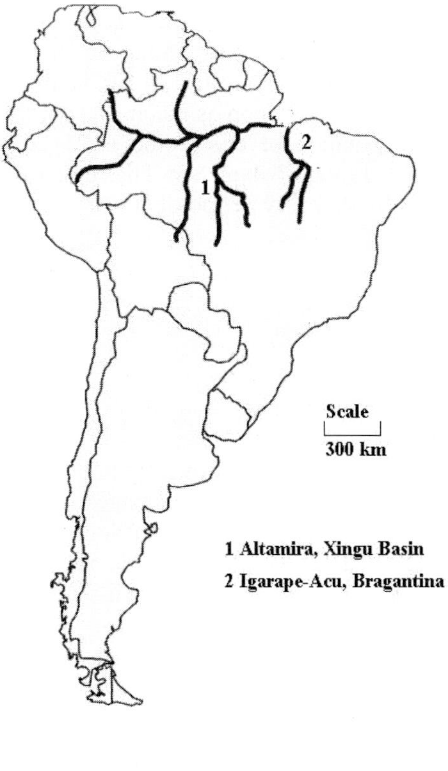

Fig. 1. Location of selected study areas

The Bragantina study area selected is located within the municipality of Igarapé-Açú in the state of Pará. The vegetation in this region is mostly composed of secondary-growth forest (*capoeira*), flooded forest (*igapo*), and a few remaining areas of dense forest. At the beginning of the twentieth century, almost one million hectares of dense tropical rain forest covered the Bragantina region; however, less than 2 percent of the forests remained by 1960. The dense forest that once surrounded the town of Castanhal had an average height of 23 meters. Heavy occupation of this region has eliminated almost all dense forests and transformed the landscape into a mosaic comprised of a variety of secondary vegetation (Tucker et al. 1998). Currently, secondary succession in all stages of regrowth dominates the landscape. The main agricultural products are passion fruit (*maracuja*), manioc, oil palm, rice, corn, beans, and cotton. Other important crops are pepper, papaya, tobacco, melon, sugar cane and beet, rubber, avocado, citrus, coconut, banana, cocoa, guarana, mango, and *caju* (Ibid.). Nutrient-poor oxisols and ultisols dominate in this area. Land use in the Bragantina region has gone through several phases, and today the dominant form is short-fallow swidden cultivation and pasture development. Cultivation of secondary-growth areas has been common for decades, and islands of mature forest are rare (Moran et al. in press). The long settlement history, high human population density, repeated land clearing (including burning) over the past century in Bragantina has degraded the landscape, leading to slower regrowth rates. In Bragantina, annual rainfall ranges from 2,200 to 2,800 mm with variation as much as +/- 1,000 mm from the mean. The average annual temperature of this region is 25–26ºC, and the dry period extends from September through November. The region is characterised by high rates of precipitation and a short dry period of at least one month with less than 66 mm of rainfall.

3 Methods

3.1 Data Collection and Analysis

A nested sampling strategy, organised by region, site, plot, and subplot, was employed to collect field data. Landsat TM images and GPS devices were used during fieldwork. The region was the categorical level representing the study area that included all sample sites. The site in this region was selected for plot sampling. In general, ten plots (10 m x 15 m) in each site were allocated in a stratified random fashion, and one randomly selected subplot (5 m x 2 m) was nested within each plot. Plots were designed to inventory trees, and subplots were used to inventory saplings, seedlings, and herbaceous species. In each plot, all individual trees with a diameter at breast height (DBH) greater than 10 cm were identified and measured for DBH, stem height (the height of the first major branch), and total height. In the subplot, all saplings (DBH between 2 cm and 10 cm), seedlings (DBH less than 2 cm), and herbaceous vegetation (percent of ground cover) were identified and counted. The diameter and total height were recorded for all individuals with DBH between 2 cm and 10 cm. At each site, soil samples were collected at 20-cm intervals to a depth of 1 m, resulting in five sets of measurements for each sample site. Soil samples were analysed at the soil laboratories in Belem, Brazil, for both chemical and physical properties.

Two models were selected for individual vegetation biomass estimation. Eq. 1 (Nelson et al. 1999) was used to calculate biomass for those trees and saplings with DBHs of less than 25 cm, and Eq. 2 (Overman et al. 1994) was used to calculate biomass for those trees with DBHs greater than or equal to 25 cm. Eq. 3 was used to calculate above-ground biomass (AGB).

$$\ln(\mathrm{DW1}) = -2.5202 + 2.14\ln(\mathrm{D}) + 0.4644 * \ln(\mathrm{H}) \quad (1)$$

$$\ln(\mathrm{DW2}) = -3.843 + 1.035\ln(\mathrm{D}^2\mathrm{H}) \quad (2)$$

$$\mathrm{AGB} = (\sum_{i=1}^{m} DW1_i + \sum_{j=1}^{n} DW2_j)/\mathrm{AP} + (\sum_{k=1}^{s} DW1_k)/\mathrm{AS} \quad (3)$$

In these equations, ln is the natural logarithm, D is diameter at breast height (cm), H is total height (m), DW1 is individual tree or sapling biomass (kg) when DBH is less than 25 cm, DW2 is the individual tree biomass when DBH is greater than or equal to 25 cm, m is the total tree number in a plot when DBH is between 10 and 25 cm, n is the total tree number in a plot when DBH is greater than or equal to 25 cm, and s is the total sapling number in a subplot area when DBH is between 2 and 10 cm. AP and AS are the plot area and subplot area (m^2), respectively, and AGB is the above-ground biomass (kg/m^2).

3.2 Landsat TM Image Processing

Landsat TM images were obtained for Altamira (image date July 20, 1991 and Bragantina (image date June 21, 1994). The TM images were geometrically rectified and then atmospherically corrected using an improved image-based DOS (dark object subtraction) model (Lu et al. in press). Four types of vegetation indices were grouped and used (Table 1):

1. Simple ratio, such as TM 4/3, TM 5/3, TM 5/7, and TM 5/4
2. Normalization ratio, such as NDVI (Normalized Difference Vegetation Index), ND32, ND54, ND53 and ND57
3. Linear combination of multiple bands, such as Albedo, MID57, VIS123, PCA (Principal Component Analysis), and KT (Tasseled Cap) transform
4. Complicated vegetation index, such as ARVI (Atmospherically Resistant Vegetation Index), ASVI (Atmospheric and Soil Vegetation Index), GEMI (Global Environmental Monitoring Index), MSAVI (Modified Soil Adjusted Vegetation Index), and SAVI (Soil Adjusted Vegetation Index)

In addition to the TM bands and different vegetation indices, four types of texture measures—mean Euclidean distance (MED), variance, skewness, and kurtosis (Table 2)—were also tested. Different texture measures combined with different window sizes (3x3, 5x5, 7x7, 9x9, and 15x15) for each Landsat TM band were tested in Altamira and Bragantina.

Table 1. Vegetation indices used in research

Index	Formula
ARVI	(NIR - 2 RED + BLUE)/(NIR + 2 RED - BLUE)
ASVI	$NIR+0.5-0.5\sqrt{(2NIR+1)^2-8(NIR-2RED+BLUE)}$
GEMI	$\xi(1-0.25\xi)-\frac{RED-0.125}{1-RED}$ $Where\ \xi=\frac{2(NIR^2-RED^2)+1.5NIR+0.5RED}{NIR+RED+0.5}$
MSAVI	$NIR+0.5-0.5\sqrt{(2NIR+1)^2-8(NIR-2RED)}$
SAVI	$\frac{NIR-RED}{NIR+RED+L}(1+L)$
ND53	(TM 5-TM 3)/(TM 5+TM 3)
ND54	(TM 5-TM 4)/(TM 5+TM 4)
ND57	(TM 5-TM 7)/(TM 5+TM 7)
NDVI	(TM 4-TM 3)/(TM 4+TM 3)
ND32	(TM 3-TM 2)/(TM 3+TM 2)
KT1	0.304TM1+0.279TM2+0.474TM3+0.559TM4+0.508TM5 +0.186TM7
KT2	-0.285TM1-0.244TM2-0.544TM3+0.704TM4+0.084TM5 -0.18TM7
KT3	0.151TM1+0.197TM2+0.328TM3+0.341TM4-0.711TM5 -0.457TM7
PC1, PC2,PC3	The constants are similar to the KT, but they are dependent on the given image of the study area.
VIS123	TM 1 + TM 2 + TM 3
MID57	TM 5 + TM 7
Albedo	TM 1 + TM 2 + TM 3 + TM 4 + TM 5 + TM 7
TM 4/3	TM 4/TM 3
TM 5/3	TM 5/TM 3
TM 5/4	TM 5/TM 4
TM 5/7	TM 5/TM 7

Table 2. Texture measures used in research

Texture	Formula	Notes
Mean Euclidean Distance (MED) (1st Order)	$\frac{\sum (X_{ij} - M)^2}{n-1}$	$X_{ij\lambda}$ is reflectance for spectral band λ and pixel (i, j) of a multispectral image.
Variance (VAR) (2nd Order)	$\frac{\lvert \sum (X_{ij} - M)^3 \rvert}{(n-1)V^{\frac{3}{2}}}$	$X_{c\lambda}$ is reflectance for spectral band λ of a window's centre pixel.
Skewness (SK) (3rd Order)	$\frac{\sum (X_{ij} - M)^3}{(n-1)V^{\frac{3}{2}}}$	X_{ij} is reflectance of pixel (i, j). M is the mean of the moving window,
Kurtosis (KU) (4th Order)	$\frac{\sum (X_{ij} - M)^4}{(n-1)V^2}$	$where\ M = \frac{\sum X_{ij}}{n}$. V is variance. N is number of pixels in a window.

3.3 Integration of Biomass and Remotely Sensed Data

Remotely sensed data are comprehensive responses of vegetation stand structure, vegetation density, and vegetation species composition. Different forest stand structures have different reflectance and texture patterns in various wavelengths, and the relationships between biomass and remotely sensed data are different. Pearson's correlation coefficient is used to analyse such relationships. It measures the strength of linear relationships between two variables. If the coefficient is close to 1, it means there is a strong relationship between them. In this research, one variable is biomass; another variable is the remotely sensed data, such as single TM band, vegetation index, and texture measure. Using biomass as a dependent variable and remote-sensing data such as TM band, vegetation indices, and texture measures as independent variables, multiple regression models are used to establish the relationships between biomass and remote-sensing data. The critical step is to find the appropriate independent variables so the combination of multiple independent variables can provide the best-estimated results. The coefficient of determination (R^2) is an indicator that can be used to determine whether or not the regression model is good, because R^2 measures the percent of variation explained by the regression model. Stepwise regression analysis was used to find the best independent variable combination.

4 Results

Linear regression models were developed using the method described. Table 3 provides a comparison of the best regression models identified, based on a single TM band, a single vegetation index, a single texture measure, and a combination of different image bands. It indicates that TM 5 is the best single band that has the highest regression coefficient in Altamira, but band TM 4 is best in Bragantina. Vegetation indices slightly improve relationships between biomass and spectral signatures in Altamira, and texture measures significantly improved the regression coefficients in Altamira, but they did not improve the regression coefficients in Bragantina. In Altamira, variance with TM 2 with a 9x9 window size is the best texture measure combination, and in Bragantina it is variance with TM 5 with a 15x15 window size. Table 3 also implies that single TM band, single vegetation index, or single texture measure did not have sufficiently high regression coefficients to develop biomass estimation models. Therefore, it may be necessary to seek two or more independent variables in order to improve the relationships between biomass and remotely sensed data.

Table 3. Comparison of regression coefficients among different models

Independent Variables		Altamira		Bragantina	
		Independent Variable	R	Independent Variable	R
Best single image band	TM	TM 5	0.627	TM 4	0.837
	VI	ND54	0.635	KT1	0.835
	Txt	VARtm2_9	0.841	VARtm5_15	0.551
Best combination	TM + Txt	TM 5 + VARtm2_7	0.874	TM 4+ SKtm4_9	0.883
	VI + Txt	KT1+ VARtm2_9	0.878	PC2+ VARtm5_15	0.851

Note: VI is vegetation index; Txt is texture measure.

Stepwise regression analysis indicates that, in the multiple regression models, if the independent variables consist of two or more TM bands, two or more vegetation indices, combinations of TM bands with vegetation indices, and two or more texture measures, the regression coefficients are not significantly improved. Only the combination of TM bands with texture measures or vegetation indices with texture measures can significantly improve the R values. This is because high correlation exists between spectral signatures or between textural signatures, or very weak relationships exist between biomass and selected image bands. However, the information associated with spectral and textural signatures is complementary. For example, in Altamira the TM reflectance or vegetation indices are not strongly related to biomass, but some specific texture measures such as VARtm2_9 (VARIANCE combined with TM 2 and 9x9 window size) are strongly related to biomass. In Bragantina, spectral signatures such as TM 4 have strong correlations with biomass, but no texture measures have strong correlations with biomass. The results (Table 3) show that integration of spectral and textural signatures improved relationships with biomass.

Table 4 summarises the best regression models that were used for biomass estimation in the Altamira and Bragantina study areas. Visually comparing the biomass estimated image with TM 4-5-3 colour composite indicates that high biomass amount corresponds to higher vegetation growth stages. For example, those areas where biomass is greater than 20 kg/m^2 are dominated by mature forest, while successional forest biomass falls in the range between 0 to 20 kg/m^2.

Table 4. Models used for biomass estimation in selected study areas

Study Area	Regression Models	R	Beta value
Altamira	122.288 - 1.078 * KT1 - 128.913 * VARtm2_9	0.878	-0.28 (sp), -0.72 (txt)
Bragantina	64.037 - 1.651 * TM4 + 1.405 * SKtm4_9	0.883	-0.76 (sp), 0.29 (txt)

Note: sp means spectral variables, such as KT1 and TM 4, used in the models; txt means texture variables used in the models.

The Beta value in Table 4 provides a means of measuring the relative changes in variables on a standard scale. It indicates how much change in the dependent variable is produced by a standardised change in one of the independent variables when the others are controlled. The Beta values confirm that texture measure is more important in a study area like Altamira with good soil conditions and fast vegetation growth rates than in a study area like Bragantina with very poor nutrient soil and very slow vegetation growth rate. Texture measure roles in biomass estimation are valuable since they offer an effective method for improving model performance based on the image itself.

5 Discussion

When the sites with very young vegetation ages (two one-year-old sites in Altamira and two two- and three-year old sites in Bragantina) were excluded, the regression coefficients and estimation errors were significantly reduced. This means the younger initial secondary succession (SS1) sites are the main cause of estimation bias. There are at least two reasons for this. First, saplings and seedlings dominate SS1 vegetation, which lacks obvious stratification of stand structure (texture measures cannot effectively extract biomass texture information from SS1 vegetation reflectance). Second, SS1 vegetation is not sufficiently dense to cover the ground, and the information captured by sensor is a mixture of soil and vegetation. The soil colour, soil moisture, and soil mineral composition can significantly influence the reflectance of these SS1 sites. This leads to wide reflectance ranges of SS1 vegetation, which result in low correlation between SS1 vegetation reflectance and SS1 biomass. Although some vegetation indices such as SAVI and MSAVI can reduce the influence of soil conditions, they are weakly related to biomass because they use near infrared (TM 4) and red (TM 3) bands that are weakly correlated with biomass.

Different soil conditions, precipitation patterns, and land-use history induce different vegetation stand structures, canopy shadows, and species compositions, resulting in different relationships between biomass and remotely sensed data (Lu 2001). In a study area with fast vegetation growth rate such as Altamira, complex stand structure and canopy shadows compromised the correlation between biomass and TM spectral signatures but enhanced relationships between biomass and texture information. In contrast, in a study area with slow vegetation growth rates, such as Bragantina, spectral signatures are more strongly related to biomass than are texture data. Thus, in the multiple regression models for biomass estimation, incorporation of texture data in the models in the Altamira study area is more important than in the Bragantina study area for improving the estimation accuracy.

A variety of vegetation indices and texture measures have been developed. A logical question to ask however, is 'which vegetation index and which texture measure can be used to establish biomass estimation models using remotely sensed data'? Previous literature has not discussed these problems to any great extent. Many applications of texture analyses to remotely sensed data are focused on improving classification accuracy by discriminating forest-cover types from non-forest land-cover types (Gordon and Phillipson 1986, Franklin and Peddle 1989, Peddle and Franklin 1991, Jakubauskas and Price 1997). Rarely has research focused on improving biomass estimation, especially for tropical successional and mature forests. Research attempting to find which texture measure—what window sizes using which wavelengths—performs better in improving biomass estimation accuracy is uncommon. Although many texture measures have been developed (Haralick et al. 1973, Haralick 1979, Marceau et al. 1990), there has been little research on how to effectively extract biomass texture information. This research indicates that texture information is a very important factor in improving model estimation performance, especially in a study area with fast vegetation growth rates and complex stand structure. The importance of texture information in a model varies. It is a difficult task to find an appropriate texture measure that is strongly related to biomass because only some texture measures with a specific window size and image band can effectively extract biomass texture information. Results can differ in different study areas. Purely textural information is not sufficient to establish a model for estimating biomass with high accuracy. More detailed research on extraction of biomass texture information is necessary in the future.

Vegetation indices can partially reduce the impacts on reflectance caused by environmental conditions and shadows. Different vegetation indices have been developed and used for classification or biomass estimation (Anderson and Hanson 1992, Anderson et al. 1993, Eastwood et al. 1997). Bannari et al. (1995) reviewed more than forty indices presented through the literature. However, not all vegetation indices are significantly correlated with biomass. In a selected study area, it is not easy to effectively and quickly determine one or more vegetation indices that are appropriate for use in the model development. Lu (2001) analysed and compared twenty-three vegetation indices and identified that linear TM band combinations (e.g., the first components of PCA or Tasseled Cap transform) have stable and good relationships with biomass. The previous analysis indicated that vegetation indices alone are not sufficient to establish an effective model for biomass estimation. This is especially true in a study area with fast vegetation growth rates and complex stand structure, largely because complex stand structure and canopy shadow negatively impacted biomass and spectral signature relationships.

Landsat TM data mainly capture the canopy information instead of individual tree information due to its spatial resolution. Other sensor data such as radar data and hyperspectral data can give new insights about biomass estimation. In a large study area such as the Amazon basin, AVHRR or MODIS data, incorporating high spatial resolution data such as Landsat TM images, can be useful in estimating biomass or carbon emission. Some high-resolution data such as IKONOS and Lidar also has the potential to estimate biomass with higher accuracy, providing a means to validate the results derived from TM images.

Many factors can influence vegetation growth vigor, stand structure, and species composition. Such factors can be terrain aspect and slope, soil types, precipitation, road construction, land-use history, population density, etc. Terrain aspect and slope can affect sun illumination and moisture distribution that directly influence plant photosynthesis and biomass accumulation. Different soil types have different physical structures and nutrient components, directly influencing the vegetation growth. Climate conditions, especially precipitation distribution in the Amazon basin, are important factors that affect vegetation vigour. For example, dry weather associated with sandy soil texture structure can wilt the leaves and reduce water content in the leaves quickly, resulting in high biomass consumption and vegetation senescence. Different human activities, such as road construction, selective logging, and mining, land-use history, and population pressures on forest resources also influence the vegetation area loss, disturbance, or fragmentation. The different biophysical conditions affect vegetation growth status and result in different vegetation reflectance captured by remote-sensing sensors. A model that incorporates remotely sensed data and associated ancillary data has the potential to improve model performance and is more applicable to a large study area. Such a model is best developed through integration of geographic information system (GIS) and remote-sensing techniques, and should be especially valuable if the required ancillary data can be captured.

6 Conclusions

Multiple regression models that consist of spectral and textural signatures are best for successional and mature forest biomass estimation. They provide better biomass estimation for those forest areas where vegetation is sufficiently dense to cover the ground. Incorporation of texture measures and spectral signatures improved estimation model performance. The models developed cannot, however, provide accurate SS1 biomass estimation. The effectiveness of texture measure in models is greatly dependent on the vegetation growth rates and vegetation stand structures. This particular finding that a model comprised of spectral and textural signatures provides better estimation accuracy is valuable in the LBA-Ecology project (Large-Scale Biosphere-Atmosphere Experiment in Amazônia). It is useful for selecting appropriate spectral and textural signatures for developing estimation models in other study areas.

The multiple regression models developed based on integration of Landsat TM images and vegetation inventory data provide a method for biomass estimation using remote-sensing data in areas beyond those defined by sample sites. These models take the advantage of remotely sensed data (digital data format, synoptic view, etc.) to update the biomass distribution image and biomass statistical data in a timely manner. This is especially important in moist tropical areas like the Amazon due to the difficulty in gathering ground-truth data representative of a large area. In a study area with rapid vegetation growth rates, selection of an appropriate texture measure is very important for improving the model performance. In a study area with slow vegetation growth rates, texture measures are relatively less important than spectral signatures. When establishing multiple regression models for biomass estimation, much attention should be paid to the selection of an appropriate texture measure.

Climate conditions, soil types, soil moisture, land-use history, and human activity influence vegetation growth rates, stand structures, species composition, and growth vigour. These factors also influence vegetation reflectance on the TM images. To accurately estimate successional and mature forest above-ground biomass using remotely sensed data, a more advanced model that includes different data sources is important. A combination of GIS data and modelling techniques would likely prove beneficial in this type of research.

Acknowledgements

The authors wish to gratefully acknowledge support from the National Science Foundation (grants 95-21918 and 93-10049), the National Institutes of Health's NICHD (grant 9710386A), and NASA's LBA program through grant N005-334 that made the data collection and analysis for this paper possible. The authors also wish to acknowledge the assistance of many collaborators in Brazil and at Indiana State University and Indiana University. None of these colleagues or funding agencies should be held responsible for the views presented herein. They are the sole responsibility of the authors.

References

Anderson GL, Hanson JD (1992) Evaluating handheld radiometer derived vegetation indices for estimating above ground biomass. Geocarto International 7:71–78

Anderson GL, Hanson JD, Haas RH (1993) Evaluating Landsat Thematic Mapper derived vegetation indices for estimating above-ground biomass on semiarid rangelands. Remote Sensing of Environment 45:165–175

Bannari A, Morin D, Bonn F, Huete AR (1995) A review of vegetation indices. Remote Sensing Reviews 13:95–120

Eastwood JA, Yates MG, Thomson AG, Fuller RM (1997) The reliability of vegetation indices for monitoring saltmarsh vegetation cover. International Journal of Remote Sensing 18:3901–3907

Franklin J, Hiernaux PYH (1991) Estimating foliage and woody biomass in Sahelian and Sudanian woodlands using a remote sensing model. International Journal of Remote Sensing 12:1387–1404

Franklin SE, Peddle DR (1989) Spectral texture for improved class discrimination in complex terrain. International Journal of Remote Sensing 10:1437–1443

Gillespie AJR, Brown S, Lugo AE (1992) Tropical forest biomass estimation from truncated stand tables. Forest Ecology and Management 48:69–87

Gordon DK, Phillipson WR (1986) A texture enhancement procedure for separating orchard from forest in Thematic Mapper imagery. International Journal of Remote Sensing 8:301–304

Haralick RM (1979) Statistical and structural approaches to texture. Proceedings of the IEEE 67(May):786–804

Haralick RM, Shanmugam K, Dinstein I (1973) Textural features for image classification. IEEE Transactions on Systems, Man and Cybernetics SMC-3:610–620

Jakubauskas ME, Price KP (1997) Empirical relationships between structural and spectral factors of Yellowstone lodgepole pine forest. Photogrammetric Engineering and Remote Sensing 63:1375–1381

Leblon B, Granberg H, Ansseau C, Royer A (1993) A semi-empirical model to estimate the biomass production of forest canopies from spectral variables, Part 1: Relationship between spectral variables and light interception efficiency. Remote Sensing Reviews 7:109–125

Lu DS (2001) Estimation of forest stand parameters and application in classification and change detection of forest cover types in the Brazilian Amazon Basin. Ph.D. diss., Indiana State University, Terre Haute, Indiana

Lu DS, Mausel P, Brondizio ES, Moran E ('No Date') Assessment of atmospheric correction methods for Landsat TM data applicable to Amazon basin LBA research. International Journal of Remote Sensing (in press)

Marceau DJ, Howarth PJ, Dubois JM, Gratton DJ (1990) Evaluation of the grey-level co-occurrence matrix method for land-cover classification using SPOT imagery. IEEE Transactions on Geoscience and Remote Sensing 28:513–519

Moran E, Brondizio E, McCracken S ('No Date') Trajectories of land use: soils, succession, and crop choice. In: Wood C et al. (eds) Patterns and Processes of Land Use and Forest Change in the Amazon. Gainesville: University Press of Florida (in press)

Nelson BW, Mesquita R, Pereira JLG, de Souza SGA, Batista GT, Couto, LB (1999) Allometric regression for improved estimate of secondary forest biomass in the central Amazon. Forest Ecology and Management 117:149–167

Nelson R, Krabill W, Tonelli J (1988) Estimating forest biomass and volume using airborne laser data. Remote Sensing of Environment 24:247–267

Nelson RF, Kimes DS, Salas WA, Routhier M (2000) Secondary forest age and tropical forest biomass estimation using Thematic Mapper imagery. Bioscience 50:419–431

Overman JPM, Witte HJL, Saldarriaga JG (1994) Evaluation of regression models for above-ground biomass determination in Amazon rainforest. Journal of Tropical Ecology 10:207–218

Peddle DR, Franklin SE (1991) Image texture processing and data integration for surface pattern discrimination. Photogrammetric Engineering and Remote Sensing 57:413–420

Roy PS, Ravan SA (1996) Biomass estimation using satellite remote sensing data – an investigation on possible approaches for natural forest. Journal of Bioscience 21:535–561

Steininger MK (2000) Satellite estimation of tropical secondary forest above-ground biomass data from Brazil and Bolivia. International Journal of Remote Sensing 21:1139–1157

Tucker JM, Brondizio ES, Moran EF (1998) Rates of forest regrowth in Eastern Amazonia: a comparison of Altamira and Bragantina regions, Para State, Brazil. Interciencia 23:64–73

Grid Computing for Real Time Distributed Collaborative Geoprocessing

Yi Shi[1], Ashton Shortridge[2] and Jon Bartholic[3]

Institute of Water Research, 115 Manly Miles Building, Michigan State University, East Lansing, MI, 48823 USA, Phone: 1-517-432-5585, shiyi1@msu.edu[1]
Department of Geography, 314 Natural Science Building, Michigan State University, East Lansing, MI, 48824 USA[2]
Institute of Water Research, 115 Manly Miles Building, Michigan State University, East Lansing, MI, 48823 USA[3]

Abstract

Grid computing has emerged as an important new field in the distributed computing arena. It focuses on intensive resource sharing, innovative applications, and, in some cases, high-performance orientation. In this paper we describe Grid computing concepts and technologies and establish compelling reasons for why they may prove useful for important geospatial research and application issues. Then we review high performance parallel geoprocessing in general. Finally, we propose an example GIS application scenario that will employ underlying Grid computing technology intensively. We report results on a prototype employing a universal soil loss equation implemented on an array of supercomputer processors.
Keywords: grid computing, parallel processing, virtual organisation, GIS, watershed modelling

1 Introduction

The Internet's exponential growth has greatly increased the significance of distributed computing technologies for information systems generally. This development did not go unnoticed by the geographic information science (GIScience) community, which early focused on identified distributed, Internet-based computing as a key research topic (UCGIS, 1998). Indeed, GIScience has long been engaged with distributed computing. Such engagement is natural:

- Computationally intensive spatial applications often tax the capacity of single machines.

- The GIS user base may itself be spatially distributed, with a need to share data, models, and results.
- Spatial operations and decisions are frequently best made in the field, requiring flexible access to a remote database.

An early representative application is Sequoia 2000, a project that linked global climate change researchers throughout the University of California system via an integrated computer environment (Stonebreaker, 1995). The project involved dozens of earth scientists and computer scientists, as well as industry representatives, working together to create a system capable of managing very large environmental databases and serving data and visualisations rapidly across a dedicated wide area network.

Experience from Sequoia 2000 and similar projects indicated the presence of both opportunities for and impediments to the employment of distributed computing. Significantly, impediments are both technological and non-technological in nature. Technological challenges include:

- integrating data from different sources, stored in different databases
- transmitting very large quantities of information over the network
- developing flexible interfaces for different users

Organisational challenges include:

- combining members of very different research communities - computer science and earth / spatial science
- communicating research objectives and requirements clearly
- orchestrating linked but independent environmental models developed in different laboratories

A variety of GIScience initiatives are intended to improve the utility of distributed computing for geospatial applications (see e.g. OGIS, 2002; Hecht, 2001; UCGIS, 1998). Generally they have focused on technical aspects, although the organisational challenges of large projects have long been acknowledged.

This paper reports on two related developments that may remove many of these impediments and greatly facilitate distributed geoprocessing. The first is the concept of the Virtual Organisation (VO). This powerful model for accomplishing group objectives in a dynamic manner independent of the geographic location of group members has gained currency in business management since the mid-1990's (Palmer & Speier, 1997). The second is the technological realisation of the Grid, a system of powerful processors and massive storage that is seamlessly knit together via the Internet by protocols and low-level programming interfaces. Grid-based technologies enable extensive Internet-based infrastructure for the VO. Section 2 describes the virtual organisation concept and Grid-based technologies that can support it. Section 3 indicates primary functions of large GIS applications and geoprocessing applications that appear to be amenable to the VO paradigm. Section 4 describes a prototypical geospatial virtual organisation and reports on the utility of Grid-based infrastructure to execute a distributed geospatial application upon it. We conclude in Section 5 with a discussion on the prospects for Grid-based geocomputation.

2 Virtual Organisation Paradigm and the Grid

A virtual organisation is a network linking geographically dispersed individuals or agencies with partially overlapping objectives in order to pool complementary core competencies (Jägers et al., 1998). The VO is a framework for accomplishing shared objectives in a flexible and efficient manner. A typology proposed by Shao et al. (1998) identifies four characteristic features of a virtual organisation:

- Connectivity, which establishes linkages where none previously existed.
- Purpose, which creates a common incentive for agents in the virtual organisation.
- Technology, which enables connectivity.
- Boundary, which separates those who are part of the VO from those who are not.

The role of trust is critical, since decisions must be made without the same degree of face-to-face interaction in more traditional settings, and because data, strategic decisions, and objectives must be shared with semi-independent participants in the VO. Because of this greater autonomy, VO participants tend to be more equal, with greater independence than is typical in a hierarchical organisation.

The VO paradigm has been highly useful for certain business activities, particularly those that are technologically oriented and require flexibility and rapid adaptation (Jägers et al., 1998). It has also been adapted in academic settings at large and small scales. At large scales, it is common for multidisciplinary, research teams with members in several academic institutions to collaborate on projects. At small scales, multiple-authored papers like this one are typically developed in concert, with portions of work done by different authors in physically different places. In either case, electronic transmission of documents and data, as well as ideas, keeps the work knit together. The network enables the rapid flow of communication, which is critical for the virtual organisation.

The Grid computing concept is intended to enable co-ordinated resource sharing and problem solving in dynamic, multi-organisational virtual organisations. The basic idea is to provide computation power to everyone who can access it, just like electric power Grid providing electricity. The sharing in this computational Grid refers to not only file exchanges but also direct access to a spectrum of computers, software, data, and other resources, as is required by a range of collaborative problem-solving and resource-brokering strategies emerging in industry, science, and engineering. This sharing is necessarily highly controlled, with resource providers and consumers defining clearly and carefully just what is shared, who is allowed to share, and the conditions under which sharing occur. A set of individual and/or institutions defined by such sharing rules form a Grid computing based virtual organisation (Foster et al., 2001).

The Grid computing based VO represents a new approach to collaborative computing and problem solving in computation- and data-rich environments. For example, the application service providers, storage service providers, computer cycle providers, university researchers and consultants engaged by a government agency might form such a VO to perform real time decision support for environmental management. Common technical concerns and requirements of different VOs include (Foster et al., 2001):

- Highly flexible sharing relationships, ranging from client-server to peer-to-peer and brokered.
- Complex and high levels of control over how shared resources are used, including fine-grained access control, delegation, and application of local and global policies.
- Sharing of varied resources, ranging from programs, files, and data to computers, sensors, and networks.
- Diverse usage modes, ranging from single user to multi-user and from performance sensitive to cost-sensitive and hence embracing issues of quality of service, scheduling, co-allocation, and accounting.

Current distributed computing technologies do not address these concerns and requirements. For example, current Internet technologies address communication and information exchange among computers but not the co-ordinated use of resources at multiple sites for computation. Business-to-business exchanges focus on information sharing (often via centralised servers). Enterprise distributed computing technologies - such as CORBA and Enterprise Java - focus on enabling resource sharing within a single organisation. Emerging "Internet computing" companies seek to harness idle computers on an international scale but, to date, support only highly centralised access to those resources. In summary, current technology either does not accommodate the range of resource types or does not provide the flexibility and control on sharing relationships needed to establish VOs (Foster et al., 2001).

Grid computing technologies were produced over the past five years to meet these concerns and requirements. They include protocols, services, and tools that address precisely the technological needs for scalable VOs. These technologies cover security solutions that support management of credentials and policies when computations span multiple organisations; resource management protocols and services that support secure remote access to computing and data resources and the co-allocation of multiple resources; information query protocols and services that provide configuration and status information about resources, organisations, and services; and data management services that locate and transport data sets between storage systems and applications (Foster & Kesselman, 1997).

We believe that the virtual organisation paradigm is broadly applicable to many large spatial management and modelling projects. This paradigm, coupled with Grid-enabling technology, concerns numerous aspects of distributed computing in such projects and changes the way that computers may be used to solve GIS problems. Perhaps more profoundly, it offers a way to reframe multi-organisational GIS and spatial modelling efforts as virtual organisations, enhancing the collaborative process. The following section considers the distributed computing applications of large GIS applications in detail.

3 Geoprocessing and the Grid

GIS is concerned with three sorts of distributed computing objectives: distribution of data, of software, and of processing. Each of these important subcategories is discussed below.

First, geographic information systems employed by public and private organisations collect, store, and maintain very large amounts of data about diverse themes (Longley et al., 2001). It is not unusual for different offices to be responsible for maintaining and updating individual layers in a regional GIS database. This division of labour has clear institutional advantages, but a significant challenge has always been maintaining the wide availability of the most current version of the entire database (Sugarbaker, 1999). While the sharing and distribution of spatial data has typically occurred only within the organisation, new Internet GIS technologies have enabled the serving of spatial information via the Internet to client organisations and the public (Coleman, 1999). A related issue is the on-line availability of massive quantities of spatial data by agencies whose purpose is distribution. Much United States base data, including digital elevation and land cover, may be downloaded at the time it is needed. Key Internet nodes serve as spatial data repositories, which can process queries and retrieval requests from users (Smith & Rhind, 1999). A key aspect of the Grid-based system for scientific research is the ability of software agents to identify and retrieve relevant data sources for processing. Significantly, the user needs not to be aware of the physical location of the data in the network. Instead, processing occurs at a (possibly remote) server that independently dispatches a search for necessary data (Foster, 2002). As a geoprocessing illustration, consider a hydrologist interested in modelling suspended sediment load in a river system. As input for the primary research, the hydrologist needs to execute an overland flow model for the watershed. The model is executed remotely, and the process itself searches for and obtains relevant elevation and landcover data, freeing the scientist from this potentially onerous search, as well as the download, pre-processing, and upload of large data quantities.

Second, a complex GIS operation may involve many autonomous steps. Traditionally, monolithic software systems have attempted to handle all possible routines. Drawbacks with the monolithic model for the user include overly complicated software design, large system elements that are never used, and inefficient processing due to the general-purpose design. An alternative approach is a distributed component GIS, in which individual spatial processing components are obtained from different sources from the network. These components are linked together to build the desired multi-step GIS procedure. Research on such software distribution falls generally under the rubric of distributed computing and interoperability, which is at least equally, concerned with open data structures. A current trend in GIS software industry reflecting this approach is to build Internet based distributed component GIS. Basic component functions are defined by OpenGIS consortium standards formed by different vendors. Actually, Grid technologies can be used as an infrastructure upon which truly distributed component GIS can be constructed. Grid concepts are designed to facilitate large data analysis tasks, with processing occurring on many different machines, and to encourage the development of scientific software portals, which provide an interface to specialised problem-solving applications on remote servers.

Third, many geoprocessing applications are extremely resource-intensive due to large volumes of data and intensive algorithms. For such applications, parallel processing enables the division of the work onto multiple machines, with potentially large reductions in time to complete the task. This goal has attracted the attention of researchers involved in diverse, high-load geoprocessing activities. Hunsaker et al. (1996) developed a parallel implementation for spatial data error simulation. The task required the production of many equi-probable realisations of a land cover data set using a complex spatial statistical model. Terrain modelling involves operations to generate terrain meshes from scattered points, detecting local surface properties, and calculating view sheds. Executing these operations on multiple processors is especially useful when grids are large or dense. Magillo & Puppo (1998) summarised progress towards implementing parallel terrain modelling. Spatial environmental modelling of snow conditions across the United States involves intensive use of resources. Research by El Haddi et al. (1996) sought to improve model performance by parallelizing the application. Spatial data mining that involves knowledge extraction from vast stores of spatial data could also benefit from parallelization. In all the cases described here, parallelization involves some combination of algorithm development and data model architecture design.

An interesting and relevant technical issue for intensive scientific computation concerns the different rates of technological change for computing power, storage, and network speed, typically measured in doubling time. These rates are roughly 18 months, 12 months, and 9 months, respectively (Foster, 2002). Foster notes that petabyte-magnitude data archives are in the planning stages based on the assumption that this trend will continue. However, processing capacity is not keeping pace. The implication is that local computer power will not be up to the task of processing this volume of data. Network speed, on the other hand, is increasing even more rapidly than storage, as is exemplified by the 40 Gbs/second rate for the Terragrid network (Benner, 2001). As a result, communication-intensive tasks will become relatively more efficient than processor-intensive tasks, encouraging network-distributed solutions for major, processor-intensive tasks.

In general, two separate problems exist for distributing geoprocessing. The first is the problem of parallelizing the relevant algorithms so that they may be employed on multiple machines. The second is concerned with identifying an optimal spatial partition of the database so that the load for each processor is roughly equivalent. Shekhar et al. (1996) and others (e.g. Hoel & Samet (1994)) developed parallel algorithms and spatial declustering techniques. Spatial operations that may be amenable to parallelization possess several qualities. The algorithms may have regularities for parallel processing so they can efficiently use idle cycles on different computers within a VO. Second, the problem may be one that can be spatially partitioned into smaller geographic units and distributed to different processors. Third, although spatial analysis may be intensive due to large data volume rather than complex processing, the results of the analysis may be much smaller than the data volume used for it. For intensive spatial analysis with parallel regularities in the algorithms, the data domain can be divided spatially. Then subsets of data are transmitted along with the (sub)algorithm to different machines. Once these smaller, more tractable subtasks are done, the outcomes are collected, reassembled spatially, and presented or stored.

A substantial challenge for the parallelization of geoprocessing applications has been implementing the low-level processing necessary for enabling multiple processors to work in concert. The Grid facilitates parallel processing by handling much of this. Toolkits have been developed that supply protocols for data handling and communications, fault detection, and cross-platform portability (Foster, 2002). This frees the programmer to concentrate on higher-level issues, such as specific parallelization implementations, improved algorithm design, or speedier heuristics.

The problems and applications that GISs address appear well suited to take advantage of the virtual organisation concept and the underlying Grid computing technology that supports it. Because the implementation is so recent, however, there have been very few publicised efforts to develop Grid-based geoprocessing applications (though some are underway (NPACI, 2002)). The next section describes our testbed for implementing a representative application, and reports on some preliminary results.

4 A Watershed Management Testbed for Grid Computing

In this section, we propose a watershed management scenario that will use underlying Grid computing technology intensively. Watershed management is a holistic activity that involves co-ordination among different organisations and collective decision-making based upon information from different sources. The ultimate goal is to optimise the use of land, water and vegetation in an area to alleviate drought, moderate floods, prevent soil erosion, and improve water quality, land use and agricultural production on a sustained basis. There is also an urgent need to educate the general public on watershed problems and to develop and implement agendas to meet these problems.

Due to its organisational characteristics and modelling goals, watershed management is an excellent theme upon which to develop a VO/Grid computing testbed. This is a problem-oriented framework that contains data and methods for facilitating decision-making in a particular geographic region of a watershed.

Fig. 1 presents a hypothetical example indicating how the Internet can be used, via Grid technologies, to develop a watershed management virtual organisation. In this illustration, a university research institute develops several computational hydrological models for a government agency on a contract basis. It sets up a GIS-based model server to provide remote modelling services to that agency. These

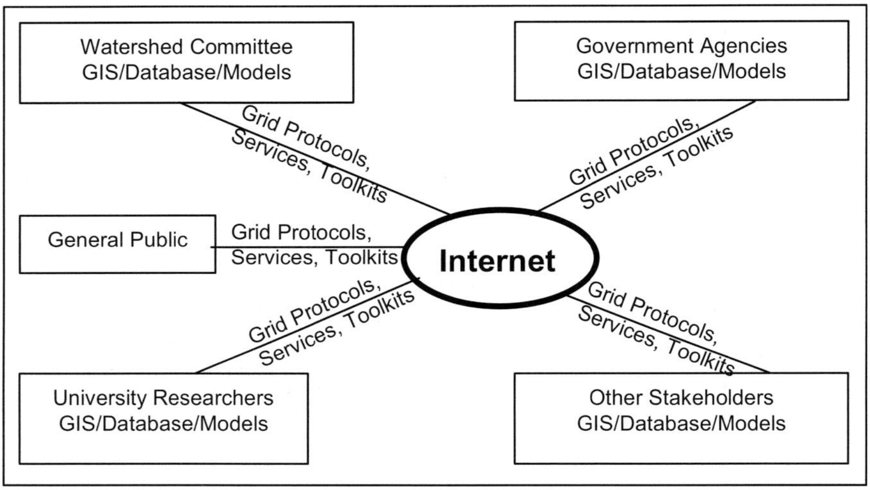

models access real time data collected from field devices maintained by many actors in the VO. In addition, for particular computation- and data-rich models, the

Fig. 1. General Internet-based integration framework for watershed management. Grid protocols handle the flow of data, processing, reports, and other information through the system

model server delegates the processing tasks to one or more of the heavy-duty supercomputers or a cluster of PCs within the VO to meet the real time requirements for the model. Each agent may selectively publish its data and geoprocessing services for the general public to access. Also a watershed management committee can set up a system to integrate data and geoprocessing services from different stakeholders and implement its own real time decision support system. The committee can also selectively publish its data, reports and geoprocessing services to the general public. All of these systems are linked through the Internet so that they can access each other's data and geoprocessing services based on certain protocols defined by Grid technologies.

This research concerns just a few aspects of the full watershed VO just described. In particular, we are interested in testing the data distribution and parallel processing capability of Grid-based technologies. We intend to test the capability of the system to distribute, process, and present results for a basic hydrologic model running on a large geographic region. Since one goal of a watershed VO might be to process results rapidly enough to enable quick scenario checking and short time scale decision making, real time geoprocessing is of particular concern. To test the potential of Grid computing technologies to meet this real time geoprocessing requirement, we developed a parallelized soil loss model using the Message Passing Interface (MPI) standard library that underlies the Grid. The model was run on both a CrayT3E supercomputer located in the University of Texas and local PC clusters we built at the Michigan State University using Grid packages. The architecture of the system is illustrated in Fig. 2.

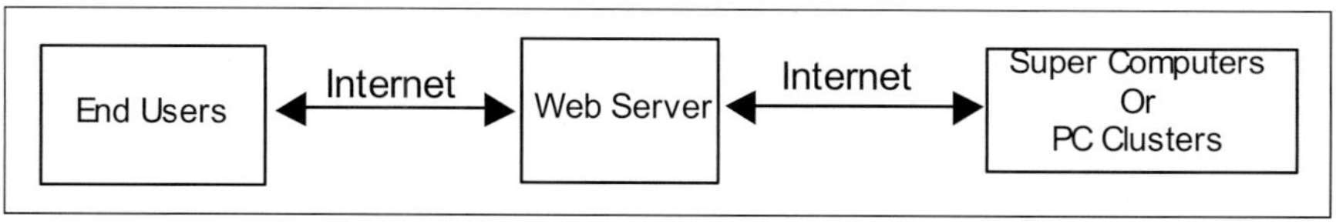

Fig. 2. Prototype system architecture for parallel processing application

The soil loss model has the following form (from Mitasova & Mitas, 1999):

$$E = R * K * LS * C * P \tag{1}$$

where E is the average soil loss, R is the rainfall intensity factor, K is the soil factor, LS is the topographic factor, C is the cover factor, and P is the prevention practices factor. Factors for a 1000x1000 cell raster grid were developed for the calculation of E. This model was run on a single processor on the Cray T3E; processing required 6.525791 seconds. The parallel model was then executed on 10 processors, which required 0.827482 seconds. Performance gain is significant, although this improvement must be weighed against network transfer time, as the following paragraphs indicate.

This spatial application employs purely local map algebra operations upon one or more input raster layers; that is, the resulting value in cell [i,j] in the output raster is solely a function of values at [i,j] in the input raster(s). In general, for applications involving local operations, processing time may be calculated for a single processor:

$$T_s = nRows \times nCols \times T_c \tag{2}$$

where T_s is the time to complete the task on a single processor, *nRows* and *nCols* are the dimensions of the raster, and T_c is the time to perform the local function on a single cell. If the operation is parallelized, processing speed is:

$$T_p = 2 \times nLayers \times \frac{(nRows \times nCols) \times bytes_c}{NetworkSpeed(bytes/s)} + \frac{nRows}{nProcs} \times nCols \times T_c \tag{3}$$

where *nLayers* is the number of input raster layers, $bytes_c$ is the number of bytes per cell value that need to be transmitted, and *nProcs* is the number of distributed processors handling spatial subsets of the operation. Processor speed is held constant for both single and parallel implementations, which is highly conservative (typically, one would take advantage of very powerful distributed processors). Setting Eqs. (2) and (3) equal to one another in order to find the break even between single and parallel processing times (T_s=T_p) and simplifying the result in the following expression:

$$T_c = \frac{2 \times nLayers \times bytes_c}{NetworkSpeed(bytes/s)} + \frac{T_c}{nProcs}. \tag{4}$$

The right-hand side of Eq. (4) consists of two elements: the added cost of parallelization, which is the time required to transfer the data, and the benefit, which is the reduction in processing time. As *nProcs* increases, the processing time drops. For a given application then, one can identify the number of processors required to offset the network speed bottleneck.

5 Discussion and Conclusions

This paper has described Grid computing as an important recent development in the general area of distributed computing. This is due to its ability to link multiple agencies with a network of shared data, software, and processors. The innovation is the degree of integration that a VO employing such technologies can achieve.

We identify three major driving factors for the implementation of Grid-based geoprocessing. As network speeds continue to increase relative to single processing power, parallel implementations become relatively efficient. Second spatial partitioning for local operators is simple and very general. Partitioning for neighbourhood operations ranges from simple (slope) to complex (flow accumulation). Complex partitioning problems restrict the utility of parallel implementations. Finally, algorithm complexity is a factor favouring parallelization, since users can take advantage of more powerful processors than are available on the local network.

The trend towards intensive multi-agency geospatial projects involving the collaboration of dozens of scientists, policy-makers, and members of the public makes the virtual organisation paradigm attractive. This research identifies the utility of real-time distributed model processing for such organisations. We maintain that Grid technologies show great promise for enhancing a variety of geospatial applications, particularly those with intensive computing requirements and a multi-organisational structure.

References

Benner J (2001) A Grid of supercomputers [online]. Wired News, August 9, 2001, Available from: http://www.wired.com/news/technology/0,1282,45977,00.html (Last checked 3/10/02)

Coleman DJ (1999) Geographical information systems in networked environments. In: Longley PA, Goodchild MF, Maguire DJ, and Rhind DW (eds) Geographic Information Systems, New York, Wiley, pp 317-329

El Haddi AA, Shekhar S, Hills R, McManamon A (1996) Mirror-image round robin spatial data partitioning: a case study with parallel SEUS. In: Third International Conference/Workshop on Integrating GIS and Environmental Modeling. January 21-25, 1996, Santa Fe, New Mexico, USA

Foster I (2002) The Grid: a new infrastructure for 21st Century Science. Physics Today February 2002, 42-47

Foster I, Kesselman C, Tuecke S (2001) The anatomy of the Grid: enabling scalable virtual organisation. Intl. J. High Performance Computing Applications 15(3): 200-222

Foster I, Kesselman C (1997) Globus: A Metacomputing Infrastructure Toolkit. Intl. J. Supercomputer Applications 11(2): 115-128

Hecht L (2001) OGC aims to enable information communities [online]. GEO World, August 2001. Available from: http://www.geoplace.com/gw/2001/0801/0801opng.asp (Last checked 3/10/02)

Hunsaker C, Ehlschlaeger CR, Davis FW, Goodchild MF (1996) Estimating spatial uncertainty as a function of scale: implications for landscape ecology. In: Third International Conference/Workshop on Integrating GIS and Environmental Modeling. January 21-25, 1996, Santa Fe, New Mexico, USA

Hoel EG, Samet H (1994) Performance of data-parallel spatial operations. In: Proceedings 20th Intl. Conf. On Very Large Data Bases. September 1994, Santiago, Chile, pp 156-167

Jägers H, Jansen W, Steenbakkers W. (1998) Characteristics of Virtual Organizations [online]. In: Organizational Virtualness: Proceedings of the VoNet Workshop, April 27-28, 1998, Bern, Switzerland pp 65-76. Available from: http://www.virtual-organization.net/files/articles/nl2-1.pdf, last checked 3/10/02.

Longley PA, Goodchild MF, Maguire DJ, Rhind, DW (2001) Geographic Information Systems and Science, New York: Wiley

Magillo P, Puppo P (1998) Algorithms for parallel terrain modeling and visualization. In: Healey R, Dowers S, Gittings B, and Mineter M (eds) Parallel processing algorithms for GIS, London: Taylor & Francis, pp 351-386

Mitasova H, Mitas L (1999) Modeling soil detachment with RUSLE 3d using GIS [online]. Available from: http://www2.gis.uiuc.edu:2280/modviz/erosion/usle.html. (Last checked 3/15/02)

NPACI (2002) SDSC Receives Six ITR Awards. EnVision 17(4):20

OGIS (2002) OpenGIS Consortium Inc. Homepage [online]. Available from: http://www.opengis.org (Last checked 3/10/02)

Palmer JW, Speier C (1997) A typology of virtual organizations: an empirical study [online]. In: Proceedings of the Association for Information Systems 1997 Americas conference. August 15-17, 1997, Indianapolis, USA Available from: http://hsb.baylor.edu/ramsower/ais.ac.97/papers/palm_spe.htm (Last checked 3/10/02)

Shao YP, Liao SY, Wang HQ (1998) A model of virtual organizations. Journal of Information Science 24(5):305-312

Shekhar S, Ravada S, Kumar V, Chubb D, Turner G (1996) Parallelizing a GIS on a Shared-Address Space Architecture. IEEE Computer 29(1):42-48

Smith NS, Rhind D W (1999) Characteristics and sources of framework data. In: Longley PA, Goodchild MF, Maguire DJ, Rhind DW (eds) Geographic Information Systems. New York: Wiley, 655-666

Stonebreaker M (1995) An overview of the Sequoia 2000 Project [online]. Digital Technical Journal 7(3) 39-48. Available from: http://www.research.compaq.com/wrl/ DECarchives/DTJ/DTJJ03/DTJJ03PF.PDF (Last checked 3/10/02)

Sugarbaker LJ (1999) Managing an operational GIS, In Longley PA, Goodchild MF, Maguire DJ, and Rhind DW (eds) Geographic Information Systems, New York: Wiley, pp 611-620

UCGIS (1998) Distributed Computing, Revised UCGIS White Papers [online]. Available from: http://www.ucgis.org/research_white/distrib.html (Last checked 3/10/02)

The Balance Between Geometry and Topology

Peter van Oosterom, Jantien Stoter, Wilko Quak and Sisi Zlatanova

Department of Geodesy, Delft University of Technology, Thijsseweg 11, 2629 JA Delft, The Netherlands, {oosterom|stoter|quak|zlatanova}@geo.tudelft.nl

Abstract

The architecture of Geographic Information Systems (GISs) is changing: more and more systems are based on the integrated architecture, i.e. storing geometric data in the Data Base Management System (DBMS) together with administrative data. The first step in building a Geo-DBMS is by having data types and operators for simple features (i.e. geometric primitives): point, line and polygon. This has reached a level of standardisation and is now implemented in several commercial DBMSs. The next step is to have support for the topologically structured features in the DBMS, i.e. complex features. The DBMS can then check and guarantee consistency. In addition, complex operations can be executed within the DBMS. Despite the fact that topologically structured models are well known and that it is not difficult to store the topological references, it still remains an unresolved issue as to how to effectively implement these models within a relational DBMS. In this paper, we describe the design and implementation of a topologically structured management at the DBMS level. Our focus is to translate topological structures into geometric primitives. It is then possible to define a DBMS view on a topological primitive, which makes this appear as a geometric primitive. This process supports the best of both worlds: on the one hand there are advantages of the topological structure (no redundancy) and on the other hand the ease of explicit geometric primitives in querying, analysis and presentation is available.
Keywords: Topology, DBMS, OpenGIS, spatial modelling, SQL queries

1 Introduction

The integrated architecture (Vijlbrief & van Oosterom1992) of storing geometric data together with administrative data in the Geo-DBMS can be contrasted to traditional GIS approaches such as the dual architecture (separate DBMSs for geometric and administrative data) and the layered architecture (all data stored in a single DBMS with spatial knowledge contained in a layer between the application and the DBMS), for example ESRI's SDE. In the integrated architecture, the DBMS is extended with spatial data types (point, polyline and polygon), and functions (overlap, distance, area and length). The first DBMSs offering these capabilities were experimental systems, such as Postgres (Stonebraker et al.1990), O2, Gral (Güting1989), and others (DeWitt et al.1990) and, of course, the functionality was not standardised in, for example, SQL (Date & Darwen1997, ISO1992). Immediately after the availability of first spatial DBMSs, the first GISs based on these DBMSs became available. These were either based on an extended (object) relational database (GEO++) or on a pure object oriented database (GeO2).

The importance of the integrated architecture was recognised by industry and the OpenGIS Consortium (Buehler & McKee1998) standardised the basic spatial types and functions, or in the OpenGIS terminology the Simple Feature Specification (SFS). The SQL/SFS implementation specification (Open GIS Consortium, Inc., 1998) will also be part of the future ISO SQL3 standard (ISO/IEC 13249-31999). In 1999 the first implementations of the OpenGIS SQL/SFS became available, which marked an important step forward in the maturing of GIS to become part of the mainstream ICT. Currently several commercial DBMSs are available with support for spatial data types (some support the OpenGIS standard): Ingres (1994), *Oracle 9i Spatial (*2001 and Hebert & Murray1999), Informix (2000) or IBM DB2 (2000). As an illustration and proof that geo-information is now part of standard ICT product, an example is shown below. In Informix the, OpenGIS compliant, SQL statement to select objects from the table `lki_boundary` where the `shape` attribute overlaps a given polygon (with two holes) provides the example:

```
select * from lki_boundary
where ST_Intersects (shape, ST_PolyFromText('polygon \
 ((103654574 460970880,104323607 460885924,104769627 460885924, \
   105523616 461013359,105544856 461395663,105061624 461741343, \
   104089976 461777967,103474041 461639912,103474041 461162032, \
   103654574 460970880), \
  (104610334 461108935,104610334 461459380,105056356 461459380, \
   105056356 461108935,104610334 461108935), \
  (103792627 461119555,103792627 461470000,104206790 461470000, \
   104206790 461119555,103792627 461119555))',28992));
```

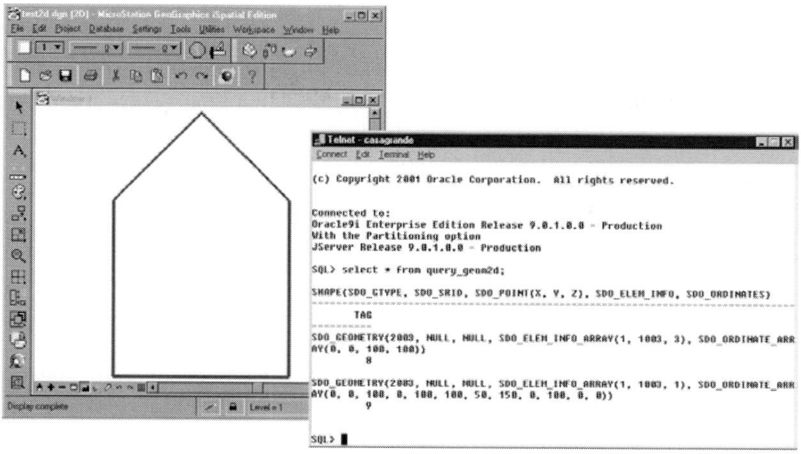

Fig. 1. Polygon stored in Oracle Spatial visualised with Bentley MicroStation

In addition, increasing numbers of commercial GIS packages support the integrated architecture: *MapInfo Professional v6.5 (*2001), *Intergraph GeoMedia Professional 4.0 (*2001) or *Bentley MicroStation GeoGraphics ISpatial edition,J 7.2.x (*2001). However, the above referenced DBMSs do not natively support topology and none of the mentioned GIS packages use topology from a DBMS. Fig. 1 shows a polygon in Bentley MicroStation, which was read from the Oracle Spatial DBMS.

The OpenGIS Consortium mentions topology several times in its abstract specifications (Open GIS Consortium, Inc.1999). What is still missing is the implementation specification for a DBMS environment. The OpenGIS abstract specification is very similar, and more or less at the same abstraction level as the work of the ISO TC 211 (ISO/DIS 19107, 2000). These two standards are now relatively well harmonized and both make a distinction between *geometric primitives* (point, curve, surface and solid) and *topological primitives* (node, edge, face and solid). The latter only have meaning within a topological structure. Note the 3D geometric and topological primitives are both called solid. We therefor propose to use the term *volume* for the 3D topological primitive instead.

Section 2 introduces the data model (options) and motivation behind topological structures. Section 3 argues why topology management should be a task of the DBMS. The design of a DBMS topology management extension is described in Section 4. A partial implementation of this design based on Oracle 9i is given in Section 5. This implementation is based on an additional topological meta data table together with PL/SQL functionality using this meta data. Special attention is paid to the aspect of spatial indexing (and clustering), which is very important when dealing with large, real world, data sets. Querying either the geometric or the topological 'side' of the system is illustrated in Section 6. Finally, conclusions, discussions and future research are described in Section 7.

2 Topological Structures

Having spatial types and operators is one part of the DBMS services required by a GIS. Two other components are 1. spatial indexing (quadtree, r-tree (Guttman1984, Samet1989)) and spatial clustering, and 2. representing and manipulating topological structures. Topological structures are used, among others, to represent planar partitions without redundancy. Another possible use of a topological structure is for the representation of a linear network. In this paper, we will focus on the topological structure for a planar partition. Topological structures for planar partitions have been well known for a long time, e.g. in TIGER (Boudriault1987, Kinnear1987), DIME (US Bureau of the Census1970), the Arc/Info system (Morehouse1989), the Netherlands Cadastre LKI (van Oosterom1997), and many other systems. They have also been studied extensively e.g. by Molenaar (Molenaar1998, de Hoop et al.1993). Relational DBMSs can effectively store the topological references: area left and right of a boundary, boundary to boundary references, treatment of islands, among other aspects., i.e. the modelling via topology. However, they do not support the topology structure management as they are not able to achieve consistency checks (area closed?, no intersecting boundaries?) and they do not provide the necessary operators, such as:

- the basic edit operations (split or merge area/insert or delete edge)
- computing the perimeter and area of topologically represented area
- solving the question: which areas are crossed by this query?
- map-overlay of two (or more) planar partitions
- selection of neighbouring features
- route planning in a linear network represented by a complex feature.

The problem with a standard relational DBMS, is that the declarative language SQL can not handle the 'navigational access' needed for the functionality described above. In SQL it is not possible to express the statement: 'Follow the *next* references of the boundary until we are back at the beginning'. Of course, this is a very simple operation in a programming language using one of the basic iterator concepts. This programming functionality is also available in every object-oriented DBMSs for implementing methods associated with classes.

Though some operations in a relational DBMS are impossible on a topologically structured area feature (e.g. compute area), this structure has many advantages:

- it avoids redundant storage (more compact than a full-polygon model);
- it is easier to maintain consistency of the data after editing;
- it is more efficient during the visualisation in some kind of front-end, because less data has to be read from disk;
- it is the natural data model for certain applications; e.g. during surveying an edge is collected (together with attributes belonging to a boundary); and
- it is efficient for certain query operations (e.g. find neighbours).

In this paper, the cadastral map will be used as a case study, because topology plays a key role in this spatial data set consisting of parcels covering the whole country of the Netherlands. The geometric data model for the cadastral *parcel* layer is based on winged-edge topology (Baumgart1975) as described in (van Oosterom1997, Lemmen & van Oosterom1995, Oosterom2001).

3 Does Topological Structure Management Belong in a DBMS?

In an implementation of topological structure management, not all functionality has to be provided by the DBMS. It is possible to provide part of the functionality in a front-end (or middleware) application. This enables the implementation to be based on standard tools without modifying the relational DBMS (server). However, as the support for complex features is quite generic, it should optimally be in the DBMS. This avoids reimplementation of the same functionality in several applications and it is the best guarantee for consistency control. Further, it also allows analysis queries on topologically structured features to be executed within the DBMS. Thus, no unnecessary data transfer to a front-end application takes place. Currently, the OODBMSs do seem to offer the most flexible platform for implementing the complex features. OODBMS offer the possibility of navigating through the database, controlled by programming the implementation methods. Their relatively weak acceptance by the market, the lack of a standard query language and the fuzzy separation between the DBMS and the application (as more and more user programming code ends up in the DBMS) form the motivation to include support for complex features in (extended/object) relational DBMSs.

Since the subject of implementing complex features in a relational DBMS is a fairly unexplored field, various issues have to be taken into account:

- The possibility of implementing a fully operational solution, that covers the total domain of all possible topological structures should be assessed. Since we will first focus on a subset of the topological structures - the planar partitions - the question will also be how far this solution will solve the total domain, and how this subset can be used as a basis to extend to other complex data types.
- The chosen solution of implementing the topology management in the database will have to be tested in an effort to establish the real benefits above other scenarios. Further, whether this functionality is best implemented in the front-end or in a middleware application, or that it be distributed over the DBMS will need to be explored
- A solution must have the possibility of being extended. Users must have the freedom to add on their own applications or variants of topology management structures, without creating conflicts with the database.
- Having an abstract standard is one important step, but the implementation forms the next required step in practice. The OpenGIS Consortium mentions

topology several times in its abstract specifications (Open GIS Consortium, Inc.1999). What is still missing is the implementation specification of topology (or sometimes also called complex features) for specific platforms. The question remains whether this could be possible for standard SQL.

- The rule of thumb whether certain functionality belongs to the DBMS or to a specific application is whenever it concerns general and reusable aspects then this belongs to the DBMS. In this section it was argued that topological management is a generic functionality and therefore belongs to the DBMS.

4 Design of DBMS Topological Structure Management

Subsection 4.1 describes an earlier attempt to extend (post)quel (a language similar to SQL, used in the Postgres DBMS) with 'prototypes' and the 'create layer' statements for the management of topology structures. The drawback of this approach is that it requires an extension of the DBMS query language, which has a serious impact on the DBMS kernel implementation. In practice (and in theory), there are many different types of topology structures used and described. Subsection 4.2 tries to categorise these different topological structures. In Subsection 4.3, this paper suggests a different approach than the 'prototype' approach for topology management having less impact on the DBMS kernel. Most relational DBMSs are currently enriched with object-oriented technology. This, among others, allows the DBMS to be extended with types and functions, which exist and run within the DBMS.

4.1 Query Language Extension Based on Prototypes

For the modelling part of topology and one of the most important operators, i.e. map-overlay, an attempt to extend (post)quel (a language similar to SQL, used in the Postgres DBMS) was described (Schenkelaars & van Oosterom1995). It was suggested to use 'prototypes' to define topological roles and the 'create layer' statement to define (declare) the roles of the different tables based on the prototypes. A translation was given from the postquel to the SQL syntax (Oosterom et al.2000),:

```
create prototype face(id=oid, bnd=edge.id[]);
create prototype edge(id=oid, line=polyline2, left=face.id,
  right=face.id);
```

The prototypes are used to model the different roles that entities can play in a topological structure. It also standardises names of the attributes with specific meaning; e.g. the object identifiers, the topological references and the base geometry. As will be described in the next subsection, there are several ways to implement topology, the method described in this subsection is based on edges and faces (no explicit nodes). There are references from a face to all its boundaries (exterior and possibly also interior) and there are references from a boundary to the left and right face. This type of topology is sometimes called, wheel topology (NNI NEN-36101995). Other topology implementations (with other restrictions) are possible (see also Subsection 4.2), but can be based on different prototypes. Based on the abstract prototypes given above, it is now possible to create specific tables that inherit the topological structure from the prototypes:

```
create table parcel(owner=text) inherit face;
create table boundary(quality=integer) inherit edge;
create layer (layer_id='cad_map', topology_type='wheel_topol',
              face_role='parcel', edge_role='boundary');
```

The last statement `create layer` defines which tables are used to implement a certain topological structure, the parameter `topology_type` specifies the type of topology. In a way the `create layer` statement could be compared to the `create index` statement, because it has important (under water) side effects later on during the updating.

It is assumed that the updates are correct and consistent with the topological rules. Otherwise a transaction can not be committed. This is checked by the DBMS as a result of the `create layer` statement. After this first layer `cad_map` is created, a second layer `soils` could be created in a similar manner. Having created two layers, it is now possible to perform the complex map-overlay operation within the DBMS. The DBMS must be extended to support this overlay operation, which returns the number of faces as a result and as a side effect is able the compute the new (topological) layer as overlay of the input layers. Again, note that the support of topology is more or less the same level in the DBMS as the support of indices or referential integrity constraints.

4.2 Topological Structures Inventory

Without claiming to be complete, this subsection tries to make an inventory of the most common topological structures. It has already been stated several times that there are many different ways to implement topology. In the previous subsection the wheel topology was discussed. Other topology implementations (with other restrictions) are possible and can be implemented in the DBMS. For example, the Dutch Cadastre uses a planar partition for the parcel layer based on edges and faces, but with different references, i.e. the chain topology (also called the winged-edge topology). The different topological structures can be characterised by the following 'parameters':

- what is the dimension of the embedding space: 2D, 2.5D, 3D, time added?
- which topological elements (primitives) are used: node, edge, face, volume?
- are the elements considered directed (oriented) or not?
- which explicit topological relationships (part_of, in, on) are stored?
- what are the topological 'rules': crossing edges allowed?, dangling elements allowed?, same topological primitive on both sides of boundary allowed?, etc.

Table 1. Inventory of different topological structures

	Topol. type	Dimension	Primitives used	Topological tables	Explicit Relationships	All tables	Rules
TIN	221	2D	Node,edge	node,edge	no	2	Planar Partition
Wing-edge	222	2D	Edge,face	edge,face	no	2	Planar Partition
Wheel (chain)	223	2D	Edge,face	edge,face	no	2	Planar Partition
3DFDS	381	3D	node,arc, edge,face	arc,edge, face	node-on-face node-in-volume arc-partof-line arc-on-face arc-in-volume	8	Space Partition
TEN	352	3D	node,arc, triangle, tetrahedron	arc,triangle, tetrahedron	tri-partof-surf arc-partof-line	5	Space Partition
Cell-tuple	313	3D	0-cell,1-cell 2-cell,3-cell	cells	no	1	Space Partition
SSS	364	3D	Node,face	face,line surface,volume	node-in-volume face-in-volume	6	Space Partition

Table 1 illustrates some 'parameters' of different topological structures.

The topological type is obtained by encoding the dimension of the data structure, the total number of topological tables and the topological 'rules'. As can be seen from in Table 1 many types of topology are possible and even more are possible; e.g. structures not for partitions such as a linear network. First of all, the dimension may be 2D or 3D. Second, not all primitives have to be used. For example, in the 2D winged-edges structure, neither the node nor the volume 'roles' are used, but only the edge and face 'roles'. In the TIN, only the node and the edge 'roles' are used. In SSS (Zlatanova & Tempfli2000) the roles that are not explicit are the edge and volume 'roles'. 3DFDS (Molenaar1998) does not use the volume 'role', while TEN (Pilouk1996) exploits the 'role' of all the primitives. It depends on the type of topology as to which roles are expected. Note, that the node table in the presented 3D topological structures contains only the coordinates of the points, i.e. links to other topological primitives are not provided. Indeed, 3D topological structures that assign topological functions to the node are also available (de la Losa & Cervelle1999).

4.3 Topology Management using Meta Information

A lighter approach than extending the DBMS kernel with topological functionality involves creating a table with meta information describing the topological structures in the DBMS. This topological meta information can than be used both within the DBMS (back-end) and outside the DBMS (front-end or middleware). In general, meta information (or system catalogs) of a DBMS contains descriptions of the data stored in the database: tables, attributes and types, and contains descriptions of the available types and operators. This enables dynamic SQL applications. A traditional example of the use of meta-data is to obtain a description of the tables within Oracle such as:

```
SELECT owner,table_name,column_id,column_name,data_type
FROM all_tab_columns
ORDER BY owner,table_name,column_id
```

Oracle spatial also uses a specific meta data table to describe geometric attributes in more detail: `USER_SDO_GEOM_METADATA`, which contains information about the number of dimensions, the extent of the domain and the resolution. In case the (relational) DBMS has to support topology management, the structural knowledge has to be stored (and be accessible for applications). For example, topological layer names, boundary tables, area tables and relevant attributes with metrical and topological information must be determined and accessible. One solution for this problem is providing prototypes as a basis for the possible topological structures; see Subsection 4.1. A drawback of this solution is that the topology elements (object ids, references and also the metric attributes) have fixed names.

An alternative is to describe this topological information in another meta data table. Again, somewhere it must be declared which tables and which attributes carry the topological information. An example of the extension of the meta information of the DBMS was given in the context of GEO++ (`geo_dyn_info table`) in (Vijlbrief & van Oosterom1992). Below the table definition of a more generic topological meta data table is shown. This model we now propose as the basis of our solution:

```
                                    | /* example entry for winged-edge
                                    |    topology of Dutch cadastre */
create table topo_meta_data(        | insert into topo_meta_data values
 topol_layer_name  varchar(32)      |   (
                     PRIMARY KEY,   |    'parcel layer',
 topol_type        integer,         |    222,
 node_table_name   varchar(32),     |    NULL,
 node_info         varchar(256),    |    NULL,
 edge_table_name   varchar(32),     |    'lki_boundary',
 edge_info         varchar(256),    |    'object_id,l_obj_id,r_obj_id,
                                    |     fr_line_id,lr_line_id,
                                    |     geo_polyline',
 face_table_name   varchar(32),     |    'lki_parcel',
 face_info         varchar(256),    |    'object_id,line_id1',
 volume_table_name varchar(32),     |    NULL,
 volume_info       varchar(256));   |    NULL);
```

The topology type is encoded in a number and stored in the meta data attribute '`topol_type`'. Knowing which tables to use for which role is not sufficient. It must be explicit about which attributes in these tables contain the needed identifier, references and other information. This depends again on the topology type. Standardising the topologically structured knowledge makes it possible to implement functionality (within the DBMS). Several categories of functionality can be considered. Initially we focus on the 'bridge' between topology and geometry by implementing a function which realises (materialises) geometric primitives from topological primitives.

5 Implementation in Oracle 9i

The Oracle 9i spatial DBMS will be used to test the topological structure management proposal. The manner in which the relational model is extended (PL/SQL, stored procedures) to support topological structure is described. The interface of a hard coded (table and attribute names are fixed) example of a PL/SQL function which translates a topological structure into a geometry is given below:

```
 CREATE OR REPLACE FUNCTION return_polygon
   (i_object_id kadtest.lki_parcel.object_id%type)
RETURN mdsys.sdo_geometry
IS
...../* body of the function */
  polygon:=mdsys.sdo_geometry(2003, NULL, NULL,
          mdsys.SDO_ELEM_INFO_ARRAY(1, 1003,1), list_coordinates);
  return polygon;
END return_polygon;
```

This function uses the 'lki_boundary' table in order to obtain the coordinates of the polygon realised for the face in the 'lki_parcel' table. In our current prototype implementation, the boundaries are accessed based on their 'object_id', on which a btree index is created. For fast access, spatial clustering of the boundary table is needed. The clustering supports fast access in case a large number of boundary records, forming one polygon are to be retrieved in a sequence and are spatially related A spatial index is not needed, per se, for this purpose. However, sometimes boundaries are spatially selected independent of the polygons and for this reason also a spatial index, an rteee, is created on the boundary table. Now, the function 'return_polygon' can be used to create a view in the following way:

```
create view lki_parcel_pgn_vw as
  select municip, osection, parcel, oarea,
    return_polygon(object_id) shape
   from lki_parcel;
```

Fig. 2 shows an example that uses the function to view topological data in a viewer that knows that the data it displays is in a topological model (Wilko Quak's Quick GIS). Currently, table names and attribute names are hard coded in the function 'return_polygon'. One might wonder what is the role of the 'from' clause in this situation, besides being necessary in a valid SQL 'select' statement. The answer is that the 'from' clause provides the iteration over the 'object_id's' from the 'lki_parcel' table.

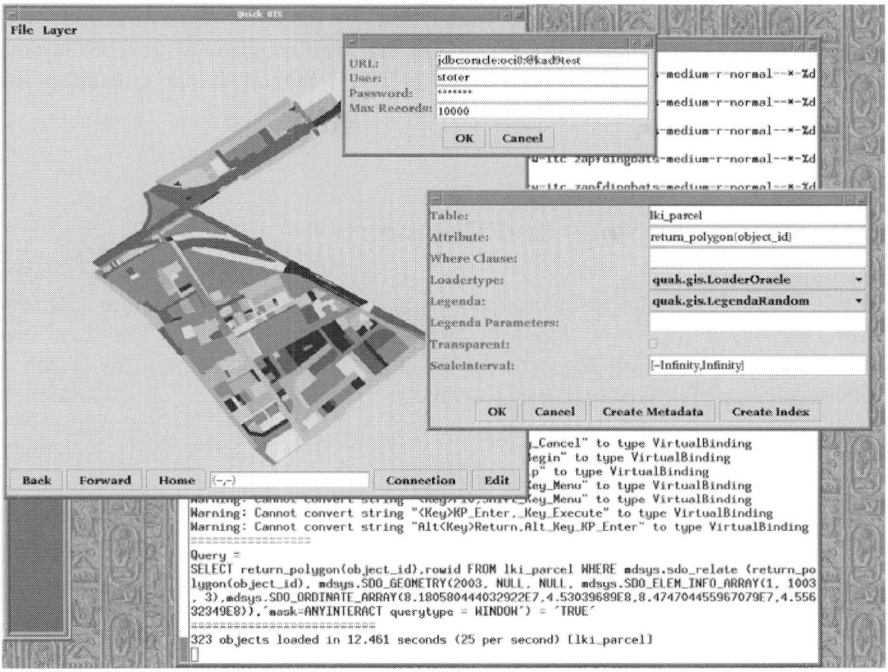

Fig. 2. Viewing topological data in a veiwer that is not topology aware

A function-based spatial index is created in order to optimise the performance. Since version 9i, Oracle offers function-based indices, i.e. an index, which is created on the return value of a function in contrast to a normal index created directly on the value of an attribute. The spatial index is created based on the pre-computed values returned by the function. This is implemented in Oracle 9i in two steps. First, by inserting a row in the table `USER_SDO_GEOM_METADATA` and then the spatial index is created by specifying the function name and parameters.

```
insert into user_sdo_geom_metadata values(
  'LKI_PARCEL', 'return_polygon(object_id)', mdsys.sdo_dim_array (
  mdsys.sdo_dim_element('X', 82291,  84261, 0.0005),
  mdsys.sdo_dim_element('Y', 453039, 455632, 0.0005)), NULL);

create index lki_parcel_idx on
  lki_parcel(return_polygon(object_id))
  indextype is mdsys.spatial_index;
```

The view that was created (`lki_parcel_pgn_vw`) can use this index and thus improve performance and functionality. Without a function-based spatial index it would not have been possible to properly index the faces. During an overlap query (or any other search query using the spatial index, for some examples see the next section), objects are filtered by means of this index. That is, using the pre-computed boxes which are stored in the R-tree. Then the '`return_polygon`' function is executed to obtain the complete geometry of filtered objects used in the exact overlap test. In Informix (Informix2000) it is possible to build an R-tree index which also includes the shape instead of only the box (as is the case in the standard R-tree), which then avoids executing the function `return_polygon`.

6 Querying Topology and Geometry

In this section examples of queries are given, which use the function to realise a geometry based on stored topology as described in section 5. The first two *geometric* queries return respectively the area and the maximum area (with a tolerance value of 0.5) of realised geometry's.

```
select municip, osection, parcel, sdo_geom.sdo_area(shape,0.5)
  from lki_parcel_pgn_vw;
select max(sdo_geom.sdo_area(shape,0.5))
  from lki_parcel_pgn_vw;
```

The next *geometric* query gives all the parcels, which are within a distance of 100m of the given rectangle (specified by lower-left and upper-right coordinates).

```
select parcel from lki_parcel
  where sdo_within_distance(
    return_polygon(object_id),mdsys.sdo_geometry(2003,NULL,NULL,
      mdsys.sdo_elem_info_array(1,1003,3),
      mdsys.SDO_ORDINATE_ARRAY(832018,453536,832824,453610)),
   'distance = 100')='TRUE';
```

Next a *topological* query selects all neighbour parcels of parcel 'X'.

```
select distinct l_obj_id neighbour_parcel
  from lki_boundary where r_obj_id='X'
union all
select distinct r_obj_id neighbour_parcel
   from lki_boundary where l_obj_id='X';
```

Another pure topological query is to find all boundaries of parcel 'X'.

```
  select distinct object_id parcel_boundary
    from lki_boundary where r_obj_id='X' or l_obj_id='X';
```

Finally, an example of a *complex* query: map-overlay, which computes the intersection of all faces from one layer with all faces from the other layer. Below, one input layer is topologically structured '`lki_parcel`' and the other layer is not topologically structured '`query_geom`'. The result is stored in a non-topologically structured table '`overlay_result`'.

```
create table overlay_result as
select object_id, tag,
  sdo_geom.sdo_intersection(return_polygon(object_id),q.shape,0.5)
    clip_geom
  from lki_parcel, query_geom q
  where mdsys.sdo_relate (return_polygon(object_id),
    q.shape,'mask=ANYINTERACT querytype = JOIN') = 'TRUE';
```

In the target list the actual geometric computation is performed in addition to selecting the identifiers from both input layers. If both input layers are partitions, the output layer will also form a partition again. To execute the map-overlay efficiently, the where clause has been added and tests for overlap (through a spatial index because a spatial operator is used).

The next step is to get the topological layer as a result of the map-overlay operation. A solution may be to make a special purpose map-overlay function (in PL/SQL), which does all the computation and populates the target topology tables.

7 Discussion and Conclusion

In this paper we presented a breakthrough in managing topology and geometry in RDBMSs based on a function, which realises (constructs) a topological primitive into its geometric counter part. The proposed solution is based on a generic topology meta data table and stored procedure (PL/SQL), which uses this meta data to realise the geometry. This makes it possible to have both the advantages of a topological structure (e.g. no redundancy) and advantages of explicit geometric primitives in querying, analysis and presentation. Any GIS package being able to display geometry from the DBMS (called `sdo_geometry` in Oracle) can now also analyse, query and display topologically structured features from the DBMS. An important issue is the connection between the features and their topology and geometry 'aspects'. Our generic approach makes it possible to integrate topology and geometry as attributes of the features or to separate those in two modelling levels. In the second approach, which is also described in (Molenaar1998), topology and geometry aspects are modelled (in tables such as node, edge and face) separately from the features, which are modelled on top of (refer to) the topology tables. This approach also fits in our generic approach and it is up to the application developer to decide on the best model for a given situation. It is our expectation that future ISO and/or OpenGIS standards with respect to topology will also fit in our generic approach.

Future work will focus on keeping the topological structure consistent under updates. This issue is related to the temporal aspect of geo-information as updating is the source of historical information (van Oosterom1997). In the same manner as there should not be gaps and overlaps in space (in a partition), there should also be no gaps and overlaps in time: a next version should start at the moment the previous version becomes outdated.

One can assume that the topological structure is correct and that updates keep the structure correct and consistent within the topological rules. Otherwise a transaction can not be committed. This must be checked by the DBMS. Two approaches can be identified. In the first solution, a group of insert/delete/updates to node/edges/face/volume tables is considered as one transaction and on the commit of this transaction a function is executed, which checks the correctness of the changes. In the second solution it is not allowed to change the node/edge/face/volume tables directly, but only through higher level edit operations, which bring the DBMS from one consistent state into the other (by changing several tables at the same time). Further research will investigate and compare these different solutions.

Other future work will try to implement the more generic solution, including support for other types of topological structures. This solution should also be applicable for other RDBMSs. The later could perhaps be achieved by not using Oracle's PL/SQL in the server, but the more generic Java/SQL. This might also be beneficial for the performance, due to the more advanced techniques of a JVM, such as just in time compilation. An interesting issue related to other topology types, raised in (de Hoop et al.1993), is whether it is possible to share the geometry in differently structured topological layers. In our model, this is not visible, although an implementation might support this. We did not investigate this, and this is again a good item for future research.

Finally, complex operations such as map-overlay of two (or more) planar partitions, resulting in a new planar partition will be investigated. Of course, as usual everything performed within the DBMS.

Acknowledgments

We would like to thank the participants in our research center, GDMC (Geo Database Management Center) for their cooperation, for making (beta) versions of their software available and for sharing their thoughts on the R&D developments in the Geo-DBMS domain. In this research project, we used Oracle 9i as our main DBMS. We owe our thanks to Tom Vijlbrief (Dutch Cadastre), Siva Ravada (Oracle), and the anonymous SDH reviewers for their valuable remarks.

References

ASK-OpenIngres (1994) INGRES/object management extension user's guide. Release 6.5, Technical report, ASK-OpenIngres

Baumgart BG (1975) A polyhedron representation for computer vision. National Computer Conference, pp 589-596

Bentley MicroStation GeoGraphics ISpatial edition (J 7.2.x) (2001)

Boudriault G (1987) Topology in the TIGER file. Auto-Carto 8:258-269

Buehler K, McKee L (1998). The OpenGIS guide - introduction to interoperable geoprocessing. Technical Report Third edition. The OpenGIS Consortium, Inc.

Date CJ, Darwen H (1997) A Guide to the SQL standard. 4th edition, Addison Wesley

DeWitt DJ, Futtersack P, Maier D, Velez F (1990) A study of three alternative workstation-server architectures for object oriented database systems. Technical Report 907, Computer Sciences Department, University of Wisconsin-Madison

Güting, RH (1989) Gral: An extensible relational database system for geometric applications. In: Proceedings of the Fifteenth International Conference on Very Large Data Bases, Amsterdam, ACM, New York, pp 33-44

Guttman A (1984) R-trees: A dynamic index structure for spatial searching. ACM SIGMOD 13: 47-57

Hebert J, Murray C (1999) Oracle Spatial User's Guide and Reference. Oracle Corporation, Redwood City, CA, USA. Part No. A77132- 01

de Hoop S, van Oosterom P & Molenaar M (1993). Topological querying of multiple map layers. In: COSIT'93, Elba Island, Italy, Springer-Verlag, Berlin, pp 139-157

IBM (2000) IBM DB2 Spatial Extender User's Guide and Reference. special web release edition

Informix (2000) Informix Spatial DataBlade Module User's Guide. Part no. 000-6868

Intergraph GeoMedia Professional 4.0 (2001) [online]. Available from: http://www.intergraph.com/gis/ gmpro/

ISO (1992) Database language SQL. Technical Report Document ISO/IEC 9075, International Organization for Standardization

ISO/DIS 19107 (2000) Geographic information - Spatial schema, Technical Report final text of CD 19107 (15046-7). International Organization for Standardization, TC211/WG2

ISO/IEC 13249-3 (1999) Information technology - database languages - SQL multimedia and application packages - part 3: Spatial. ISO/IEC JTC 1 SC 32, Technical Report Part 3, International Organization for Standardization

Kinnear C (1987) The TIGER structure. In: Auto-Carto 8. pp 249-257

Lemmen C, van Oosterom P (1995) Efficient and automatic production of periodic updates of cadastral maps. In: JEC'95, Joint European Conference and Exhibition on Geographical Information, The Hague, The Netherlands, pp 137-142

de la Losa A, Cervelle B (1999) 3D topological modelling and visualisation for 3D GIS. Computer & Graphics 23

MapInfo Professional v6.5 (2001) [online]. Available from: http://dynamo.mapinfo.com/products/index.cfm

Molenaar M (1998) An Introduction to the theory of spatial object Modelling for GIS. Taylor and Francis

Morehouse S (1989) The architecture of Arc/Info. In: Auto-Carto 9. pp 266-277

NNI NEN-3610 (1995) Terreinmodel vastgoed. Termen. Technical report, Nederlands Normalisatie-instituut (In Dutch, English version also available)

van Oosterom P (1997) Maintaining consistent topology including historical data in a large spatial database. In: Auto-Carto 13. pp 327-336

van Oosterom P, Lemmen C (2001) Spatial data management on a very large cadastral database. Computers, Environment and Urban Systems 25(4-5):509-528

van Oosterom P, Verbree E, Kap A (2000) Storing and manipulating simple and complex features in database management systems. In: Proceedings of the 3rd AGILE Conference on Geographic Information Science, Helsinki/Espoo, Finland, pp 178-182

Open GIS Consortium, Inc. (1998) OpenGIS simple features specification for SQL. Technical Report Revision 1.0, OGC

Open GIS Consortium, Inc. (1999) The OpenGIS abstract specification, topic 1: Feature geometry. Technical Report Version 4 (99-101.doc), OGC

Oracle 9i Spatial (2001) [online]. Available from: http://otn.oracle.com/products/spatial/

Pilouk M (1996) Integrated Modelling for 3D GIS. Ph.D. thesis, ITC Enschede, Netherlands

Samet H (1989) The Design and Analysis of Spatial Data Structures. Addison-Wesley, Reading, Mass.

Schenkelaars V, van Oosterom P (1995) Map-overlay within a geographic interaction language. In: Auto-Carto 12, Charlotte, NC, pp 281-290

Stonebraker M, Rowe LA, Hirohama M (1990) The implementation of Postgres. IEEE Transactions on Knowledge and Data Engineering 2(1): 125-142

US Bureau of the Census (1970) The DIME geocoding system. Technical Report 4, Census Use Study, US Department of Commerce, Bureau of the Census, Washington, DC

Vijlbrief T, van Oosterom P (1992) The GEO++ system: An extensible GIS. In: Proceedings of the 5th International Symposium on Spatial Data Handling. Charleston, South Carolina, pp 40-50

Zlatanova S, Tempfli K (2000) Modelling for 3D GIS: spatial analysis and visualisation trough the web. Proceedings of the XIX congress of ISPRS Com. VI/2. 16-23 July, Amsterdam, Vol. XXXIII, B4/3, Comm. IV, pp 1257-1264

STOQL: An ODMG-Based Spatio-Temporal Object Model and Query Language

Bo Huang[1] and Christophe Claramunt[2]

Department of Civil Engineering, National University of Singapore, 10 Kent Ridge Crescent, Singapore 119260, Tel: (65) 874-2158 Fax: (65)779-1635, cvehb@nus.edu.sg[1]
Naval Academy Research Institute, Lanveoc-Poulmic, BP 600, 29240 Brest Naval, France, Tel: 33 2 98 234206 Fax: 33 2 98 233857, claramunt@ecole-navale.fr[2]

Abstract

Spatio-temporal databases are a topic of growing research interest but there is still little work reported on spatio-temporal query languages in line with the Object Database Management Group (ODMG) standard. This paper presents the design of such a language, called STOQL, by adding spatial and temporal dimensions to OQL. This language is based upon a data model that extends the ODMG Object Model with basic spatial and temporal types as well as a parameterised type, **Temporal<T>**, which lifts spatial types into spatio-temporal types to support the representation of the history of an object. We show that STOQL can support the formulation of various spatio-temporal queries in relation to historical states of spatial objects as well as spatial changes including spatial type substitution.
Key words: spatio-temporal database, ODMG, object model, OQL, STOQL

1 Introduction

Spatial and temporal databases have long been studied on parallel, but relatively separate tracks. As natural and man-made objects are often associated with both space and time, there has been a growing interest in the integration of spatial and temporal components (Langran 1992, Worboys 1994, Claramunt and Thériault 1995, Peuquet and Duan 1995), leading to the emerging research field of spatio-temporal databases (Güting *et al.* 2000, Spery *et al.* 2001).

Compared with considerable research on temporal relational databases (e.g. Tansel *et al.* 1993, Snodgrass 1995a), a smaller amount of work has been reported in the context of temporal object databases (Snodgrass 1995b). This has similarly occurred in spatio-temporal object databases. One of the important reasons has been the lack of an agreement for a common object model and query language (Huang *et al.* 2000). However, a consensus is now emerging, with the Object Data Management Group (ODMG) effort on standardising Object-Oriented Database (OODB) and a recent release of ODMG 3.0 (Cattell and Barry 2000). The proposed ODMG standard encompasses an object model (ODMG object model), an object definition language (ODL) and an object query language (OQL). This paper addresses the spatio-temporal data model and an associated query language. We propose the modelling of spatial and temporal data types using a novel parameterised temporal type **Temporal<T>** that lifts spatial types into spatio-temporal types to support the representation of the history of an object. The potential of this data type is illustrated by the application of spatio-temporal queries on a simplified case study.

Object-oriented databases (OODBs) provide powerful data abstractions and modelling facilities. The challenge is, however, to augment an object data model to capture the history of objects and to augment an object query language to permit queries over this history. The history could be represented at the object or attribute level (Fegaras and Elmasri 1998). As the attributes of an object may change asynchronously, attribute level time-stamping is preferred in this study. To represent the history of an attribute, we extend the ODMG Object Model with a parameterised type (a type generator just like **List<T>**), named **Temporal <T>**, which lifts a type **T** into a temporal type that contains the history of all states of **T**. The parameter type **T** can be any spatial type such as **Point**, **LineString**, **Polygon**, **Points**, **LineStrings** and **Polygons**, and so spatial types can be promoted to spatio-temporal types. Corresponding operations within the **Temporal** class are defined for traversing the history of objects. Thus, the implementation of temporal types is actually the implementation of the parameterised **Temporal** type, which, in turn, can meet the requirement of spatio-temporal data modelling and queries.

In this paper, we discuss the design of a spatio-temporal object query language, called STOQL, by extending OQL with some syntactic constructs to uniformly manipulate historical information of both spatial and non-spatial objects. The underlying model of STOQL establishes a parameterised type to support the definition of spatio-temporal types, facilitating the representation of the history of spatial objects and spatial changes including spatial type substitution. The reminder of this paper is organised as follows. Section 2 briefly summarises the principles of the underlying ODMG model and introduces the spatial, temporal and parameterised types. Section 3 describes how these data types can be manipulated at the query language level and illustrates the potential of our model using a simplified case study. Section 4 presents some related work. Finally Section 5 draws some conclusions.

2 Data Model

ODMG provides neither spatial types nor temporal types except Temporal Structured_literals (e.g., **Date**, **Time** and **Timestamp**). The objective of this paper is to extend the ODMG Object Model with spatial, temporal and parameterised types and to illustrate the interest of these extensions at the query language level. The ODMG standard provides an object definition language, ODL, for defining the object types that conform to the ODMG Object Model. STODL extends ODL with a number of class definitions and syntactical constructs that accommodate spatio-temporal information. STODL is hereafter used in class definitions.

2.1 Basic Spatial and Temporal Data Types

Spatial data types and operations have been defined in (OGC 1999). We primarily use the following geometric types: **Geometry**, **Point**, **LineString**, **Polygon**, **GeometryCollection**, **Points**, **LineStrings** and **Polygons**. **Geometry** is the root class of all these types, and is an abstract (non-instantiable) class. It includes operations that access properties of objects (e.g. **envelope**), determine topological relationship (e.g. **overlaps**, **meets**, and **disjoint**), and support spatial queries (e.g. **distance**, **buffer**, **intersection**, and **union**) (Figure 1). There are also other methods defined in the instantiable classes like **length** for the data type **LineString** and **area** for the data type **Polygon**; see (OGC 1999) for more details.

```
Class Geometry {
  //dimension 0-Point, Points
  //1-LineString, LineStrings
  //2-Polygon, Polygons
  attribute Unsigned Short dimension;

  //accessing spatial properties
  Struct MBR (Double minx; Double miny;
  Double maxx; Double maxy;);
  MBR envelope(in Geometry g);

  //test topological relationship
  Boolean disjoint (in Geometry g);
  Boolean intersects (in Geometry g);
  Boolean crosses (in Geometry g);
  Boolean contains (in Geometry g);
  Boolean overlaps (in Geometry g);
  ...
  //create new spatial objects
  ...
  Geometry intersection (in Geometry g);
  Geometry difference (in Geometry g);
}
```

```
Class TimeInterval {
  Attribute Timestamp start;
  Attribute Timestamp end;

  Unsigned Long duration();

  TimeInterval intersection (in
  TimeInterval tv);
  TimeInterval union (in TimeInterval tv);

  Boolean equals (in TimeInterval tv);
  Boolean contains (in TimeInterval tv);
  Boolean meets (in TimeInterval tv);
  Boolean overlaps (in TimeInterval tv);
  Boolean before (in TimeInterval tv);
  Boolean starts (in TimeInterval tv);
  Boolean finishes (in TimeInterval tv);
}
```

Fig. 1. The Geometry and the TimeInterval classes

In the context of temporal databases, two time dimensions, i.e., valid-time and transaction-time, are commonly represented (Jensen *et al.* 1992). Currently STODL only supports valid time although it can be easily extended to transaction time (valid-time is the time of a real-world event while a transaction-time gives the time of a database event). Based on the period definition in TSQL2 (Snodgrass 1995a) rather than the time interval definition in Allen (1983), the **TimeInterval** class is defined using STODL in Fig. 1.

The class **TimeInterval** has two attributes **start** and **end**, both of ODMG type Timestamp. Let **TI** be a variable of type **TimeInterval**. **TI** is constructed as **[TI.start, TI.end]**, which represents all the time points between and including **TI.start** and **TI.end**. If **TI.start** equals **TI.end**, **TI** contains just one time point, simply represented as **[TI.start]**. **[TI.start]** is treated as an equivalent of a time interval, i.e. **[TI.start, TI.start]**. This class defines a number of elementary temporal operations for comparing two time intervals such as **equals**, **contains** and **before**, as well as those for the intersection and union of two time intervals.

For a spatio-temporal data model, the critical types should include the integration of spatial and temporal types into spatio-temporal types. These are tackled by a parameterised type, or a type generator as described below.

2.2 The Temporal<T> Type

The ODMG Object Model includes type generators, collection types, and collection instances.

The collections supported by the ODMG Object Model include **Set<T>**, **Bag<T>**, **List<T>**, **Array<T>** and **Dictionary<T, V>**. Each of these is a type generator, parameterised by the type **T**. All the elements of a collection object are of the same type **T**. Note that collection type generators are represented as Template classes in the C^{++} binding of the ODMG standard.

In the class definition of these types (see Cattell and Barry, 2000 for more details), the ODL type **Object** has been chosen to represent type parameters. As a time-varying object can contain a list of historical states, and the value of this object can be of any type, a parameterised type (type generator), **Temporal<T>**, is needed to promote any type to a temporal type so that the history of this object can

```
Struct State {Object val; TimeInterval vt;};

interface TemporalFactory: ListFactory{
   Temporal new_of_size(in long size);
}
```

```
class Temporal: List {
    attribute temporal<t> value;

   //get entire history of the attribute
   List<State> getHistory();

   get state index number
   Unsigned Long getStateIndex(in Object val);
            raises (no_such_Tvalue)
}
```

Fig. 2. The Temporal<T> type

be represented and manipulated. The **Temporal** type, an extension of the **List** type, is specified and illustrated in Fig. 2.

In the ODMG Object Model, objects are created by invoking creation operations on **factory** interfaces provided on factory objects supplied to the programmer by the language binding implementation. The **new_of_size** operation in the **TemporalFactory** interface creates a collection with the given amount of initial storage allocated, where the given size is the number of elements for which storage is to be reserved.

A **Temporal** object is a temporally ordered collection of **val-vt** pairs:

$$\{(val_1, vt_1), (val_2, vt_2), \ldots, (val_n, vt_n)\}$$

where $val_1, \ldots, val_n$ are legal values of type **T**, $vt_1, \ldots vt_n$ are time intervals such that $vt_i \cap vt_j = \emptyset$, $i \neq j$ and $1 \leq i, j \leq n$, and each pair, i.e., (val_i, vt_i), is an instance of the **State** structure (Fig. 2). An attribute of the **Temporal <T>** type is called a *historical attribute*.

The parameter type **T** can be any ODMG type, and so a ODMG type is lifted into a temporal type. **T** can also be any spatial type, which lifts a spatial type (e.g., **Polygon**) into a spatio-temporal type (e.g., **Temporal <Polygon>**).

All the operations defined in the **List** are valid for **Temporal** types when a **State** is specified as the argument. In addition, the **Temporal** class defines two additional operations. The **getHistory** operation retrieves the entire history of a historical attribute, and the **getStateIndex** operation retrieves an index number associated with a state.

After accessing the entire history of a historical attribute, the value (**.val**), validtime (**.vt**) and index number of each state can be obtained. Operations on the value of a particular type, e.g., **Geometry**, can then be carried out according to application purposes.

2.3 Type Substitution

A spatial change implies a change of an object's spatial attribute value during a given time interval. Spatial objects may change their location, orientation, shape, and size over time, and so their topological relationships with other objects may also change.

Attributes, particularly spatial attributes, can change not only their values, but also their types within a supertype. For example, a fire projected on the 2D plane may be modelled as a point at time t_1, but as a polygon at time t_2, and finally as a polygons at time t_3. To tackle such a problem, an alternative is to use the supertype of relevant types, e.g. **Geometry** in this case. This benefits from the rich modelling capabilities of object-orientation. When instantiating the air pollution class, different geometric types such as **Point**, **Polygon** and **Polygons** are used such that the shape attribute of the air pollution is represented by

$$\{(\text{pointObj}, vt_1), (\text{polygonObj}, vt_2), \ldots, (\text{polygonsObj}, vt_n)\}$$

in which, pointObj, polygonObj, …, polygonsObj are instances of the **Geometry** type, and vt_1, vt_2, …, vt_n are time intervals of type **TimeInterval**.

In general, the key to handling type substitution lies in the definition of operations in a supertype that should be able to deal with different subtypes. In the above case, the operations defined in the **Geometry** type can be applied to these instances.

3 Query Language

Although there is no consensus for the taxonomy of temporal queries, some common requirements have been identified (Snodgrass 1995a) such as temporal selection, join and aggregation, as well as valid-time projection. Besides these, as spatio-temporal databases also target spatial changes over time, the spatio-temporal join (Güting *et al.* 2000) is incorporated in STOQL as well.

3.1 Syntactical Constructs in STOQL

In order to illustrate the expressiveness of the STOQL language, we introduce an example database that has been used in previous works to evaluate some geographical data models and query languages (Claramunt 1999, Spery *et al.* 2001). This example describes a land use database, relationships between some parcels and buildings, protected areas (whose areas have been extended over time) and a fire. The STODL specification of these classes is defined in Fig. 3.

In the following classes, the geometry of either a parcel, a building or a protected area may change over time. However, type substitution does not occur to these three classes. It might occur to the class **fire** under the **Geometry** type. Type substitution refers to the change of geometric type, in the context of our example a fire is first located by a point that approximates its location and then by a polygon that varies in time and eventually disappears.

class **parcel** (extent parcels key identifier)	class **building** (extent buildings key identifier)

```
{
 attribute String identifier;
 attribute Temporal<String> landuse;
 attribute set Temporal<String>
                                    owners;
 attribute Temporal<Polygon> geo;
}
```

```
{
 attribute String identifier;
 attribute Temporal<String> building-
                                       use;
 attribute Temporal<Polygon>  geo;
}
```

```
Class protected-area
 (extent protected-areas
  key name)
{
 attribute String name;
 attribute String level;
 attribute Temporal <Polygon> geo;
}
```

```
class fire
 (extent fires
  key name)
{
 attribute String name;
 attribute Temporal<Geometry>
                                         geo;
}
```

Fig. 3. Example database

STOQL extends OQL facilities to retrieve spatial temporal information. The states in histories are extracted through iteration in the OQL *from-clause*. Constraints in the *where-clause* can then be applied to the value, timestamp and index number of a state through corresponding operations that have been defined in the above model part. Finally the result is obtained through the projection operation in the *select-clause*. Clearly if the value is of a spatial type (i.e. **Point**, **LineString**, **Polygon**, **Points**, **LineStrings** and **Polygons**), then spatial operations will be employed.

Given the above principle, STOQL provides some syntactical extensions to OQL to traverse a history that is of type **Temporal<T>**. In Table 1, **time**$_1$ and **time**$_2$ are expressions of type **Timestamp**, **e** is an expression of type **Temporal<T>**, and **es** is an expression denoting a state within a history. These constructs facilitate the access to each state in a history.

Table 1. Syntactical constructs in STOQL

STOQL	OQL	Type
[time$_1$, time$_2$]	Struct(start: time$_1$, end: time$_2$)	TimeInterval
e!	e.getHistory()	List
es.val	es.val	T (any ODMG type and basic spatial types)
es.vt	es.vt	TimeInterval
es.index	e.getStateIndex(ev) (es in e)	Unsigned Long

Consider an example of class **parcel**. Let **parcel**$_1$ be an object of the class **parcel**. Let **parcelgeo_3** be the fourth state in **parcel**$_1$**.geo**'s history. **parcel**$_1$**.geo!** refers to the entire history of **parcel**$_1$**.geo**, which is, in fact, the whole list of **parcel**$_1$**.geo**'s states. **parcelgeo_3.index** retrieves the index number of the state **parcelgeo_3**, i.e., 4 in this case. **parcelgeo_3.val** retrieves the geometric value of **parcelgeo_3**, and **parcelgeo_3.vt** the valid-time associated with **parcelgeo_3.val**.

The default value of $\mathbf{parcel_1.geo}$ is the value of $\mathbf{parcel_1.geo}$ at the current time (denoted as **[now]**), if no time value is given.

3.2 Query examples

The following examples illustrate different types of spatio-temporal queries such as temporal selection, index-based information retrieval, temporal and spatio-temporal joins, valid-time projection and type substitution. These queries are applied to self-explanatory classes. We assume that the granularity for all time-varying attributes in classes **parcel**, **building** and **protected-area** is the year, while the granularity for the geometrical attribute of the class **fire** is the day.

Query 1 (Temporal selection). Display graphically all the parcels of land use 'agricultural' that existed over the year 1980.

```
Select p-geo.val
From   parcels As parcel, parcel.geo! As p-geo, parcel.landuse! As p-landuse
Where p-landuse.vt.contains([1980]) and
       p-geo.vt.contains([1980]) and
       p-landuse.val = 'agricultural'
```

In the above query, **parcel.geo!** returns the entire geometric history of **parcel**, and **parcel.landuse!** the entire landuse type history. Variables **p-geo** and **p-landuse** are states ranging over these two histories, respectively. The value of state **p-geo** can be viewed as a pair of a **Polygon** and **TimeInterval**. Similarly, the value of state **p-landuse** can be regarded as a pair of a **String** and **TimeInterval**. Time constraints on **p-geo** and **p-landuse** are respectively specified in the **Where** clause.

The next example illustrates how to retrieve the ith state of an object, and how to synchronize events in STOQL (also called **temporal join**).

Query 2 (Index-based information retrieval and temporal join). Display graphically the first land use type of the parcel identified 'L1' from 1980 to 1990, and at that time where was the protected area of the river 'River1'.

```
Select p-landuse.val, p-areageo.val
From   parcels As parcel, parcel.landuse! As p-landuse,
       protected-areas As p-area, p-area.geo! As p-areageo
Where parcel.identifier = 'L1' and
        p-landuse.index <=
           All (Select ap-landuse.index
                From parcel.landuse! ap-landuse
                Where ap-landuse.vt.overlaps([1980, 1990])) and
        p-areageo.vt.overlaps(p-landuse.vt) and
        p-area.name = 'River1'
```

The first land use type of the parcel identified as 'L1' during a period may not simply be represented by '**p-landuse.index = 0**'. Instead it is the minimal index number during that period. Thus a subquery in the Where clause is employed. **p-**

areageo.vt.overlaps(p-landuse.vt) explicitly synchronizes events in relation to **p-landuse** and **p-areageo**. The valid-time of **p-areageo** is not necessarily [1980, 1990], but possibly a subset of this period, e.g. [1980, 1982].

As shown in the next example, spatial changes reflected in spatial properties and topological relationships of objects can be expressed by using two variables of the same class extent as well as different operations. This example also involves qualification over both space and time, and a **join** between different classes of objects.

Query 3 (Spatio-temporal join). Who were the owners of the parcels, which intersected the protected area of the river 'River1' over the year 1990, while they were away from that protected area over the year 1980.

```
Select  parcel.owners
From    parcels As parcel, parcel.geo! As parcelgeo1 parcelgeo2,
        protected-areas As p-area, p-area.geo! As p-areageo1 p-areageo2
Where   p-area.name = 'River1' and
        p-areageo1.vt.contains([1980]) and parcelgeo1.vt.contains([1980]) and
        p-areageo1.val.disjoint(parcelgeo1.val)  and
        p-areageo2.vt.contains([1990]) and parcelgeo2.vt.contains([1990]) and
      p-areageo2.val.intersects(parcelgeo2.val)
```

As each state of an historical attribute can be accessed, spatial operations in the **Geometry** class such as **disjoint** and **intersects**, and temporal operations such as **contains** and **intersects** are employed naturally. In this sense, spatial and temporal operations are harmonised in the language.

Let us introduce a fourth query example that retrieves the time associated with a spatial change:

Query 4 (Validtime projection). When did the protected area of the river 'River1' first touched the parcel identified 'L1' ?

```
element
      (Select  (p-areageo.vt.intersection(p-geo.vt)).start
       From    protected-areas As protected-area,
               protected-area.geo! As p-areageo,
               parcels As parcel, parcel.geo! As p-geo
       Where   protected-area.name = 'River1' and parcel.name = 'L1' and
               p-areageo.val.touches(p-geo.val)
      ).requires[0]
```

As both the protected area and the parcel may spatially change, the **intersection** of the two relevant time intervals is used. The use of **element…requires[0]** obtains the first time that satisfies the conditions.

Temporal aggregation in STOQL is achieved using the standard OQL aggregators. The following query also uses the method **duration** to retrieve duration of a time interval before an aggregation:

Query 5 (Duration). For how long was the parcel identified 'L1' separated from the protected area of the river 'River1'?

```
Sum
(Select  p-geo.vt.duration()
 From    parcels As parcel, parcel.geo! As p-geo,
             protected-areas As p-area, p-area.geo! As p-areageo
 Where  parcel.identifier = 'L1' and p-area.name = 'River1' and
             p-geo.val.disjoint(p-areageo.val)
)
```

STOQL can also deal with queries involving spatial type substitution. A type casting might be used if necessary.

Query 6 (Type substitution). What is the total area of buildings that were affected by the area of the fire named 'Fire1' on May 25, 1992?

```
Sum
(Select  (Polygon)(f-geo.val).intersection(b-geo.val).area()
 From    buildings As building, building.geo! As b-geo,
         fires As fire, fire.geo! As f-geo
 Where   f-area.name = 'Fire1' and
         f-geo.vt.overlaps([25/5/1992]) and
         b-geo.vt.start = f-geo.vt.start.year() and
         b-geo.vt.end  = f-geo.vt.end.year() and
         b-geo.val.overlaps(f-geo.val)
)
```

This query returns the intersection area of the buildings with the fire identified 'Fire1' for a given time interval. The **b-geo.val.overlaps(f-geo.val)** expression in the **Where** clause evaluates if two objects overlap. If this relationship holds, the spatial **intersection** operation is performed, otherwise not. The **year()** operation assigns the valid-time of **f-geo** with the granularity of one year.

The use of **Polygon** casting before **f-geo.val** indicates that the compile-time type checker is told via the downcast that the intersection result must be of the **Polygon** type and the query is accepted as type correct. Note that at runtime, each occurrence of the intersection geometry in the select clause will be checked for its type.

4 Related Work

Previous work relevant to this paper is primarily associated with the **Temporal<T>** type and spatio-temporal object query languages. The notion of the **Temporal<T>** type was introduced in (Bertino *et al.* 1997), but there was no ODL-like definition for this type or a query language. Fegaras and Elmasri (1998) specified the definition of a similar type, which was, however, taken as a parametric class that is not supported by the current version of the ODMG standard (see also Alagic, 2001). Furthermore, since we apply this type in a different query language, operations defined in this type are distinguished from theirs.

Güting *et al.* (2001) explored the representation and querying of moving objects by defining a set of Abstract Data Types (ADT), and ensuring consistency and closure of data types. They defined temporal types such as **mreal**, **mpoint**, **mpoints** and **mregion**. These types obtained through the moving type constructor are functions, or sets of pairs (instant, value). While these types are similar to type **Temporal <T>**, the former are instant based, and each temporal type like **mreal**, **mpoint**, **mpoints** and **mregion** requires a class definition, or code rewriting. In addition, this data model does not touch the representation and handling of spatial type substitution.

STOQL is motivated by the temporal object language proposed by Fegaras and Elmasri (1998), but differs greatly from the latter in the expression of time and temporal projections. In STOQL, such expressions are specified as time constraints in the **Where** clause, similar to the treatment of other constraints in OQL. STOQL also employs fewer syntactical constructs to make the language easier to understand especially for a novice user. More importantly, STOQL has attempted to express spatio-temporal queries and temporal queries in a uniform way. Spatio-temporal queries involve spatial properties, as well as spatial changes.

Cheng and Gadia (1993) have also proposed an object-oriented spatio-temporal structured query language (OOSTSQL), an extension of SQL, which supports both spatial and temporal data. OOSTSQL employs a proprietary clause, **Restricted to**, and temporal expressions to handle different constraints, and it focuses on temporal aspects rather than spatial aspects. Thus it does not address a rich set of spatial data types together with spatial operations.

Let us also mention the recent release of SQL-99 (formerly known as SQL 3). While OQL is an attempt to bring the best of SQL into the object-oriented world, SQL-99 can be characterised as bringing the best of object-orientation into the relational world (cf. Ullman and Widom 1997). Since we adopt an object database approach instead of object-relational, the ODMG standard is preferred in this study. There are also a number of standardisation efforts in the area of geographic and spatial information such as OGC (Open GIS Consortium), ISO TC211 and SQL/MM Spatial (Kottman 1998). Our extensions to the ODMG standard with respect to spatial and temporal operations have been addressed to comply with the core of these proposed standards.

5 Conclusion

This paper has presented an extension to the ODMG Object Model that incorporates spatial and temporal dimensions into database objects. The temporal extension is achieved by a parameterised type that is an orthogonal property of data types. On top of the model, we have also presented the design of a query language, STOQL, designed as a minimal extension to OQL. This supports an homogeneous manipulation of historical information for spatial and non-spatial attributes. Benefited from the object-oriented paradigm, type substitution can also be represented and queried in STOQL.

In contrast to other related work, the distinct features of STOQL rely on that it builds upon existing standards (e.g. ODMG and OGC) and provides a flexible mechanism to deal with synchronous and asynchronous changes of spatial and aspatial properties. STOQL has been implemented on top of the commercial GIS package ArcView through an object-oriented scripting language. Further work concerns the support of transaction time and the integration of multiple time granularities.

Acknowledgements

Part of the work in this paper was done when the first author was working for the joint TRIPOD project between Keele University and Manchester University, UK. Discussions with Mike Worboys, Keith Mason, Chris Johnson, John Stell, Norman Paton, Alvaro Fernandes and Tony Griffiths were helpful in shaping the contribution of this paper. Part of the second author's work was carried out while he was working at the Nottingham Trent University.

References

Allen JF (1983) Maintaining knowledge about temporal intervals. Communications of the ACM 26: 832-843

Alagic S (1999) Type checking OQL queries in the ODMG type systems. ACM Transactions on Database Systems 24: 319-360

Bertino E, Ferrari E, Guerrini, G (1997) T_Chimera: a temporal object data model. Theory and Practice of Object Systems 3: 103-125

Cattell R, Barry D (eds) (2000) The Object Data Standard: ODMG 3.0. San Francisco, Morgan Kaufmann Publishers Inc.

Cheng TS, Gadia SK (1993) An object-oriented model for temporal databases. In: Snodgrass RT (ed) Proceedings of the International Workshop on an Infrastructure for Temporal Databases. Arlington, TX, June

Claramunt C, Thériault M (1995) Managing time in GIS: an event-oriented approach. In: Clifford J, Tuzhilin A (eds) Recent Advances in Temporal Databases. Springer-Verlag

Claramunt C, Parent C, Spaccapietra S, Thériault M (1999) Database modelling for environmental and land Use changes. In: Geertman S, Openshaw S, Stillwell J (eds) Geographical Information and Planning: European Perspectives. Springer-Verlag

Fegaras L, Elmasri R (1998) A temporal object query language. In: Fifth International Workshop on Temporal Representation and Reasoning (TIME-98), May 1998, Sanibel Island, Florida

Griffiths T, Fernandes A, Paton N, Mason K, Huang B, Worboys M, Johnson C and Stell J (2001) Tripod: a comprehensive system for the management of spatial and aspatial historical objects. ACM GIS 2001

Gueting RH, Böhlen MH, Erwig M, Jensen CS, Lorentzos NA, Schneider M, Vazirgiannis M (2000) A foundation for representing and querying moving objects. ACM Transactions on Database Systems 25: 1-42

Hornsby K, Egenhofer M (2000) Identity-based change: a foundation for spatio-temporal knowledge representation. International Journal of GIS 14: 207 – 224

Huang B, Worboys M, Johnson C, Stell J, Mason K (2000) A spatio-temporal object model and query language (extended abstract). In: GIScience 2000: First International Conference on Geographic Information Science. 28-31 Octover, Savannah, Georgia, USA

Jensen CS, Clifford J, Gadia SK, Segev A, Snodgrass RT (1992) A glossary of temporal database concepts. SIGMOD Record 21: 35-43

Kottman C (1998) Geoprocessing standards connections: OGC and TC/211, TC/204, SQL/MM, CORBAgis, W3C [online]. Availalbe from: http://www. opengis.org/info/newsletter/ 9810/16.htm.

Langran G (1992) Time in Geographic Information Systems. Taylor & Francis, London

Open GIS Consortium Inc. (OGC) (1999) OpenGIS Simple Features Specification for SQL, (Revision 1.1) [online]. Available from: http://www.opengis.org/techno /specs/99-049.pdf.

Peuquet D, Duan N (1995) An event-based spatio-temporal data model (ESTDM) for temporal analysis of geographical data. International Journal of GIS 9: 7-24

Snodgrass R (ed) (1995a) The TSQL2 Temporal Query Language. Kluwer Academic Publishers

Snodgrass R (1995b) Temporal object-oriented databases: a critical comparison. In Kim W (ed) Modern Database Systems: The Object Model, Interoperability, and Beyond. Addison-Wesley, pp 386-408

Spery L, Claramunt C, Libourel T (2001) A spatio-temporal model for lineage metadata. Geoinformatica 5:51-70

Tansel U, Clifford J, Gadia S, Jajodia S, Segev A, Snodgrass R (eds) (1993) Temporal Databases: Theory, Design, and Implementation. The Benjamin/Cummings Publishing Company

Ullman JD, Widom J (1997) A First Course in Database Systems. Prentice Hall Inc.

Worboys M (1994) A unified model for spatial and temporal information. The Computer Journal 37: 26-34

A Semantics for Version Queries In GIS Databases

Hanna Kemppainen

Finnish Geodetic Institute, Geodeetinrinne 2, FIN-02430 Masala, Finland, hanna.kemppainen@fgi.fi

Abstract

A formal semantics for spatio-temporal version queries is presented. A version query specifies criteria against which change information is obtained from a database. The query criteria may concern spatial and temporal characteristics and indexed states of objects. The semantics is given using an object calculus, where the calculus terms refer to versioned geographical objects, their changing states and spatial and temporal relationships. The purpose of specifying the semantics of queries is to aid in understanding the exact meaning of spatio-temporal queries better. The work provides a framework for future research in modelling of complex identity-affecting mutations of objects and related queries.
Keywords: spatio-temporal, object-oriented, change, query

1 Introduction

Implementation of a GIS that supports various aspects of object-based change is founded on models of change itself and models of queries to get information of the stored change data. A model of change can be based on an object-oriented model of geographical information where a geographical object consists of spatial, aspatial and temporal components (Worboys, 1992, 1994b). A change in an object may involve the identity, spatial or aspatial attribute values or the thematic content of the object. A software system that supports change should not only be able to store data of the changing states and identities of the objects. It should also provide a query language to specify criteria against which change information is obtained from a database. A query language should provide support for queries where the user wants to find out what happened to an entity at a given time or during a given time interval. The user should also be able to ask what happened to an entity when a related entity changed at a given time, or how an entity changed between two consecutive changes of a related entity. Such analysis capabilities are not provided by current commercial GIS.

A model of extracting data from a database is often given as an SQL-like query language. Several languages exist that augment the standard SQL with temporal constructs (Snodgrass, 1992). A query language that integrates both spatial and temporal aspects is needed for geographical applications (Egenhofer, 1994), and such languages have been developed (Griffiths *et al.* 2001, Kemppainen, 2001). This paper does not present another SQL-like query language, but examines how the *semantics* of change queries can be expressed. Starting with queries expressed in an SQL-like language, the objective is to look for a semantics that aids in finding a solution to implement them. This can be accomplished if the semantics is given in terms that are near to the concepts used in implementing systems, for example objects and methods. A formal semantics also aids in understanding the exact meaning of the queries.

Specification of the semantics of change queries is studied here using a logical language called spatio-temporal version calculus. The calculus, as any formal language, consists of syntax and semantics. The syntax specifies the symbols that may be used to express spatio-temporal version queries and how those symbols may be arranged to create well-formed query formulas. The semantics of the calculus specifies how meaning is ascribed to the symbols and the well-formed query formulas. The calculus is designed such that the terms of the calculus correspond to geographical objects and their spatial and temporal operations. Given a set of database objects I, the semantics of a query q specifies which objects are to be included in the query result. The calculus approach to specifying semantics of queries is adapted from the field of database theory for relational query languages (Abiteboul *et al.* 1995, Gallaire *et al.* 1984). An object calculus is also suggested as a query language for geographic databases in (Clementini *et al.* 1993).

The remainder of the paper is organized as follows. Related work is discussed in section 0. A note is made in section 0 as to how change data can be stored in GIS systems. The paper presents a semantics for some typical query types (section 0) based on a conceptual model of version data (section 0). The semantics is similar to the relational query semantics, which is reviewed in section 0. The main contribution of the paper is the object calculus and its use in expressing version queries and their semantics. This is the topic of the remaining chapters.

2 Related Work

Geographical change and temporal GIS have been popular research topics for over a decade (Armstrong, 1988, Langran, 1992, Peuquet, 1994, Frank *et al.* 2000). Conceptual models of change have been proposed for object-based representations, where the object mutates to another object or the object state changes (Claramunt *et al.* 1995). Hornsby & Egenhofer developed a change description language that can be used to graphically represent operations on identity of simple and composite objects (Hornsby *et al*, 1997, 1998). Medak (1999) developed a theory of lifestyles in terms of operations affecting object identity. The theory is presented as an algebra, i.e. as types, operations and axioms that define their behaviour.

Implementation models have been proposed to store change-related information in GIS databases, relational or object-oriented (Langran, 1989, Kemp *et al.* 1992, Tryfona, 1998). Change that does not affect the object identity can be modelled as object versions (Claramunt *et al.* 1995, Wachowicz *et al.* 1994). A unified model to space and time and its implementation using simplicial complexes was introduced in (Worboys, 1994a).

Research in query languages for changing geographical objects has been somewhat less intense compared to the development of conceptual models of change. Typologies of spatio-temporal queries are presented in (Peuquet, 1994, El-Geresy *et al.* 2000). A model of extracting data from a database is often given as an SQL-based query language. Several query languages have been proposed for temporal (Snodgrass, 1992, Chomicki, 1994) and spatial applications (Egenhofer, 1994, Herring, 1987). A recent contribution augments the syntax of an object oriented query language OQL (Cattell *et al.* 2000) with spatial and temporal operations and index numbers on object states (Griffiths *et al.* 2001). An earlier work by the author of this paper developed a similar but SQL-based query language that also contains constructs to refer to the version history of an object (Kemppainen, 2001).

3 Support for Change in GIS

It has been stated that implementation of a temporal GIS is still problematic and support for tracking changes in spatio-temporal databases is not effective (Peuquet, 2001, Griffits *et al.* 2001). Current GISs provide two kinds of mechanisms that support storing change-related data: user-defined attributes and version management. Users may time-stamp their data by storing a temporal value as one of an object's attributes. To track the evolution of a database object one can chain different objects representing the same real world entity. Some systems implement *database versions* rather than *object versions* to provide users with simultaneous access to data. Version management systems (Easterfield *et al.* 1990) consider change as a tree of database alternatives. A database alternative is a database view that corresponds to some state S_i of the database, and each sub-alternative S_{ii} derived from S_i provides another view to the database by updating S_i independently of other sub-alternatives. Object's representation in each of the database alternatives corresponds to a version of an object.

The purpose of such a versioning mechanism is to allow different users to work simultaneously with the same data rather than to provide a representation of the evolution of objects. If, however, database alternatives are used to store evolution data, a specific meaning is assigned to each database alternative. For example, there might be a cadastral database alternative, utilities alternative, etc. To track the “evolution” of a single object in different alternatives, the alternative tree has to be traversed to extract the object information in each of the alternatives. It is left to the application programmers to extract the version data from different alternatives and make analysis as to what changes the entity has experienced, and how the different versions of the entity relate to other entities.

4 Examples of Queries on Time-Varying Geographical Objects

The starting point for this work is provided by a typology of spatio-temporal queries by Peuquet (1994). The queries are not given with respect to any data model, and no indication is made as to how they should be implemented. The formulation of the queries suggests a natural approach to consider changing geographical phenomena, for example, when making observations of the landscape by eye, or studying maps or plans.

The first class of queries addresses changes in an *object* or *feature*: "How has this object changed or moved over the last five years" and "Where was this object two years ago". These queries involve some real world entity ("this object") that is to be monitored at different points in time. Such an entity is thus identifiable, but not necessarily well defined. The problem is that we should identify exactly the same entity at different time instants.

The second class of queries involves changes in the spatial distribution of an object or a set of objects; for example, "What agricultural areas in January 1980 changed to residential land use areas as of December 1989" and "Did any land use changes occur in this drainage basin between t_1 and t_2". Entities may not be well defined here either. As far as land use areas are concerned, it may be questioned whether a land use area should be considered as an identifiable object at all; rather, such areal phenomena might be abstracted as fields. If they are considered as discrete objects, it should be possible to model *class changes* of objects and provide support for such change queries.

Class III queries address the temporal relationships among multiple geographic phenomena, for example, "Which areas experienced a landslide within one week of a major storm event?" In such a query, both the thematic attribute and the spatial characteristics of objects are related.

Based on this framework, geographical real world entities are considered here as spatio-temporal objects that are elements of some class. The name of the class reveals the thematic content of the object (for example, Road class contains road objects). Objects have aspatial, spatial and temporal characteristics. A change may occur that concerns any apatial or spatial value of the object, or the class of the object may change.

5 Version Model and Model of Version Queries

5.1 Conceptual version model

Modelling of changing geographical entities as versioned objects is suggested in e.g. (Wachowicz *et al.* 1994). A versioned object consists of a sequence of versions. The versioned object denotes the changing real world entity, and each version in the version sequence denotes an entity state during some time interval. Change of an object to another object, or other mutations affecting object identity (Claramunt *et al.* 1995) are not supported. Using this model, a road object *Road-A* would be represented as a versioned object, whose changing states correspond to versions *Road-A-1*, *Road-A-2*, etc. The version scenario is depicted in Fig. 1, where versioned objects are depicted as small squares, connected to a sequence of circles that denote versions.

Possible queries for versioned objects are identified by examining potential relationships between versioned objects and their versions. Different ways of associating versioned objects are depicted using large rectangles in Fig.1. Given versionable objects O_1 and O_2, with a set of versions V_1 and V_2, respectively, a query may involve the location of an element of V_1 in the version sequence (similarly for V_2). Such a query extracts versions using expression such as "get the latest version of the object" or "get the first and the second version of the object", to determine how the entity has changed. Another category of queries relates versions based on their aspatial, spatial, and temporal characteristics.

The limitations of this version scenario are the following: 1) Many geographical phenomena are difficult to identify so that exactly the same entity or phenomenon can be detected, say, years later. 2) Creating new versions causes data duplication, which may result in inefficient systems unless the storage problem can be resolved efficiently.

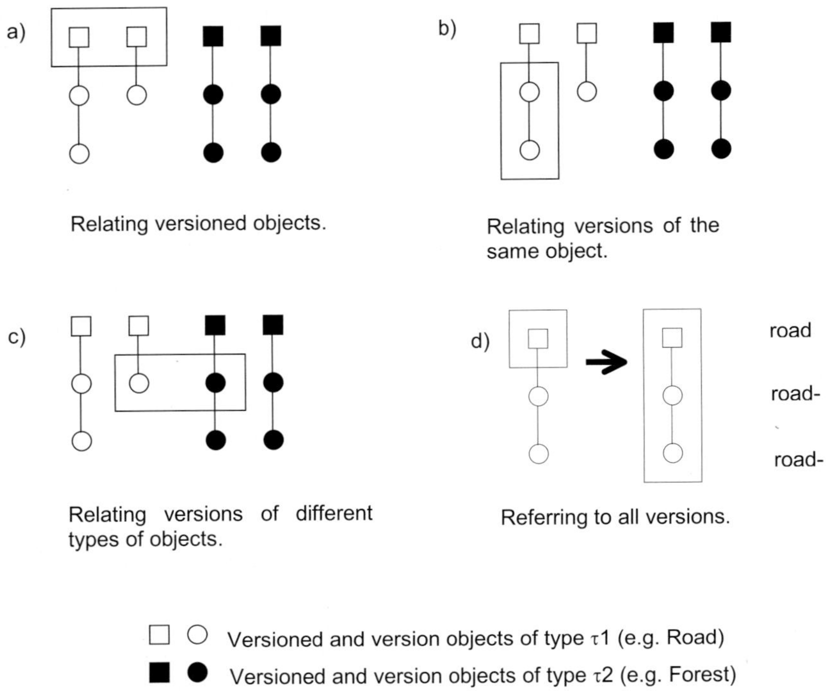

Fig. 1. Relationships between versioned objects

5.2 SQL-Like Version Queries

Version queries are expressed using a *select-from-where* expression, which is the basic construct of SQL, the standard relational query language (explained in e.g. Elmasri *et al.* 1989). Similar expression, but with a somewhat different syntax, is also used in OQL, a standard for object database queries (Cattell *et al.* 2000). The SQL-like version query language developed in (Kemppainen, 2001) is used to express queries whose semantics is studied in this paper.

Let us suppose the data about the versioned objects is stored in a relational database. There are two tables: one for versioned objects and one for versions. A version query would be expressed as:

select versioned_exp | version_exp
from relation_name [,relation_name*]
where select-condition

Here *relation_name* refers to the tables storing data of versioned and version objects. The *select* clause identifies the data that will comprise the query result; for example, *versioned_exp* is an expression that refers to the whole versioned object. In this case, the select clause would be formulated as "*select versioned*", to select all the versions that exist of the object. The expression might be used in queries such as "Did something change about this object?" The query would return all the versions of the versioned object, for which conditions specified in the select condition apply. A *version_exp* may be of the form 1) "*select version.aspatial*", 2) "*select version.spatial*" or 3) "*select version.temporal*", where *aspatial*, *spatial* and *temporal* denote the corresponding attributes of the version, or 4) "*select version*" if the whole version is asked for.

The select condition is expressed as a combination of version sequence clauses and aspatial, spatial and temporal clauses. An example of a version sequence clause is "*version in (last –1, last)*", to denote the latest version and the one preceding it. A spatial clause uses spatial operators (Egenhofer *et al.* 1991) to evaluate relationships between spatial components of versions, for example, "*version_1.spatial inside version_2.spatial*" to relate spatial components of the two versions using is a spatial operator *inside*. In a similar manner, we express aspatial clauses using standard comparison operators $<, \leq, >, \geq, =, \neq$, and temporal clauses using standard temporal operators (Halpern *et al.* 1991). When relating temporal characteristics of objects, we use temporal expressions to refer to time instants or time intervals given with respect to some calendar value, or *now*. For example, to denote a time interval that started two years ago and ends at the current moment, we would specify "*last (“year”, 2, now, “interval”)*".

Examples of queries and their semantics are given in section 9.

6 Specification of the Semantics of Queries

The semantics of queries is discussed in terms of relational database theory in (Abiteboul *et al.* 1995, Gallaire *et al.* 1984). These principles are extended to object-oriented databases for versioned geographical objects.

From a logic perspective to databases, one considers a relational database as an interpretation of a first-order language, where the atomic constructs of the language are *n*-ary predicates $P(x_1,\ldots, x_n)$, and equality operator =. A predicate is understood as a tuple in a relation, and is deemed true in the interpretation if the corresponding tuple exist in the database. Queries are expressed as logical formulas. Their truth-value can be determined when we know how the constants of the language are interpreted, and how the variables occurring in the formulas are valuated. The truth of a formula is derived with respect to the values that are stored in the database. An answer to a query *q*, presented as a logical formula φ, is obtained by finding a valuation of variables in φ that makes φ true.

To put this formally, we consider a relational database with *n*-ary relations R_i. A fact over R_i is an expression of the form $R(a_1,\ldots,a_n)$, where a_i is an alphanumeric value. A relation instance over *R* is a finite set of facts over *R*. A database schema **R** consists of the relation names, $\mathbf{R} = \{R_1, \ldots, R_n\}$ and a database instance is a finite set **I** that is the union of relation instances over R_i, for $R_i \in \mathbf{R}$.

Let L be a first order language whose domain is **dom**, an infinite set of alphanumeric values. In standard logic, the constants of the language are used to denote elements of the domain, but it is customary in the database context to consider elements of the domain as the constants of the language also. Predicate symbols $R_1,\ldots, R_n$ denote real predicates among the individuals of the domain. In this setting, a database instance is understood as a finite interpretation **I** of L. This means that the tuples in the database are interpreted as true facts about the universe, and negation of a fact can be derived if the corresponding tuple is not found in the database. The naming of the domain elements is arranged such that each element of the domain is named, and individuals with different names are different.

Queries would be expressed in the relational calculus in the form $q = \{e_1,\ldots, e_m \mid \varphi\}$ where $e_1,\ldots, e_m$ are variables that are free in φ. The semantics of a query is given as an image of database instance **I** under q relative to **d**:

$q_{\mathbf{d}}(\mathbf{I}) = \{\nu([e_1,\ldots, e_n]) \mid$ where ν is the *valuation* over free variables of φ with range contained in **d**, and **I** satisfies φ for $\nu\}$. The underlying domain **d**, with $\mathbf{d} \subseteq$ **dom**, is used to permit us to talk of different underlying domains. The set of all constants occurring in **I** is contained in **d**.

The semantics of spatio-temporal version queries shall be given here in the form of an object calculus. The calculus consists of 1) appropriate domains for versioned geographical objects, 2) an alphabet and rules to use the alphabet to build well-formed query expressions, and 3) specification of truth-values of the calculus expressions with respect to the set of objects that exist in the database.

7 Basic Notions for the Object Calculus

A query q is expressed as $q = \{x_i \mid \varphi\}$, where φ is an object calculus formula and x_i denotes a free variable in φ. The calculus is designed such that the atoms in φ denote meaningful constructs in the geographical application domain. The individuals of the domain are abstracted as complex values and objects.

The calculus is similar to the complex value calculus presented in (Abiteboul *et al.* 1995), extended to incorporate object-oriented characteristics.

7.1 Domains

The domain of the calculus consists of the standard domains **integer**, **string**, **bool** and **float** whose disjoint union is **dom**. Complex value domains are defined recursively using the following abstract syntax: $\tau =$ **dom** $\mid [B_1 : \tau,\ldots, B_k : \tau] \mid \{\tau\}$, where $[B_1 : \tau,\ldots, B_k : \tau]$ denotes a tuple value, and $\{\tau\}$ is a set value. Objects are referred to using elements of **obj**, which is a finite set of object identifiers, **obj** $= \{o_1,\ldots, o_n\}$.

Spatial objects are elements of the spatial domain, with **spatial** $\subseteq$ **obj**. Given an OID assignment π for spatial classes *Point*, *Line* and *Area*, a finite set of spatial objects is denoted by $\pi(Point) \cup \pi(Line) \cup \pi(Area)$. Temporal values are contained in temporal domain **temp**, whose objects can denote time-stamps given in valid time or transaction time. We assume **versioned**, the domain for the object identifiers of versioned objects, with **versioned** $\subseteq$ **obj**, and **version**, the domain for the object identifiers of **version** objects, with **version** $\subseteq$ **obj**. The domains **spatial**, **temp**, **version** and **versioned** are pair-wise disjoint.

The underlying object model abstracts a changing geographical entity as a *versioned object*, whose changing states are described as *versions*. Each object version may have spatial, aspatial and temporal components. For example, changing real world roads would be modelled using *Versioned_Road* class. An object of *Versioned_Road* would consist of an ordered set of *Road_Version* objects, each of which is an aggregation of values coming from the **spatial**, **temporal** and **aspatial** domains. Objects of *Versioned_Road* and *Road_Version* come from **versioned** and **version**, respectively.

7.2 Predicates for Version Queries

Relationships between objects are modelled as predicates $P(o_1,\ldots, o_n)$. For the purposes of defining the semantics of queries, for a given predicate P, it is sufficient to know whether P holds for some valuation of the object variables standing for $o_1,\ldots, o_2$.

To formulate spatial and temporal selection conditions we assume a set of spatial predicates *SP* and temporal predicates *TP*. Comparisons between numeric and textual values are included in the calculus operations. The definition of spatial and temporal relationships is discussed in e.g. (Egenhofer *et al.*, 1991, Allen, 1984, Halpern *et al.* 1991).

$$SP = \{D, SM, SO, COV, COVB, I, CON, SE\}$$
$$TP = \{B, TE, TM, TO, D, S, F, AS, AE\}$$

Spatial predicates *SP* refer to spatial operators disjoint (*D*), spatial_meet (*SM*), spatial_overlap (*SO*), covers (*COV*), covered_by (*COVB*), inside (*I*), contains (*CON*), and spatial_equal (*SE*). Temporal predicates correspond to temporal operators before (*B*), temporal_equal (*TE*), temporal_meet (*TM*), temporal_overlap (*TO*), during (*D*), starts (*S*), finishes(*F*), at_start (*AS*) and at_end (*AE*).

8 The Object Calculus

The object calculus for expressing version queries is based on the complex value calculus of (Abiteboul *et al.* 1995), extended with some object-oriented notions.

The alphabet of the object calculus (as that of any formal language) consists of constants, variables, logical connectives and quantifiers, and of rules that govern the formulation of well-formed formulas.

Constants and variables For each domain, a countable finite set of constants and variables of that domain is assumed.

Terms are interpreted as standing for an individual of the domain. Constants and variables are terms. An object variable stands for an object, and a tuple variable denotes a complex value that is a tuple. A field A of a tuple variable x (of sort [A : **dom**]) is referred to as $x.A$.

Atomic formulas:

a) A predicate $P(t_1,\ldots, t_n)$ is an atom ($t_1,\ldots, t_n$ are terms).

b) Atoms for simple values: $t = t'$ and $t < t'$ are atoms (t and t' are terms).

c) Atoms for complex values: $t \in t'$, $t \subseteq t'$, t and t' are terms and the appropriate sort restrictions apply ($\in, \subseteq$ are many sorted).

d) Object-oriented characteristics of the calculus are supported by 1) value equality $o = o'$, and 2) identity equality $o =_{\mathrm{id}} o'$ for object variables o and o'. To compare values returned by methods we have: $t = m(t_1,\ldots, t_n)$ and $m_1(t_1,\ldots, t_n) = m_2(t_1,\ldots, t_m)$ where $t, t_1,\ldots, t_n ,\ldots,t_m$ are terms and m, m_1 *and* m_2 denote method names. The value of $m(t_1,\ldots, t_n)$ is obtained by evaluating the implementation of m under a variable assignment μ. that associates values to $t_1,\ldots, t_n$. The method signature reveals the type of the value which m returns.

Formulas are created from atomic formulas using the logical connectives $\neg$, $\wedge$, $\vee$, $\rightarrow$ and quantifiers $\forall$ and $\exists$.

8.1 Semantics of the calculus

The truth-value of an atomic formula is defined like the standard relational calculus. For example, an equality atom $\varphi = (s = s')$ is true in an interpretation **I** if the valuations of the two variables are equal, i.e. $\nu(s) = \nu(s')$. We say that **I** satisfies φ for ν, relative to **d**, denoted by $\mathbf{I} \vDash_{\mathbf{d}} \varphi[\nu]$. If φ is formed using logical connectives for example if $\varphi = (\psi \wedge \xi)$, $\mathbf{I} \vDash_{\mathbf{d}} \varphi$ if $\mathbf{I} \vDash_{\mathbf{d}} \psi[\nu']$ and $\mathbf{I} \vDash_{\mathbf{d}} \xi [\nu']$, where ν' is a restriction of ν to free variables of the formula. For other forms of φ, we shall only discuss the interesting features that are related to complex values and objects. Given a database instance **I**, **I**(C) denotes an instance of class C, and **I**(P) denotes an instance of relation P. The truth of a formula φ in **I** is given as follows:

(a) $\varphi = C(o)$ is true if $\nu(o) \in \mathbf{I}(C)$ and ν is a valuation of object variables.

(b) $\varphi = (s = s')$ is true if $\nu(s) = \nu(s')$. s,s' denote variables of any domains (value equality)

(c) $\varphi = (s =_{\mathrm{id}} s')$ is true if $\nu(s) =_{\mathrm{id}} \nu(s')$. s,s' denote variables of object domains (identity equality)

(d) $\varphi = (s = m(s_1,\ldots,s_n))$ if $\nu(s) = m(\nu(s_1),\ldots, \nu(s_n))$ and $\nu(s)$ complies with the method signature of m under ν, especially we require that $\nu(s)$ is of the correct type. Similarly, $m_1(s_1,\ldots, s_n) = m_2(s_1,\ldots, s_m)$ if $m_1(\nu(s_1),\ldots, \nu(s_n)) = m_2(\nu(s_1),\ldots, \nu(s_m))$ and the values returned by m_1 and m_2 are of the same type.

(e) $\varphi = (s \in s')$ and $\nu(s) \in \nu(s')$

(f) $\varphi = (s \subseteq s')$ and $\nu(s) \subseteq \nu(s')$

(g) $\varphi = P(o_1,\ldots,o_n)$ if $[\nu(o_1),\ldots, \nu(o_n)] \in \mathbf{I}(P)$

9 The Semantics of Version Queries

The version queries are formalised as $q = \{x \mid \varphi\}$, where φ is an object calculus formula and x is free in φ. For example, to select a versioned object from the database we use a query language statement "*select versioned*" which corresponds to a query $\{r \mid \varphi(r)\}$, where r is a variable ranging over **versioned**, and r is free in φ.

The object calculus formulation of some of the queries discussed in section 0 requires a sample database schema and a set of objects of the schema classes. Let *Versioned_Road, Road_version, Versioned_landuse, Landuse_version* and *Line* denote classes of a schema *S*. A database instance *I* consists of objects belonging to classes of *S*. Methods *M* are used to extract attribute values and versions of objects. A method definition is given as a signature $m : c \times \tau_1 \times \ldots \times \tau_{n-1} \rightarrow \tau_n$, where c is a class name and τ_i is a type over C, the set of all class names (Abiteboul *et al.* 1995). This signature is associated with class c; we say that method applies to objects of class c. $\tau_1,\ldots,\tau_{n-1}$ denote parameter types and the method returns a value whose type is τ_n. A coded definition of m is the piece of program code to implement the method.

For example, for class *Road_version* with attributes *aspatial*, *spatial* and *temporal*, method *spatial* gets the value of the spatial attribute: *spatial* : *Road_version* → *Spatial*. Temporal component is returned by a method whose signature is *temporal* : *Road_version* → *Temporal*. Method *version*(*Versioned*, {**int**}) → {version} extracts versions of an instance of a versioned class The method returns a set of versions specified by a set of integers as a method parameter.

A query "Has Road A moved in the last two years?" would be formulated as an SQL-like query as follows:

select version
from *Versioned_Road, Road_version*
where root = "road-A" **and version.temporal during last** ("year", 2, **now**, "interval")

This query extracts the versions of Road A, whose time stamp indicates that the version existed during a time interval specified by **last** ("year", 2, **now**, "interval"), for example [1.1.1999,31.12.2000], if the query was performed at 31.12.2000.

The query extracts versions whose spatial components have to be compared to determine whether a change occurred during the specified interval of time.

The following calculus query gets all versions of Road A:

$$\{v \mid \exists r\, (Versioned\_Road(r) \wedge name(r) = \text{"}road\text{-}A\text{"} \wedge v \in versions(r)\}$$

We restrict the query result further by requiring that the temporal value associated with the version overlaps a temporal constant tc_1, where tc_1 denotes a time interval starting two years ago and ending *now*.

$$\{v \mid \exists r\, (Versioned\_Road(r) \wedge name(r) = \text{"}road\text{-}A\text{"} \wedge v \in versions(r) \wedge D(temporal(v), tc_1)\,)\}$$

Here variable r denotes an object for which predicate $Versioned\_Road(r)$ should hold. Answer to the query is obtained by a valuation for r such that r denotes an object from the *Versioned_Road* class. Let o be such an object. The query also requires that the method *name* applied to o returns "*road-A*". A free variable v is used to denote an object that belongs to the set of versions of o. Finally, temporal predicate D (for "during") is used to determine whether the temporal component of version v takes place during the time interval denoted by a temporal constant tc_1. Once we have found a valuation for r and v such that the formula holds, we have an answer to the query.

In a similar manner, "Where was Road B two years ago?" would be formulated SQL-like as

select spatial
from *Versioned_road*, *Road_version*
where root = "Road B" **and version.temporal temporal_overlap last** ("year", 2, **now**, "instant")

This query extracts the spatial component of the version of Road B that existed two years ago. This is a direct answer to the original query, given as a spatial object.

Formally:

$$\{s \mid \exists r,v\, (Versioned\_Road(r) \wedge name(r) = \text{"}Road\ B\text{"} \wedge v \in versions(r) \wedge D(temporal(v), tc_1) \wedge spatial(v) = \mathrm{s})\}$$

A query "Did any land use changes occur in this drainage basin between Jan 1, 1980 and Dec 31, 1989?" might be interpreted as involving a change in the area extent of the land use areas, or a change in their classification attribute, or whether the particular area object disappeared or new areas were born. The following query finds versions of a land use area 12345 where the value of its the classification attribute changes. The select condition first specifies that we are interested in areas that lie inside "this drainage basin"; that area is referred to using a spatial constant "REF". A negated aspatial clause is then used to specify that the versions whose classification attribute values are different looked for.

select versioned, version_1, version_2

from Versioned_Landuse, Landuse_version, Landuse_version
where versioned.id = "12345" **and**
    version_1.spatial inside "REF" **and**
    version_2.spatial inside "REF" **and**
    not (version_1.classification = **version_2**.classification)

In this query, the class *Versioned_Landuse*, and methods defined similarly as in the previous example are assumed. Spatial constant sc_1 represents the reference area REF. I is a spatial predicate for "inside". The query is formalised as:

$$\{v_1, v_2 \mid \exists r_1, v_1, v_2, s_1, s_2 (\\ Versioned\_Landuse(r_1) \wedge id(r_1) = \text{"12345"} \wedge \\ v_1 \in versions(r_1) \wedge v_2 \in versions(r_2) \wedge \neg(v_1 =_{\text{id}} v_2) \wedge \\ spatial(v_1) = s_1 \wedge spatial(v_2) = s_2 \wedge \\ I(sc_1, s_1) \wedge I(sc_1, s_2))\}$$

10 Concluding Remarks

The purpose of specifying a formal semantics for SQL-like spatio-temporal queries is to aid in understanding the meaning of the queries better, and to outline a solution for their implementation. The semantics of spatio-temporal version queries is developed here using an object calculus. The alphabet of the calculus consists of concepts that can be easily implemented: classes, objects and methods. Given a set of database objects, the answer to a query is obtained by finding a proper valuation for query variables such that the query criteria are fulfilled.

A question then arises, what can we describe using the calculus. The calculus can be used to refer to objects from different classes, and methods to extract their values and other characteristics. Predicates can be used to make statements about relationships between objects Spatial predicates denote spatial conditions between object versions, and temporal predicates are used for evaluating temporal relationship between them. It is thus possible to describe the relationships between versions of objects in terms of their spatio-temporal characteristics, and determine the semantics of such a description with respect to what data exists in a database. What first-order predicate calculus cannot do is to state something about sets of objects as a whole. For that purpose, we should be able to quantify over predicates, not just individual elements of domains.

The formalisation of queries and their semantics provides a framework for research that allows the development of models of discrete change further, for example, to model complex mutations of objects.

Acknowledgements

This work has been carried out under supervision of Professor Tapani Sarjakoski, whose advice and support is gratefully acknowledged. The author is indebted to the anonymous reviewers for the comments that helped in shaping the contribution of the paper.

References

Abiteboul A, Hull R, Vianu V (1995) Foundations of databases. Addison-Wesley

Allen JF (1984) Towards a general theory of action and time. Artificial Intelligence 23(2):123-154

Armstrong M (1988) Temporality in spatial databases. *Proceedings GIS/LIS'88*, San Antonio, Texas, pp 880-889.

Cattell R, Barry DK, Berler M, Eastman, J, Jordan D, Russell C, Schadow O, Stanienda T, Velez F (2000) The Object Data Standard: ODMG 3.0. Morgan Kaufmann Publishers

Chomicki J (1994) Temporal query languages: a survey. In: Proceedings of the First International Conference Temporal Logic. Bonn, Germany, pp 506-534

Claramunt C, Theriault M (1995) Managing time in GIS: an event-oriented approach. In: Proceedings of the International Workshop on Temporal Databases. 17-18 September 1995, Zürich, Switzerland, pp 23-42

Clementini E, Di Felice P (1993) An object calculus for geographic databases. In: Proceedings of the 1993 ACM/SIGAPP Symposium on Applied Computing: States of the Art and Practice, Indianapolis, IN, pp 302-308

Easterfield M, Newell R, Theriault D (1990) Version management in GIS – applications and techniques. In: First European Conference on Geographical Information Systems. Amsterdam, The Netherlands, pp 288-297

Egenhofer M (1994) Spatial SQL: a query and presentation language. IEEE Transactions on Knowledge and Data Engineering 6(1): 86-95

Egenhofer M, Franzosa R (1991) Point set topological relations. International Journal of Geographical Information Systems 5(2):161-174

El-Geresy B, Jones C (2000) Spatio-Temporal Models and Queries in GIS. In: Atkinson P, Martin D (eds) Innovations in GIS 7, Taylor and Francis, pp 27-39

Elmasri R, Navathe S (1989) Fundamentals of Database Systems. Menlo Park, CA, Benjamin/Cummings

Frank AU, Raper J, Cheylan, J-P (eds) (2000) Life and Motion of Socio-Economic Units. GISDATA Volume 8, London, Taylor & Francis.

Gallaire H, Minker J, Nicolas J-M. (1984) Logic and databases: a deductive approach. ACM Computing Surveys 16(2):153-185

Griffiths T, Fernandes A, Paton N, Mason K, Huang B, Worboys M (2001) Tripod: A Comprehensive Model for Spatial and Aspatial Historical Objects. In: 20th International Conference on Conceptual Modeling. Yokohama, Japan, pp 84-102

Halpern J, Shoham Y (1991) A propositional modal logic of time intervals. Journal of the ACM 38(4):935-962

Herring J (1987) TIGRIS: Topologically integrated geographical information system. Proceedings of AUTO-Carto 8. March 1987, Baltimore, MD, pp 282-291

Hornsby K, Egenhofer M (1997) Qualitative representation of change. In: Hirtle S, Frank A (eds) Spatial Information Theory: A Theoretical Basis for GIS, Proceedings of the International Conference COSIT '97. pp 15-33

Hornsby K, Egenhofer M (1998) Identity-Based Change Operations for Composite Objects. In: Proceedings of the 8th International Symposium On Spatial Data Handling. Vancouver, Canada, pp 202-213

Kemp Z, Thearle R (1992) Modeling relationships in spatial databases. In: Proceedings 5th International Symposium on Spatial Data Handling. Charleston SC, pp 313-322

Kemppainen H (2001) A data model and query language for versioned spatio-temporal objects. Submitted to *Geoinformatica* (under review)

Langran G (1989) A review of temporal database research and its use in GIS applications. International Journal of Geographical Information Systems 3(3): 215-232

Langran G (1992) Time in Geographic Information Systems. Taylor & Francis

Medak (1999) Lifestyles – an algebraic approach to change in identity. Spatio-Temporal Database Management, International Workshop STDBM'99. Edinburgh, Scotland pp 19-38

Peuquet D (1994) It's about time: a conceptual framework for the representation of temporal dynamics in geographic information systems. Annals of the Association of American Geographers 84(3):441-461

Peuquet D (2001) Making space for time: issues in space-time data representation. GeoInformatica 5(1):11-32

Snodgrass R (1992) Temporal Databases. International Conference GIS - From Space to Territory: Theories and Methods of Spatio-Temporal Reasoning. Pisa, Italy, pp 22-64

Tryfona N (1998) Modeling Phenomena in Spatiotemporal Databases: Desiderata and Solutions. In: Database and Expert Systems Applications, 9th International Conference, DEXA '98. Vienna, Austria, pp 155-165

Wachowicz M, Healey R (1994). Towards temporality in GIS. In: Worboys MF (ed) Innovations in GIS 1. Taylor and Francis, London, pp 105-115

Worboys MF (1992) A Generic Model for Planar Geographic Objects. International Journal of Geographical Information Systems 6(5): 353-372

Worboys MF (1994a) A unified model of spatial and temporal information. The Computer Journal 37(1):26-34

Worboys MF (1994b) Object-oriented approaches to geo-referenced information. International Journal of Geographical Information Systems 8(4):385-399

A Differential Spatio-temporal Model: Primitives and Operators

Giorgos Mountrakis, Peggy Agouris, Anthony Stefanidis

Dept. of Spatial Information Science and Engineering, National Center for Geographic Information and Analysis, 348 Boardman Hall, University of Maine, Orono, ME 04469-5711, USA, {giorgos, peggy, tony}@spatial.maine.edu

Abstract

In this paper, we present a differential change-oriented model for the storage and communication of spatio-temporal information. The focus is on the development of model primitives and operators to support the aggregation of change over time and the propagation of change across resolution. Our investigation is motivated by recent advances in image-based geospatial databases, with constantly increasing update frequencies, and diverse user communities performing queries of various levels of detail. The primitives and operators presented here extend existing qualitative operators to support the management of quantitative and geometric information within a change-oriented spatio-temporal environment. We also show how the design of our model results from 'change semantics' at different granularities. By doing so, advanced communication operations can be addressed within our model in an efficient way.
Keywords: spatio-temporal, change, differential, queries, granularity

1 Introduction

Spatio-temporal applications are increasingly the focus of research activities in the geospatial and database communities. The complexity behind the combination of spatial and temporal representations is well documented (Erwig et al, 1999; Worboys, 1994; Yeh and de Cambray, 1995; Theodoridis et al, 1998), but an efficient solution has yet to emerge (Peuquet, 2001). In early approaches, representations of objects (states) were stored at different time instances. In this case change information was handled indirectly, as it was not stored but could be calculated using the stored data. More recently we saw the introduction of change-oriented approaches, focusing mostly on qualitative attributes of geospatial entities (Peuquet and Wentz, 1995) and variations of these attributes (Hornsby and

Egenhofer, 2000). Approaches addressing the recording of positional information for mobile objects proceeded by reducing these objects to point data and ignoring their spatial extent (Pfoser et al, 2000).

In this paper we present a differential change-oriented model for the storage and communication of spatio-temporal information. We focus on the development of model primitives and operators to support the aggregation of change over time and the propagation of change across resolution. Our investigation is motivated by recent advances in image-based geospatial databases that have constantly increasing update frequencies, and increasingly diverse user communities. The primitives and operators presented here extend the qualitative operators presented by (Worboys, 1992; Hornsby and Egenhofer, 2000) to support the management of qualitative and geometric information within a change-oriented spatio-temporal environment. Thus, they complement the general framework outlined by (Peuquet and Wentz, 1995; Langran, 1993) to support the management of raster and vector geospatial data within a spatio-temporal environment.

Our paper begins with a brief description of our differential change model, followed by a more detailed presentation of change primitives and operators (Section 2). We continue by discussing the use of these operators within a differential spatio-temporal image-based model (Section 3) to demonstrate the resulting improvement in expressiveness and redundancy minimisation.

2 Differential Change Model

Developments in sensor technologies have enabled the continuous collection of geospatial data, and the constant updates of geospatial databases. This supports complex spatio-temporal analysis, but at the same time imposes interesting challenges on detecting changes in geospatial objects and managing this change information. Addressing these challenges we have developed the model of a differential spatio-temporal gazetteer, and differential image-based change detection approaches to populate this model (Agouris et al., 2000; Agouris et al., 2001). We use the term *differential* in our approach to reflect the emphasis put on change as the explicit information that is both captured and stored in our spatio-temporal model. We proceed by storing an initial state of an object (in essence change from non-existence) and all subsequent changes (Fig. 1). An object representation at any instance t_n is obtained through a multi-dimensional aggregation operator of t_0 and all subsequent changes $\Delta t_{i-1,i}$ for i=1,…n.

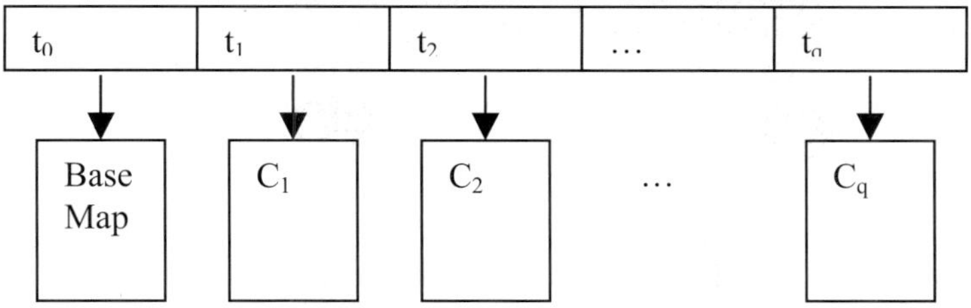

Fig. 1. A snapshot change representation model based on (Peuquet and Wentz, 1995)

This is in accordance with the conceptual model presented by (Peuquet and Wentz, 1995). In addition to being an actual implementation of this model, our approach extends it by including 'change semantics' in multiple granularities. We expand the method to work on non-grid based geometric information, since a significant part of GIS information is in vector format by making use of (Langran, 1993) representation. In addition, we present a set of change primitives and operators that can be used to represent and extract information at an index and a qualitative level within our model. These operators are naturally expressive and extend previous approaches (Worboys, 1992; Hornsby and Egenhofer, 2000) by handling quantitative and geometric information.

2.1 Primitives

Let O_j be an object from the set of objects (O), and that this object is observed in a subset T_n of the set of time instances (T). So we have:

$$\vec{O}_j \in O, T_n \in T \text{ with } T_n = [0, t_1, t_2, t_3, \ldots, t_n] \tag{1}$$

The representation of an object at a time instance t_n can be expressed as the (continuous) accumulation of changes that appeared from the time that the object was created (t=0) until time t_n. Or mathematically:

$$\vec{O}_j^{t_n} = \int_0^{t_n} \partial \vec{O}_j \tag{2}$$

where the integration is performed over time with limits 0 and t_n($\int_0^{t_n}$). The $\partial \vec{O}_j$ vector expresses the change of object O_j.

The efficient modelling of geospatial change is a challenging issue due to the inherently diverse nature of change itself. Towards this goal we proceed with a multi-dimensional, multi-resolution decomposition of geospatial change. Change dimensionality reflects a *semantic* analysis of the composition of geospatial entities, while resolution expresses various levels of *scale* and *abstraction* when analysing geospatial information. Our goal is to express the inherently multi-dimensional change by a set of 1-D elements, in essence the attributes stored in our change-oriented differential database.

Let's assume that change exists over a multi-dimensional set R_n. We decompose change as an aggregation of j subset dimensional spaces,

$$R^n = [C_1^{a_1}, C_2^{a_2}, ..., C_j^{a_k}] \tag{3}$$

where C_1, C_2,…,C_j define multidimensional subspaces of change, and a_1, a_2,.., a_k are the corresponding dimensions of each subspace. For example C_1 might refer to geometric change, C_2 to thematic state, etc. In this case a_1 would describe the number of dimensions necessary to describe the geometric space (of change), and a_2 would relate to the dimensions of thematic form (e.g. a_2=2, if the only thematic attributes we monitor are colour and use of an object). In a way these subspaces act as the basis of the multi-dimensional change space (R_n).

Following the above analysis we can decompose the multi-dimensional object vector $\vec{O}_j$ to its n 1-D dimensions or basis of change:

$$\vec{O}_j = [O_j^1, O_j^2, ..., O_j^n] \tag{4}$$

So the $\partial\vec{O}_j$ change vector would correspond to:

$$\partial\vec{O}_j = [\delta O_j^1, \delta O_j^2, ..., \delta O_j^n] \tag{5}$$

The δO_j 1-D element of vector $\partial\vec{O}_j$ represents the change that occurred in that dimension. Here we should mention that δO_j, like $\partial\vec{O}_j$, could not be defined without a temporal interval. Throughout the following analysis, we replace consciously $\int_{t_1}^{t_2} \delta O_j$ with δO_j to simplify things.

In order to explore the values of this element, we will define fundamental unary predicates. These predicates hold in a specific temporal interval, but for the time, we omit this.

is_positive (δO_j) = True	if and only if in δO_j addition exists
is_negative (δO_j) = True	if and only if in δO_j subtraction exists
is_empty (δO_j) = True	if and only if δO_j has no change information
is_null (δO_j) = True	if and only if δO_j has information that leads to no change

Based on the above predicates δO_j can have the following value types:

$$\delta O_j = \begin{cases} + & | \text{ is\_ positive } (\delta O_j) = \text{True} \\ - & | \text{ is\_ negative } (\delta O_j) = \text{True} \\ \mp & | \text{ is\_ negative } (\delta O_j) = \text{True } \wedge \text{ is\_ positive } (\delta O_j) = \text{True} \\ * & | \text{ is\_ empty } (\delta O_j) = \text{True} \\ \varnothing & | \text{ is\_ null } (\delta O_j) = \text{True} \end{cases}$$

The first three value types can contain further qualitative or quantitative descriptors, while the last two are defined explicitly though the value types. For example a {+} type change element might be {+3 feet} if that dimension refers to the width of the road. The third value type $\{\mp\}$ applies when more than one predicate holds true for the specified temporal interval. A good example of this case would be a qualitative dimension, the departments that are using one building. In this case, the building is the object under examination. Within a temporal interval, maybe one department has moved out and a new one came in. In this case, we would have:

$$\int_{t_1}^{t_2} \delta O_j = \{\text{- Computer Science Dept, + Civil Engineering Dept}\} \tag{6}$$

2.2 Handling Change Within our Model

The above defined primitives can be used to create a multi-resolutional change representation (Fig. 2). At the coarser level, we have a general description of change. Change is treated as a whole without going into its' specifics and very general change semantics are used. Values of that level can be one of the following:

$$C_{q,q-1} = [+,-,*,\mp,\varnothing,\rightarrow,\leftarrow] \tag{7}$$

The first five values follow the same concept as the one introduced in the δO_j 1-D elements. More on how these values are defined is provided in the operator's section of this paper. To facilitate computations two new elements are included that act as objects and not change descriptors. The [→] sign denotes the birth of an object and the [←] represents the end of existence.

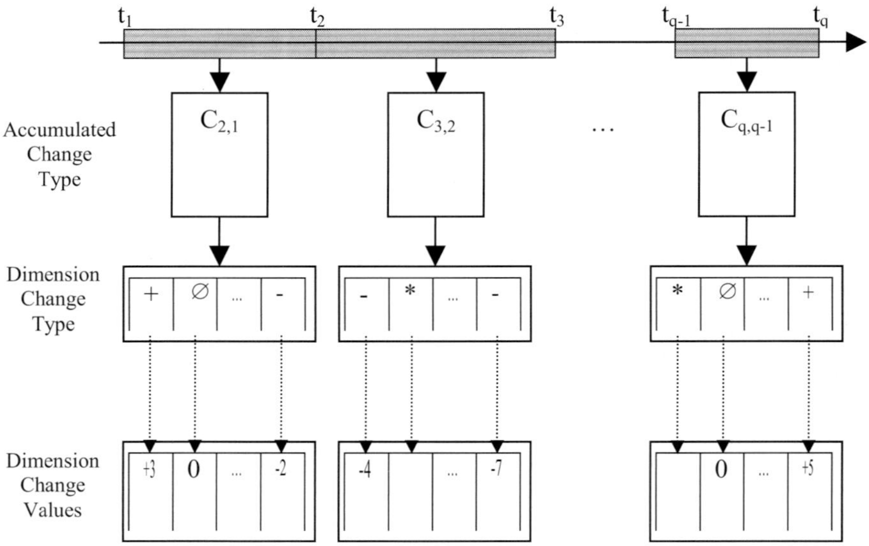

Fig. 2. Proposed change representation model

At the second level of our model, we use the value types of predicate δO_j to provide a summary (index) of change in every dimension. This level can also be described as a fundamental change semantics description in each dimension. Such indexing structures are especially important in distributed environments (Dolin et al, 1997). Since change type is indexed we can also provide a multi-resolutional query approach (Mountrakis et al, 2000) where change is treated as a binary query at first and through propagation rules the requested fields are accessed. By doing so we facilitate faster change information extraction when the specifics of change are not important, just the type/semantics of change.

At the most detailed level, we are storing the values of change in every dimension. The hierarchical structure of our model reduces the access frequency of this level, since only detailed change information triggers such access. In addition, our multi-resolutional environment can support distributed systems where the first two levels can act as indices/pointers to other databases.

2.3 Operators

In our model, we distinguish two types of operators, the ones that function horizontally and the ones that work vertically within the structure of Fig. 2. Horizontal operators *aggregate* change over time, while vertical operators *propagate* change across different resolutions for a specific temporal interval.

2.3.1 Multi-dimensional Change Value Aggregator Operator

The next step in our analysis is to provide a mechanism to aggregate changes over time. We do so by disintegrating the integral of Eq. 2 based on the discrete subset $T_n = [0, t_1, t_2, t_3, ..., t_n]$. Accordingly, we have:

$$\vec{O}_j^{t_n} = \int_0^{t_1} \partial\vec{O}_j \oplus \int_{t_1}^{t_2} \partial\vec{O}_j \oplus ... \oplus \int_{t_{n-1}}^{t_n} \partial\vec{O}_j \tag{8}$$

We make use of a multi-dimensional operator $\oplus$ that allows change aggregation in each dimension separately. Aggregation can be logical or metric, depending on the nature of each dimension. It differs from a common multi-dimensional vector aggregator, by having elements that can be qualitative or quantitative and have one or multiple instances. It compares every dimension of $\partial\vec{O}_j = [\delta O_j^1, \delta O_j^2, ..., \delta O_j^n]$ separately and groups the result as follows:

$$\begin{bmatrix} O_j^1 \\ O_j^2 \\ ... \\ O_j^n \end{bmatrix} = \begin{bmatrix} \int_0^{t_1} \delta O_j^1 \\ \int_0^{t_1} \delta O_j^2 \\ ... \\ \int_0^{t_1} \delta O_j^n \end{bmatrix} \oplus \begin{bmatrix} \int_{t_1}^{t_2} \delta O_j^1 \\ \int_{t_1}^{t_2} \delta O_j^2 \\ ... \\ \int_{t_1}^{t_2} \delta O_j^n \end{bmatrix} \oplus ... \oplus \begin{bmatrix} \int_{t_{n-1}}^{t_n} \delta O_j^1 \\ \int_{t_{n-1}}^{t_n} \delta O_j^2 \\ ... \\ \int_{t_{n-1}}^{t_n} \delta O_j^n \end{bmatrix} \tag{9}$$

Due to sensor limitations and information availability, most spatio-temporal changes are expressed as snapshots in time. In this case the continuous interval can be substituted by a discrete summation function showing the discrete rather than continuous nature of change:

$$\vec{O}_j^{t_n} = \sum_0^{t_1} \partial\vec{O}_j \oplus \sum_{t_1}^{t_2} \partial\vec{O}_j \oplus ... \oplus \sum_{t_{n-1}}^{t_n} \partial\vec{O}_j \tag{10}$$

Based on the capturing method and requested accuracy a discrete representation can be considered as continuous (complete).

Here we should note that this aggregator can function at any "horizontal" level of our model. If it is applied at the coarser level, a vector is created since there is only one dimension, at the other levels a matrix-type representation is returned.

If users wish to collapse change over time then some rules have to be defined on how change is summarised. For a metric attribute, this can be straightforward. For example in cases such as a geometric description, a simple vector overlay would be sufficient. In non-metric attributes, however, change summarisation rules would have to be defined based on user needs. Conceptually it would be hard to apply summarisation rules at the two coarser levels and it is beyond the scope of this paper. The most appropriate solution would be to apply it on the most detailed level, obtain the summarised results and then present them incoarser detail following the propagation operator described below.

2.3.2 Multi-resolutional Change Propagation Operator

Information flow in our model would commonly be bottom-up. Detailed change information propagates upwards to update the corresponding indices. In order to do so we introduce a *change propagation operator* $\wr$. If we define as $\partial\vec{O}_j = [\delta O_j^1, \delta O_j^2, ..., \delta O_j^n]$ the change value vector expressing the change of object O_j in n dimensions and we omit the temporal interval of application to simplify the representation and label $\partial\vec{O}_j index$ the change value type vector we have:

$$\partial\vec{O}_j index = \wr\, \partial\vec{O}_j = \wr\, [\delta O_j^1, \delta O_j^2, ..., \delta O_j^n] \tag{11}$$

Since there is a one-to-one relation between the dimensions of the index (value type) and the actual values, we can apply the operator separately in each dimension:

$$\partial\vec{O}_j index = [\wr\, \delta O_j^1, \wr\, \delta O_j^2, \ldots, \wr\, \delta O_j^n] \tag{12}$$

In the complex case of propagating change to the coarser level of the one-dimensional accumulated change, the problem of grouping dimensions to produce a single result arises. We reserve this for future work within our environment. For now, we provide the general framework to incorporate this in our model.

2.4 Operations

Based on the above change representation we can apply a variety of operations within our model. First we discuss operations that only require access to coarser levels of change representation and then we show how more detailed ones are applied by using actual change values.

2.4.1 Index-based Operations

We begin our discussion by showing how our representation supports fundamental index-based operations as introduced in (Worboys, 1992). These operations (e.g. birth, expansion and death) were later extended by (Hornsby and Egenhofer, 2000). We will show how these queries can be addressed within the content of the coarser two levels of our approach.

- Birth/Death

The creation of an object can be returned directly by querying the coarser level and return the temporal value of the [→] birth pointer. Similar use of the [←] value shows the end of existence.

- Expansion and Reduction

These two operations can be addressed in a qualitative and a quantitative level. If the user requests information about the presence of expansion or reduction the second level would be enough. Value types such as [+, -, $\mp$] and their corresponding temporal pointers can answer this type of query sufficiently.

- Advanced operations

In addition to the above operations that were introduced in the past we also support new, more complex change information retrieval. For example in large geospatial systems, an automated process can be supported to facilitate future information acquisition. Change information gaps can be detected easily by making use of our [*] value type. In other cases, the absence of change might be of importance for some applications such as video compression/summarization. The [∅] value can directly point to the unchanged objects within the database for a specified temporal interval. In more advanced scenarios a temporal change pattern match operation can be triggered, for example show me when this dimension changed like that and that dimension like this after time t. Such issues, however, are beyond the scope of this paper, although a general framework of support is provided.

2.4.2 Value-based Operations

- Detailed change retrieval

At this level detailed change, information is available. Change-oriented queries can be applied on single or multiple objects. For example for a single object we support information retrieval such as:

- "Show me Boardman's largest expansion/reduction"
- "Has Boardman ever showed a specific shape of change (e.g. Π-shape)?"
- "Was there an expansion at the North side?"

We can also combine multiple objects and summarise results:

- "In this area (on campus) show me the largest expansion decade"
- "Return the most popular expansion direction (e.g. North)"
- "Has the campus ever changed following this pattern (where pattern might be a combination of spatial/thematic dimensions over time)?"

- Consistency operations

In order to provide the user with valid results, some consistency checks are introduced. In the first category, we can find operations that apply to all dimensions. Such operations might look for validity of a subtraction. The idea is that the system cannot subtract something that does not exist.

Let's assume that a subtraction in dimension w of object O_j takes place at t_k .We perform a one-dimensional aggregation in $[0, t_k]$ through the $\oplus$ operator:

$$|_0^{t_k} \vec{O}_j^w = \int_0^{t_1} \partial O_j^w \oplus \int_{t_1}^{t_2} \partial O_j^w \oplus \ldots \oplus \int_{t_{k-1}}^{t_k} \partial O_j^w \tag{13}$$

The $|_0^{t_k} \vec{O}_j^w$ shows the current state of object Oj in dimension w in time t_k. Assuming $|_{t_k}^{t_{k+1}} \delta O_j^w$ is the one-dimensional change element then

$$\text{is\_ negative}\,(|_{t_k}^{t_{k+1}} \delta O_j^w) = \text{True} \tag{14}$$

If we define consistency unary predicate as *cons_subtraction* and A=$|_0^{t_k} \vec{O}_j^w$, while B=negative descriptor(s) of $\{|_{t_k}^{t_{k+1}} \delta O_j^w\}$ then we have:

$$cons\_subtraction\,(\text{A, B})= \left\{ \begin{array}{l} \text{if } A \cap (-B) = (-B) \text{ then True} \\ \\ \text{else False} \end{array} \right\} \tag{15}$$

In the second category of consistency checks we find validity operations that depend on the dimension. For example if one dimension is the "building outline" then a consistency check might be that it is a closed polygon. Another important operation would be to compare change dimensions with expected behaviour to filter out inconsistencies.

3 Implementation in a Differential Gazetteer

In Agouris et al (2000) we proposed a Spatio-temporal Gazetteer (STG) as an efficient model to store and retrieve spatio-temporal information. We will use this prototype to demonstrate the practical use of the operators introduced in this paper. In the following figures we show the original dataset in the STG that was provided for change detection in Boardman Hall.

Fig. 3. Boardman Hall in t_1 (1932)

Fig. 4. Boardman Hall in t_2 (1971)

Fig. 5. Boardman Hall in t_3 (1985)

Fig. 6. Boardman Hall in t_4 (1997)

After analysing the above dataset, this is how change information and specifically the building's outline is represented in our model. At the coarser level a general change description is stored. At the second indexing level, the δO_j data types are stored and at the more detailed level the metrics of change are shown.

Table 1. Building's outline change representation example

Accumulated Change Type	$\rightarrow$ +		$\varnothing$	$\mp$
Dimension Change Type	$\varnothing$	+	$\varnothing$	$\mp$
Dimension Change Values	$\varnothing$	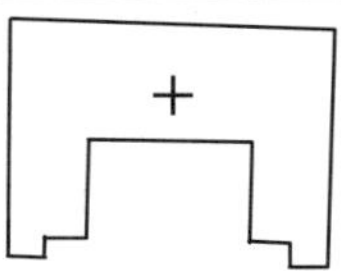	$\varnothing$	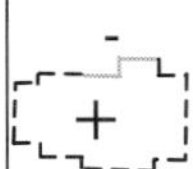
Time Line	$0 < t < t_1$	$t_1 < t < t_2$	$t_2 < t < t_3$	$t_3 < t < t_4$

In a different example this is how the building area (using information from complementary datasets like blueprints and maps) would be stored in a state-based structure as opposed to our change-oriented one using the primitives defined before.

Table 2. Building's area state representation example

Time (years)	Building Area (sq. meters)
1925	∅
1932	∅
t_1	∅
t_2	1043
1979	1043
1982	1043
t_3	1043
t_4	1239
2000	1239
2001	1239

Table 3. Building's area change representation example

<table>
<tr><td>Accumulated Change Type</td><td colspan="2">→ +</td><td>∅</td><td>+</td></tr>
<tr><td>Dimension Change Type</td><td>∅</td><td>+</td><td>∅</td><td>+</td></tr>
<tr><td>Dimension Change Values</td><td>∅</td><td>+1043</td><td>∅</td><td>+196</td></tr>
<tr><td>Time Line</td><td>$1925 < t < t_1$</td><td>$t_1 < t < t_2$</td><td>$t_2 < t < t_3$</td><td>$t_3 < t < t_4$</td></tr>
</table>

The main points demonstrated by these two examples are that the minimal redundancy and clear expressiveness of the method can be achieved. With this approach, we can ensure *redundancy minimisation* in most cases by reducing a multi-dimensional problem to its minimal modified dimensions. There are however, some cases where a change-oriented approach might require a larger volume of storage (e.g. constantly changing qualitative dimensions). Nevertheless, for common geospatial applications, where we have numerous instances in which a monitored object remains unchanged the gain of a differential model over a state-based one becomes substantial.

Regarding *expressiveness,* it can be easily seen that a differential STG model directly supports numerous types of object and scene queries. They range from object to scene queries, and can address any level of resolution within the model of Fig. 2. This allows for example queries on the index level (e.g. how many times has a building changed over the last 10 years?), the object level (e.g. what is the largest expansion of this building during the last 5 years?), and even the scene level (e.g. which building within this area has expanded the most in the last decade?).

4 Conclusions

In this paper we presented a differential change-oriented model for the storage and communication of spatio-temporal information. We discussed the development of model primitives and operators to support the aggregation of change over time and the propagation of change across resolution. By making use of these primitives the result is a multi-resolutional change model that captures the semantics of change from the coarser level to the most detailed one. Our primitives and operators presented here extend existing qualitative operators to support the management of quantitative and geometric information within a change-oriented spatio-temporal environment. The major advantage of this approach lies in the minimisation of redundancy, and its superb expressiveness in the communication process. A GIS implementation prototype is discussed to reveal the effectiveness of our change model.

Acknowledgements

This work is supported by the National Science Foundation through grants number CAREER IIS-9702233, DG-9983445, and ITR-0121269.

References

Agouris P, Beard K., Mountrakis G, Stefanidis A (2000) Capturing and Modeling Geographic Object Change: A Spatio-Temporal Gazeteer Framework. Photogrammetric Engineering and Remote Sensing 66(10):1224-1250

Agouris P, Stefanidis A, Gyftakis S (2001) Differential Snakes for Change Detection in Road Segments. Photogrammetric Engineering and Remote Sensing 67(12):1391-1399

Dolin R, Agrawal D, El Abbadi A, Dillon L (1997) Pharos: A Scalable Distributed Architecture for Locating Heterogeneous Information Sources. In: Nevada F, Golshani, Makki K (eds) Proceedings 6th ACM Int. Conference on Information and Knowledge Management. Las Vegas, pp 348-355

Erwig M, Guting R, Schneider M, Vazirigiannis M (1999) Abstract and Discrete Modeling of Spatio-Temporal Data Types. Geoinformatica 3(3):269-296

Han J, Dong G, Yin Y (1999) Efficient mining of partial periodic patterns in time series database. In: Proceedings Int. Conf. Data Engineering (ICDE'99). Sydney, Australia

Hornsby K, Egenhofer M (2000) Identity-based change: a foundation for spatio-temporal knowledge representation. International Journal of Geographical Information Science 14(3): 207-224

Langran G (1993) Issues of Implementing a Spatio-temporal System. International Journal of Geographical Information Systems 7(4): 305- 314

Mountrakis G, Agouris P, Stefanidis A (2000) Navigating Through Hierarchical Change Propagation in Spatio-temporal Queries. In: IEEE Press, Time 2000 Workshop. Cape Breton, NS, Canada, pp 123-131

Peuquet D (2001) Making Space for Time: Issues in Space-Time Data Representation. GeoInformatica 5(1): 11-32

Peuquet D, Wentz E (1995) An Event-Based Spatio-temporal Data Model (ESTDM) for Temporal Analysis of Geographic Data. International Journal of Geographic Information Systems 9: 7-24

Pfoser D, Jensen C, Theodoridis Y (2000) Novel Approaches to the Indexing of Moving Object Trajectories. VLDB. 395-406

Theodoridis Y, Sellis TK, Papadopoulos A, Manolopoulos Y (1998) Specifications for Efficient Indexing in Spatio-temporal Databases. In: Intl. Conference on Scientific and Statistical Database Management. Capri, Italy, pp 123-132

Worboys M (1992) A Model for Spatio-Temporal Information. In: Proceedings of the 5th International Symposium on Spatial Data Handling. Charleston, South Carolina, USA pp 602-611

Worboys M (1994) A Unified Model of Spatial and Temporal Information. Computer Journal 37:26-34

Yeh TS, de Cambray B (1995) Modeling Highly Variable Spatio-Temporal Data. In: 6th AustraliAsian Database Conference. pp 221-230

A QTM-based Algorithm for Generation of the Voronoi Diagram on a Sphere

Xuesheng Zhao[1,2], Jun Chen[3], Zhilin Li[1]

Department of Land Surveying and Geo-Informatics , The Hong Kong Polytechnic University, Kowloon, Hong Kong, Tel: (+852) 2766 5960, Fax: (+852) 2330 2994, lsllz@polyu.edu.hk[1]
China University of Minging and Technology (Beijing), D11 Xueyuan Road, Beijing, China, 100083, zxs@mail.cumtb.edu.cn[2]
National Geometrics Center of China, 1 Baishengcun, Zizhuyuan, Beijing, China, 10004, chenjun@nsdi.gov.cn[3]

Abstract

To efficiently store and analyse spatial data at a global scale, the digital expression of the Earth's data must be global, continuous and conjugate, i.e., a spherical dynamic data model is needed. The Voronoi data structure is the only published attempt and only solution (which is currently available) for dynamic GIS. The complexity of the Voronoi algorithms for line and area data sets in a vector-based context limits its application in dynamic GISs. As yet, there is no raster-based Voronoi algorithm for objects (including points, arcs and regions).

To overcome this deficiency, an algorithm for generating a spherical Voronoi diagram, that is a Voronoi diagram on a spherical surface, is presented based on O-QTM (Octahedral Quaternary Triangular Mesh). The basic idea is to apply the dilation operation developed in mathematical morphology to objects on the sphere in an effort to produce the effect of distance transformation. The distance contours of objects will form the Voronoi boundaries of the spherical objects.

The algorithm presented in this paper can handle point, line and area objects. Additionally, it has been tested and concluded that the processing time required for this algorithm with point, arc and region data is proportional to the levels of complexity of the spherical surface tessellation. The difference (error) between the *great circle distance* and the QTM cells distance is related to the spherical distance.

Keywords: O-QTM, neighbour triangular, Voronoi diagram on sphere, recursive dilation

1 Introduction

It is been argued by Li et al (1999) that the Voronoi data structure is the only possible solution (which is currently available) for dynamic GISs. This view is also similar to that suggested by Wright and Goodchild (1997), who point out that the Voronoi methods are the only published attempts of which we are aware that are well suited to achieving a dynamic GIS. This is because Voronoi diagrams (VD) have many excellent properties in spatial analysis (Gold 1992, Edwards 1993), dynamic operation (e.g. to add or delete objects without destroying the bubble structure of the cells) (Gold and Condal 1995; Gold and Mostafavi 2000), and computational geometry (Aurenhammer 1991; Okabe et al. 2000), etc.

So far, the Voronoi diagram on a spherical surface has been applied to some areas, such as global spatial indexing (Lukatela 1987, 2000), interpolation on a sphere (Watson 1988,1998) and dynamic operations (Gold 1997; Gold and Mostafavi 2000), *etc.* For example, Lukatela (1987) sets up a digital geo-positioning model and develops an operational software package that provides geometrical and geo-relational functions to applications that manipulate spatial objects. A Voronoi tessellation is used as a base for a highly efficient indexing system to increase the speed of data manipulation (Lukatela 1987) (Fig. 1a). Watson (1988,1998) develops the MODEMAP system and uses a point-set Voronoi diagram for interpolation on a spherical surface (Fig. 1b). Gold and Mostafavi (2000) attempt to develop a global dynamic data structure with the Voronoi diagram. In this type of structure, the Voronoi diagram is a basic data model that dynamically maintains spatial relationships.

(a) (b)

Fig. 1. The Voronoi diagram of spherical objects, **(a)** as a spatial index (Lukatela 1987), and **(b)** for interpolation (Watson 1989)

It is evident that the Voronoi diagram is a spatial data structure, which has become increasingly important. Considerable efforts have been spent on the development of the algorithms, however, most of them are applicable for vector data and are based on point sets in a planar surface. The algorithms for generating the Voronoi diagrams of line and area sets in vector data are very complex. This complexity has greatly limited the application of the Voronoi data model and only limited advances have been achieved. On the other hand, only a few algorithms are available for generating Voronoi diagrams for point-sets, but no algorithms for arc-sets (or curve-face sets) have yet been thoroughly demonstrated for a spherical surface.

This paper presents a method for computing a spherical Voronoi diagram for arbitrary input objects (including points, arcs and curve faces) based on the Quaternary Triangular Mesh (QTM). It makes use of the dilation operation through the neighbour triangles. It is simplified by using their labelling codes. Advantages of the QTM-based method are that the QTM data structure is seamless, hierarchical and numerically stable everywhere on the surface of the sphere. Its hierarchical data structure can be used to efficiently manage multi-resolution global data and it allows spatial characteristics be studied at different levels of detail in a consistent fashion across extensive regions of the sphere (Lee and Samet 2000).

The next section critically reviews a selection of algorithms. Section 3 will introduce the selection of tessellation methods on a spherical surface as a reference system. Section 4 will present in detail, a searching method of spherical neighbour triangles.. In section 5, an algorithm for generating the Voronoi diagram for objects on the sphere by means of a dilation operation in QTM is presented..

2 A Critical Examination of Algorithms for Generation of the Voronoi Diagram on a Sphere

The Voronoi diagram has been one of the research highlights in the area of computational geometry since it was introduced to the computer domain by Shamos and Hoey (see Okable et al. 2000) as an efficient data structure. Most methods of the Voronoi diagram generation are based on point sets in planar space, such as the incremental method, the divide and conquer method, the indirect generating method and the parallel method (Aurenhammer 1991; Li et al 1999; Okabe et al. 2000). There are only a few algorithms for generating spherical Voronoi diagrams (Geyer, 2000). In this section, such algorithms will be examined.

2.1 Voronoi Diagrams on a Sphere: Basic Concepts

The distance on a sphere is different from that of Euclidean space. As a result, the Voronoi diagram on a sphere will have a different definition (Okabe, et al. 2000; Lee and Samet 2000): Let $P = \{ p_1, p_2, ..., p_n \}$ $(2 \leq n < \infty)$ be distinct points on a sphere $\boldsymbol{S}$ with the unit radius centred at the origin, and $\boldsymbol{X}$ and $\boldsymbol{X}_i$ be the location vectors of $p \in \boldsymbol{S}$ and $p_i \in \boldsymbol{S}$, respectively. The shortest distance from p to p_j on $\boldsymbol{S}$ is defined by the length of the shortest arc on the great circle (the circle whose centre is at the Center of $\boldsymbol{S}$) passing through p and p_i. Mathematically, this distance is written as

$$d_{gc}(p, p_i) = \arccos\left(X^T X_i\right) \leq \pi \tag{2.1}$$

This distance is called the *great circle distance*. The bisector defined with the great circle distance is given by the great circle that *perpendicularly* passes through the mid-point of the great circular arc combining p_j and p (*perpendicularly* means that sufficiently small segments of the two great circles around the mid-point are orthogonal). This bisector divides the sphere $\boldsymbol{S}$ into two disjoint hemispheres. Thus the bisector defined with the great circle distance is well-behaving, and

$$V(p_i) = \left\{d_{gc}(p, p_i) \leq d_{gc}(p, p_j), j \in I_n \setminus \{i\}, p \in S\right\} \tag{2.2}$$

gives a non-empty set in $\boldsymbol{S}$. This set is called the spherical Voronoi polygon associated with p_i. The set of resulting spherical Voronoi polygons gives a generalised Voronoi diagram, which is called the spherical Voronoi diagram defined by p_i on $\boldsymbol{S}$. Fig. 2 shows a spherical Voronoi diagram.

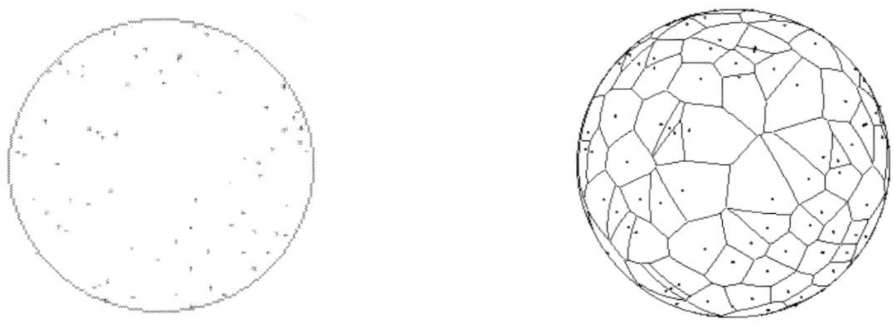

(a) (b)

Fig. 2. A Voronoi diagram on a sphere based on points sets (Geyer et al 2000), **(a)** discrete points on sphere, **(b)** a Voronoi diagram on spherical points sets

A *Voronoi edge* is the boundary between two Voronoi regions and a *Voronoi vertex* is the intersection of three or more Voronoi edges.

2.2 Voronoi Diagrams on a Sphere: Algorithms

Augenbaum (1985) gives an insertion method for computing the Voronoi diagram of a set of n points on a sphere with time complexity $O(n^2)$, and Robert (1997) presents an incremental algorithm, which can be constructed with time complexity $O(nlogn)$. As pointed out by (Gold 1992; Gold and Condal 1995), the vector-based methods are good for point sets, but complex for line or area input sets. This is also true as far as the spherical surface is concerned. This deficiency impedes the Voronoi data structure from being widely applied in dynamic GISs.

To solve this problem, Yang and Gold (1996) presented a *point-line* model. In their model, the complex objects are decomposed to points and lines. Voronoi diagrams for points and lines are generated first, and then translated to Voronoi diagrams of the complex objects by removing the Voronoi edges between points or lines of the same object. More recently, Gold and Mostafavi (2000) extend this model onto a spherical surface. The advantage of this method is that it can generate a Voronoi diagram of relatively complex vector-based objects and deals with dynamic changes of topological relations. However, this result can only be reached through many additional steps, such as *decompose*, *calculation*, *remove, compose*, etc. As a result, their algorithm and the data structure is relatively complex. More importantly, the data structures of these vector-based methods lack hierarchical expression. Therefore, it is very difficult to deal with the hierarchical expression of a large volume of spherical data.

Due to the complexity of vector-based methods, alternative methods, i.e. raster-based methods, on planar have been approached (Dehne 1989; Embrechts and Roose 1996; Okabe et al 2000). These algorithms make use of distance transformations, such as city block, chessboard, and Octagon, etc. It is such that the approximation of these raster distances to the Euclidean distance becomes poorer and poorer when the distances becomes larger. To solve this problem, Li et al (1999) present a method based on raster dynamic distance transformation, which employs the dilation operator which was developed in mathematical morphology. In their algorithm, the distance errors are confined to one pixel.

These raster-based algorithms, however, are limited to planar surfaces and cannot be translated onto spherical surfaces directly, simple because planar and spherical spaces are not homoeomorphic. In this paper, a spherical mesh-based method is presented for the generation of the Voronoi diagram on a spherical surface. In this method, the spherical surface is subdivided into triangles by QTM tessellation, which is similar to the planar raster. The spherical hexagon distance (to be discussed in section 5.1) is used for the dilation operation to compute the spherical Voronoi diagram of the arbitrary spatial objects (including points, arcs, and curve faces).

3 Selection of a Tessellation Method for a Spherical Reference System

To generate a Voronoi diagram on a sphere, a tessellation method for spherical surfaces and labelling schemes should be selected first. The selection criteria are based on the transformation efficiency between triangle code and its spherical coordinates and the encoding method, which best suits a search of neighbouring triangles.

3.1 Selection of a Tessellation Method for a Sphere's Surface -- O-QTM

Originally, the concept of spherical surface tessellation was presented by Fuller, a German cartographer, for studying the mapping projection in the 1940s (Dutton 1996). Since then, many researchers have approached this problem to project, analyse and index global data. Many methods are based on *inscribed polyhedron*, such as tetrahedron, cube (Snyder 1992), *octahedron* (Dutton 1996, 1999a, 1999b; Goodchild and Yang 1992, 1992; Otoo and Zhu 1993; Clarke and Mulcahy 1995), *dodecahedron* (Wickman and Elvers, 1974), *icosahedron* (Fekete 1990; White et al. 1992; Lee and Samet 2000), as shown in Fig. 3. Edges of the polyhedron are projected to the spherical surface and form the edges of spherical triangles.

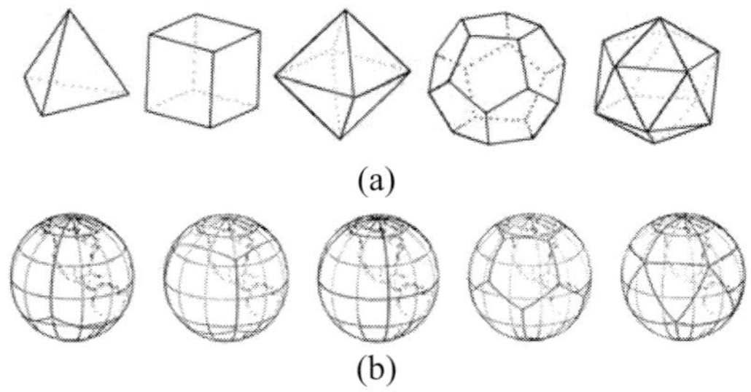

Fig. 3. Spherical surface tessellation based on inscribed polyhedrons (White et al 1992), **(a)** five polyhedrons, **(b)** projected to the spherical surface

In this study, the octahedron is chosen as a basis for an O-QTM. The reason for this selection is that it can be readily aligned with the conventional geographic grids of longitude and latitude. When this is done, its vertices occupy cardinal points and its edges assume cardinal directions, following the equator, the prime meridian, and the 90th, 180th and 270th meridians, making it simple to determine which facet a point on the sphere occupies (Dutton, 1996). In addition, each facet is a right spherical triangle and one subdivision line of each face is parallel to the equator.

There are some methods of recursive tessellation that satisfy the different requirements. It is well recognised that the transformation efficiency between triangle code and its spherical coordinates (latitude, longitude) is an important function in an application system (White and Kimerling 1998). As such the *Latitudes and longitudes average method* is selected in this study, as done by Dutton (1996). Fig. 4 illustrates levels 1, 2 and 3.

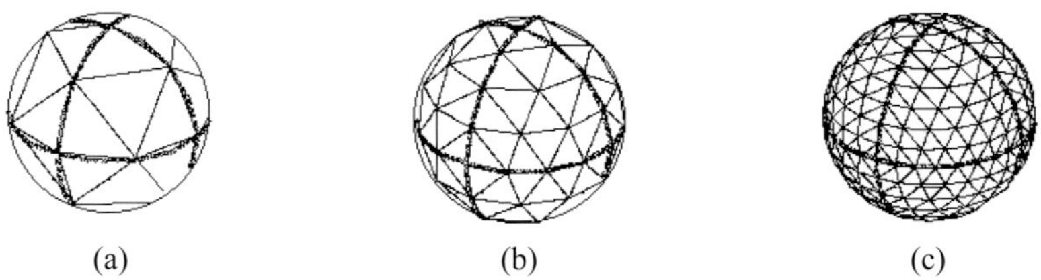

Fig. 4. Hierarchical tessellation of the spherical facet based on octahedron (Dutton 1996), **(a)** level 1, **(b)** level 2, and **(c)** level 3

3.2 Selection of an Encoding Method

The addressing methods and data structures of triangular regions are related to the operation and indexing on spherical surfaces. Some methods, such as SQT (Sphere Quadtree) (Fekete 1990), SQC (Semi-quadcode) (Otoo et al 1993), THDS (Triangular Hierarchical Data Structure) (Goodchild et al 1991) and QTM-IDs (Dutton 1996, 1999) have been applied.

The triangle labelling method selected in our study is similar to the one used for the octahedron (Goodchild et al 1991). The address code of each QTM cell consists of an octant number (from "0" to "7") followed by up to 30 quaternary digits (from "0" to "3"), which names a leaf-node in a triangular quadtree rooted in the given octant. At the *k-th* level of decomposition, the triangle address A is represented by: $A = a_0a_1a_2a_{3...}a_k$, where a_1 to a_k are k quaternary digits and a_0 is an octal digit representing the initial octahedral decomposition at level 0.

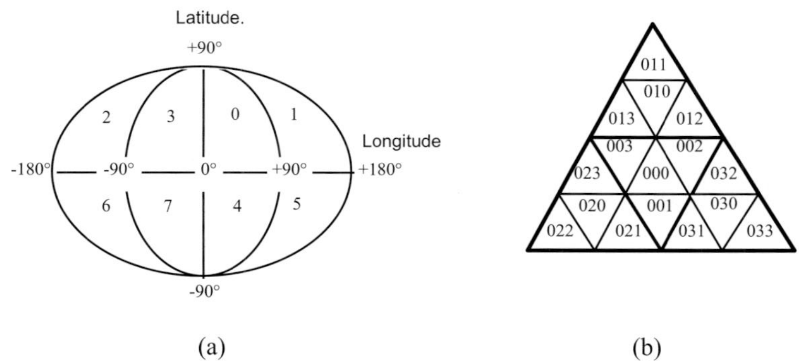

Fig. 5. Surface partition and encoding, **(a)** original partition of the earth's surface, **(b)** ecnoding in 0 unit

4 A Method for Searching Neighbours in QTM

4.1 Different Types of Neighbours

The neighbours with shared edges are called *edge-neighbour-triangles.* Those only with common vertices are called *vertex-neighbour-triangles*, as shown in Fig. 6a. In the data structure of O-QTM, triangle neighbours, that are located in two adjacent octants, must also be specifically considered since the search methods for different locations of *border triangles* are different. From Fig. 7b, it can be seen that, if a triangle has edge(s) at the border of an octant, the triangle will have *edge-neighbour-triangle*(s) and *vertex-neighbour-triangles* in its neighbouring octant(s). On the other hand, if a triangle has vertices(s) at the border of an octant, the triangle will only have *vertex-neighbour-triangle*(s) in its neighbouring octant(s). Border triangles can be classified into 4 categories: *edge, sub-edge, corner* and *sub-corner triangles* and can be defined as follows (Goodchild et al 1991), see Fig. 6b:

- *Edge triangle (1)* --if it has exactly one edge-neighbour-triangle in the adjacent octant.
- *Sub-edge triangle (2)* --if it has exactly three vertex-neighbour-triangles in the adjacent octant.
- *Corner triangle (3)* --if it has exactly two edge-neighbour-triangles in the adjacent octant.
- *Sub-corner triangle (4)* --if it has exactly six vertex-neighbour-triangles in the adjacent octant.

(a) (b)

Fig. 6. Definitions of neighbour triangles and border triangles, **(a)** edge-neighbour-triangles and vertex-neighbour-triangles, **(b)** classification of a border triangle

4.2 A Method of Searching Edge-Neighbour-Triangles

All triangles have three edge-neighbour-triangles in QTM on a sphere. We use the codes *t, l, r* to represent the three edge-neighbour-triangles with common top, left and right edges for a given triangle *U* inside an octant, the code *T, L, R* to represent the edge-neighbour-triangle of a top, left and right edge triangle lying in the adjacent octant. Different searching methods used for the triangles at the different locations are in one octant. Border triangles can be classified into 7 categories and are shown in Fig. 7.

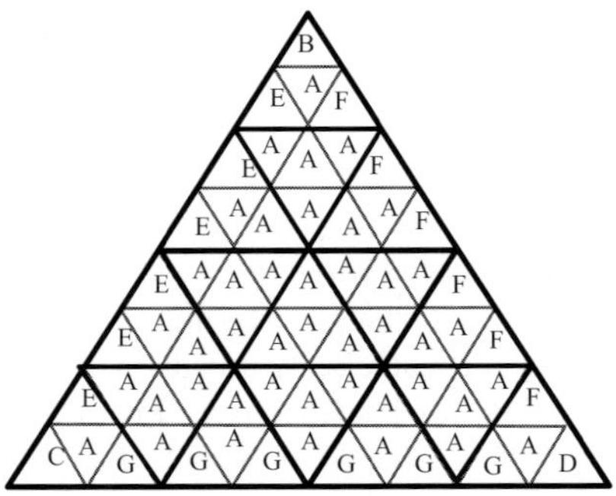

Fig. 7. Categories of triangles by different searching algorithm of edge-neighbour-triangle

The data strings *t, l, r, T, L* and *R* can be obtained by the triangular address *U*. The details can be seen in (Goodchild et al 1991).

4.3 A Method of Searching Vertex-Neighbour-Triangles

For all triangles in an octant, a corner triangle has 7 vertex-neighbour-triangles (Fig. 8-a), and each of the others have 9 vertex-neighbour-triangles, as shown in Fig. 8-b. There are several methods to search vertex-neighbour-triangles. In our approach, they have been obtained through their edge-neighbour-triangles.

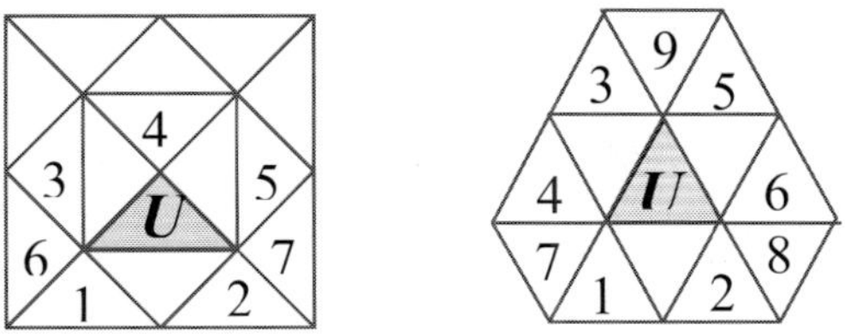

(a) (b)

Fig. 8. Angle-neighbour-triangles, **(a)** corner triangle, **(b)** no-corner triangle

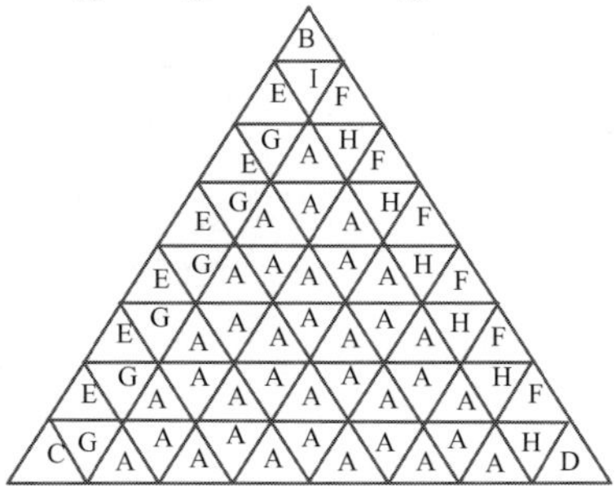

Fig. 9. Nine different types of searching methods of angle-neighbour-triangles

The searching algorithm of vertex-neighbour-triangles varies with the locations of *U* in the octant. In particular, the border triangles are much more complex. These triangles can be classified into 9 different types (shown as Fig. 9).

5 A QTM- Based Algorithm for Generating the Voronoi Diagram

The algorithm for the generation of the Voronoi diagram on a sphere is based on the dilation operator of spherical triangles. In this section, we will develop the triangle-dilation-operator in QTM according to the principle of the raster dilation operator in mathematical morphology. From this and the neighbour searching method described in section 4, the *triangle cell distance* diagram, which consists of a number of distance contours, radiated from each object on the sphere, is obtained. The most distant contours form the approximate boundaries of spherical the Voronoi diagram.

5.1 Distance Contours in QTM Generated by a Dilation Operator

Dilation and erosion are two of the basic operators in mathematical morphology and have wide applications in digital image processing as well as geographical information science (Su et al. 1997, Li et al 1999). These two basic operators in QTM cells can be defined as follows:

$$\begin{array}{ll} \textit{dilation} & A \oplus B = \cup b \in BA_b \\ \textit{erosion} & A \ominus B = \cap b \in BA_b \end{array} \quad (5.1)$$

Where A is an original region on sphere and B is a structuring element, examples of dilation and erosion are given in Fig. 10:

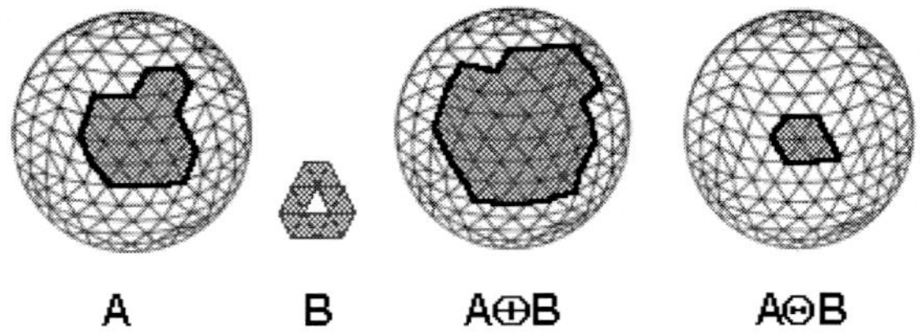

Fig. 10. Dilation and erosion operations based on QTM

Now the concept of distance becomes important. In vector mode, the *distance* on sphere means the great circle (or arc or geodesic) distance. The distance between two points $\boldsymbol{X_1}$ (λ_1, ϕ_1) and $\boldsymbol{X_2}$ (λ_2, ϕ_2) on sphere is defined as formula (5.2):

$$L_{X_1X_2} = R\theta = R\cos^{-1}(X_1 \bullet X_2) \quad (5.2)$$

Where θ is an angle between $\boldsymbol{X_1}$ and $\boldsymbol{X_2}$, and the range of $cos^{-1}\theta$ is taken to be $[0, \pi]$. In QTM, the distance in the integer number is more desirable and thus normally employed. Accordingly, the order of neighbours could be the best candidate to be used as the QTM distance to approximate the *great circle distance* (e.g. hexagon structuring element in Fig. 11a).

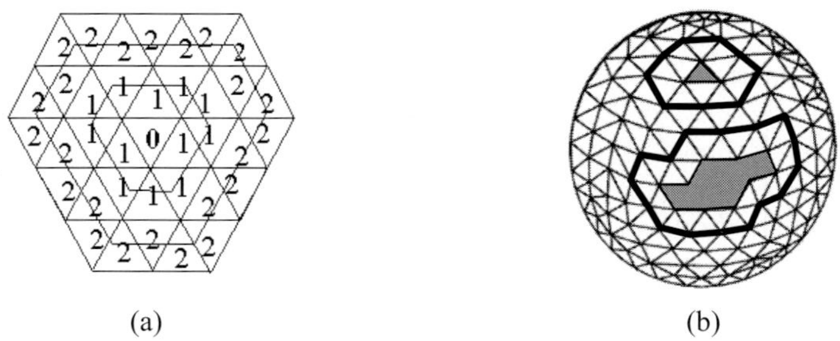

Fig. 11. Hexagon structuring element and dilation on a sphere, **(a)** the hexagon structuring element, **(b)** the triangular dilation on a spherical surface

For an inside triangle, the region expanded is a hexagon with three edges of length $(m\text{-}1)\times l$ and three edges of length $m\times l$ if the region does not cross to another octant. Where m is the number of times the procedure is repeated and l is the edge length of the triangle at the given level. The form of the region changes if the region crosses the edge of an octant as shown in Fig. 11-b. The distance between the border of the dilated region and the nearest edge of a given triangle (point) varies from $nl\sqrt{3}/2$ *to nl*, a factor of 0.866, which is larger than in a rectangular raster where the ratio of the edge to the diagonal of a square is 0.717 (Goodchild et al. 1991). The error of forming the dilation region by hexagon structuring element in QTM is smaller than the chess structuring element in rectangular cells in planar space. In addition, the topological and metrical properties of the region are preserved:

- The region generated is connected and there exists no hole.
- The region dilated each time is a stripe region surrounding the old region with the width of $l\sqrt{3}/2$.

The neighbouring triangle search algorithm described in section 4 can be used directly for dilation since the dilation operation requires searching all neighbours (edge neighbours and vertex neighbours) of a point, arc or region.

5.2 The Principle of Generating the Voronoi Diagram on a Sphere in QTM

The Algorithm for generating the Voronoi diagram on a spherical surface is based on the principle of dilation operation in mathematical morphology. In the QTM, a point is represented by a triangle, an arc by a series of neighbour triangles and a region by a series of neighbour triangles on and within its boundary trace. The dilation operation of an arc or region can be simply done by dilation of all triangles by which the arc or region is described. Thus, the process of generating the Voronoi diagram is as follows: First, determine the *edge* and *vertex neighbour triangles* around the object (such as points, arcs and regions) by using the algorithm of searching neighbour triangles presented in section 4. Second, remove all duplicate triangles and generate the dilation trace of the object. The spherical distances are approximate equal from the outer boundary of the dilation trace to the boundary of the object. Next, repeat the dilation operation and stop when the dilation trace is intersected with the other dilation trace. The intersecting trace is just the Voronoi edge between two objects.

5.3 Algorithm

Input: tessellation level N and an object data set on a spherical surface: $\Gamma=\{A_1, A_2, A_3, ..., A_n\}$
Output: Voronoi diagram of input data set Γ stored in the file *VoronoiData*.

CsphVoronoiView::OnCaculateVoronoi( )
{

```
    step1: LongLatitude_to_QTMcode(Γ);
    step2: For every objects A_i in Γ
    {
        step2.1: For every QTMcode Q_j in A_i;
            {
                Adjact12(Q_j); //searching neighbour triangles
                if Adjact12(Q_j) are copy code
                    Delete_copyQTMcode(Q_j);
Else  Dialation_A[i] ← Adjact12(Q_j)
            }
step2.2: For every Dialation_A[i]
            {
                For every QTMcode Q_im in A[i] and every QTMcode Q_jk in A[j], i≠j
                    {
                        if (Q_im = Q_jk)
                        VoronoiData←Q_jk;
                    }
            }
    }
    step3: QTMcode_to_LongLatitude(VoronoiData);
    step4: Output( );
}over
```

Based on the algorithm described above, a prototype system has been developed on OpenGL with VC++ language.

A number of experimental tests have been conducted but for the sake of brevity will not be discussed here. An example of the Voronoi diagram for arbitrary objects positioned on a sphere are shown in Fig. 12).

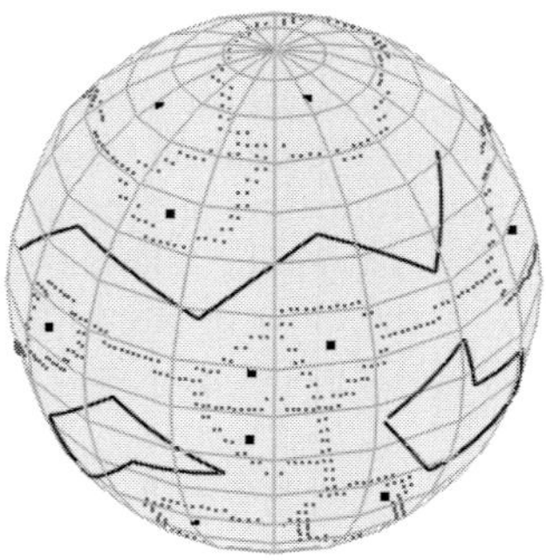

Fig. 12. Voronoi diagrams of arbitrary objects on a sphere based on QTM

6 Conclusions

In this paper, a new algorithm for generating a Voronoi diagram on the sphere is developed by the recursive dilation operation in QTM (Quaternary Triangular

Mesh). This method can easily handle arbitrary composite objects (including arcs and regions).The dilation operation, developed in mathematical morphology, was applied to objects on the sphere, in an effort to provide the effects of a distance transformation. The distance contours of objects will be used to form the boundaries of Voronoi regions of spherical objects. In this case, the principle of dilation is extended to spherical surfaces. A method for spherical distance transformation based on QTM is developed and a detailed algorithm is presented. This algorithm can handle point, line and area objects. It has also been tested and it was observed that the time consumption of this algorithm with input points, arcs and regions are equal, and is proportional to the levels of the spherical surface tessellation. The difference (error) between *great circle distance* and QTM cells distance is related to the spherical distance (better than the raster dilation in the planar space), and is related mainly to the locations of the generating points.

Acknowledgements

The work described in this paper was supported by the National Natural Science Foundation of China (under grant No.69833010) and by the Research Grants Council of the Hong Kong Special Administrative Region (Project No. PolyU 5048/98E).

References

Augenbaum M (1985) On the Construction of the Voronoi Mesh on a Sphere. Computational Physics 59: 177-192

Aurenhammer F (1991) Voronoi Diagram-A Survey of a Fundamental Geometric Data Structure. ACM Computing Survey 23(3): 345-350

Clarke KC, Mulcahy KA (1995) Distortion on the Interrupted Modified Collignon Projection. In: Proceddings of GIS/LIS 95. Nashville, TN, pp 175-181

Dehne F, Hassenklover A, Sake J (1989) Computing the configuration space for a robot on a mesh-of-processors. Parallel Computer 12(2):221—231

Dutton G (1996) Encoding and Handling Geospatial Data with Hierarchical Triangular Meshes, In: Kraak, MJ and Molenaar M (eds) Proceeding of 7th International Symposium on Spatial Data Handling. Netherlands, pp 34-43

Dutton G (1999) A hierarchical Coordinate System for Geoprocessing and Cartography. Lecture Notes in Earth Sciences, Springer-Verlag

Edwards G (1993) The Voronoi Model and Cultural Space: Applications to the Social Sciences and Humanities. In: Frank AU, Compari I (eds) Spatial Information Theory: A Theoretical Basis For GIS: European Conference, COSIT'93, Marciana Marina, Elba Island, Italy, pp 202-214

Embrechts H, Roose D (1996) A Parallel Euclidean Distance Transformation Method. Computer Vision and Image Understanding 63:15-26

Fekete G (1990) Rendering and Managing Spherical Data With Sphere Quadtree, In: Proceedings of Visualization '90. IEEE Computer Society, Los Alamitos, CA, pp 176-186

Geyer C (2000) Voronoi diagram on the surface of the sphere [online]. Available from: http://www.cis.upenn.edu/~cgeyer/sphr-vor.html

Gold CM (1992) The Meaning of Neighbour. In: Frank AU, Campari I, Formentini U (eds) Theories and Methods of Spatio-Temporal Reasoning in Geographic Space, Lecture Notes in Computing Science, 639, Springer-Verlag, pp 220-235

Gold CM (1997) The Global GIS. In: Proceeding of the International Workshop on Dynamic and Multi-Dimension GIS. Hong-Kong, China, pp 80-91

Gold CM, Condal AR (1995) A Spatial Data Structure Integrating GIS and Simulation in a Marine Environment. Marine Geodesy 18: 213-228

Gold CM, Mostafavi M (2000) Towards the Global GIS. ISPRS Journal of Photogrammetry and Remote Sensing 55(3): 150-163

Goodchild MF, Yang Shiren (1992) A Hierarchical Data Structure for Global Geographic Information Systems. Computer Vision and Geographic Image Processing 54(1): 31-44

Goodchild MF, Yang Shiren, Dutton G (1991) Spatial Data Representation and Basic Operations for a Triangular Hierarchical Data Structure. NCGIA report, 91-8

Lee M, Samet H (2000) Navigating through Triangle Meshes Implemented as Linear Quadtree. ACM transactions on Graphics 19(2): 79-121

Li C, Chen J, Li Z (1999) Raster-based Methods for the Generation of Voronoi Diagrams for Spatial Objects. International Journal of Geographic Information Science 13(3): 209-225

Lukatela H (1987) Hipparchus Geopositioning Model: An Overview. In: Proceedings of the Eighth International Symposium on Computer-Assisted Cartography. Baltimore, Maryland, 87-96

Lukatela H (2000) Ellipsoidal Area Computations of Large Terrestrial Objects [online]. Available from: http://www.ncgia.ucsb.edu/globalgrids/papers

Okabe A, Boots B, Sugihara K, Chiu S (2000) Spatial Tessellations: Concepts and Applications of Voronoi Diagrams (2nd Edition). John Wiley and Sons Ltd

Otoo E, Zhu H (1993) Indexing on Spherical Surfaces Using Semi-Quadcodes. In: Abel J, Beng CO (eds) Advances in spatial Databases 3th International Symposium, SSD`93. Singapore, pp 509-529

Robert JR (1997) Delaunay Triangulation and Voronoi Diagram on the Surface of a Sphere. ACM Transactions on Mathematical Software 23(3):416-434

Snyder JP (1992) An Equal-Area Map Projection for Polyhedral Globes. Cartographica 29(1):10-21

Su B, Li Z, Lodwick G, Muller JC (1997) Algebraic Models for the Aggregation of Area Features Based upon Morphological Operators. International Journal of Geographical Information Science 11(3): 233-246

Watson DF (1988) Natural Neighbour Sorting on the N-Dimensional Sphere. Pattern Recognition 21(1): 63-67

Watson DF (1998) Modemap: An Implementation of Natural Neighbour Interpolation on the Sphere [online]. Available from: http://members.iinet.net.au/~watson/modemap.html

White D, Kimmerling AJ (1998) Comparing Area and Shape Distortion on Polyhedral Based Recursive Tessellations of the Sphere. International Journal of Geographical Information Science 12(8): 805-827

White D, Kimmerling J, Overton WS (1992) Cartographic and Geometric Components of a Global Sampling Design For Environment Monitoring, Cartography & Geographical Information Systems 19(1): 5-22

Wickman FE, Elvers E (1974) A system of domains for global sampling problems. Geografiska Annaler 56(3/4): 201-212

Wright D, Goodchild MF (1997) Data from Deep: Impliications for the GIS Community. International Journal of Geographical Information Science 11(6):523-528

Yang W, Gold C (1996) Managing Spatial Objects With the VMO-Tree, In: Kraak, MJ, M Molenaar (eds) Proceeding of 7th International Symposium on Spatial Data Handling. Netherlands, pp 15-31

The Dimensional Model: A Framework to Distinguish Spatial Relationships

Roland Billen[1], Siyka Zlatanova[2], Pierre Mathonet[3], and Fabien Boniver[4]

Geography Department, University of Liege, Allée du 6-Août 17, 4000 Liège, Belgium, +32-(0)4/3665752, +32-(0)4/36656 93, Roland.billen@ulg.ac.be[1]
Department of Geodesy, Delft University of Technology, Thijsseweg 11, 2629 JA Delft, The Netherlands, +31 (0)15 278 2714, +31 (0)15 278 2745, S.Zlatanova@citg.tudelft.nl[2]
Math. Dept., University of Liege, Grande Traverse 12, 4000 Liège, Belgium, +32-(0)4/3669480, +32-(0)4/366 9547, P.Mathonet@ulg.ac.be[3]
Math. Dept., University of Liege, Grande Traverse 12, 4000 Liège, Belgium, +32-(0)4/3669417, +32-(0)4/366 9547, F.Boniver@ulg.ac.be[4]

Abstract

A unique characteristic of GIS as compared to other information systems, is their capacity to manage *spatial* relationships, such as connections or interrelations among objects in the geometric domain. A number of frameworks use topology as a basic mechanism to define spatial relationships. The OpenGIS consortium has adopted one of them, i.e. the 9-intersection model. In this paper, a new framework for representing spatial relationships - the *Dimensional model* - is introduced. The model was first developed for convex spatial objects and is now extended to topological n-manifolds. It is based on two major concepts, i.e. the *dimensional elements* of spatial objects and the *dimensional relationships, i.e.* the relationships existing between dimensional elements. The model addresses a substantial group of spatial relationships and provides a flexible framework to consider either generalised or specialised types of associations..
Keywords: spatial relationships, spatial model, convexity, topological manifolds

1 Introduction

The development of a mathematical theory for categorising relationships has been identified as an essential task in tackling the diversity and incompleteness of spatial-relationships' representations (see Boyle *et al.* 1983, NCGIA 1989). Among all the approaches (e.g. metrics, topology, ordered sets), topology seems the most appreciated (see Egenhofer *et al.* 1994, Egenhofer 1989, Egenhoher and Sharif 1998, Clementini *et al.* 1993, Kainz *et al.* 1993, Molenaar 1998). The OpenGIS consortium has adopted one of the topologically based frameworks, referred to as the 9-intersection model, as a generic mechanism for implementation and development.

In this paper, a new framework for representing spatial relationships named the *Dimensional Model* (DM) is introduced. The model was first developed for convex spatial objects and is now extended to topological n-manifolds. The paper is organised in the following sections. First, the order concept for convex bodies is presented. Then, the order concept for a topological manifold is developed. A formal description of the Dimensional Model follows. After a brief comparison with the 9-intersection model, some spatial relationships are depicted using the DM. Finally; the implementation of the DM is overviewed.

2 An Order Formula for Convex Bodies

2.1 Affine Subspaces and Convex Sets

Geographers and mathematicians use co-ordinates to describe location in space. Whenever d is a positive integer, we denote by $\mathbf{R}^d$ the set of tuples $(\alpha_1,\ldots,\alpha_d)$ of real numbers $\alpha_1,\ldots,\alpha_d$. As Euclidean vector (or affine) space, $\mathbf{R}^d$ is the natural framework in which geometry can formally be studied. It is also the first example of Euclidean topological space. A detailed introduction to general topology and manifolds can be found in Alexandrov (1998) and Lee (2000).

A subset A of $\mathbf{R}^d$ is an *affine subspace* if, for any distinct points x, y belonging to A, the (infinite) straight line defined by x and y lies in A. Points, straight lines, planes, and $\mathbf{R}^3$ itself are the only affine subspace of $\mathbf{R}^3$. Their respective *dimensions* are 0, 1, 2 and 3. An affine subspace of dimension d-1 of $\mathbf{R}^d$ is named *hyperplane*. For instance, the hyperplanes are merely straight lines in $\mathbf{R}^2$ and planes in $\mathbf{R}^3$. A hyperplane divides the whole space in two regions, called *halfspaces*.

A subset A of $\mathbf{R}^d$ is *convex* if, for any two points x,y belonging to A, the segment $[x,y]$ lies in A.

A *supporting hyperplane* M of a convex set C is a hyperplane such that

- C is included in one of the halfspaces defined by M,
- $M \cap C \neq \varnothing$.

Fig. 1 shows examples of convex sets and supporting hyperplanes in $\mathbf{R}^2$ and $\mathbf{R}^3$.

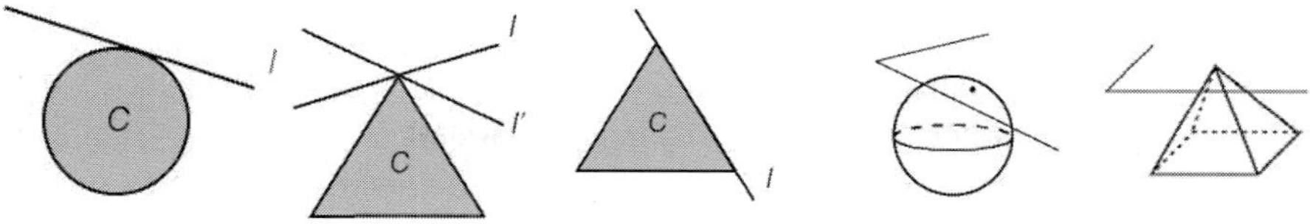

Fig. 1. Examples of closed convex sets and hyperplanes

2.2 Order of Points in Closed Convex Set

Let C be a convex set, closed with respect to the Euclidean topology. Each point of C has an *order*, whose definition is given in Berger (1978, p. 50), and can be stated as follow.

Let C be a closed convex set in $\mathbf{R}^d$ and $x \in C$. The order of x in C, denoted by $o(x,C)$, is the dimension of the intersection of all supporting hyperplanes containing x.

In particular, if no supporting hyperplanes contains x, then x has order d. One can prove that those points with order d are exactly the interior points of C with respect to the Euclidean topology.

Fig. 2 shows points with various orders in a triangle, a drop, and a segment. For instance, the top point of the triangle has order 0. Indeed, infinitely many supporting hyperplanes (only two are represented) go through it and intersect in that point itself. Fig. 3 shows examples in $\mathbf{R}^3$.

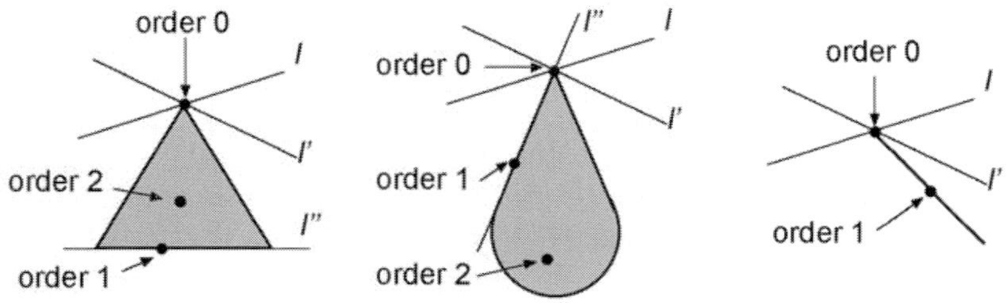

Fig. 2. Order of points of closed convex sets in $\mathbf{R}^2$ (with some hyperplanes)

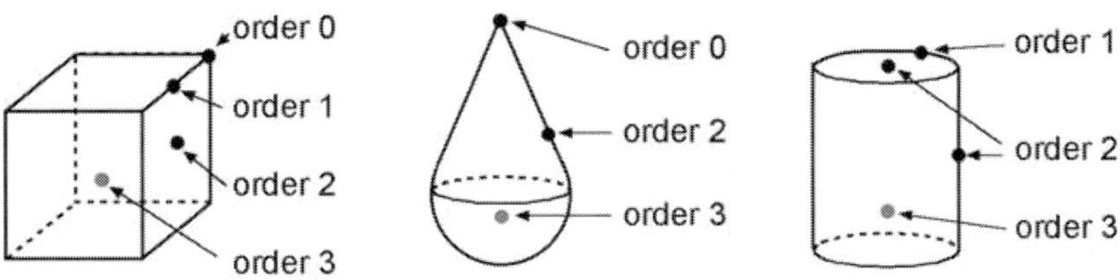

Fig. 3. Order of points of closed convex sets in $\mathbf{R}^3$ (without hyperplanes)

3 An Order Formula for Manifolds

More often than not, objects in GIS application environments are not convex. It is therefore reasonable to seek for a suitable formula for more general objects. Those will be *topological manifolds*. Our approach goes as follow: given a point x in a topological manifold, we create a closed convex neighbourhood of x and compute its order with respect to this neighbourhood. The key point is that, intuitively, the order of a point is a "local" property: it actually depends only on the manifold in a neighbourhood of x

Therefore, the steps to find the order of points are: 1) determine if the point of interest stands on the interior or the boundary of the manifold; 2) centre a ball on this point with a given radius r; 3) take the intersection between the ball and either the manifold or the boundary of the manifold; 4) determine the convex hull of this intersection; 5) determine the order of the point regarding the convex object created (the convex hull); 6) repeat the operation 2 to 6 with a smaller radius, until getting a minimum value for the order.

3.1 Topological Manifolds

Let n be a positive integer. We denote by $\mathbf{R}^n_+$ the subset of tuples $(\alpha_1,\ldots,\alpha_n)$ with $\alpha_n \geq 0$. A subset A of $\mathbf{R}^d$ is a *topological n-manifold with boundary* if each point $x \in A$ has neighbourhood, which is homeomorphic to an open subset of $\mathbf{R}^n_+$.

Let A be such n-manifold. It is the disjoint union of its *interior*, A° , and its *boundary*, ∂A. The points belonging to A° are named *interior points*. A point is interior if it has a neighbourhood homeomorphic to an open subset of $\mathbf{R}^n$. It is worth noticing that those definitions do not coincide with the Euclidean topological definitions of interior and boundary.

Two additional notations will be useful. If B is a subset of $\mathbf{R}^d$, we denote by *conv(B)* the *convex hull* of B, which is the smallest convex subset of $\mathbf{R}^d$ containing B. If r is a positive real number and $x \in \mathbf{R}^d$, we denote by $b_{x,r}$ the closed ball with centre x and radius r.

3.2 The General Formula

We can now write the general formula that encapsulates the algorithm presented above. Let X be a topological manifold. Assume furthermore that X is closed as a subset of $\mathbf{R}^d$. If x is an interior point of X, then its *order* is defined as:

$$\lim_{r \to 0, r \neq 0} o(x, conv(b_{x,r} \cap X)) . \quad (1)$$

Now, if $x \in \partial X$ (x is a boundary point of X), then its *order* is:

$$\lim_{r \to 0, r \neq 0} o(x, conv(b_{x,r} \cap \partial X)) . \quad (2)$$

In both cases, the limit exists and equals the minimum of the values of the function *o*.

Fig. 4 illustrates this approach for 1-manifold. Figs. 5 and 6 show some more examples for 1-manifold (curve line) and 2-manifold.

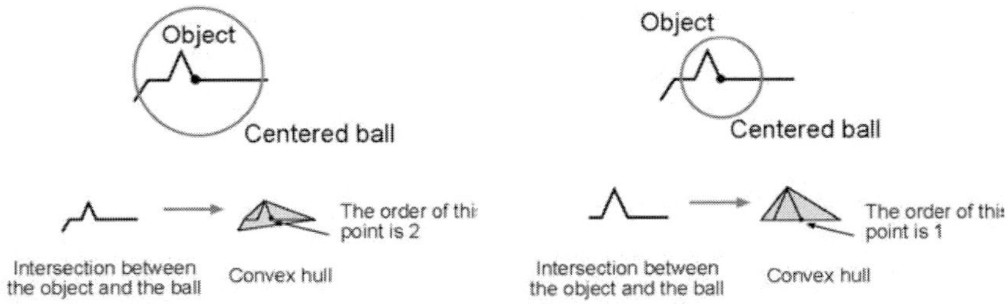

Fig. 4. Determination of the order of a broken line's point with different ball radius

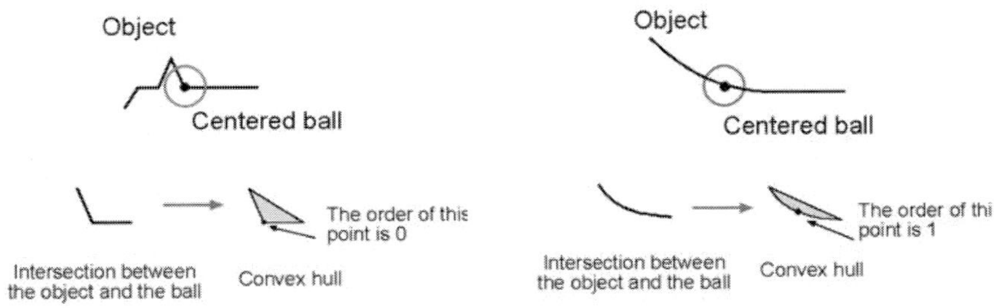

(a) (b)

Fig. 5(a) Determination of the minimum order of a broken line's point, **(b)** determination of the minimum order of a curve line's point

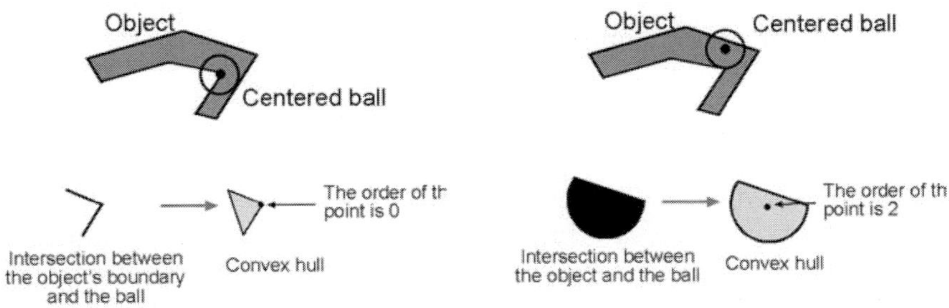

(a) (b)

Fig. 6(a) Determination of the minimum order of a polygon's boundary point, **(b)** determination of the minimum order of a polygon's interior point

4 The Dimensional Model

The Dimensional Model (DM) is a (conceptual) framework to describe both spatial objects and spatial relationships. The spatial objects are composed by *dimensional elements,* which are based on order's points of object. The spatial relationships between spatial objects are described in terms of *dimensional relationships*, i.e. relationships that exist between the dimensional elements of the objects.

4.1 Spatial objects in DM

In our model, *a simple spatial object of dimension d is equivalent to a topological d-manifold.* They are called simple because it is possible to apply directly the order formula to them, and therefore determine their dimensional elements. We also define a complex spatial object (as a combination of simple spatial object), which will not be discussed here (see Fig. 7).

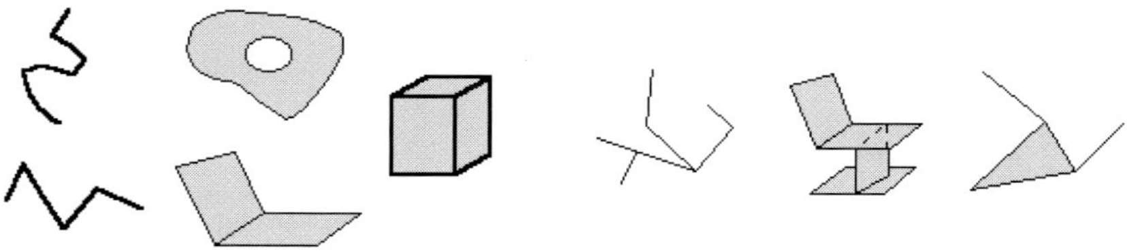

Simple spatial objects (manifold)

Complex spatial objects (aggregation of simple objects)

Fig. 7. Examples of spatial objects

4.2 Dimensional elements of DM

The *dimensional elements* are associated with different parts (or points) of a spatial object according to their order.

1. The α–dimensional element (denoted αD-element) of a spatial object C (which has at least dimension α), corresponds to the set of all the points (or parts) of C which have order 0 to α.
2. The αD-element of a spatial object C has an **extension** and may have a **limit**.
3. The extension is the subset of C formed by its points of order α, and the limit is the subset of C formed by its points of order 0 to order (α-1).

Thus, if the αD-element has a limit, this limit corresponds to a lower (α-1)D-element. The 0D-element does not have a limit by definition. Fig. 8 illustrates the dimensional elements of a polygon. First, the order of all the points is determined.

This convex is composed of 2D, 1D and 0D-elements. The different extension and limits are presented in the Fig. 8. In the case of an ellipse, the 1D-element does not have a limit. It should be noted that there is one and only one *x*D-element for this object.

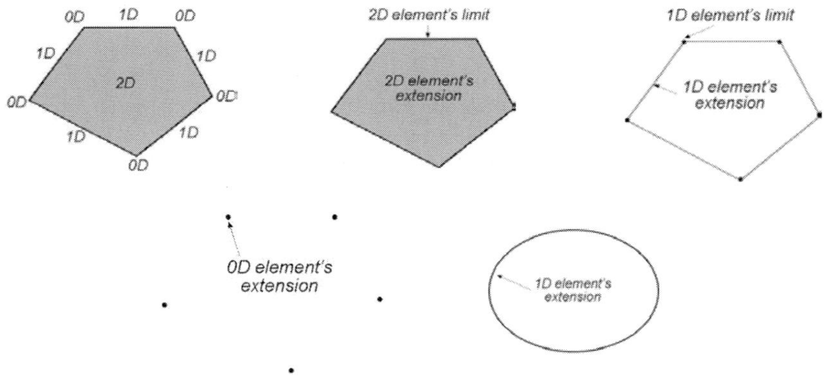

Fig. 8. Order of points and dimensional elements of a polygon and of an ellipse

4.3 Dimensional Relationships

The *dimensional relationships* are defined as the relationships existing between dimensional elements. These relationships can either be total, partial or non-existent, and are oriented (from one element to an other one).

1. A dimensional element is in **total relation** with another dimensional element if their intersection is equal to the first element, and if the intersection between their extensions is not empty.
2. A dimensional element is in **partial relation** with another dimensional element if their intersection is not equal to the first element, and if the intersection between their extensions is not empty.
3. A dimensional element is in **no relation (non-existent)** with another dimensional element if the intersection between their extensions is empty.

Fig. 9 illustrates the three types of dimensional relationships for 2D-elements.

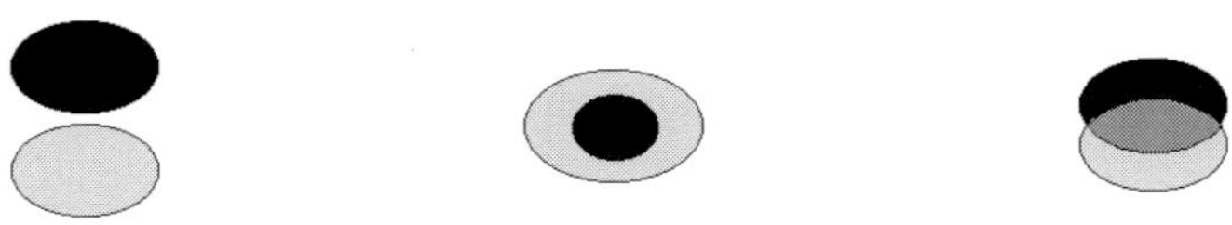

No relation (non-existent) Total relation Partial relation

Fig. 9. The different types of dimensional relationships between two 2D-elements (from black element to grey element)

4.4 The Dimensional Model for Investigation of Spatial Relationships

The dimensional elements (with their limits and extensions) and the dimensional relationships (i.e. total, partial and non-existent) are the basic tools to decode spatial relationships. The spatial relationship between two objects can be expressed by the dimensional relationships that exist between the dimensional elements of both objects. For example, let us consider a polygon *A* (with 2D, 1D, 0D elements) and a line *B* (with 1D and 0D-elements). The dimensional relationships between the spatial object *A* and the spatial object *B* can be identified in the following sequence: first, check the dimensional relationship between the 2D-element of *A* and all the dimensional elements of spatial object *B*; then, check the dimensional relationship between the 1D-element of *A* and all the dimensional elements of spatial object *B*, etc. The dimensional relationships between *B* and *A* can be found following the same approach. Three groups of dimensional relationships can be distinguished following this approach, i.e. the *simplified*, the *basic* and the *extended* relationships.

A dimensional relationship is coded *RnDy*, using the notations *R* for relationships, *nD* for dimension of the element of the first object, and *y* dimension of the element of the second object. R2D1 represents the dimensional relationships between the 2D-element of the first object and the 1D-element of the second object. Furthermore, a numeric code for the three types of dimensional relationships, i.e. 0 for non-existent, 1 for total and 2 for partial, is specified.

The basic relationships. This group contains all the relationships between every possible combination of dimensional elements. For example, the spatial relationship between a polygon (considering 2D, 1D, 0D elements) and a line (considering 1D and 0D elements) can be expressed by basic dimensional relationships as follows:

R2D1	R2D0	R1D1	R1D0	R0D1	R0D0
{0,1,2}	{0,1,2}	{0,1,2}	{0,1,2}	{0,1,2}	{0,1,2}

where 0, 1, 2 correspond to the possible dimensional relationships, i.e. non-existent, total and partial.

The extended relationships. The partial relation can be further investigated for the dimension of the intersection. For example, in $\mathbf{R}^3$, a 2D-element and a 1D-element may have a 1D- or a 0D-intersection. If the intersection has the same dimension as the lowest dimensional element in the relation, it keeps the code 2 (e.g., if a 2D-element and a 1D-element have a 1D-intersection, it would be noted as R2D1 2). If the dimension of the intersection is just inferior, then it would have code 3 (a 0D-intersection in the example, R2D1 3). Our example of the extended dimensional relationships between a polygon and a line becomes:

R2D1	R2D0	R1D1	R1D0	R0D1	R0D0	
{0,1,2,3}		{0,1,2}	{0,1,2,3}	{0,1,2}	{0,1,2}	{0,1,2}

The simplified dimensional relationships. In many cases the dimensional elements of the second object is not relevant. For example, it might be interesting to know if the 2D-element of a polygon has a relationship with another object independently of its dimensional elements. In such cases, some dimensional relationships can be aggregated. The complete aggregation rules will not be exposed here. Our example of the simplified dimensional relationship becomes:

R2D	R1D	R0D
{0,1,2}	{0,1,2}	{0,1,2}

Fig. 10 illustrates how the spatial relationships between a polygon and a line are represented according to the different groups in the Dimensional Model.

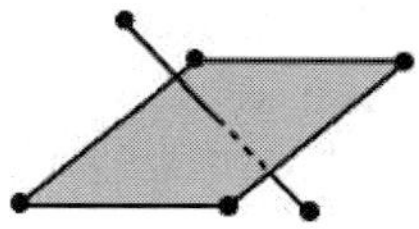

Basic relationships

R2D1	R2D0	R1D1	R1D0	R0D1	R0D0
2	0	0	0	0	0

Extended relationships

R2D1	R2D0	R1D1	R1D0	R0D1	R0D0
3	0	0	0	0	0

Simplified relationships

R2D	R1D	R0D
2	0	0

Fig. 10. The Dimensional Model applied to the relationship between polygon and line

5 The Dimensional Model Versus the 9-intersection Model

As mentioned, the 9-intersection model is a one of the topological relationships standard. We want to clearly expose the difference that exists between the 9-intersection model and the Dimensional Model.

5.1 The 9-intersection Model (9i Model)

The framework is based on the assumptions of spatial objects (represented by 0,1,2,3-cells) without holes and intersecting parts. If the spatial objects A and B are defined in the same topological space, their *boundary, interior* and *exterior* are denoted by $\partial A, A°, A$-$,\partial B, B°$ and B^- (see Egenhofer, M. and J. Herring, 1990). The binary relationship *R(A,B)* between the two objects is then identified by composing all the possible set intersections of the six topological primitives, i.e. $\partial A \cap \partial B$, $A° \cap B°$, $\partial A \cap B°$, $A° \cap \partial B$, $A^- \cap B^-$, $A^- \cap \partial B$, $A^- \cap B°$, $\partial A \cap B^-$ and $A° \cap B^-$, and detecting empty (0) or non-empty (1) intersections. For example, if two objects have a common boundary, the intersection between the boundaries is non-empty, i.e. $\partial A \cap \partial B = 1$; if they have intersecting interiors, then the intersection $A^\circ \cap B^\circ$ is not empty, i.e. $A^\circ \cap B^\circ = 1$. To represent the relationships a decimal coding is adopted here (see Kufoniyi 1995). That is to say, that the binary number (obtained from a particular ordering of all the intersections) is converted into a decimal number. For example, the relationship that has a binary number 000011111 (considering the ordering given above) equals decimal number 31 and hence the relationship is R031 (disjoint). Thus, all the decimal codes are between R000 and R512

5.2. Differences Between 9i Model and DM

The differences between the 9-intersection model and DM can be summarised as follows:

1. The definition space of 9i model is topological when it is an affine one for DM.
2. In the 9i model, the spatial relationships are determined looking to intersections of the topological primitives. In the DM, the spatial relationships are determined looking to intersections of the dimensional elements.
3. The 9i model is related to the cell approach. The dimension of the cells are not (always) equivalent to order of points (see Fig. 11). Therefore, the union of *d*-cells of an object is not equivalent to its *d*D element, except for some types of objects (as polytopes).

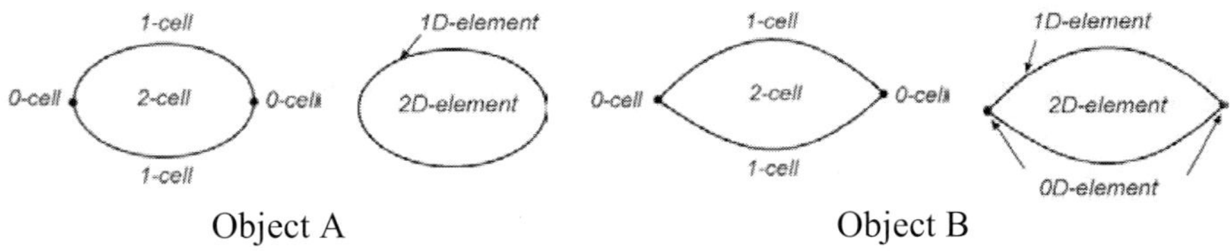

Fig. 11. Examples of differences between cell decomposition and dimensional elements

4. The DM allows grouping spatial relationships at different level of complexity: more general groups than with 9i model, same kind of groups or more specific ones (see the next section).

6 Possible Spatial Relationships

As mentioned, three groups of dimensional relationships can be used to express the spatial relationship between objects. Furthermore, one has the choice to take into consideration only relevant dimensional elements in a geographical perspective. For example, a particular geographical phenomena represented by a polygon may not need a distinction between the 1D-element and the 0D-element which form its border. In such a case, although the 0D-element exists in the object's definition, it would not be taken into account in the determination of the spatial relationship.

The number of potential relationships between two objects depends, with respect to the Dimensional Model, on: 1) the dimensional nature of the objects (given by dimensional elements), 2) the semantic dimension of the object (only the "relevant" dimensional elements from a semantic point of view) and 3) the group of dimensional relationships. Similarly to the 9-intersection model, only a small number of the theoretical relationships can be realised in reality (see Zlatanova, 2000 for examples in $\mathbf{R}^3$). The same approach is adopted, i.e. elimination of impossible relationships by negative conditions. All the possible relationships between line-line, line-surface, line-body, surface-surface, surface-body and finally body-body have been established and studied for the different criterion mentioned above. Note that in this study, some elements have been simplified. For example the 0D element of a line corresponds only to its extremities (no broken lines). Table 1 portrays simplified, basic and extended relationships for all levels of dimensional relevancy.

Table 1. Possible relationships according to the Dimensional Model

D element	Dim. Rel.	Line-Line	Surface-line	Body-body	Surface-surface	Body-Line	Body-surface
(n)D	S.	5	3	5	5	3	3
	B.	5	3	5	5	3	3
	E.	7	5	5	11	3	3
(n)D	S.	11	10	8	15	6	8
&(n-1)D	B.	33	31	8	43	19	19
	E.	61	?	15	?	43	48
(n)D	S.		?	?	?	19	?
&(n-1)D	B.		?	?	?	?	?
&(n-2)D	E.		?	?	?	?	?
(n)D	S.			?		?	?
&(n-1)D	B.			?		?	?
&(n-2)D	E.			?		?	?
&(n-3)D							

With Dim. Rel. = Dimensional relationship; S. = simplified; B. = basic; E. =extended

? non determinate

(dark grey) impossible case

33 possible relationships according to the 9-intersection model

Considering the highest and the second highest dimensional element and using the basic relationship, the relationships reported by Zlatanova (2000) are found. They are shown in bold font in table 1 (light grey row). Most of the equivalent topological cases can be "refined" using some of the more complex criteria in the Dimensional Model (i.e. everything that is below the shaded line in table 1). Further, more aggregated relationships can be found with simpler criterions (i.e. everything that is above this line). Fig. 12 presents the extended dimensional solutions (0D elements are not taken into account) to the topological equivalence R095 and R287 (between a body *A* and a surface *B*).

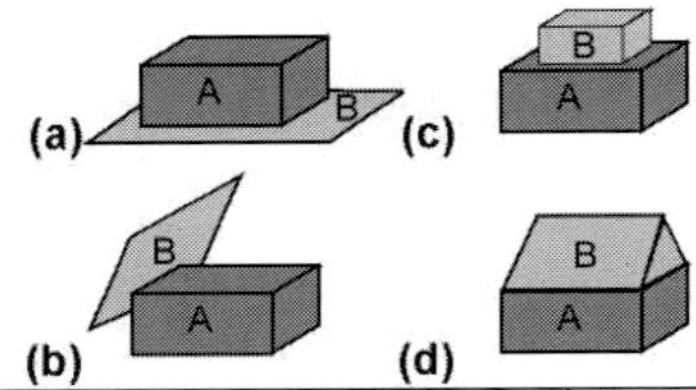

	R3D2	R3D1	R2D2	R2D1	R1D2	R1D1
(a)	0	0	2	0	2	0
(b)	0	0	0	0	2	0
(c)	0	0	0	2	0	0
(d)	0	0	0	0	0	2

Fig. 12. Dimensional relationships for, **(a)** and **(b)** topological equivalence R095, and **(c)** and **(d)** R287

Another interesting example concerns the "meet" relationship between two bodies. All the spatial situations in Fig. 13 are equivalent to R287 according to the 9-intersection model, while all of them can be differentiated with the Dimensional model. Solutions are given only for cases *a, b, c* and *d.* The cases *e* and *f* need to consider the 0D-element.

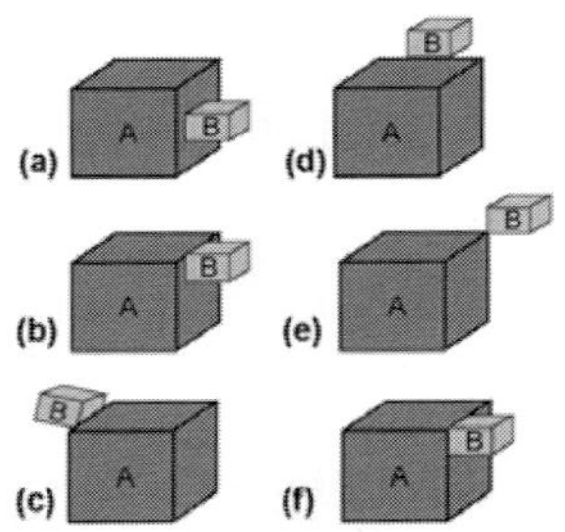

	R3D3	R3D2	R3D1	R2D3	R2D2	R2D1	R1D3	R1D2	R1D1
(a)	0	0	0	0	2	2	0	0	0
(b)	0	0	0	0	2	2	0	0	2
(c)	0	0	0	0	0	0	0	2	2
(d)	0	0	0	0	0	0	0	0	2

Fig. 13. Dimensional relationships for topological equivalence (R287)

7 Implementation of the DM

The DM is oriented spatial relationships. . It is possible, however, to use it to enrich spatial data structures and some convenient tests have been done. The "purest" solution is to use D elements instead of topological primitives. Of course, some problems may occur when objects are transformed – the D elements are not invariant through all the homeomorphic transformations. A more pragmatic solution is to keep a topological data structure and compute the dimensional elements during the determination process of dimensional relationships. This computation uses colinearity and coplanearity algorithms. The search for an adapted implementation of DM has revealed some very interesting concepts, such as the possibility of relinquishing the *single-valued map* approach (see Molenaar, 1998). This is still being considered and will not be addressed here.

8 Conclusions

In this paper, we have presented a new framework, i.e. the Dimensional Model, to distinguish spatial relationships of spatial objects. The Dimensional Model supports the presentation of a very large group of spatial relationships. It provides a flexible framework that allows either generalised or specialised types of relationships to be considered. The freedom in choosing geographically relevant dimensional elements and groups of dimensional relationships allows one to decide on a particular complexity of spatial relationships. It covers a large range of spatial objects (topological n-manifold for simple spatial objects and combinations of simple spatial objects for complex spatial objects).

Acknowledgements

The authors would like to thank Bernard Cornélis for his critical proof reading and constructive comments.

References

Alexandrov PS (1998) Combinatorial topology, Vol. 1, 2 and 3. Dover Publications Inc., Mineola, New York

Boyle A, Dangermond J, Marble D, Simonett D, Smith L, Tomlinson R (1983) Final report of a conference on the review and synthesis of problems and directions for large scale geographic information system development. Technical report, NASA NAS2-11246

Berger M (1978) Géométrie 3 : convexes et polytopes, polyèdres réguliers, aires et volumes. Paris, Cedic/Fernand Nathan

Clementini E, di Felice P, van Oosterom P (1993) A small set of formal topological relations suitable for end-user interaction. In: Proceedings of the 3rd International Symposium on Large Spatial Databases. Springer-Verlag, Berlin, pp 277-295

Egenhofer M (1989) A formal definition of binary topological relationships. In: Proceedings of the 3rd International Conference on Foundation of Data Organisation and Algorithms. pp 457-472

Egenhofer M, Clementini E, di Felice P (1994) Topological relations between regions with holes. International Journal of Geographical Information Systems 8(2):129-144

Egenhofer M, Herring J (1990) A mathematical framework for the definition of topological relationships. In: Proceedings of Fourth International Symposium on Spatial Data Handling. Zurich, Switzerland, pp 803-813

Egenhofer M, Shariff B (1998) Metric details for natural-language spatial relations. ACM transactions on information systems 16(4):295-321

Kainz W, Egenhofer M, Greasley (1993) Modelling spatial relations and operations with partially ordered sets. International Journal of Geographical Information Systems 7(3):215-229

Kufoniyi O (1995) Spatial coincidence modelling, automated database updating and data consistency in vector GIS. Ph.D. thesis, ITC, The Netherlands

Lee J (2000) Introduction to topological manifolds. New-York, Springer-Verlag

Molenaar M (1998) An introduction to the theory of spatial objects modelling. London, Taylor&Francis

NCGIA (1989) The research plan of the National Center for Geographic Information and Analysis. International Journal of Geographical Information Systems 3(2):117-113

Zlatanova S (2000) On 3D topological relationships. In: Proceedings of the 11th International workshop on Database and Expert System applications. Greenwich, London, pp 913-919

A Quantitative Description of Spatial Configurations

Francisco Godoy and Andrea Rodríguez

Departamento de Ingeniería Informática y Ciencias de la Computación,
Universidad de Concepción,Casilla 160-C, Correo 3, Concepción, Chile.
{ fgodoyf , andrea } @ udec.cl

Abstract

The description of spatial configurations plays a fundamental role in content-based retrieval and in the design of user interfaces, as this description is often the basis for the specification of query constraints. In this paper we present measures that capture the content description of spatial configurations as quantitative values that reflect characteristics of individual objects and relations among objects. These content measures are evaluated in terms of their ability to distinguish topological relations, and they are used to compare spatial configurations. Such content measures are suitable for similarity-based retrieval and indexing schemas of spatial configurations.
Keywords: spatial configurations, similarity-based retrieval, content measures, content-based retrieval

1 Introduction

The description of spatial configurations plays a fundamental role in content-based retrieval and in the design of user interfaces, as this description is often basis for the specification of query constraints. These spatial configurations consist of sets of objects that are spatially arranged. They are described by the spatial relations among objects (*i.e.*, topological relations, distance relations, and orientation relations) as well as by the geometric characteristics of objects (*e.g.*, shape, size, and density), and by attributes specifying the semantics of the spatial objects (*i.e.*, entity-type classification).

Previous studies have investigated the content description of images in the context of images retrieval (Ankerst *et al.*, 1999, Ahmad and Grosky, 1997, Park and Golshani, 1997, Petrakis and Faloustos, 1997, Flickner *et al.*, 1995, Faloustsos *et al.*, 1994). In those studies, the similarity between images relies on judgments in terms of visual descriptions, such as shape, size, texture, and color. For spatial configurations, on the other hand, the spatial arrangement of the object becomes the subject of comparison. These object arrangements are typically expressed by a set of constraints dealing with orientation (*e.g.*, north and south), topology (*e.g.*, inside, overlap), and distance (*e.g.*, 5 miles).

Many studies in the domain of image databases have compared object arrangements based on variations of *2D-strings*. 2D-strings represent configurations with a sequential structure for each encoded dimension (Costagliola *et al.*, 1992, Lee and Hsu, 1992, Lee *et al.*, 1992). Query processing with this structure is carried out as string matching, which is only possible when users specify queries by the schema of the relations according to which 2D-strings are built, and images are composed of a predefined set of objects. Another binary string representation codifies topological, orientation, and distance relations between objects that are seen as Minimum Boundary Rectangles (MBRs) (Papadias *et al.*, 1998b). Relations between MBRs are treated as interval relations in two dimensions, and a similarity function is defined as inversely proportional to the number of changes needed to make two strings equivalent. In a similar way, a 3x3 matrix was used to determine the orientation relation by calculating the proportional area in the quadrants defined by the orthogonal projection of the reference object's MBR (Goyal and Egenhofer, 2001). In general, methods based on 2D-strings and their variations handle changes in scale and translation, but they are sensitive to rotation (El-Kwae and Kabuka, 1999). Using a different approach, some studies represent configurations and queries using Attribute Relation Graphs (ARGs) (Berretti *et al.*, 2000, Papadias *et al.*, 1998b, Petrakis and Faloustos, 1997). In these graphs, spatial relations are represented quantitatively by the distance and angle between centroids, and qualitatively by the symbolic representation of spatial relations, such as the topological relations defined by Egenhofer and Franzosa. (Egenhofer, 1994, Egenhofer and Franzosa, 1991). Similarity is then defined between quantitative values as the inverse of their differences (Berretti *et al.*, 2000, Petrakis and Faloustos, 1997) and between qualitative relations as the inverse of the distance in a conceptual neighborhood structure (Papadias *et al.*, 1998a, Bruns and Egenhofer, 1996).

This work focuses on the definition of content measures that describe the object arrangement of spatial configurations. The definition of such content measures uses a simple strategy for comparing spatial configurations that should be independent of rotation and scaling and that can be used in current spatial information systems. This work uses the simplified and common representation of objects (*i.e.*, MBRs) in current spatial indexing schemas of Geographic Information Systems (GISs). Although this simplification of objects lacks details, it is broadly used for its desirable computational properties, and it can be used as an approximation in a similarity-based retrieval. This work uses a quantitative approach to characterize object arrangements in terms of topology and relative size. In this sense, it follows closely the ideas derived from Egenhofer and Shariff's work (1998) that provide metric refinement of topological relations. However, instead of defining measures to refine individual topological relations, we pursue the definition of a single and continuous content measure that distinguishes topological relations using metric refinement. Although the problem of determining spatial similarity is not a new one, to the best of our knowledge, none of the previous studies has attempted to define a content measure that continuously distinguishes spatial relations, making this single content measure suitable for comparing transitions as objects move and change their spatial relations. Furthermore, our proposed content measures can be used for content-based indexing schemas instead of complementing current indexing schemas (*e.g.*, R-Tree) with heuristics based on similarity functions (Papadias *et al.*, 1999, Papadias *et al.*, 1998b).

The organization of this paper is as follows. Section 2 defines parameters that characterize MBRs, whereas Section 3 defines the content measures that are used to describe arrangements of MBRs. Section 4 compares the behavior of these content measures in terms of their ability to distinguish topological relations. Content measures are used to compare spatial configurations in Section 5. Finally, conclusions and future research directions are presented in Section 6.

2 Characterising Objects' Arrangements

This work defines content measures of object arrangements based on basic parameters that characterise individual MBRs and pairs of MBRs. In doing so, it takes an incremental approach to characterising spatial configurations composed of an arbitrary number of MBRs; that is, the aggregation of content measures that describe a pair of MBRs determines a global characterisation of spatial configurations.

Two basic parameters allow us to characterize MBRs as single dimensional values: areas and diagonals, such that they are simplified views of MBRs.

A second set of parameters characterizes a pair of MBRs. The idea is to create parameters that reflect the relationship between MBRs rather than the properties of individual MBRs. These parameters are the area and diagonal of the MBRs created by the *union* and *intersection* of pairs of individual MBRs, and the *external* (d_e) and *internal* (d_i) *minimum distances* between MBRs' boundaries (δA and δB) (Fig. 1).

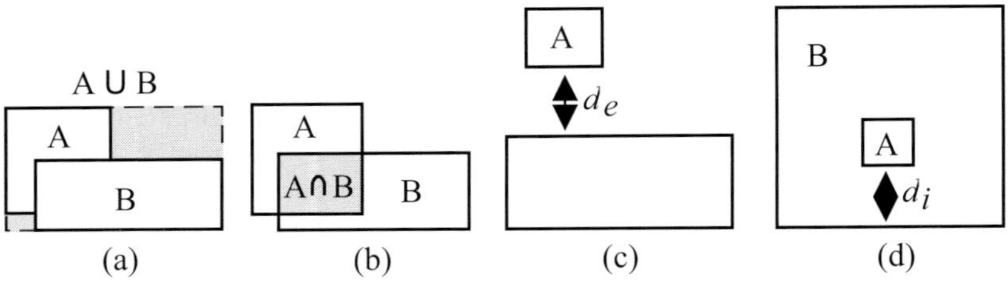

Fig. 1. *Union* **(a)**, *intersection* **(b)**, *external minimum distance* **(c)**, and *internal minimum distance* **(d)**, derived from the combination of two individual MBRs

We explain in the following section how the parameters described above are used to define content measures of objects' arrangements.

3 Defining Content Measures

Two basic ideas guide the definition of content measures of objects' arrangements:

Making metric refinement of topological relations so that we want distinguish degrees of disjointedness or overlap. In Fig. 2, for example, while both pairs of objects A-B and C-D are disjoint by the same distance *d*, the pair A-B should be considered more disjoint than the pair C-D, thus making the configurations independent of scale (Egenhofer and Shariff, 1998).

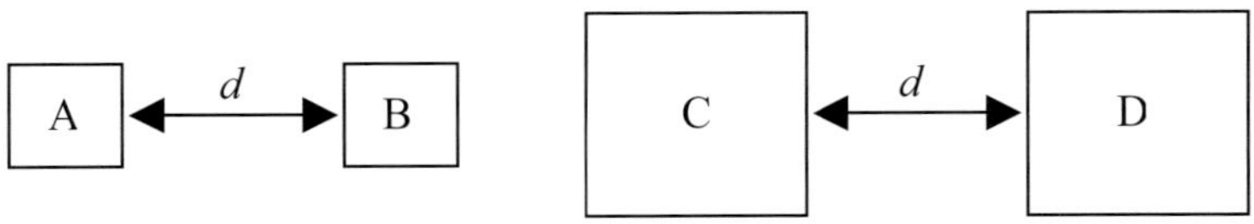

Fig. 2. Differences in the degree of disjointedness

Making explicit the influence of each object in the relation; that is, being able to distinguish between large and small objects as well as between objects as containers and objects as containments in a contains/inside relation. In Fig. 3, for example, objects A and B are disjoint, but A is more disjoint with respect to B than vice versa. In other words, the degree of disjointedness is asymmetric with respect to the size and distance between MBRs. Therefore, describing the arrangement of two objects needs two values, one in each direction of the relation.

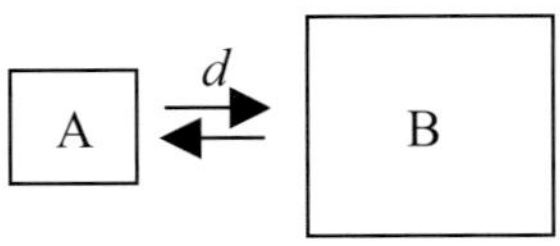

Fig. 3. Asymmetric property of topological relations

We explore three different content measures:

The first measure (F_a) takes areas of individual object MBRs and normalizes them by the area of the union of objects' MBRs, leading to values that are larger than 0 and less or equal than 1 (in the range between ~0 and 1) (Eqs. 1). The ratio between areas represents a relative size relation, while the area of the union of objects' MBRs may indicate whether or not objects share common areas.

$$F_a(A,B) = \frac{\text{area}(A)}{\text{area}(A \cup B)} \quad , \quad F_a(B,A) = \frac{\text{area}(B)}{\text{area}(A \cup B)} \tag{1}$$

Like the first measure, the second measure (F_d) uses diagonals instead of areas of MBRs (Eqs. 2), yielding to values between ~0 and 1.

$$F_d(A,B) = \frac{\text{diagonal}(A)}{\text{diagonal}(A \cup B)} \quad , \quad F_d(B,A) = \frac{\text{diagonal}(B)}{\text{diagonal}(A \cup B)} \tag{2}$$

Making further distinctions among topological relations, the last measure (F_m) uses the diagonal and area of the base MBR[1], the minimum internal or external distances (d_i and d_e, respectively) between MBRs's boundaries (indicated with the δ symbols), and the area of intersection between MBRs (Eqs. 3). The underlying idea of this function is that *distance* between objects is a basic parameter for the refinement of *disjointedness*, while *area* is for *overlapping*.

$$F_m(A,B) = \frac{\text{area}(A) - 2\text{area}(A \cap B)}{\text{area}(A)} + \frac{\text{distance}(\delta A, \delta B)}{\text{diagonal}(A)} \quad ,$$

$$F_m(B,A) = \frac{\text{area}(B) - 2\text{area}(A \cap B)}{\text{area}(B)} + \frac{\text{distance}(\delta B, \delta A)}{\text{diagonal}(B)} \tag{3}$$

where :

$$\text{distance}(\delta A, \delta B) = \begin{cases} d_e(\delta A, \delta B) \text{ if } A \cap B = \varnothing \\ -d_i(\delta A, \delta B) \text{ if } A \cap B \neq \varnothing \end{cases}$$

[1] We consider the *base* object as the first parameter and the *target* object as the second parameter of the function *F*.

4 Properties of Content Measures

The first two content measures (F_a and F_d) use the same structures, that is, areas or diagonals over both individual MBRs and the union of MBRs. Unlike the first measure, the second measure is less sensitive to orientation. For example, Fig. 4 has configurations (a) and (b) with equivalent MBRs separated by the same distance. The difference determined between configurations using an area-based measure is larger than it is using a diagonal-based measure. Although we differentiate these configurations, this difference should not be very large, since in topological terms, the MBRs are disjoint by the same distance.

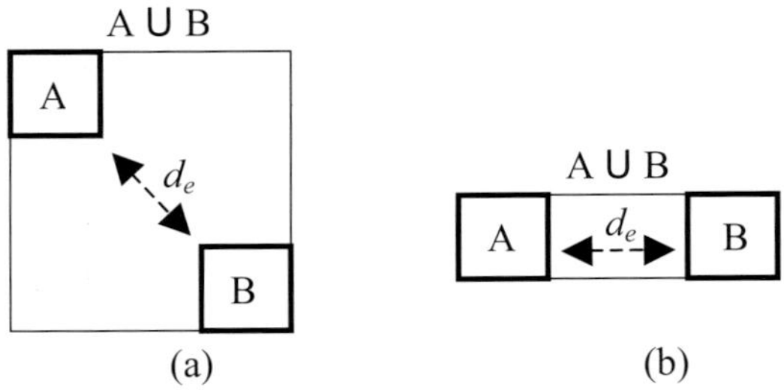

Fig. 4. *Diagonals* are less sensitive than *areas* to orientation.

The three content measures defined above are continuous functions whose values depend on the objects size and arrangement, such that they are independent of scale. Values of F_a and F_d tend to 1 when the MBRs overlap completely; whereas, these values approximate to zero as the MBRs' separate and, therefore, the area and diagonal of the MBRs' union increase. The last measure F_m ranges from negative to positive values, being equal to zero when objects overlap and the area of the *base* object inside of the *target* object is equivalent to the area outside of the *target* object.

Table 1 shows the ranges of values for each measure and each of the topological relations between regions. As table 1 indicates, these measures are asymmetric, and it is the combination of the evaluations in both directions (*e.g.*, *F*(A,B) and *F*(B,A)) that provides some indication of the type of topological relation between objects. The measures F_a and F_d distinguish 3 of the 8 relations, whereas values of F_m can differentiate all 8 topological relations.

As an example of how these proposed content measures are capable of distinguishing topological relations, consider Fig. 5, where the same two objects have different topological relations in transition states.

In Fig. 5, the values of the measures based on *areas* and *diagonals* increase as the degree of overlap increases. These measures are, however, incapable of distinguishing between *inside* and *covered_by* relations, since the MBR that contains both individual MBRs is the same. The third measure, on the other hand, changes from positive to negative values as the smaller object moves inside the larger one. Although the last measure detects differences between the *inside* and *covered_by* relations, it is unable to detect different degrees of *meet*.

The proposed measures do not address explicitly *orientation*; however, a synergy occurs when constrains that related more than one pair of objects are satisfied. Thus we can distinguish between configurations whose objects have the same relative orientation, making the content description independent of the rotation of the whole configuration. For instance, consider three objects with the same topological relation that are arranged in different locations.

Table 1. Range of values for the three content measures (according to 8 topological relations)

Content Measure	Relations	Values
F_a	Disjoint	$0 < F_a(A,B) + F_a(B,A) < 1$
	Meet	$0 < F_a(A,B) + F_a(B,A) \leq 1$
	Overlap	$0 < F_a(A,B) < 1,\ 0 < F_a(B,A) < 1$
	Contain	$F_a(A,B) = 1,\ 0 < F_a(B,A) < 1$
	Cover	$F_a(A,B) = 1,\ 0 < F_a(B,A) < 1$
	Inside	$0 < F_a(A,B) < 1,\ F_a(B,A) = 1$
	Covered_by	$0 < F_a(A,B) < 1,\ F_a(B,A) = 1$
	Equal	$F_a(A,B) = 1,\ F_a(B,A) = 1$
F_d	Disjoint	$0 < F_d(A,B) < 1,\ 0 < F_d(B,A) < 1$
	Meet	$0 \leq F_d(A,B) + F_d(B,A) \leq \sqrt{2}$
	Overlap	$0 < F_d(A,B) + F_d(B,A) < 2$
	Contain	$F_d(A,B) = 1,\ 0 < F_d(B,A) < 1$
	Cover	$F_d(A,B) = 1,\ 0 < F_d(B,A) < 1$
	Inside	$0 < F_d(A,B) < 1,\ F_d(B,A) = 1$
	Covered_by	$0 < F_d(A,B) < 1,\ F_d(B,A) = 1$
	Equal	$F_d(A,B) = 1,\ F_d(B,A) = 1$
F_m	Disjoint	$1 < F_m(A,B),\ 1 < F_m(B,A)$
	Meet	$F_m(A,B) = 1,\ F_m(B,A) = 1$
	Overlap	$\lvert F_m(A,B) \rvert < 1,\ \lvert F_m(B,A) \rvert < 1$
	Contain	$\lvert F_m(A,B) \rvert < 1,\ F_m(B,A) < -1$
	Cover	$\lvert F_m(A,B) \rvert < 1,\ F_m(B,A) = -1$
	Inside	$F_m(A,B) < -1,\ \lvert F_m(B,A) \rvert < 1$
	Covered_by	$F_m(A,B) = -1,\ \lvert F_m(B,A) \rvert < 1$
	Equal	$F_m(A,B) = -1,\ F_m(B,A) = -1$

In Fig. 6, both configurations are composed of the same three objects separated by the same external distances between each other. They differ only on the rotation of the whole configuration such that the MBRs of the unions of pairs of objects are equivalent in size but not in rotation. The content measures of corresponding pairs of objects are, therefore, the same. We have the same results when the configurations differ by a mirror effect; *e.g.*, when the objects are swapped with respect to a horizontal or vertical axis.

Transition State	$F_a(A,B)$, $F_a(B,A)$	$F_d(A,B)$, $F_d(B,A)$	$F_m(A,B)$, $F_m(B,A)$
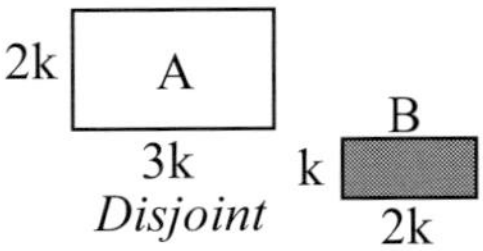 *Disjoint*	(0.33 , 0.11)	(0.54 , 0.33)	(1.28 , 1.45)
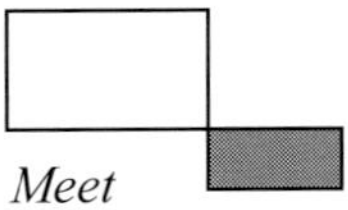 *Meet*	(0.40 , 0.13)	(0.62 , 0.38)	(1.00 , 1.00)
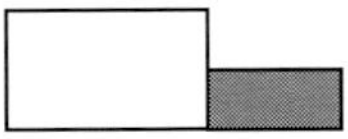 *Meet*	(0.60 , 0.20)	(0.67 , 0.42)	(1.00 , 1.00)
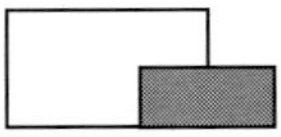 *Overlap*	(0.75 , 0.25)	(0.81 , 0.50)	(0.67 , 0.00)
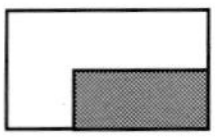 *Covers, covered_by*	(1.00 , 0.33)	(1.00 , 0.62)	(0.33 , -1.00)
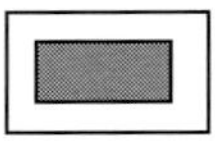 *Contains, inside*	(1.00 , 0.33)	(1.00 , 0.62)	(0.19 , -1.22)

Fig. 5. Distinguishing different transition states of topological relations between two objects

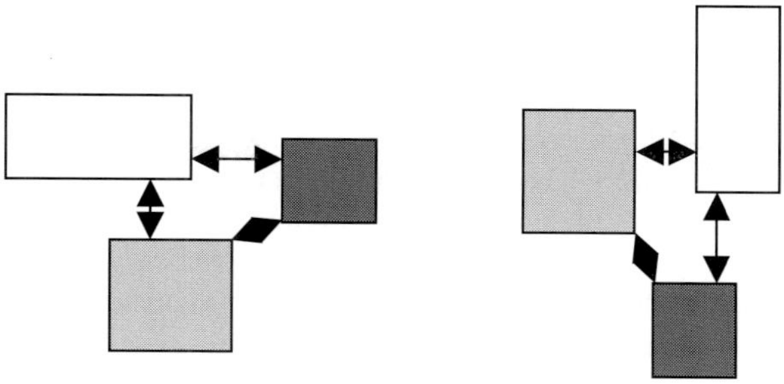

(a) (b)

Fig. 6. Capturing global orientation of spatial configurations

5 Comparing Spatial Configurations

This section checks the performance of the three proposed content measures when comparing different spatial configurations. In this context, this paper concentrates just on the question how *sensible* and *consistent* the results of these measures are. By *sensible* we mean that the content measures find correct topological relations, by *consistent* we mean that all three measures yield *similar*.

The experiments use a spatial database composed of 10,000 configurations with 3 objects of different classes created randomly. The definition of a general searching algorithm for comparing configurations with an undefined number of objects can be built upon these measures and is left for future work. The number of occurrences of different types of topological relations in the database is as follows: 33550 *disjoint*, 68 *meet*, 24512 *overlap*, 28 *covers*, 907 *contains*, 28 *covered_by*, 907 *inside*, and zero *equal*. We describe configurations as points in a 6D space (*i.e.*, a space composed of two values for each pair of relations between MBRs). The similarity is determined as inversely proportional to the Euclidean distance in this 6D space.

Fig. 7 shows results of two queries. The first query includes *overlap* and *disjoint* relations, the two most common relations in the database. The second query is composed of *overlap*, *meet*, and *disjoint*. Since there are few occurrences of *mee*t in the database, this constraint is highly significant to discriminate configurations in topological terms. For both queries we run all three measures and we give the 3 best matchings with their respective rankings for each of the similarity measures.

For all three content measures, the best configurations have objects whose relations closely satisfy the query's topological constraints. The worse results are obtained with the measure F_d , where we obtain configurations that have objects with different topological relations with respect to the query. A reason is that neither F_a nor F_d differentiates the topological relations as F_m does, and that the Euclidean distance combines each dimension without ensuring the satisfaction of any topological constraints.

In terms of consistency, our results indicate that content measures may result in different rankings of similarity for spatial configurations; however, results obtained from measure F_m are closer to the best results derived from F_a and F_d. Some reasons to the low consistency in the rankings given by the different measures are that the query has objects with very common topological relations in the database and that each measure captures different aspects of spatial configurations (*i.e.*, *area* versus *diagonal*, *area* versus *distance*). If there were few configurations in the database that satisfy the query constraints, there would be better chances that the three measures give the same results.

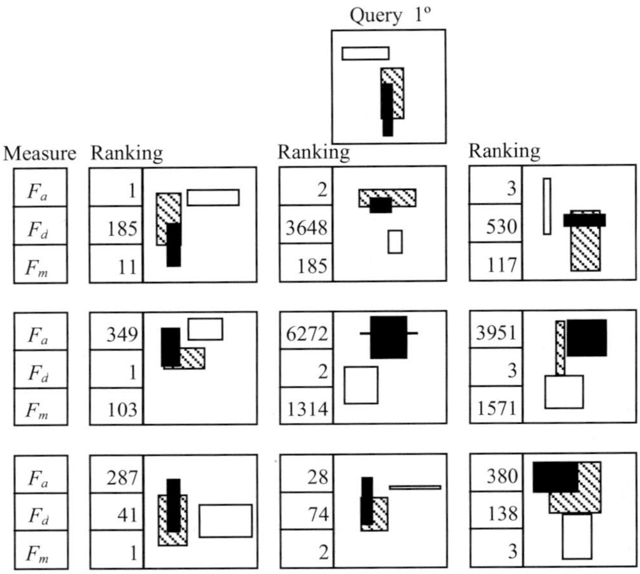

(a)

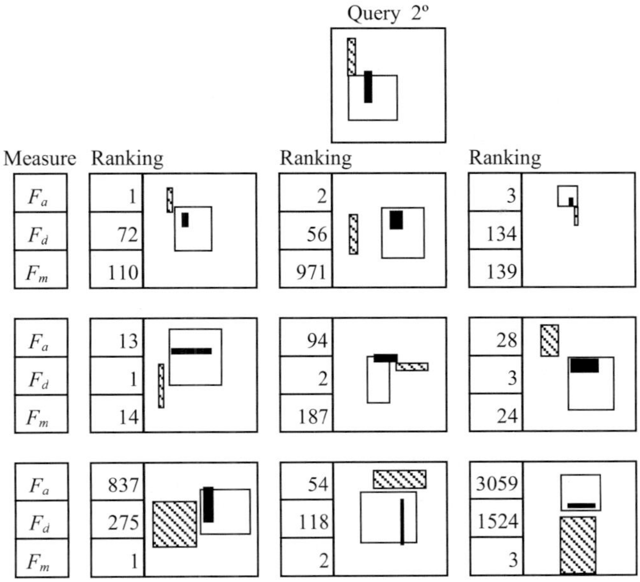

(b)

Fig. 7. Answer to first and second queries using each different measure

6 Conclusions

We have proposed three content measures to characterise spatial configurations. They are based on parameters that describe objects by MBRs' basic dimensions (*i.e.*, areas and diagonals) and objects' relations by MBRs' interaction (*i.e.*, areas and diagonals of MBRs' union and intersection, and internal and external distances between MBRs' boundaries).

Our experimental results indicate that these simple content measures are able to distinguish topological relations at different levels of detail. Therefore, these content measures allow us to quantitatively compare spatial configurations. In particular, the last measure F_m that combines basic dimension of MBRs with distances between MBRs' boundaries was seen to be the best of our content measures for comparing spatial configurations.

We left for future work the analysis of different aggregation functions that define a similarity measure. Likewise, we want to explore different weight schemas for similarity measures of topological relations. For example, considering a query composed by objects that *meet* and objects that are *disjoint*, we would like to perform the search process giving more importance to the *meet* constraints than to the *disjoint* constraints. We will also pursue the definition of a content-based indexing schema that avoids the exhaustive search over the whole database.

Acknowledgements

This work has been partially funded by CONICYT under FONDECYT project 1010897 and by Fundación Andes, Chile.

References

Ahmad I, Grosky WI (1997) Spatial Similarity-Based Retrievals and Image Indexing by Hierarchical Decomposition. In: International Database Engineering and Applications Symposium (IDEAS '97). pp 269-278

Ankerst M, Kastenmüller G, Kriegel H-P, Seidl T (1999) 3D Shape Histograms for Similarity Search and Classification in Spatial Databases. In: Güting RH, Papadias D, Lochovsky F (eds) Advances in Spatial Databases-6th International Symposium (SSD '99). Hong Kong, China. Springer-Verlag, Berlin, pp 207-226

Berretti S, Bimbo AD, Vicario E (2000) The Computational Aspect of Retrieval by Spatial Arrangment. In: International Conference on Pattern Recognition. pp 1:1047-1051

Bruns HT, Egenhofer M (1996) Similarity of Spatial Scenes. In: Kraak JM, Molenaar M (eds) Seventh International Symposium on Spatial Data Handling (SDH '96). Taylor & Francis, London, pp 173-184

Costagliola G, Tortora G, Arndt T (1992) A Unifying Approach to Iconic Indexing for 2-D and 3-D Scenes. IEEE Transactions on Knowledge and Data Engineering 4(3):205-222

Egenhofer M (1994) Deriving the Composition of Binary Topological Relations. Journal of Visual Languages and Computing 5(2):133-149

Egenhofer M, Franzosa R (1991) Point-set topological spatial relations. International Journal for Geographical Information Systems 5(2):161-174

Egenhofer M, Shariff A (1998) Metric Details for Natural-Language Spatial Relations. ACM Transactions on Information Systems 16(4):295-321

El-Kwae E, Kabuka M (1999) A Robuts Framework for Content-Based retrieval by Spatial Similarity in Image Databases. ACM Transactions on Information Systems 17(2):174-198

Faloustsos C, Barber R, Flickner M, Hafner J, Niblack W, Petrovic D, Equitz W (1994) Efficient and Effective Querying by Image Content. Journal of Intelligent Information Systems 3:231-262

Flickner M, Sawhney H, Niblack W, Ashley J, Huang Q, Dom B, Gorkani M, Hafner J, Lee D, Petrovik D, Steele D, Yanker P (1995) Query by Image and Video Content: The QBIC System. IEEE Computer 28(9):23-32

Goyal R, Egenhofer M (2001) Similarity of Cardinal Directions. In: Jensen C, Schneider M, Seeger B, Tsotras V (eds) Seventh International Symposium on Spatial and Temporal Databases Lectures Notes in Computer Science. Springer-Verlag, 2121:36-55

Lee S, Hsu F (1992) Spatial Reasoning and Similarity Retrieval of Images Using 2D C-Strings Knowledge Representation. Pattern Recognition 25(3):305-318

Lee S, Yang M, Chen J (1992) Signature File as a Spatial Filter for Iconic Image. Journal of Visual Language and Computing 3:373-392

Papadias D, Arkoumanis D, Karacapilidis N (1998a) On the Retrieval of Similar Configurations. In: Poiker T, Chrisman N (eds) 8th International Symposium on Spatial Data Handling. International Geographical Union, Vancouver, pp 510-521

Papadias D, Mamoulis N, Delis V (1998b) Algorithms for Querying Spatial Structure. In: 24th VLDB Conference. pp 546-557

Papadias D, Mantzourogiannis M, Mamoulis N, Ahmad I (1999) Content-Based Retrieval using Heuristic Search. In: ACM Conference on Information Retreival (SIGIR). pp 168-175

Park Y, Golshani F (1997) ImageRoadMap: A New Content-Based Image Retrieval System. In: Haneurlaim A, Tjoa A (eds) Database and Expert Systems Applications DEXA'97. Springer Verlag, Toulouse, France, pp 225-239

Petrakis G, Faloustos C (1997) Similarity Searching in Medical Image Databases. IEEE Transactions on Knowledge and Data Egineering 9(3):435-447

Implementing Topological Predicates for Complex Regions

Markus Schneider

University of Florida, Department of Computer and Information Science and Engineering, Gainesville, FL 32611, USA, mschneid@cise.ufl.edu

Abstract

Topological predicates on spatial objects have been a main area of research in spatial data handling, reasoning, and query languages. The focus of research, has, to a large extent, been on the design of and reasoning with these predicates, whereas implementation issues have been somewhat neglected. The goal of this paper is to show how an abstract design of topological predicates for *complex* regions can be efficiently implemented. The resulting algorithms are based on the *realm* concept, which is well known in the spatial database community.
Keywords: complex region, topological predicate, implementation, realm, ROSE algebra

1 Introduction

In recent years, significant achievements have been made on the design of topological predicates for spatial objects. Almost all results have been based on either the 9-intersection model (Egenhofer *et al.*, 1990), which rests on point sets and point set topology, or the RCC model (Cui *et al.*, 1993), which employs spatial logic. The foundation for this work is the first model, since the goal of the paper is to address the *efficient* implementation of topological predicates for regions in a spatial database or GIS context.

Whereas the predicates in the aforementioned two models operate on simplified abstractions of spatial objects like simple regions, we are interested in the design and implementation of topological predicates for *complex* regions, which may consist of several components (faces) and which may have holes.

Implementation issues for these predicates, regardless of whether they operate on simple or complex geometries, have so far been somewhat neglected. Hence, it is especially the goal of this paper to demonstrate how an abstract design of topological predicates for complex regions can be efficiently implemented.

Section 2 discusses related work. Section 3 presents an abstract, formal model of topological predicates for complex regions. This model is called *abstract*, because it makes clean and simple definitions in terms of infinite point sets and point set topology, without worrying whether finite representations of these sets exist. Section 4 describes a corresponding discrete model. This model is called *discrete*, because it takes into account *finite representations* only available in computers. In this sense it is nearer to implementation. *Realms* (Güting *et al.*, 1993; Schneider, 1997) will be the basis of this model. Section 5 discusses the implementation level. We will show a realm-based data structure for complex regions and algorithms for the predicates operating on this data structure. Finally, Section 6 draws some conclusions.

2 Related Work

In the past, a number of data models and query languages for spatial data have been proposed with the aim of formulating and processing spatial queries in databases. *Spatial data types* (Schneider, 1997) like *point*, *line*, or *region* are the central concept of these approaches and provide fundamental abstractions for modelling the structure of geometric entities, their relationships, properties, and operations. Whereas in older models the geometric structure of spatial data has been restricted (only simple regions, continuous lines, single points), in the meantime a few models (Güting *et al.*, 1995; Clementini *et al.*, 1996) also allow complex spatial objects which may consist of several disjoint components. Additionally, a region object may have holes, and a component of a line object may be a connected collection of curves. The reasons for defining complex geometric structures are closure properties for spatial operations and application-specific requirements. The OpenGIS Consortium (OGC) has incorporated similar generalised geometric structures, called *simple features*, into their OGC Abstract Specification (OGC, 1999) and into the *Geography Markup Language* (*GML*) (OGC, 2001), which is an XML encoding for the transport and storage of geographic information. These geometric structures are called *MultiPoint*, *MultiLineString*, and *MultiPolygon*. Implementation descriptions for these structures have so far not been given.

A well known, abstract model for characterising topological relationships between simple regions is the 9-*intersection model* (Egenhofer *et al.*, 1989). For two simple regions, eight meaningful configurations have been identified which lead to the eight predicates of the set T_{sr} = {*disjoint*, *meet*, *overlap*, *equal*, *inside*, *contains*, *covers*, *coveredBy*}. Each predicate is determined by a unique 9-intersection matrix representing the nine intersections of the boundary, interior, and exterior of the first region with the corresponding parts of the second region. All predicates are mutually exclusive and cover all topological situations. However, a generalisation of spatial data types also necessitates a generalisation of the corresponding topological predicates. It is surprising that such predicates have until recently not been defined. In (Clementini *et al.*, 1995) the so-called TRCR

(Topological Relationships for Composite Regions) model only allows sets of disjoint simple regions without holes. In (Egenhofer *et al.*, 1994) only topological relationships of simple regions with holes are considered; multi-part regions are not permitted. Recently, a general (Behr *et al.*, 2001) and a limited model (Schneider, 2001) have been presented for topological predicates on complex regions. This paper will be based on the limited model (see Section 3).

So far, implementation issues for topological predicates have been almost completely neglected in the literature. In (Clementini *et al.*, 1994) query processing strategies are discussed for topological predicates on simple regions. The central idea is to find a minimal but still unique subset of the nine intersections for each predicate. Only this subset has then to be evaluated, and the hope is that this reduction leads to a better performance. It is assumed that the computational cost of each of the nine possible intersections is equal. Although this approach is interesting from a theoretical point of view, it is doubtful from an implementation perspective, because the computation of all nine intersections can be and has to be performed by the same algorithmic scheme (see Section 5).

The implementation of spatial objects and predicates is impeded by the assumption of a Euclidean space and an infinite-precision arithmetic in spatial data models which conflicts with the reality of finite-precision number systems available in computers (Schneider, 1997). This leads inevitably not only to numerical but especially to topological errors and thus to wrong query results in database systems. A solution for this problem is the *realm* concept (Güting *et al.*, 1993). A realm replaces the Euclidean space with a discrete geometric basis and is intended to represent the entire underlying geometry of an application. It is based on a finite resolution computational geometry and consists of a finite set of points and *non-intersecting* line segments which are defined over a discrete point grid and which form a spatially embedded planar graph. We will see that on top of realms complex regions and topological predicates operating on them can be easily implemented.

3 The Abstract Model

In this section we review the limited abstract model for topological predicates on complex regions, which has been described in (Schneider, 2001). The objective of this model is not to find all possible topological relationships between two complex regions but to generalise the eight topological relationships for simple regions to complex regions in a straightforward way. This may be regarded as an ad hoc approach leading to too coarse predicates. For many spatial applications, however, this predicate collection is practicable enough, and a more fine-grained differentiation is even not desired.

3.1 Complex Regions

First, we informally describe the structure of complex regions, which results in the spatial data type *region*. A formal definition can be found in (Schneider, 2001) and is based on point set theory and point set topology. Regions are embedded into the two-dimensional Euclidean space IR^2 and are thus point sets.

A *simple region* has a connected interior, a connected boundary, and a single connected exterior. Hence, it does not consist of several components, and it does not have holes.

A *simple region with holes F*, which is also called a *face* (Fig. 1c-e), is a simple region with an outer boundary containing several other simple regions, which are called *holes*. Holes represent areas (simple regions) that do not belong to the region object but are enclosed by it. A hole is allowed to touch the outer boundary of F or the boundary of another hole in at most single points. To permit holes to have a partially common border with other holes makes no sense because then adjacent holes could be merged to a single hole by eliminating the common border (similarly for adjacency of a hole with the outer boundary). A face is atomic and cannot be decomposed into two or more faces. For example, the configuration shown in Fig. 1a, must be interpreted as two faces with two holes and not as a single face with four holes.

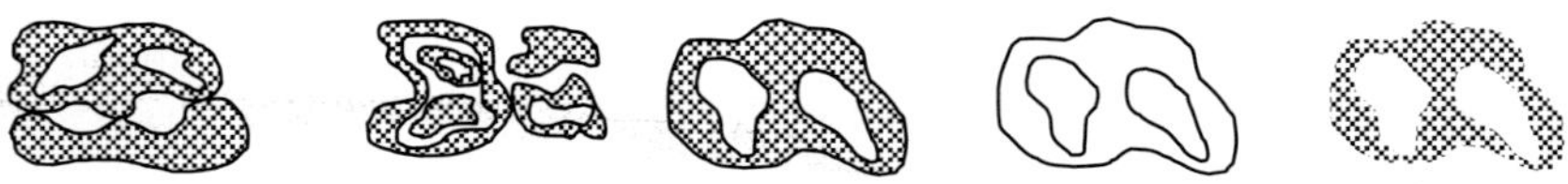

(a) (b) (c) (d) (e)

Fig. 1. Unique representation of a face **(a)**, a complex region with five faces **(b)**, a simple region with two holes **(c)**, its boundary **(d)**, and its interior **(e)**

Finally, a *complex region* (Fig. 1b) is a set of faces. A face has to be disjoint to another face, or to meet another face in one or several single boundary points, or to lie within a hole of another face and possibly share one or several single boundary points with the boundary of the hole. Faces having common connected boundary parts with other faces or holes are disallowed. The argumentation is similar to that for the face definition.

3.2 Topological Predicates on Simple Regions with Holes

In the following we use the notation $(P_1|P_2|\ldots|P_n)(F,G)$ as a syntactical simplification for the term $P_1(F,G) \vee P_2(F,G) \vee \ldots \vee P_n(F,G)$ where P_i : *region* × *region* → *bool* is a topological predicate for each $1 \leq i \leq n$.

As a first step to a general definition of topological predicates for complex regions we consider such predicates for simple regions with holes and base their definition on the well known semantics of the topological predicates T_{sr} for simple regions (Section 2), as they have been derived from the 9-intersection model.

Let F and G be two simple regions with holes, F consisting of the simple region with the outer boundary F_0 and the holes $F_1, \ldots, F_n$, and G consisting of the simple region with the outer boundary G_0 and the holes $G_1, \ldots, G_m$. The topological predicates operating on F and G are defined as follows:

$$
\begin{aligned}
\mathit{disjoint}_{srh}(F, G) &:= \mathit{disjoint}(F_0, G_0) \vee \\
&\quad\; (\exists\, 1 \leq i \leq n : \mathit{inside}(G_0, F_i)) \vee \\
&\quad\; (\exists\, 1 \leq j \leq m : \mathit{inside}(F_0, G_j))
\end{aligned}
$$

$$
\begin{aligned}
\mathit{meet}_{srh}(F, G) &:= \mathit{meet}(F_0, G_0) \vee \\
&\quad\; (\exists\, 1 \leq i \leq n : \mathit{coveredBy}(G_0, F_i)) \vee \\
&\quad\; (\exists\, 1 \leq j \leq m : \mathit{coveredBy}(F_0, G_j))
\end{aligned}
$$

$$
\begin{aligned}
\mathit{inside}_{srh}(F, G) &:= \mathit{inside}(F_0, G_0) \wedge \\
&\quad\; (\forall\, 1 \leq j \leq m : \mathit{disjoint}(F_0, G_j) \vee \\
&\quad\quad\quad (\mathit{inside}(G_j, F_0) \wedge (\exists\, 1 \leq i \leq n : \mathit{inside}(G_j, F_i)))
\end{aligned}
$$

$$
\mathit{contains}_{srh}(F, G) := \mathit{inside}_{srh}(G, F)
$$

$$
\begin{aligned}
\mathit{equal}_{srh}(F, G) &:= \mathit{equal}(F_0, G_0) \wedge n = m \wedge \\
&\quad\; \exists\, \pi : \{1, \ldots, n\} \rightarrow \{1, \ldots, n\}, \pi \text{ bijective}, \\
&\quad\; \forall\, 1 \leq i \leq n : \mathit{equal}(F_i, G_{\pi(i)})
\end{aligned}
$$

$$
\begin{aligned}
\mathit{coveredBy}_{srh}(F, G) &:= \neg\, ((\mathit{inside}_{srh}|\mathit{equal}_{srh})(F, G)) \wedge \\
&\quad\; (\mathit{inside}|\mathit{coveredBy}|\mathit{equal})(F_0, G_0) \wedge \\
&\quad\; (\forall\, 1 \leq j \leq m : ((\mathit{disjoint}|\mathit{meet})(F_0, G_j) \vee \\
&\quad\quad (\exists\, 1 \leq i \leq n : (\mathit{inside}|\mathit{coveredBy}|\mathit{equal})(G_j, F_i)))
\end{aligned}
$$

$$
\mathit{covers}_{srh}(F, G) := \mathit{coveredBy}_{srh}(G, F)
$$

$$
\begin{aligned}
\mathit{overlap}_{srh}(F, G) &:= \neg\, ((\mathit{disjoint}_{srh}|\mathit{meet}_{srh}|\mathit{coveredBy}_{srh}|\mathit{covers}_{srh}| \\
&\quad\quad \mathit{inside}_{srh}|\mathit{contains}_{srh}|\mathit{equal}_{srh})(F, G))
\end{aligned}
$$

In (Schneider, 2001) we have shown that the set $T_{srh} = \{\mathit{disjoint}_{srh}, \mathit{meet}_{srh}, \mathit{coveredBy}_{srh}, \mathit{covers}_{srh}, \mathit{inside}_{srh}, \mathit{contains}_{srh}, \mathit{equal}_{srh}, \mathit{overlap}_{srh}\}$ provides a complete coverage of topological relationships for two simple regions with holes, and its elements are mutually exclusive. The set T_{srh} of topological predicates on simple regions with holes is in two ways compatible with the set T_{sr} of topological predicates on simple regions obtained by the 9-intersection model. First, if both F and G do not have holes, then T_{srh} and T_{sr} coincide. Second, each of the eight topological predicates on simple regions with holes has the same boolean results for the nine intersections as the corresponding predicate on simple regions.

3.3 Topological Predicates on Complex Regions

With the aid of the topological predicates on simple regions with holes we are now able to define the corresponding predicates on complex regions. Let F and G be complex regions, F consisting of the simple regions with holes $F_1, \ldots, F_n$, and G consisting of the simple regions with holes $G_1, \ldots, G_m$. We define the following predicates:

$$
\begin{aligned}
disjoint_{cr}(F, G) &:= \forall\, 1 \le i \le n\; \forall\, 1 \le j \le m : disjoint_{srh}(F_i, G_j) \\
meet_{cr}(F, G) &:= \neg disjoint_{cr}(F, G) \wedge \\
&\quad\; (\forall\, 1 \le i \le n\; \forall\, 1 \le j \le m : \\
&\qquad\qquad (disjoint_{srh}|meet_{srh})(F_i, G_j)) \\
inside_{cr}(F, G) &:= \forall\, 1 \le i \le n\; \exists\, 1 \le j \le m : inside_{srh}(F_i, G_j) \\
contains_{cr}(F, G) &:= inside_{cr}(G, F) \\
equal_{cr}(F, G) &:= n = m \wedge \exists\, \pi : \{1, \ldots, n\} \rightarrow \{1, \ldots, n\}, \pi \text{ bijective}, \\
&\qquad\qquad \forall\, 1 \le i \le n : equal_{srh}(F_i, G_{\pi(i)}) \\
coveredBy_{cr}(F, G) &:= \neg\, ((inside_{cr}|equal_{cr})(F, G)) \wedge \\
&\quad\; (\forall\, 1 \le i \le n\; \exists\, 1 \le j \le m : \\
&\qquad (inside_{srh}|coveredBy_{srh}|equal_{srh})(G_j, F_i)) \\
covers_{cr}(F, G) &:= coveredBy_{cr}(G, H) \\
overlap_{cr}(F, G) &:= \neg\, ((disjoint_{cr}|meet_{cr}|coveredBy_{cr}|covers_{cr}| \\
&\qquad inside_{cr}|contains_{cr}|equal_{cr})(F, G))
\end{aligned}
$$

With similar arguments as in Section 3.2 we can recognise that two complex regions satisfy exactly one of these topological predicates. In other words, the topological predicates of the set T_{cr} = $\{disjoint_{cr}$, $meet_{cr}$, $inside_{cr}$, $contains_{cr}$, $equal_{cr}$, $coveredBy_{cr}$, $covers_{cr}$, $overlap_{cr}\}$ are mutually exclusive and complete.

4 The Discrete Model

The discrete model is the interface between the abstract model and the implementation level. It provides a discrete, finite representation based on the realm concept for the complex *region* type of the abstract model with its infinite point set representation. Similarly, the topological predicates of the abstract model are mapped to corresponding predicates of the discrete model and operate now on the finite representations. The discrete model still "abstracts" from implementation aspects (that is, from concrete data structures and algorithms) and gives a definition of the *region* type and the predicates on a discrete geometric basis. Efficiency does not play a role here. The challenge is to ensure numerical robustness (due to rounding errors) and topological consistency of predicates. We show how these requirements can be satisfied. Needed realm concepts are briefly reviewed in Section 4.1. Topological predicates on discrete regions are defined in Section 4.2.

4.1 Needed Realm Concepts

In Section 2 we briefly addressed the problems of implementing spatial objects and topological predicates in a numerically robust and topologically consistent way. Further, we mentioned the *realm* concept (Güting *et al.*, 1993; Schneider, 1997) as a successful approach to solving these problems. In this subsection we will informally give some needed realms definitions. The formal definitions can be found in (Güting *et al.*, 1993).

The underlying space of a realm is a finite discrete grid $N \times N$ with $N = \{0, \ldots, m-1\} \subset \mathbb{N}$. As primitive objects, points and line segments with co-ordinates in N are defined over this grid. A realm represents a finite, user-definable set of points and *non-intersecting*[1] line segments over this grid and is characterised by the following features: (i) Each point and each end point of a line segment of a realm is a grid point, (ii) each end point of a realm segment is also a point of the realm, (iii) no realm point lies properly within a realm segment, and (iv) no two realm segments intersect except at their end points. There is an obvious interpretation of a realm as a *spatially embedded planar graph* with the realm points as nodes and the realm segments as edges. For the formal realm definition *robust geometric primitives* on realm points and realm segments are needed whose definition can also be found in (Güting *et al.*, 1993). Their essential feature is that they can be implemented in a numerically precise way based on integer arithmetic.

In Section 3.1 we have used a bottom-up strategy at the abstract model to (informally) define “abstract” complex regions, and we have started with *simple regions* as the fundamental and simplest kind of regions. In (Schneider, 2001) all more complex kinds of regions are derived from this notion by *finite* means, because in their definitions we only employ finite quantifications and already known predicate specifications. Hence, at the discrete level, it is sufficient to define the finite analogy of a simple region in the realm context. Discrete versions of simple regions with holes and of complex regions are then straightforward! The analogy of a simple region is called *cycle* alluding to the graph interpretation of a realm and describes the well known linear approximation of a region as a simple polygon. A cycle c is defined by a set of realm segments $S(c) = \{s_0, \ldots, s_{m-1}\}$ such that any two consecutively indexed segments (also s_0 and s_{m-1}) meet, that is, share exactly one common end point, and such that no more than two segments from $S(c)$ meet in any point. Cycle c partitions all grid points into the three finite subsets $P_{in}(c)$ containing the grid points lying inside c, $P_{on}(c)$ containing the grid points lying on c, and $P_{out}(c)$ containing the grid points lying outside c. Let $P(c) :=$

[1]The task of transforming an application’s set of *intersecting* line segments into a realm’s set of *non-intersecting* line segments is performed by the concept of *redrawing* (Güting *et al.*, 1993).

$P_{in}(c) \cup P_{on}(c)$. Furthermore, let *region* be the type of all discrete, complex regions.

4.2 Topological Predicates on Discrete Regions

Since the topological predicates derived from the 9-intersection model are defined on "abstract" simple regions based on infinite point sets and real numbers, they cannot be directly implemented. However, even if their geometry would be based on floating-point numbers, the simple test whether a point lies on a segment is doomed to fail due to rounding errors. If two computed segments are intended to meet in a common point, a corresponding *meet* predicate will probably yield the wrong result because the equality test of the two common end points is error-prone due to presumably slightly different point representations. In the realm context all these computations are numerically precise and robust and do not cause any problems.

We now give the definitions of the eight topological predicates for simple polygons (cycles), which correspond to the 9-intersection model but are specified on a discrete geometric domain. Let F and G be two simple polygons. Then we define:

$$
\begin{aligned}
disjoint(F, G) &:= P(F) \cap P(G) = \varnothing \wedge S(F) \cap S(G) = \varnothing \\
meet(F, G) &:= P_{in}(F) \cap P(G) = \varnothing \wedge P(F) \cap P_{in}(G) = \varnothing \wedge \\
&\quad\;\; (P_{on}(F) \cap P_{on}(G) \neq \varnothing \vee S(F) \cap S(G) \neq \varnothing) \\
equal(F, G) &:= P(F) = P(G) \wedge S(F) = S(G) \\
inside(F, G) &:= P(F) \subseteq P(G) \wedge P_{on}(F) \cap P_{on}(G) = \varnothing \wedge \\
&\quad\;\; S(F) \cap S(G) = \varnothing \\
contains(F, G) &:= inside(G, F) \\
coveredBy(F, G) &:= P(F) \subseteq P(G) \wedge \\
&\quad\;\; (P_{on}(F) \cap P_{on}(G) \neq \varnothing \vee S(F) \cap S(G) \neq \varnothing) \\
covers(F, G) &:= coveredBy(G, H) \\
overlap(F, G) &:= \neg\, ((disjoint|meet|coveredBy|covers| \\
&\quad\;\; inside|contains|equal)(F, G))
\end{aligned}
$$

In Section 3 we have used a bottom-up strategy to define topological predicates on abstract complex regions. We have based their definitions on the predicates on simple regions derived from the 9-intersection model and, as an intermediate step, on the predicates on simple regions with holes. Since the definitions of the predicates on simple regions with holes and on complex regions in addition only employ logical connectives and finite quantifications, we can simply adopt these definitions for topological predicates on discrete complex regions! Consequently, there is not more to do at this point.

5 Data Structure and Algorithms

It is obvious that, in this context, computing with finite points sets is not very efficient in practice. Hence, an efficient data structure for the complex data type *region* (Section 5.1) as well as efficient algorithms for the topological predicates operating on this data structure are needed (Section 5.2). For the data structure a special version of the classical linear boundary representation is used.

5.1 A Data Structure for Complex Regions

Rather than describing the data structure directly in terms of arrays, records, etc., we introduce a higher level description which offers suitable access and construction operations to be used in the algorithms. In a further step, one can then easily design and implement the data structure itself.

Let $N = \{0, \ldots, m-1\} \subset$ IN, and let $P_N = N \times N$ denote the set of all *N-points* (grid points). We define an (x, y)-lexicographical order on P_N, which is defined for $p = (x_1, y_1)$ and $q = (x_2, y_2)$ as $p < q \Leftrightarrow x_1 < x_2 \vee (x_1 = x_2 \wedge y_1 < y_2)$. Let $S_N = P_N \times P_N$ denote the set of *N-segments*. We normalise S_N by the requirement that $\forall s \in S_N : s = (p, q) \Rightarrow p < q$. This enables us to speak of a *left* and a *right end point* of a segment.

The crucial idea for the representation of *region* objects is to regard them as *ordered sequences of halfsegments* (Güting *et al.*, 1995). Let $H_N = \{(s, d) \mid s \in S_N, d \in \{left, right\}\}$ be the set of *halfsegments* where flag d emphasises one of the N-segments' end points, which is called the *dominating point* of h. If $d = left$, the left (smaller) end point of s is the dominating point of h, and h is called *left halfsegment*. Otherwise, the right end point of s is the dominating point of h, and h is called *right halfsegment*. Hence, each N-segment s is mapped to two halfsegments $(s, left)$ and $(s, right)$. Let dp be the function which yields the dominating point of a halfsegment.

For two distinct halfsegments h_1 and h_2 with a common end point p, let α be the enclosed angle such that $0° < \alpha \leq 180°$ (an overlapping of h_1 and h_2 is excluded by the realm properties). Let a predicate *rot* be defined as follows: $rot(h_1, h_2)$ is *true* if and only if h_1 can be rotated around p through α to overlap h_2 in counterclockwise direction. We can now define a complete order on halfsegments which is basically the (x, y)-lexicographical order by dominating points. For two halfsegments $h_1 = (s_1, d_1)$ and $h_2 = (s_2, d_2)$ we obtain:

$$h_1 < h_2 \quad \Leftrightarrow \quad dp(h_1) < dp(h_2) \vee (dp(h_1) = dp(h_2) \wedge ((d_1 = right \wedge d_2 = left) \vee (d_1 = d_2 \wedge rot(h_1, h_2)))$$

A value of type *region* (an example is given in Fig. 2) is now defined as an ordered sequence $\langle(h_1, a_1), \ldots, (h_n, a_n)\rangle$ of n halfsegments[2] where each halfsegment h_i has an attached set a_i of *attributes*. Attribute sets are used in algorithms to attach auxiliary information to segments. For example, those halfsegments of a *region* object F carry an associated attribute *InsideAbove* where the area of F lies above or left of its segments.

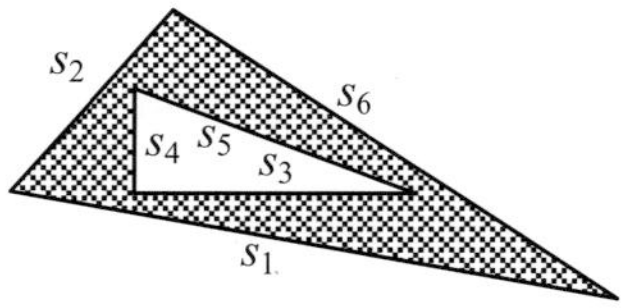

Fig. 2. Example of the halfsegment sequence of a region. If $h_i^l = (s_i, left)$ and $h_i^r = (s_i, right)$ denote the left and right halfsegments belonging to the segments s_i for $1 \leq i \leq 6$, the halfsegment sequence is $\langle(h_1^l, \{I\}), (h_2^l, \varnothing), (h_3^l, \varnothing), (h_4^l, \{I\}), (h_4^r, \{I\}), (h_5^l, \{I\}), (h_2^r, \varnothing), (h_6^l, \varnothing), (h_5^r, \{I\}), (h_3^r, \varnothing), (h_6^r, \varnothing), (h_1^r, \{I\})\rangle$. I stands for the attribute *InsideAbove*.

Simple implementations for type *region* would represent a sequence of n halfsegments in a linked list or sequentially in an array. Unfortunately, in the latter case, inserting a halfsegment at an arbitrary position then needs $O(n)$ time. Alternatively, an AVL-tree embedded into an array can be used whose elements are additionally linked in sequence order. An insertion then requires $O(\log n)$ time.

5.2 Algorithms for Topological Predicates

We now study the algorithms for our set T_{cr} of topological predicates on complex regions. These algorithms are *realm-based* in the sense that the regions as their arguments are defined over the *same* realm, that no two segments intersect within their interiors and that no point lies within a segment. The description of the algorithms requires three important concepts, namely *realm-based plane sweep*, *parallel object traversal*, and *overlap numbers*, which are explained first in the following.

5.2.1 Realm-Based Plane Sweep

The algorithmic scheme we employ for all predicates is the same and is based on the popular *plane sweep* technique (Preparata *et al.*, 1985) of Computational Geometry. A vertical *sweep line* sweeping the plane from left to right stops at

[2]Note that this data structure only stores halfsegments and attributes. The algorithms using this structure are responsible for constructing only halfsegment sequences that maintain the region properties.

special points called *event points* which are generally stored in a queue called *event point schedule*. The event point schedule must allow one to insert new event points discovered during processing; these are normally the initially unknown intersections of line segments. The state of the intersection of the sweep line with the geometric structure being swept at the current sweep line position is recorded in vertical order in a data structure called *sweep line status*. Whenever the sweep line reaches an event point, the sweep line status is updated. Event points which are passed by the sweep line are removed from the event point schedule. Note that in general an efficient fully dynamic data structure is needed to represent the event point schedule and that in many plane-sweep algorithms an initial sorting step is needed to produce the sequence of event points in (x, y)-lexicographical order.

In the special case of realm-based algorithms where no two segments intersect within their interiors, the event point schedule is static (because new event points cannot exist) and given by the ordered sequence of halfsegments of the operand objects. No further explicit event point structure is needed. Also, no initial sorting is necessary, because the plane sweep order of segments is the base representation of *region* objects anyway.

We assume some operations on the sweep line status which later make the description and understanding of the algorithms easier. Each segment stored in this structure is associated with a set of attributes. A new sweep line status structure is initialised by the operation *new_sweep*. If a left (right) halfsegment of a *region* object is reached during a plane-sweep, the operation *add_left* (*del_right*) stores (removes) its segment component into (from) the segment sequence of the sweep line status sorted by the order relation *above*. A segment s lies *above* a segment t if the intersection of their x-intervals is not empty and if for each x of the intersection interval the y-co-ordinate of s is greater than the one of t (except possibly for a common end point where the y-co-ordinates are equal). The operation *pred_exists* (*common_point_exists*) checks whether for the segment currently considered in the sweep line status a predecessor (a neighboureded segment with a common end point) exists. The operation *set_attr* (*get_pred_attr*) sets (gets) a set of attributes for (from the predecessor of) the segment currently considered in the sweep line status.

For the sweep line status an efficient internal dynamic structure like the AVL-tree can be employed which realises the operations *add_left* and *del_right* each in time $O(\log n)$ and the other operations in constant time.

5.2.2 Parallel Object Traversal

Four further operations make the algorithms more readable. The access to the current element of the halfsegment sequence of a *region* object is given by the operation *get_hs*. The operation *get attr* yields the pertaining attribute set of this halfsegment.

For the evaluation of a predicate we have to perform a parallel traversal through the ordered halfsegment sequences of both operand objects. To simplify the description of this parallel scan, two operations are provided. The operation *rr_select_first*(F, G, *object*, *status*) selects the first halfsegment of each of the *region* objects F and G and positions a logical pointer on both of them. The parameter *object* with possible values {*none*, *first*, *second*, *both*} indicates which of the two object representations contains the smaller halfsegment. If the value of *object* is *none*, no halfsegment is selected, since F and G are empty. If the value is *first* (*second*), the smaller halfsegment belongs to F (G). If it is *both*, the first halfsegments of F and G are identical. The parameter *status* with possible values {*end_of_none*, *end_of_first*, *end_of_second*, *end_of_both*} describes the state of both halfsegment sequences. If the value of *status* is *end_of_none*, both objects still have halfsegments. If it is *end_of_first* (*end_of_second*), F (G) is empty. If it is *end_of_both*, both object representations are empty. The operation *rr_select_next*(F, G, *object*, *status*) searches for the next smaller halfsegment of F and G; parameters have the same meaning as for *rr_select_first*. Both operations together allow one to scan in linear time two object representations like one ordered sequence.

5.2.3 Overlap Numbers

Overlapping of region parts plays an important role for computing their topological relationships. For this purpose we introduce the concept of *overlap numbers*. A point of the realm grid obtains the overlap number 2 if it is covered by (or part of) two *region* objects. This means that for two intersecting simple polygons the area outside of both polygons gets overlap number 0, the intersecting area gets overlap number 2, and the other areas get overlap number 1. Since a segment of a region separates space into two parts, an inner and an exterior one, each segment is associated with a pair (m/n) of overlap numbers, a lower (or right) one m and an upper (or left) one n. The lower (upper) overlap number indicates the number of overlapping *region* objects below (above) the segment. In this way, we obtain a *segment classification* of two *region* objects and speak of (m/n)-segments. Obviously, $m, n \leq 2$ holds. Of the nine possible combinations only seven describe valid segment classes. This is because a (0/0)-segment contradicts the definition of a *region* object, since then at least one of both regions would have two holes or an outer cycle and a hole with a common border. Similarly, (2/2)-segments cannot exist, since then at least one of the two regions would have a segment which is common to two outer cycles of the object. Hence, possible (m/n)-segments are (0/1)-, (0/2)-, (1/0)-, (1/1)-, (1/2)-, (2/0)-, and (2/1)-segments. Examples of (m/n)-segments are given in Fig. 3.

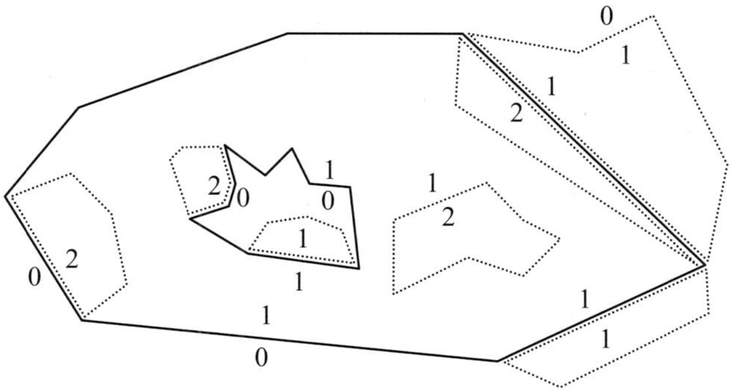

Fig. 3. Example of the segment classification of two *region* objects

5.2.4 Algorithms

The algorithmic scheme, which is common to all topological predicates, is demonstrated by the algorithm for $coveredBy_{cr}(F,G)$. For this predicate all segments of F must lie within the area of G but no segment (and hence no hole) of G may lie within F. If we consider the objects F and G as halfsegment sequences together with the segment classes, the predicate $coveredBy_{cr}$ is *true* if (i) all halfsegments that are *only* elements of F have segment class (1/2) or (2/1), since only these segments lie within G, (ii) all halfsegments that are *only* elements of G have segment class (0/1) or (1/0), since these definitely do not lie within F, and (iii) all *common* halfsegments have segment class (0/2) or (2/0), since the areas of both objects lie on the same side of the halfsegment. We have to require that F and G share at least a common segment or a common point. In the case of a (1/1)-segment the areas would lie side by side so that F could not be covered by G. In the algorithm, whenever a segment is inserted into the sweep line status, first the pair (m_p/n_p) of overlap numbers of the predecessor is determined (it is set to $(*/0)$ if no predecessor exists). Then the overlap numbers (m_s/n_s) for this segment are computed. Obviously $m_s = n_p$ must hold; n_s is also initialised to n_p and then corrected. We can now formulate the algorithm $coveredBy_{cr}$:

algorithm $coveredBy_{cr}$
input: Two *region* objects F and G
output: *true*, if F is covered by G
false, otherwise

begin
S := *new_sweep*();
inside := *true*;
common_segment := *false*;

```
common_point := false;
rr_select_first(F, G, object, status);
while (status ≠ end_of_first) and inside do      /* Let h = (s, d). */
  if (object = first) or (object = both) then h := get_hs(F);
  else h := get_hs(G);
  endif;
  if d = right then S := del_right(S, s);
  else
    S := add_left(S, s);
    if not pred_exists(S) then (m_p/n_p) := (*/0)
    else {(m_p/n_p)} := get_pred_attr(S);
    endif;
    m_s := n_p; n_s := n_p;
    if ((object = first) or (object = both)) and (InsideAbove ∈ get_attr(F))
      then n_s := n_s+1
      else n_s := n_s−1
    endif;
    if ((object = second) or (object = both)) and
          (InsideAbove ∈ get_attr(G))
    then n_s := n_s+1
    else n_s := n_s−1
    endif;
    S := set_attr(S, {(m_s/n_s)});
    if object = first then
      inside := ((m_s/n_s) ∈ {(1/2), (2/1)});
      if common_point_exists(S) then common_point := true endif
    else if object = second then inside := ((m_s/n_s) ∈ {(0/1), (1/0)})
    else inside := ((m_s/n_s) ∈ {(0/2), (2/0)}); common_segment := true
    endif;
  endif;
  rr_select_next(F, G, object, status);
endwhile;
return inside and (common_segment or common_point);
end coveredBy_cr.
```

If F has l and G m halfsegments, the while-loop is executed at most $n = l+m$ times, because each time a new halfsegment is visited. The most expensive operations within the loop are the insertion and the removal of a segment into and from the sweep line status. Since at most n elements can be contained in the sweep line status, the worst time complexity of the algorithm is $O(n \log n)$.

The other predicates mostly require slight modifications of the algorithm above. The predicate $covers_{cr}$ is symmetric to $coveredBy_{cr}$. The predicate *insidecr* forbids common (0/2)- and (2/0)- segments and also common points. The same holds for the symmetric predicate $contains_{cr}$. The predicate $disjoint_{cr}$ yields *true* if both objects do not share common areas, common segments, and common points.

Hence, it only allows (0/1)- and (1/0)-segments and forbids common points. The predicate $meet_{cr}$ is equal to $disjoint_{cr}$ but additionally requires the existence of at least one (1/1)-segment or a common point. The predicate $overlap_{cr}$ is *true* if two regions have a common area which means that there exist segments of the segment classes (1/2)) or (2/1).

6 Conclusions

Based on a formal and coherent definition of complex regions and their topological predicates both at the abstract and the discrete level, we have demonstrated how regions and predicates on them can be implemented in a numerically robust and topologically consistent manner. Spatial query languages can now also be employed to pose queries using topological relationships on more complex regions.

References

Behr T, Schneider M (2001) Topological Relationships of Complex Points and Complex Regions. In: Int. Conf. on Conceptual Modeling. pp 56-69

Clementini E, Di Felice P (1996) A Model for Representing Topological Relationships between Complex Geometric Features in Spatial Databases. Information Systems 90:121–136

Clementini E, Di Felice P., Califano G (1995) Composite Regions in Topological Queries. Information Systems 20(7):579–594

Clementini E, Sharma J, Egenhofer MJ (1994) Modeling Topological Spatial Relations: Strategies for Query Processing. Computers and Graphics 18(6):815–822

Cui Z, Cohn A G, Randell DA (1993) Qualitative and Topological Relationships. In: 3rd Int. Symp. on Advances in Spatial Databases, LNCS 692, pp 296–315

Egenhofer M J, Frank A, Jackson J P (1989) A Topological Data Model for Spatial Databases. In: 1st Int. Symp. on the Design and Implementation of Large Spatial Databases. LNCS 409, pp 271–286

Egenhofer M J, Herring J (1990) A Mathematical Framework for the Definition of Topological Relationships. In: 4th Int. Symp. on Spatial Data Handling. pp 803–813

Egenhofer MJ, Clementini E, Di Felice P (1994) Topological Relations between Regions with Holes. Int. Journal of Geographical Information Systems 8(2):128–142

Güting R H, Schneider M (1993) Realms: A Foundation for Spatial Data Types in Database Systems. In: 3rd Int. Symp. on Advances in Spatial Databases, LNCS 692, pp 14–35

Güting R H, Schneider M (1995) Realm-Based Spatial Data Types: The ROSE Algebra. VLDB Journal 4:100–143

OpenGIS Consortium (OGC) (1999) OGC Abstract Specification [online]. Available from: http://www.opengis.org/techno/specs.htm.

OpenGIS Consortium (OGC) (2001) OGC Geography Markup Language (GML) 2.0 [online]. Available from: http://www.opengis.net/gml/01-029/GML2.html.

Preparata F P, Shamos M I (1985) Computational Geometry. Springer Verlag

Schneider M (1997) Spatial Data Types for Database Systems–Finite Resolution Geometry for Geographic Information Systems. In: LNCS 1288, Springer-Verlag

Schneider M (2001) A Design of Topological Predicates for Complex Crisp and Fuzzy Regions. In: Int. Conf. on Conceptual Modeling. pp 103-116

Route Specifications with a Linear Dual Graph

Stephan Winter

Institute for Geoinformation, Vienna University of Technology, Gusshausstr. 27-29, 1040 Vienna, Austria, winter@geoinfo.tuwien.ac.at

Abstract

The objective often explored in Web-based route planners is to find an optimal route in a network for a given mode of transport. Usually this involves searching for the cheapest route corresponding to some cost function. For many types of trips, not all desired route properties can be satisfied in this way. Following is a proposed solution for planning hiking trips. The method can easily be transferred to tourist guides in urban areas, for advice for taking a drive, or can be used in related contexts. It is anticipated that these applications will become increasingly relevant in our mobile leisure society.

Planning a hiking route is based primarily on the intended length of the trip, and then on the decision about whether start and destination points need to coincide. As a result, planning trips represents neither a shortest path nor a travelling salesman problem.

The problem with this approach lies in the fact that hikers require trip proposals that cannot necessarily be provided by a shortest path algorithm: round tours, routes including loops, or routes where a route segment which has to be passed in both directions are more consistent with route planning for hikers. A surprisingly simple solution for these requirements appears to be a linear dual graph, which applies a *k* shortest path algorithm to the linear dual graph. This paper outlines this approach and demonstrates that it achieves realistic and practical route planning.
Keywords: route planning, shortest path, linear dual graph

1 Introduction

Web-based route planners or mobile route guides solve the task of finding an optimal route in the network of a given mode of transport, e.g., by car, public transport, aeroplane, bike, or on foot. They typically search for the cheapest route corresponding to any cost function, like time, fuel, or number of changes. At least for pedestrians (but in principle for all travel modes) not all desired route properties can be satisfied by a simple graph and a shortest path algorithm. Some of the incompatible properties are an intended duration of a trip, or the option of round tours. We are investigating these problems for planning hiking trips (Cziferszky 2002), and propose a linear dual graph approach (not to be confused with the dual graph of a partition). The following discussion can easily be transferred to tourist guides in urban areas, for guidance in taking a drive, or for similar types of contexts. These applications will become increasingly relevant in our mobile leisure society.

Let us consider a person asking advice from a web service concerning her hiking trip next Sunday. Her main criterion may be the intended duration of the trip. It is clear that planning routes for trips may not only be a shortest path problem. The trip planning service needs to be able to search for routes of a predetermined duration. A planning service needs also the flexibility to provide both, routes from a start to a destination and round tours. Finding a round tour can be a boundary condition for instance for hikers who want to come back to their car at the end of the trip. Another required flexibility lies in the fact that hikers accept to visit nodes more than once. Consider for example a loop through a park, added to an otherwise linear trip. Hikers often walk a route segment in both directions, e.g., if this segment is the only one to a mountain's peak, or the only connection between a car park and an attractive hiking area. There are many other route selection criteria, for example, maximal slope, most scenic path, or possible stops at inns. These criteria can likely be discovered, provided that a number of predetermined or preconsidered interests could be established.. However, we discuss neither the introduction of these additional criteria nor the ranking of routes in this paper.

The round tour problem is also not entirely like a travelling salesman problem. The travelling salesman is looking for the shortest round tour that visits a set of places. Hikers on the other hand, who look for a round tour may not necessarily ask for the shortest route, even if they specify points of interests to be visited on that trip.

In principle, the posed problem of route planning for hiking trips requires a combination of all possible route segments in a region, with a set of possible start and destination nodes (hiker parking places, bus stops, and similar). The constructed routes have to be selected that best fit to the specified criteria. This is an exponential combinatorial problem. A known efficient approach is by exploiting algorithms for the k shortest paths. These algorithms yield an ordered sequence of the k shortest routes (some of them yield even loopless shortest paths). Instead of calculating k shortest routes, the algorithms can run recursively, stopping when the intended route duration is exceeded. Consequently, the set of routes satisfying the duration criterion can then be evaluated for other properties. Evaluation could rank the routes to recommend the ‘best’ one first.

The k shortest routes show all combinatorial possibilities of extending a shortest path. Many of the routes presented, however, are not appropriate and would not usually be acceptable to hikers. The question to pose is 'how does one specify route properties requested by hikers such that the artificial or inappropriate solutions can be rejected'. A surprisingly simple solution appears to be found through using a linear dual graph, as presented in the following sections of this paper. The hypothesis presented in this paper explores the applicability of the *k* shortest path algorithm on the linear dual graph to determine if it can produce realistic route tips, including round trips, trips with loops, and trips with segments to be walked in both directions – but never twice in the same direction.

The paper is structured as follows. In section 2, the problem is discussed in relation to the state of the art. section 3 introduces the conceptual model of a dual linear graph and a dual path. In section 4, the application of the linear dual graph to trip planning is presented. Section 5 discusses the further exploitation of the conceptual model in the calculation of trip proposals. We close with a discussion and outlook (Section 6).

2 State of the Art

The requirements for creating hiking trip proposals is based on a user-centred approach to model the route planning (Cziferszky 2002), and is supported by general research in human wayfinding (Golledge 1995; Golledge and Stimson 1997) and navigation (Denis et al. 1999; Frank 2001). According to a survey of Cziferszky most people planning an excursion first specify the duration of the trip.

Shortest path algorithms, among them the most prominent (Bellman 1958; Dijkstra 1959) yield minimum spanning trees from a start node which through execution navigates to all other nodes in a network (Chvatal 1983; Cormen et al. 1990), i.e., exactly one path from a start to any destination node. Different trip duration can only be found by varying the start and destination points. Common for all shortest path algorithms is a graph search strategy for local optima such that a node, registered once, will never be visited again. This property excludes loops.

The *k* shortest path algorithms (Bellman and Kalaba 1960; Azevedo et al. 1990; Eppstein 1994; Ruppert 2000) vary the shortest path, finding not only the shortest, but also the second shortest, and so on. The variations are produced either by route labelling methods, or by route deletion. There are loopless and non-loopless *k* shortest path algorithms. Loops extend the length of a route but then continue with already known parts of a path. Loops also include edges that can be travelled in both directions.

The travelling salesman algorithms find shortest round tours to a set of nodes in the graph. Some variation has been applied to shortest round tours through development of the sub-tour problem presented by (Stille 2001). In this case, a subset of the available points of interest (inns, views, and peaks) is combined to (shortest) round tours. However, this option is still not flexible enough to find tours of a predetermined duration.

The linear dual graph was originally proposed to model graphs with edge *and* node costs (Caldwell 1961; Knödel 1969). It can be used to express relations between consecutive edges (Winter 2001). Although the idea is appealing for its elegance, it did not find much attention in the graph theory literature nor in applications. The only implementation of which we are aware is reported by

(Anez et al. 1996). Current network analysis tools deal with relations differently (Zeiler 1999).

Due to the complexity of the trip planning problem current services use mostly a set of predefined routes (Davies et al. 2001) or, alternatively, have to restrict the problem to a very limited set of route segments.

3 The Linear Dual Graph

In the following sections we use undirected and directed weighted graphs, paths, and the linear dual graph. These concepts are introduced in the following sections.

3.1 Graphs

A graph G consists of a set of nodes N and a set of edges E. The edges represent binary relations between nodes, $e: N \rightarrow N$. We distinguish the start node f, $f: E \rightarrow N$, and the end node l, $l: E \rightarrow N$. We treat an undirected edge u as a short form of a pair of bi-directional edges: $u = \{e_1\ (f(u), l(u)),\ e_2\ (l(u), f(u))\}$. Typically edges have some costs attached, $w: E \rightarrow \Re^+$; default value is 1. Then a graph is given by the tuple $G(N,E)$ with f_E, l_E, and w_E.

The total number of edges ending in a node n is called the *indegree* of n, $\text{indegree}(n) = \sum_{i=1}^{|E|} e_i \mid l(e_i) = n$, and the number of edges starting in n is its *outdegree*, $\text{outdegree}(n) = \sum_{i=1}^{|E|} e_i \mid f(e_i) = n$.

A path p from node $f(e_1)$ to $l(e_k)$ is the sequence of edges $p=(e_1,\ldots,e_k)$ such that $l(e_i)=f(e_{i+1})$, $i=1,\ldots,k-1$, and $e_i \neq e_j$ for $i \neq j$. The path include loops if there exist any pairs $f(e_i)=l(e_j)$ for $i < j$. The length of a path is k; the total cost of a path is $W(p) = \sum_{i=1}^{k} w(e_i)$.

3.2 Linear Dual Graphs

A *linear dual graph D* is defined by the following definition (see also Fig. 1):

Definition: Given a (primal) graph $G(N_G,E_G)$ with f_G, l_G, w_G, the graph $D(N_D, E_D)$ with f_D, l_D, w_D with the following properties is called its *linear dual graph*:

- For each edge e_i in G there is a node v in D, assigned by a function d: $v_i=d(e_i)$. d is a bijective function so that $d^{-1}(v_i)=e_i$. Thus, $N_D=d(N_G)$.
- For each pair of edges (e_i, e_j) in G with $l(e_i)=f(e_j)$ there is an edge ε in D between the corresponding nodes $v_i=d(e_i)$, $v_j=d(e_j)$, such that $f(\varepsilon)= v_i$ and $l(\varepsilon)=v_j$. $E_D = \bigcup_i \varepsilon_i$.

- A cost function w_D: $E_D \rightarrow \Re^+$.

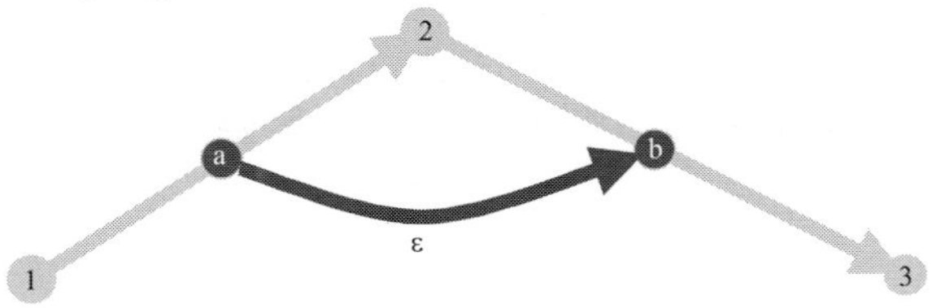

Fig. 1. A directed primal graph *G* (gray) and its linear dual *D(G)* (dark)

For simplicity we will speak of a *dual graph* in the following, always referring to the linear dual graph. Note that the more common notion of a dual graph to a partition is completely different.

The size of the dual graph *D* is given by the following equations:

- The number of nodes in *D* equals the number of edges in *G*: $|N_D| = |E_G|$.
- The number of edges in *D* equals the number of paths of length 2 in *G*. For each $n_i \in G$ there are $p(n_i)=indegree(n_i) \times outdegree(n_i)$ such paths. With route graphs being sparse, we can estimate the upper limit of such paths assuming a maximal number of ingoing (*inmax*) and outgoing edges (*outmax*) for all nodes. Then the number of edges in *D* is $\leq |E_G|$ *min (inmax, outmax)*.

Theorem: For each primal path with a length $k \geq 2$ there exists a corresponding path in the dual graph.

Proof. For each neighboured pair of edges in the primal path there exists per definition a dual edge. Thus, a sequence $(e_1,..., e_k)$ corresponds to a sequence $(\varepsilon_1, ..., \varepsilon_{k-1})$ such that $f(\varepsilon_1)=d(e_1)$, $l(\varepsilon_1)=d(e_2)$, and so forth till $l(\varepsilon_{k-1})=d(e_k)$. q.e.d.

Corollary: For a primal path of length *k* the length of the corresponding path in the dual graph is k-1.

3.3 Dual Paths

The properties of paths in the dual graph are not promising for the tasks at hand: in the context of trips paths with only one primal edge are valid paths; but they cannot be represented as paths in the dual graph. For that reason we propose to change to a *dual path*:

Definition: A dual path π to a given path $p=(e_1, ..., e_k)$ is the sequence of nodes $v_1=d(e_1), ..., v_k=d(e_k)$.

For each path in the primal graph exists a dual path in the dual graph. The length of the dual path is *k*, which is equal to the length of the path. The costs of the primal edges can be transferred to the corresponding dual nodes by the relation

$v(v_i)=w(e_i)$. In the context of this paper[1] we limit the total costs of a dual path to $V(\pi)=\sum_{i=1}^{k} v(v_i)$.

The dual path consists of a subset of elements of the dual graph, of a primal path. Only if the primal path contains no loops, the dual path consists of all elements of the dual graph.

4 Route Planning with the Linear Dual Graph

4.1 Start and Destination in the Dual Path

The dual graph is introduced in this section and discussed in the context of specifying routes for trip planning.

In the primal graph, a node means a state and an edge a transition from its start to its end node. For hiking trips we deal with embedded graphs, i.e., the state described by a node is its (or a visiting agent's) position in a plane. With the restriction that trips start and end at graph node positions, a path in the primal graph can directly be translated into a trip route, and constructed by a sequence of route segments.

In the dual graph a node shall mean a movement along the corresponding primal edge from its start to end node. An edge shall mean a reorientation in the corresponding primal node with continuation of hiking along another primal edge. Hence, a dual path $\pi=(v_1, \ldots, v_k)$ in the dual graph needs to be translated carefully into a trip route.

- The start position s of a route is (with the above restriction) $f(d^{-1}(v_1))$.
- The destination t is accordingly $l(d^{-1}(v_k))$.

This leads to a slightly different planning problem.

4.2 Routes and the Dual Path

We transfer the route planning problem to the dual graph.

A trip-planning problem identifies among the primal nodes a start and a destination of the route. Then its goal is to find a path with costs close to the intended trip duration. In case of round trips, start and destination are identical.

- A hiker always leaves the (primal) start node taking an outgoing edge. We call all the dual nodes v_i for which $f(d^{-1}(v_i))=s$ the *start gateways* of the route planning problem.

[1] In other contexts it can make sense to consider the costs of the dual edges also, see for example Winter, S., 2001: Weighting the Path Continuation in Route Planning. In: Aref, W.G. (Ed.), 9th ACM International Symposium on Advances in Geographic Information Systems. ACM Press, Atlanta, GA, pp. 173-176..

- A hiker always reaches the (primal) end node along an incoming edge. We call all the dual nodes v_i for which $l(d^{-1}(v_i))=t$ the *destination gateways* of the route planning problem.

For any path from node s to node t the dual path contains at least one element of the start gateways and one element of the destination gateways. Both sets can overlap (if a path is of length 1), hence, the element from the start gateways and the element of the destination gateways may coincide.

The route planning problem in the dual graph is then to find a dual path between any node from the start gateways and any node of the destination gateways such that its costs are close to the intended duration of the trip.

4.3 Extended Modelling Capabilities with the Dual Path

As indicated previously, a shortest path algorithm shall be applied for the calculation of the route planning problem.. Shortest path algorithms cannot include loops (by definition). This is certain limitation for planning trips, and can be solved by the extended expression of the dual path.

- A round trip is a popular trip type. Paths may have loops, but no shortest path algorithm allows visiting a node twice. The algorithm will stop after initialisation.

  For the dual path the problem is formulated differently: for a round trip start and destination gateways are disjoint (a loop of length 1 is excluded in the graph definition), and thus, the length of a round trip is at least 2

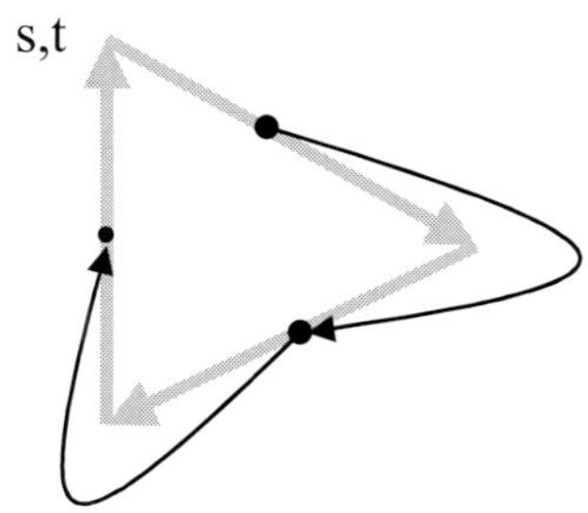

Fig. 2. A round trip does not require visiting any dual path node more than once

- Hiking routes may include loops in between. Shortest path algorithms will never visit a node more than once. Again, in the dual path no node is visited more than once (Fig. 3).

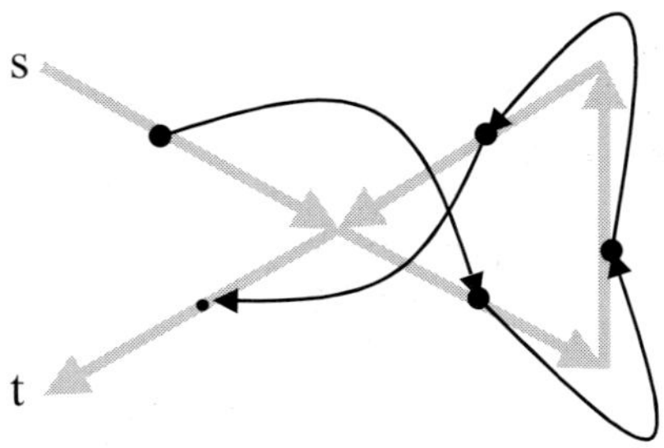

Fig. 3. Loops do not require visiting any dual path node more than once

- Hiking routes may include segments that have to be hiked in both directions. Again, in the dual path no node is visited more than once (Fig. 4).

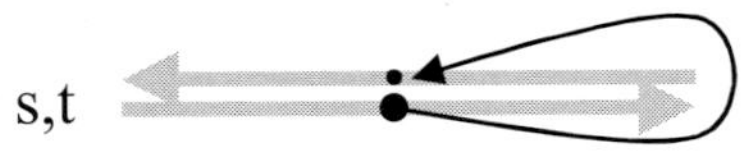

Fig. 4. Segments travelled in both directions do not require visiting any dual path node more than once

- The argument is even valid for combined cases (Fig. 5).

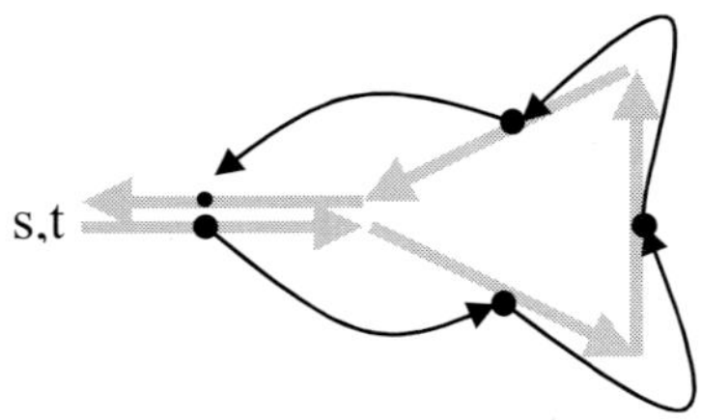

Fig. 5. A combination of previous cases

That means, common shortest path algorithms work in some central cases only on the dual graph. The specific properties of the dual graph would avoid hiking a segment a second time in the same direction. We expect that this property is conforming to the expectations of hikers on trip proposals, although we have not checked that.

5 Routes of Intended Duration

A shortest path algorithm between start and destination always yields the shortest path, however, we are looking for paths close to a predefined cost. One possibility to consider is the variation of start and destination. A route planning service will make use of this possibility, however, it is limited in most areas to a few numbers of hiker parking places, bus stops or similar points of interests.

The idea is to therefore systematically vary the shortest path, as done in k shortest path algorithms. The result is an ordered list of k routes from start to destination with monotonic increasing costs. Setting k large and storing the found routes in a database allows off-line calculation. The route candidates have only to be updated if the network changes. A request for a trip with intended duration x can access the database selecting all route candidates with a duration close to x, let us say, $0.85x \ldots 1.15x$.

6 Discussion and Outlook

This paper has introduced the linear dual graph and a dual path as a conceptual model for route specifications in trip planning. It has been demonstrated that this model can express critical planning cases, which were not achievable in primal graphs using shortest path algorithms.

Currently a k shortest path algorithm has been implemented to prove the concept. A subsequent paper will report on the algorithm, experiences with test cases, and computing performance of the calculations. In Winter, 2001, the linear dual graph for modelling turning costs and turn restrictions has been demonstrated.. In the current paper the specific needs of trip planning, can be accomplished, as well. Hence, one can expect that the potential of the linear dual graph has not been exhausted.

Acknowledgements

The idea for this paper came from discussions with Andreas Ciferszky, Technical University Vienna, who is working on multi-criteria optimisation for hiking routes.

References

Anez J, de la Barra T, Perez B (1996) Dual Graph Representation of Transport Networks. Transportation Research 30(3):209-216

Azevedo JA, Silvestre Madeira JJER, Martins EQV, Pires FMA (1990) A shortest paths ranking algorithm. In: Proc. of the Annual Conference of the Operations Research Society of Italy (AIRO '90). Associazione Italiana Ricerca Operativa, pp 1001-1011

Bellman RE (1958) On a Routing Problem. Quart. Appl. Math 16:87-90

Bellman RE, Kalaba R (1960) On *k*th best policies. Journal of SIAM 8:582-588

Caldwell T (1961) On finding minimum routes in a network with turn penalties. Communications of the ACM 4(2): 107-108

Chvatal V (1983) Linear Programming. W. H. Freeman & Co.

Cormen TH, Leiserson CE, Rivest RL (1990) Introduction to Algorithms. MIT Electrical Engineering and Computer Science Series. MIT Press, Cambridge

Cziferszky A (2002) Routenplanung anspruchsvoller Wanderungen. Diploma thesis in progress, Institute for Geoinformation, Technical University Vienna, Austria

Davies N, Cheverst K, Mitchell K, Efrat A (2001) Using and Determining Location in a Context Sensitive Tour Guide. IEEE Computer Journal 34(8):35-41

Denis M, Pazzaglia F, Cornoldi C, Bertolo L (1999) Spatial Discourse and Navigation: An Analysis of Route Directions in the City of Venice. Applied Cognitive Psychology 13:145-174.

Dijkstra EW (1959) A note on two problems in connexion with graphs. Numerische Mathematik 1:269-271

Eppstein D (1994) Finding the *k* Shortest Paths. In: IEEE Symposium on Foundations of Computer Science. pp 154-165

Frank AU (2001) Pragmatic Information Content - How to Measure the Information in a Route Description. GeoInformatica (accepted)

Golledge RG (1995) Path Selection and Route Preference in Human Navigation: A Progress Report. In: Frank AU, Kuhn W (eds) Spatial Information Theory - A Theoretical Basis for GIS. Lecture Notes in Computer Science, 988. Springer, Berlin, pp 207-222

Golledge RG, Stimson RJ (1997) Spatial Behavior: A Geographic Perspective. The Guildford Press, New York

Knödel W (1969) Graphentheoretische Methoden und ihre Anwendungen. Ökonometrie und Unternehmensforschung, 13. Springer, Berlin

Ruppert E (2000) Finding the *k* Shortest Paths in Parallel. Algorithmica 28:242-254

Stille W (2001) Lösungsverfahren für Prize-Collecting Traveling Salesman Subtour Probleme und ihre Anwendung auf die Tourenplanung im Touristeninformationssystem Deep Map. Diploma thesis, Fakultät für Mathematik und Informatik, Ruprecht-Karls-Universität Heidelberg, Germany

Winter S (2001) Weighting the Path Continuation in Route Planning. In: Aref WG (ed) 9th ACM International Symposium on Advances in Geographic Information Systems. ACM Press, Atlanta, GA, pp 173-176

Zeiler M (1999) Modeling Our World. ESRI Press, Redlands

Projecting a Regular Grid onto a Sphere or Ellipsoid

Rune Aasgaard

SINTEF Applied Mathematics, PO.Box 124, N-0314 Oslo, Norway,
rune.aasgaard@sintef.no

Abstract

This paper presents a new map projection intended for mapping regular Level-of-Detail structures like quad-trees, wavelets or Lindstrom/ROAM triangulations on a spherical/ellipsoidal earth. The projection maps the squares of regular structures to near-squares on the globe, even near the poles. The projection gives simple equations, visually pleasing graphic output and works well with familiar algorithms and data structures.
Keywords: global grid, cartographic projection, level-of-detail, spatial reference system

1 Introduction

In many applications, geographical data is stored as a regular quadratic grid. The data unit is a near quadratic cell. When only a limited area is handled, this grid is usually oriented to a map projection.

For global applications however, we have the problem that no projection handles the whole earth very well without having areas where the scale shows strong variation. There are discontinuities and shape or area distortions.

For a global grid of dimension COLS×ROWS a common method is to use a simple uniform scaling from the geographical coordinates, giving grid positions:

$$x = \frac{COLS}{2\pi}\lambda \qquad y = \frac{ROWS-1}{\pi}(\phi + \lambda/2) \tag{1}$$

Where the latitude domain is: $\phi \in [-\pi/2, \pi/2]$, and for the longitude: $\lambda \in [0, 2\pi\rangle$ where λ wraps around at 2π.

This method has some advantages and several disadvantages. It is very simple, and for equatorial areas, it gives reasonable results. However, as we approach the poles the cells on the sphere are increasingly narrower and more trapezoidal in shape.

In a Level-of-Detail (LOD) environment (for Lindstrom triangulations, terrain texture pyramids, wavelets, quad-trees etc. ..) it is often important that the aspect ratio of the cells is as close to 1 as possible. The length of the longest axis of the cell usually determines the detail level for a cell. Where the ratio between the cell dimensions is very far from 1 we would get an unnecessarily high data density in the direction where the cells are shorter.

In spite of this, the uniform scaling is frequently used in systems for global mapping. Examples are the tessellations used for GeoVRML (Reddy M, Leclerc YG, Iverson L, Bletter N (1999) or VGIS(Lindstrom P, Koller D, Ribarsky W, Hodges LF, Op den Bosch A, Faust N (1997). It is also used as the storage coordinate system of many publicly available global data sets:

- GTOPO30 (http://edcdaac.usgs.gov/gtopo30/gtopo30.html),
- Globe (http://www.ngdc.noaa.gov/seg/topo/globe.shtml).

Using a conformal map projection would correct some of this, but a conformal projection cannot map the whole earth in one projection system without having areas where the scale approaches infinity. However, continuous LOD methods allows the model to adapt to changing detail requirements over the area, and compensates for some of the large variations of scale implied with conformal projections.

Some large scale (but not global) data sets use conformal projections:

- the Smith-Sandwell bathymetric model uses the Mercator projection between 72°N and 72°S (http://topex.ucsd.edu/marine_topo/mar_topo.html), and,
- the IBCAO arctic ocean bathymetric model uses a polar stereographic projection: (http://www.ngdc.noaa.gov/mgg/bathymetry/arctic/arctic.html).

Other authors have gone in the opposite direction (Tobler W, Chen ZT (1986) and (Ottoson P (2001), defining an equal-area co-ordinate system for preserving the area of the cells instead of the shape. This approach has the advantage of having an equal data volume in each cell at the same level of detail, given a uniform complexity.

There have been other attempts to develop global grid systems based on approximating the sphere to a regular polyhedron. A prominent example is octahedral Quaternary Triangular Meshes (Dutton G (1990) and several other papers of the same author). It is based on a hierarchical decomposition of equal-sided triangles where a triangle is split into four smaller triangles for each level of detail. Other authors, for example: (Goodchild MF, Shiren Y (1992) and

(Bartholdi JJ, Goldsman P (2001) have explored the QTM with emphasis on its use for spatial indexing and its geometrical properties.

The QTM gives a very homogeneous hierarchical mesh, but its triangle based split pattern makes it difficult to integrate with the commonly used square based methods. In addition, if it should be used for continuous view dependent LOD we will need non equal-sided triangles for the connections between triangles at different levels. This breaks much of the elegance of the original model.

For global wavelets both (λ, ϕ) parameterisations (Freeden W, Windheuser U (1994) and QTM methods (Schröder P, Sweldens W (1995) have been used.

In 2000 the U.S. National Center for Geographic Information and Analysis arranged a conference on global grid structures: International Conference on Discrete Global Grids (http://www.ncgia.ucsb.edu/globalgrids/index.html). Several of the mentioned methods where addressed there, but there does not seem to be any general agreement on a universally "best" method.

2 Mercator's Projection

An infinitesimal change in geographic position is mapped via Mercator's projection to an infinitesimal distance on the surface of the earth ellipsoid by the following equations:

$$du = N(\phi)\cos(\phi)\,d\lambda, \quad dv = M(\phi)\,d\phi$$

Where $M(\phi)$ is the Meridial radius of curvature and $N(\phi)$ is the Normal radius of curvature:

$$N(\phi) = \frac{r_{eq}}{W(\phi)}$$

$$M(\phi) = \frac{r_{eq}\left(1-\varepsilon^2\right)}{W(\phi)^3} = N(\phi)\frac{1-\varepsilon^2}{W(\phi)^2}$$

$$W(\phi) = \sqrt{1-\varepsilon^2\sin^2\phi}$$

Where r_{eq} is the equatorial radius of the ellipsoid and ε is the eccentricity.

In Mercator's projection the east/west transform is defined as $x = r_{eq}\,\lambda$ and the north/south transform is defined so that the projection is conformal ($dx/dy = du/dv$):

$$dy = \frac{dv}{du}dx = \frac{M(\phi)\,d\phi}{N(\phi)\cos(\phi)\,d\lambda}r_{eq}d\lambda = \frac{r_{eq}\left(1-\varepsilon^2\right)}{\left(1-\varepsilon^2\sin^2\phi\right)\cos\phi}d\phi \tag{2}$$

In the spherical case $\varepsilon = 0$ and the equation simplifies to:

$$dy = \frac{r_{eq}}{\cos\phi} d\phi \tag{3}$$

The projection transform functions are found by integrating Eqs. 2 or 3.

3 A Finite Near-Conformal Regular Grid Projection

When mapping a rectangular grid to the sphere we still want the meridians to be mapped to vertical grid lines:

$$dx = c \cdot d\lambda \tag{4}$$

In the north/south direction we want a transform that to some degree grows like the Mercator's projection, but maps to a finite range. We try a polynomial expansion of Eq. 3:

$$dy = \frac{a \cdot d\phi}{1 - (b\phi)^2} \tag{5}$$

This can be integrated to find a relation between y and ϕ:

$$y = \int_0^{\phi} \frac{a \cdot d\dot{\phi}}{1 - (b\dot{\phi})^2} = \frac{a}{2b} \ln\left|\frac{b\phi + 1}{b\phi - 1}\right| \tag{6}$$

The parameters a and b are determined to let the projection have the properties we desire for scaling and range. From Eq. 5 we see that $dy/d\phi \rightarrow \infty$ if $b\phi \rightarrow \pm 1$. To avoid asymptotic behaviour within the ordinary range of ϕ we want $| b\phi | < 1$ for all $\phi \in [-\pi/2, \pi/2]$. Therefore $b \in <-2/\pi, 2/\pi>$. As the equations are symmetric for positive and negative b we concentrate on $b \geq 0$.

The case $b = 0$ results in $dy = a \cdot d\phi$, and is trivial. For $b \rightarrow 2/\pi$ the scaling $dy/d\phi$ gets increasingly steeper and approaches the behaviour of the Mercator's projection. We still want a finite range for the projection, and expect a value of b that is slightly smaller than $2/\pi$.

In the following we use a domain $\phi \in [-\pi/2, \pi/2]$ and a range of $y \in [-y_{\pi/2}, y_{\pi/2}]$. We insert the values for the extremes and use a substitution for b:

$$b = \tfrac{2}{\pi}(1 - \delta)$$

$$y_{\pi/2} = \frac{a\pi}{4(1 - \delta)} \ln\left|2/\delta - 1\right|$$

Considering the expected domain of b we would similarly expect $\delta \in <0, 1]$, with the most useful value near 0. Then we can remove the absolute value symbol, and rewrite the equation as:

$$\delta = \frac{2}{1 + e^{\frac{4(1-\delta)}{a\pi} y_{\pi/2}}} \tag{7}$$

which is easily solved numerically by fixed-point iteration starting with $\delta_0 = 0$. Since the exponent over e is always positive the denominator is always greater than 2, and the restriction on δ is followed. We should note that this relation always has a solution for $\delta = 1$. Whether it also has a solution for $\delta \in <0, 1>$ depends on the value of $4y_{\pi/2}/a\pi$. It is relatively easy to prove that this solution exists and is unique when:

$$y_{\pi/2} > a\pi/2 \tag{8}$$

The inverse projection is easily computed when we know that $|b\phi| < 1$:

$$\begin{aligned} e^{\frac{2b}{a}y} &= \frac{1+b\phi}{1-b\phi} \\ \phi &= \frac{e^{\frac{2b}{a}y}-1}{e^{\frac{2b}{a}y}+1}\left(1/b\right) = \tfrac{1}{b}\tanh\left(\tfrac{b}{a}y\right) \end{aligned} \tag{9}$$

3.1 Scale

The dimension of the cells are defined by du/dx and dv/dy:

$$du/dx = \frac{N(\phi)\cos(\phi)d\lambda}{c \cdot d\lambda} = N(\phi)\cos(\phi)/c \tag{10}$$

$$dv/dy = \frac{M(\phi)d\phi}{\frac{a \cdot d\phi}{1-(b\phi)^2}} = M(\phi)\left(1-(b\phi)^2\right)/a = \frac{M(\phi)}{a \cdot \cosh^2\left(\frac{b}{a}y\right)} \tag{11}$$

In the following dv/dy is called the y-scale: s_y

The aspect ratio of a cell is then:

$$\theta(\phi) = \frac{dv/dy}{du/dx} = \frac{cM(\phi)\left(1-(b\phi)^2\right)}{aN(\phi)\cos\phi} = \frac{c\left(1-\varepsilon^2\right)\left(1-(b\phi)^2\right)}{a\left(1-\varepsilon^2\sin^2\phi\right)\cos\phi}$$

The aspect ratio tells us how close the cells are to be quadratic. A value of 1 represents a quadratic cell, higher values indicate a "narrow" cell and lower values a "wide" cell. If we want the aspect ratio to be 1 at $\phi=0$ we get $a = c\ (1-\varepsilon^2)$.

3.2 Distance, Direction and Area

The length of an infinitesimal line element is given as:

$$dS^2 = du^2 + dv^2 = \left(\frac{N(\phi)\cos\phi}{c}dx\right)^2 + \left(\frac{M(\phi)\left(1-(b\phi)^2\right)}{a}dy\right)^2$$
$$= \frac{M(\phi)^2\left(1-(b\phi)^2\right)^2}{a^2}\left(\frac{dx^2}{\theta(\phi)^2} + dy^2\right) = s_y(\phi)^2\left(\frac{dx^2}{\theta(\phi)^2} + dy^2\right) \tag{13}$$

The direction (azimuth, angle between the north direction and the line) of the line element:

$$\tan(Az) = {du}/{dv} = \frac{aN(\phi)\cos\phi}{cM(\phi)\left(1-(b\phi)^2\right)}\frac{dx}{dy} = \frac{1}{\theta(\phi)}\frac{dx}{dy} \tag{14}$$

The area of a rectangular element:

$$dA = du \cdot dv = \frac{N(\phi)\cos\phi \cdot M(\phi)\left(1-(b\phi)^2\right)\cdot dx \cdot dy}{c \cdot a}$$
$$= \frac{M(\phi)^2\left(1-(b\phi)^2\right)^2}{a^2} \cdot \frac{dx}{\theta(\phi)} \cdot dy = s_y(\phi)^2 \cdot \frac{dx}{\theta(\phi)} \cdot dy \tag{15}$$

3.2.1 Distance Approximation

Returning to the distance calculation we introduce Eq. 14 into Eq. 13 to eliminate *dx*.

$$dS = s_y(\phi)\sqrt{\frac{dx^2}{\theta(\phi)^2} + dy^2} = s_y(\phi)\sqrt{\tan^2(Az)+1}\cdot|dy| = \frac{s_y(\phi)}{|\cos(Az)|}|dy|$$

To develop a simplified equation for the length of a geodetic line we introduce some approximations that are valid for smaller distances, where some of the properties are considered constant along the line:

- Azimuth is constant along the line, and approximated as: $\tan(Az) = {\Delta x}/{\theta(\phi)\cdot\Delta y}$ then also $|\cos(Az)| = {|\Delta y|}/{\sqrt{{\Delta x^2}/{\theta(\phi)^2}+\Delta y^2}}$
- $M(\phi)$ is constant for small changes of latitude.
- $\theta(\phi)$ is nearly constant for small changes of latitude, except near the poles.

$$S = \int_{p_1}^{p_2} dS = \int_{y_1}^{y_2} \frac{s_y(\phi)}{|\cos Az|}|dy| = \int_{y_1}^{y_2} \frac{M(\phi)}{a|\cos Az|\cosh^2\left(\frac{b}{a}y\right)}|dy|$$

$$\approx \frac{M(\phi)}{a|\cos Az|}\int_{y_1}^{y_2}\frac{|dy|}{\cosh^2\left(\frac{b}{a}y\right)} = \frac{M(\phi)}{b|\cos Az|}\left|\tanh\left(\tfrac{b}{a}y_2\right) - \tanh\left(\tfrac{b}{a}y_1\right)\right|$$

This is again expanded to the second degree in a Taylor series around $y = y_0$ (and $\phi=\phi_0$), and the approximation for $\cos(Az)$ is inserted. By selecting $y_0=(y_1+y_2)/2$ the second degree terms disappear and we get:

$$S = \frac{M(\phi_0)}{a\cdot\cosh\left(\frac{b}{a}y_0\right)^2}\sqrt{\Delta x^2 \big/ \theta(\phi_0)^2 + \Delta y^2} \tag{16}$$

3.2.2 Area Approximation

The area of a grid cell is found by integrating Eq. 15 and using the same simplifications as in section 3.2.1:

$$A = \int_{x_{\min}}^{x_{\max}}\int_{y_{\min}}^{y_{\max}} \frac{M(\varphi)^2}{a^2\cosh^4\left(\frac{b}{a}y\right)}\frac{dx}{\theta(\phi)}dy \approx \frac{M(\phi)^2\Delta x}{a^2\theta(\phi)}\int_{y_{\min}}^{y_{\max}}\frac{dy}{\cosh^4\left(\frac{b}{a}y\right)}$$

When using a Taylor series expansion around y_0 in the same way as above we get:

$$A = \frac{M(\phi)^2\Delta x\cdot\Delta y}{a^2\theta(\phi)\cosh^4\left(\frac{b}{a}y_0\right)} \tag{17}$$

4 Examples

In this section we will develop the equations further for a $2^n \times 2^m+1$ grid. Here we get this range: $0 \le x<2^n$, $-2^m/2\le y\le 2^m/2$. For simplicity we assume a spherical earth, the elliptical equations are developed with slightly more work from the full equations in section 3.

The east/west transform is:

$$x = \frac{2^n}{2\pi}\lambda$$

If we set the aspect ratio at equator to 1 we get $a=c=2^n/2\pi$. Given the range of y we have $y_{\pi/2}=2^m/2$.

The fix point recurrence relation is then:

$$\delta_0 = 0, \delta_{k+1} = \frac{2}{1+e^{(1-\delta_k)\cdot 2^{(m-n+2)}}}$$

and has a solution other than $\delta=1$ if (from Eq. 8):

$$1 < \frac{2y_{\pi/2}}{a\pi} = \frac{2\cdot 2^m/2}{\frac{2^n}{2\pi}\pi} = 2^{m-n+1} \Leftrightarrow m > n-1$$

If the grid has relative dimensions of $m = n\text{-}1$ we get $b = 0$ and a uniform scaling as in Eq. 1.

For $m>n\text{-}1$ the other equations are:

$$b = \tfrac{2}{\pi}(1-\delta)$$

$$dy = \frac{2^{n-1}d\phi}{\pi-(b\phi)^2}$$

$$y = \frac{2^{n-3}}{1-\delta}\ln\frac{1+b\phi}{1-b\phi}$$

$$\phi = \frac{1}{b}\cdot\tanh\left(\frac{1-\delta}{2^{n-2}}y\right)$$

$$\theta(\phi) = \frac{1-(b\phi)^2}{\cos\phi}$$

4.1. Aspect ratio

As the aspect ratio $\theta(\phi)$ is important for the shape of the grid cells it is studied in more depth. For the limit case $m=n+\infty$, $b = 2/\pi$ the aspect ratio is:

$$\theta_{m=n+\infty}(\phi) = \frac{1-\left(\frac{2}{\pi}\phi\right)^2}{\cos\phi}$$

This can be considered the "best" aspect ratio that can be obtained by this method. It has its maximum at $\theta(\pi/2) = 4/\pi$. Naturally, for a conformal projection the aspect ratio would be constant 1 for the whole domain.

Some computations of aspect ratios are summarised in Fig. 1. Except for the limit case the aspect ratio rises to infinity at the poles. The size of the grid cells (the projection scale) would also change dramatically from cell to cell as we approach the pole. As expected, the projection is not well suited if the area of interest is very close to the poles.

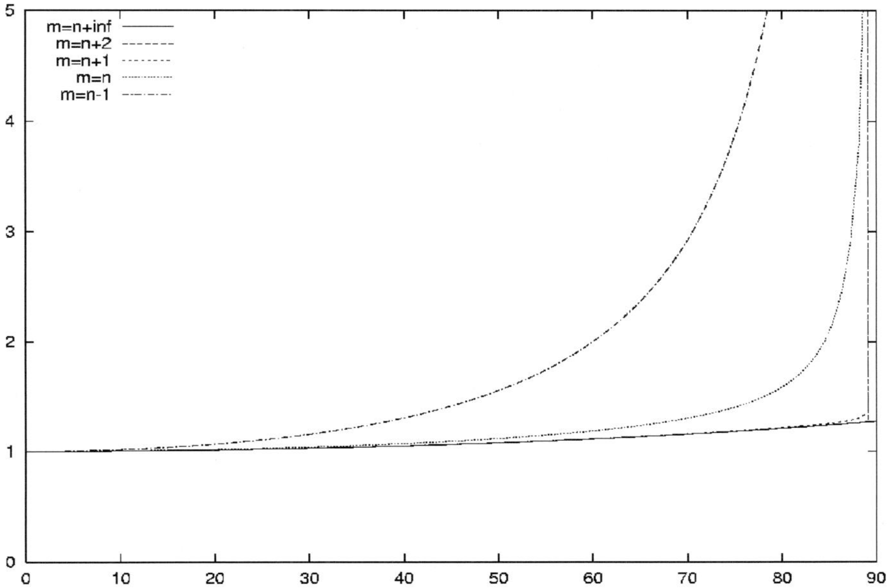

Fig. 1. Aspect ratios for different values of m=n+x

However, for $m = n+1$ the aspect ratio is reasonable even a few ten kilometers from the pole and for $m = n+2$ it is still less than 1.5 at around 100m from the pole.

For many applications an aspect ratio of around 2 at 85° is acceptable, at least much better than an aspect ratio of 2 at 60° as the uniform scaling would give. At the same time the quadratic coordinate range we get when $m=n$ may have advantages by using the coordinate values most efficiently.

The improvement given by going from $m=n+1$ to $m=n+2$ may not be worth the extra narrowing of the x coordinate range. Instead it may have some advantages to change the scaling (parameter a) so that the aspect ratio at the equator is slightly less than 1, to distribute the error more evenly.

4.2 Approximation Errors

When using the approximated equations for distance or grid cell area we would like to know their area of acceptable accuracy. This is not necessarily easily computed, as it may be difficult to estimate the impact of the simplifications in the development of the approximated equations.

The errors are found by comparing distances computed with the approximated equations and "exact" geodetic line methods. A series of computations is performed with successively larger distances until the error reaches a given threshold. Fig. 2. shows the maximum distance in kilometers before the error threshold is reached, at different latitudes.

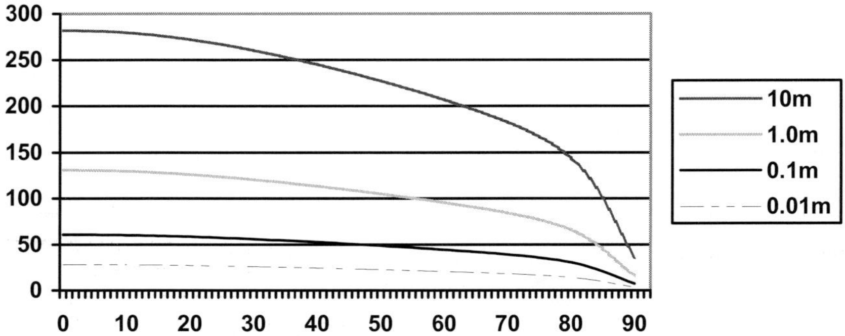

Fig. 2. Approximation error thresholds in distance computation, m = n grid

As expected, the maximum distance for a given error tolerance is reduced with increasing latitude. In this case, the distances are computed for a line with azimuth 135°, for longer distances, the error is direction dependent, and using another direction would give slightly different results. Similar tables can also be computed for the area approximation, comparing the results with data from the Lamberts equal-area projection.

5 Conclusion

For data visualisation purposes, the global regular grid projection is capable of mapping a regular grid to the globe with acceptable distortions. It has two singularities - at the poles - and most of the projection distortions are concentrated there. As the scale increases towards the poles a grid cell covers successively smaller area. This implies an increasing resolution towards the poles. For most purposes, however, the interesting data is located in the lower latitudes, and the projection should preferably be combined with a LOD structure for balancing the scale with the need for resolution.

The projection is relatively simple to integrate with many well-known methods and data structures. More complex methods may give better results, but are usually considered too difficult for practical use. Most projects therefore fall back to using a uniform scaling to latitude/longitude. For them, this method could be a good alternative.

For limited regions, various geometrical properties like distance and area can be computed directly from projected grid coordinates. The approximation error thresholds are not easily found, and a table area size, where the error is within acceptable limits, should be computed before the approximation functions are used.

The projection is used in a "virtual globe" project on the Internet where a Lindstrom/ROAM -like regular triangulation is mapped to the earth, described in (Aasgaard R, Sevaldrud T (2001). The methods are demonstrated on http://globe.sintef.no.

As an example two regular partition triangulations are shown in Fig. 3.

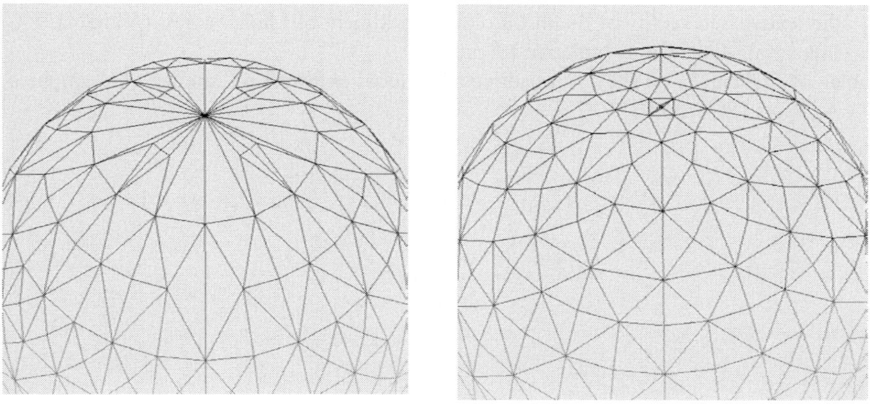

Fig. 3. Uniform scaling (left) vs. m=n+1 projection (right).

Acknowledgements

The Research Council of Norway through the DynaMap project and the European Union through the TellMaris project have supported this work.

References

Aasgaard R, Sevaldrud T (2001) Distributed handling of level of detail surfaces with binary triangle trees. Proceedings of ScanGIS'2001

Bartholdi JJ, Goldsman P (2001) Continuous indexing of hierarchical subdivisions of the globe. International Journal of Geographical Information Science 15(6): 489-522

Bugajevsky LM, Snyder JP (1995) Map Projections, A Reference Manual. Taylor & Francis

Dutton G (1990) Locational properties of quaternary triangular meshes. Proc. Spatial Data Handling Symp. 4. July, 1990, Dept. of Geography, U. of Zurich, pp 901-910

Freeden W, Windheuser U (1994) Spherical Wavelet Transform and its Discretization. Tech. Rep. 125, Universität Kaiserslautern, Fachbereich Mathematik

Goodchild MF, Shiren Y (1992) A hierarchical spatial data structure for global geographic information systems. Computer Vision, Graphics and Image Processing 54(1): 31-44

Lindstrom P, Koller D, Ribarsky W, Hodges LF, Op den Bosch A, Faust N (1997) An Integrated Global GIS and Visual Simulation System. Technical report GIT-GVU-97-07, March 1997

Ottoson P (2001) Retrieval of geographic data using elliptical quadtrees. Proceedings of ScanGIS'200

Reddy M, Leclerc YG, Iverson L, Bletter N (1999) TerraVision II: Visualizing Massive Terrain Databases in VRML. IEEE Computer Graphics and Applications (Special Issue on VRML) 19(2): 30-38

Schröder P, Sweldens W (1995) Spherical wavelets: Efficiently representing functions on the sphere. University of South Carolina Department of Mathematics: Columbia, S.C. Industrial Mathematics Initiative 1995:01

Tobler W, Chen ZT (1986) A quadtree for global information storage. Geographical Analysis 18(4): 360-371

Creating Spatial Information: Commissioning the UK Land Cover Map 2000

Alexis Comber, Peter Fisher and Richard Wadsworth

REVIGIS Project, Department of Geography, University of Leicester, Leicester, LE1 7RH, UK, Tel: +44 (0) 116 252 3859, ajc36@le.ac.uk

Abstract

Environmental datasets are subject to ontological differences in the way that information is recorded and reported. One of the implications is that environmental monitoring and the identification of loci of change is difficult. This generic problem is explored with reference to land cover mapping in the UK. The decennial land cover projects (LCMGB and LCM2000) are introduced and their differences described. Their different scientific, policy and commission contexts contribute to the overall change in methodology. Increased user participation in the specification of the 2000 dataset is the result of structural changes in the way policy is conducted. The Broad Habitat classification system was determined by many factors and not solely the responsibility of the scientists producing the data. The drive to integrate data that have fundamentally different characteristics and properties is examined and the botanical background to one dataset being used to provide quality measures for another is described. The value of combining two fundamentally different approaches and presenting them as an integrated product may be misleading. Each survey offers a different approach in terms of the spatial detail, extent and data primitives that are recorded.
Keywords: ontology, integration, Land Cover Map 2000

1 Introduction

There is a generic problem of discordance faced by workers seeking to integrate different spatial data which at least seem to report the same or similar information (Fisher, 1999). This can be exemplified through land cover. Many different datasets provide information about land, land cover and vegetation. It is often difficult to translate or relate the information contained in one dataset to that of another, despite the fact that they record similar thematic attributes and are spatially or temporally coincident. This is because surveys are subject to different methodologies, even when they are repeated or updated. Whilst more accurate recording of environmental features during an inventory is important, often we would like to be able to address specific aspects such as monitoring change or to make more informed decisions based on sound environmental information. These objectives are hindered by the difficulty in translating and relating the information captured in *Survey1* to that captured in *Survey2*. This problem is typical of very many spatial databases. Two major issues are entwined in this problem, the first is the technical methods by which the information may be translated, and the second is the institutional-social conditions that cause it to happen.

We explore this generic problem with reference to land cover mapping in the UK and in particular to the Land Cover Map of Great Britain (LCMGB) produced in 1990 and the UK Land Cover Map 2000 (LCM2000) produced a decade later[1], both derived from satellite remote sensing. These two surveys provide country-wide inventories of land cover for a range of activities and a large number of end users. Whilst there is much interest in the change detection and environmental monitoring, the stated objectives of LCM2000 are to "provide an enhanced product, rather than pursuing the measurement of change through a direct repeat of the LCMGB" (Fuller *et al.*, forthcoming). However, the fact is that both products are called the "Land Cover Map" (LCM), this suggests to users that LCM2000 is the second in the series, in spite of methodological differences in how the information was derived and presented.

In this paper, we document the process of commissioning the datasets, and the resulting uncertainties produced by a shift in ontologies consequent to many developments especially a change in policy context. We demonstrate that one contribution to this uncertainty in spatial information is embedded in the socio-political process of creating the data.

2 LCMGB and LCM2000

The 1990 LCMGB recorded the stock of land cover across the country and details can be found in Fuller *et al.* (1994). The LCMGB raster dataset was the first complete high-resolution survey of the land cover of Britain to be mapped from satellite imagery. It was produced using a per pixel supervised maximum likelihood classification (MLC) of Landsat Thematic Mapper data with a derived pixel size of 25 x 25 m, recording 25 "Target" land cover types. LCMGB was financed by the British National Space Centre and NERC with funds from the Department of the Environment (DoE, later DETR). It provided a demonstration of the viability of using satellite imagery, and was incorporated into a wider umbrella project, the Countryside Survey 1990 (CS90; Barr *et al.*, 1993). CS90 included a sample-based field survey of detailed vegetation composition and land cover among other data from 508 $1km^2$ squares. The field survey and the LCMGB

[1] Note GB is Great Britain, and includes England, Scotland and Wales, while the UK (United Kingdom) includes these countries as well as Northern Ireland.

were both produced by CEH (the Centre for Ecology and Hydrology, then known as the Institute for Terrestrial Ecology, ITE).

LCMGB was the first inventory of England and Wales since Coleman (1968), and of GB since Stamp (1939). The value of the LCMGB survey is illustrated by the extent to which it has been used in many and various applications. LCMGB data have been supplied under license to ~500 academic, policy and commercial users, with some 300 further users of more generalised data through the Countryside Information System (Haines-Young *et al.*, 1994). Its use and application has been reported in a variety of academic papers.

LCM2000 represents a significant methodological departure from LCMGB. It combines a parcel-based supervised MLC with ancillary geographical data and records land cover information in vector format rather than raster (pixel). Vector format was to increase structural integrity by applying methodological developments made through CLEVER-Mapping (Smith and Fuller, 2001). In addition, it allows attributes to be attached to the parcel (Smith and Fuller, in press) LCM2000 provides an enhanced product, i.e. with explanatory meta information, and increased thematic detail. A knowledge-base of about 600 rules (Knowledge Based Corrections; KBCs) were developed under the method proposed by Groom and Fuller (1996), and applied to parcels that were classified with low confidence and / or with classes out of their natural context (Fuller *et al.*, forthcoming). Extensive parcel attributes describe class probabilities, within parcel heterogeneity, and each parcel's processing history, as well as recording the parcel's land cover class. LCM2000 is distributed as part of the Countryside Survey 2000, (CS2000) which includes a field survey of 569 1km^2 sample locations. Further details about LCM2000 can be found at http://www.ceh.ac.uk/data/lcm/LCM2000.shtm

3 LCMGB Issues

The aim of this paper is to illustrate the generic problem of ontological changes that occur as the understanding advances in a science that supports a process of national inventory, and to describe the origins of such change. The value of LCMGB, as the first census survey of the land cover of Britain, cannot be over-stated and is reflected by the number of end users, applications and reported academic investigations. A number of themes emerged post LCMGB and in the cycle of reports and papers, that inevitably accompany such an important undertaking:

- Integration LCMGB information with the field survey component of CS90;
- Lack of understanding in *some* less technically informed users about the limitations of maps in general, and by-pixel classification, in particular;
- Reviews of CS90 and recommendations for a future CS2000.

3.1 Integration of Field Survey and Satellite Map

In 1993, the field survey data and the LCMGB were presented as part of an integrated project CS90 (Barr *et al.*, 1993). Comparisons of the land cover information showed them to be different (Wyatt, *et al.*, 1993; Fuller *et al.*, 1998). For instance, the satellite derived LCMGB recorded the area of the key cover type "Open shrub heath / moor" as 8% of the country while the field survey identified 5% (Fuller *et al.*, 1998: Table 7, p.113). This is an inevitable consequence of the different methodologies. Studies that have compared the two approaches such as Cherrill *et al.* (1995), Fuller *et al.* (1998) and Wyatt *et al.* (1993), have consistently described differences due to:

- *meaning* (class labels and their nomenclature, in terms of class recognition on the ground, and spectral separability in the imagery);
- *timing of data capture* (effects of: different years, growth cycles, seasonal differences, and diurnal effects);
- *spatial factors* (scale, geo-registration, heterogeneity, boundary pixels);
- *statistical issues* (training data collection).

Wyatt *et al.* (1993) and Fuller *et al.* (1998) go to considerable effort to describe the origins of the different "answers" provided by the two approaches. They acknowledge the difficulty in defining any kind of "truth" in mapping and Fuller *et al.* (1998) comment on users failing to accommodate differences between maps.

3.2 Perceptions of LCMGB

The LCMGB does not capture the botanical detail of the field survey, but does provide a comprehensive survey covering all of Britain. The first thing many users of LCMGB did was to compare the data with their local knowledge (Andrew Stott, pers. comm.). The trade-off between an extensive and comprehensive survey is increased spatial and thematic generalisation, and in spite of local errors the database is largely correct, Fuller *et al.* (1998) comment that "spectral misclassification caused most of the errors in the Land Cover Map" and "might have been removed with more sophisticated knowledge-based procedures" (p123).

As part of CS90 objectives, information from the field survey and satellite land cover maps were compared. This comparison was used to provide a measure of thematic quality for LCMGB whilst stressing the lack of "truth" in mapping and the nature of different classifications. The user community, however, found the descriptions of LCMGB quality based on a comparison with the field survey confusing (Haines-Young and Swanwick, 1996; Wyatt *et al.*, 1993).

Better-informed users, who could view the dataset through "Thematic Mapper Eyes", were able to understand the data and accommodate these errors much more easily than others (Neil Veitch, pers. comm.). They were able to appreciate the value of LCMGB due to its spatial extent and used it to provide a first cut for directing more detailed user activity resulting in targeted efforts and further analyses.

3.3 Reviews of CS90 and Recommendations for CS2000

Reviews of CS90 have noted the issues raised by integrating information from the field survey with that from the satellite map (Fuller *et al.*, 1998; Haines-Young *et al.*, 2000). Most of the differences between the two approaches reflect the different paradigms or views of the world embodied in them. Others are more subtle, such as the confusion between land cover and land use. The field survey often recorded the land *use*, whilst the LCMGB necessarily recorded the dominant *cover*. For instance, juvenile trees were recorded as "woodland" in the field survey and as the understory vegetation in the LCMGB (Fuller *et al.*, 1998).

Some of the perceived discrepancies reflect the fact that the two surveys were rolled together after their inception, rather than being planned and designed together. LCMGB was a demonstrator project. The independence of the surveys is illustrated by the names and meaning of the land cover classes they use. For instance, the reporting of the land cover stocks was for 59 *Reporting Classes* in the field survey and 25 *Target Classes* in LCMGB, and they employed different definitions (Wyatt, *et al.*, 1993). The classes did aggregate to a common classification of 17 classes, however.

Recommendations for the proposed repeat land cover map LCM2000 concentrated on the discrepancies between the field survey and the satellite data (e.g. Fuller *et al.,* 1998: p123, state that "there is likely to be inconsistency if one is used to validate the other") and the issue of better integration. For instance, amongst many points Fuller *et al.* (1998) conclude their paper by saying: "A planned update in 1998-2000 will address many of the problems of the CS90, will upgrade the detail and accuracy and will ensure even closer integration between the field and satellite surveys." (p125).

4 Policy Background for LCM2000: A Changed World

The policy climate within which national environmental monitoring efforts were taking place had changed considerably by the mid 1990s. This was due to: the Earth Summit in Rio in 1992 (Secretariat on the Convention on Biological Diversity, 2000); European Union *Habitats Directive* 92/43 (CEC, 1992); "*Biodiversity: the UK Action Plan*" (BAP) (UK Biodiversity Steering Group, 1994).

The Convention on Biological Diversity (Article 7) requires that each country identifies and monitors national components of biodiversity. The EU Habitats Directive promotes the maintenance of biodiversity and demands that surveillance, identification and monitoring of the natural habitat and species be undertaken (Article 11). The UK BAP was implemented as a result of these commitments. UK legislation delegates the duty of implementing the Directive to all organisations and agencies that have or could have an impact on the areas designated under the Directive. This was a significant increase in the role of environmental agencies and the legislation emphasised the role of locally prepared management plans. In this context there were specific references to the potential role of and modifications to initiatives such as the Countryside Survey series whose "repeat surveys provide a mechanism for monitoring the biodiversity of the wider countryside" (UK Biodiversity Steering Group, 1994: 9.19).

5 LCM2000

Land Cover Map 2000 was undertaken within the umbrella of the CS2000 (Haines-Young *et al.*, 2000) to update and upgrade LCMGB. It sought to provide an enhanced survey and aimed from the outset to be a complete revision rather than a repeat using the earlier methods (Robin Fuller, pers. comm.). To this end, an improved vector based methodology was adopted. It is possible to delineate two distinct spheres of influence, firstly a drive for "better" science and secondly a demand for policy relevant map products.

5.1 Scientific Influence

The objective concern of most scientific research in the Descartian tradition is greater certainty and understanding through the application of scientific reason in an objective manner in order to better explain and predict. In the LCM2000 project this activity can be seen in:

- The development and adoption of improved techniques in LCM2000 using technological developments made in CLEVER-Mapping (Smith and Fuller, 2001). Using a parcel representation allows processing history and heterogeneity attributes to be attached to the parcel;
- Application of the lessons learnt from the previous LCMGB. One of the key issues to emerge from the LCMGB project was the problem created by the fundamental differences between the field survey and satellite map. To this end explicit statements about improved methodological integration had been made as the LCM2000 project was beginning (Fuller *et al.*, 1998).

The nature of these influences are implicit in most academic scientific activity, where improved answers are sought by applying the lessons learnt from previous exercises and by developing new techniques.

5.2 Policy Influence

The second sphere of influence on the LCM2000 project was the increased role of policy as exercised by external agencies. Here a number of inter-linking factors combined:

- The devolution of direct responsibility for local environmental monitoring away from government to environmental agency accountability;

- The links to policy were explicit in that the stated objectives of CS2000 were to enable the UK to fulfil its international obligations to monitor biodiversity;
- The concept of Broad Habitats within which biodiversity monitoring could take place. These provide the framework within which the UK seek to meet their *Priority Habitat* obligations and were specified by governmental departments and agencies funding the LCM2000 (Fuller *et al.*, forthcoming);
- The increased role of environmental agencies such as English Nature and the Environment Agency. They were legally required to fulfil national policy objectives locally.

5.3 Broad Habitats

The two spheres of influence came together in the LCM2000 steering group. The steering group played a much more active role than its predecessor in 1990.

The LCM2000 steering group was constituted from potential users who had increased policy obligations and data producers with improved scientific knowledge. The agencies wanted information about Priority Habitats, but the scientists did not believe that these could be mapped directly from remotely sensed imagery. In the end, a compromise was made to try to map Broad Habitats, even though the scientific view was that some of these could not be mapped directly.

The steering group had the unenviable task of balancing the policy desires of the agencies with the expert scientific knowledge of what features could be reliably determined by their reflectance characteristics. The remote sensing paradigm requires that identifiable features have distinct spectral characteristics, but many of the Broad Habitat classes do not have distinct spectral signatures. This problem was compounded by the late delivery of formal Broad Habitats definitions in the overall CS2000 project (Jackson, published in July 2000). Both the field survey and land cover map projects had to use their own *working definitions*. Nonetheless, the customer's wishes were accommodated, which was a time-consuming exercise as it meant identifying thousands of spectral classes, which were aggregated into class Variants, Subclasses and ultimately Target classes. Broad Habitats were constructed from these. In order to meet requirements of the steering group external data were obtained and used particularly for grassland and bog covers.

The result was that the specification of the LCM2000 could be seen to be driven by two general influences: science and policy. The satisfaction of both of these influences resulted in a compromise between what was scientifically possible and what was politically desirable.

5.4 Integration in the Production of LCM2000

One of the themes arising from CS90 that was the focus of much work was the need for the field survey and land cover map components to be better integrated in

CS2000. Integration can be addressed from many positions, for instance spatial factors, meaning, timing and statistical issues. Some of these are easier to address than others. However, other differences are fundamental to each technique and their integration may be irreconcilable. This section explores integration from the thematic perspective of land cover reporting.

At one level, greater integration is illustrated by the reporting of land cover stocks in terms of Broad Habitats, using a set of similarly defined and labelled classes. Examination of the definitions of the Broad Habitat classes used by each approach shows that the classes are described similarly, principally on their botanical and bio-geographic properties. Descriptions can be found at:

http://www.cs2000.org.uk/Report_pdf/appen.pdf
http://www.ceh.ac.uk/data/lcm/lcmleaflet2000/leaflet3.pdf.

However the nature of data capture in each approach is fundamentally different in terms of the data primitives recorded by the field survey as compared to the land cover map. Necessarily the Broad Habitats identified by LCM2000, were defined by their spectral and reflectance properties, rather than their botanical ones. The field survey by contrast records features based on their botany and ecology such as species composition and number. These differences are important when trying to present information of thematic equivalence from the two components.

To address the differences in perception, staff preparing the LCM2000 attended a training exercise in the field with the field surveyors of the CS2000 to develop a shared understanding of land cover concepts.

The steering group for LCM2000 was much more active as a forum for decision making than was its predecessor. Indeed the advisory group for LCMGB was perceived by some not to have existed because: the data and approach had been determined before the group had been formed. For LCM2000 there was increased explicit dialogue between users and scientists before the survey was started and the nature of involvement of these stakeholders was different (come to that the word stakeholder had been invented).

The LCM2000 steering group, typical for a large publicly funded exercise, was set up by the government (DETR). The initial objectives were to gather financial support for production and to drive the LCM2000 project towards policy. The LCM2000 steering group had other roles: it provided a forum by which user specifications could be articulated and solutions to problems could be discussed, and most importantly the existence of a steering group demonstrated that the project had wide support. This in turn allowed the development of a collection of possible funding organisations under the aegis of the steering group. The need for answers to different questions to those in 1990 resulted in different specifications. There is some difference in opinion about the precise role of the LCM2000 steering group in resolving issues of specification and approach. Steering group members experienced different degrees of involvement in the decision making process. Some thought that they exercised influence over decisions; they were able to modify the specification to enable them to make some use of the outputs in support of the advice that government seeks from them. Others believed that by the time the steering group was formed, scientific decisions about specification and data had already been made, and that therefore their role was to rubber stamp these decisions.

6 Discussion

The preceding sections have shown the origins of difference between surveys: scientific development, applying lessons learnt from experience and the impact of national and international policy. The resultant methodological change common to many repeated national inventories: the previous project is implicitly supported by the desire for a repeat, and scientists will always try to improve their methodology.

The use of the Broad Habitats, defined primarily by their botany, illustrates the strong influence of policy on the LCM2000 project. Broad Habitats were agreed as the reporting classes for both components of CS2000. However, there were greater difficulties in defining these classes using the relatively coarse satellite image data compared to the finer granularity of information captured by the field survey. For the field survey there was considerable post survey work, such as the ECOFACT series (Bunce *et al.*, 2000), which established that it could deliver information in terms of "Broad Habitats" as demanded by policy. Roy Haines-Young (pers. comm.) stated that the field survey had "learnt to speak the language of policy and [so] got funded". In contrast, despite having many hundreds of applications and end uses, the land cover map did not have the equivalent to the ECOFACT work.

Umbrella projects, such as the Countryside Surveys of 1990 and 2000, seek to present a unified message from rolling together the two different world views embedded in the Land Cover Map and the Field Survey. This can be confusing and result in tension if one approach is presented as implicitly "better" than the other. If policy demands are being satisfied by the field survey *and* satellite land cover map then the assumption is that both approaches are providing a meaningful resource inventory. The description of LCM2000 shows an attempt to accommodate the "ecological view" as embodied by the field survey. There is no equivalent to assess the field survey in terms of the remote sensing view of the world. Thus, there is an implicit assumption that field survey paradigm is superior. The official integrated product will probably be released in the next couple of years, until then users will have to accommodate different approaches in terms of the spatial detail, extent and data primitives that are recorded.

The origins of dataset differences between LCMGB and LCM2000 can be attributed to one of three general types: data capture, information reporting and representation, and in labelling semantics. The reasons for the methodological changes within time series land survey projects may be scientific (technological and scientific developments) or political. That is to say, that changes in the political and social context within which the survey is being commissioned means that the questions to which answers are being sought are different. One of the net results of changing the question is to reduce the capacity to track changes over time.

Standardisation, is frequently cited as a panacea to inter-operability, but misses the point that different users have different objectives at different times, in different institutions. Presenting users with a single land cover database assumes an homogeneity of interest that does not exist. Alternatively, presenting the user with the raw data in a standardised way is also problematic: it assumes users have the technical skills to construct their own models. This is not the case, different users will create similarly named land cover features that are in turn assumed to mean the same thing, and the problem expounded in this paper persists.

The UK Land Cover 2000 dataset seeks to address this paradox by including extensive attributes at the polygon level. These allow the user to apply specific local analyses without having to assume that the data quality is evenly distributed. The philosophy of the production team has been to explicitly retain and report as much information as is feasible, by publishing meta-information about the polygon processing history and the top 5 spectral sub classes within the parcel. However, due to changes in approach the LCM2000 comes with an explicit warning against the direct measurement of change. This is essential, as users will compare their local knowledge with the LCMGB and LCM2000 datasets.

7 Conclusion

We have documented the changing pressures on the development of a land cover map of the UK, in its decennial reporting in 1990 and 2000. The changes in policy between those dates have made one contribution to the dissimilarity in the products and the changing methodology has made another, providing tensions between what is technically possible, organisationally expedient and scientifically objective. The two land cover activities of the CS2000 have so far had limited integration, despite persistent identification of the problems of integration post CS90. The majority of the integration has yet to take place and is about to start as we write in March 2002, some four years after data capture began.

Although the specific case we document here is unique to the Countryside Surveys, we believe that the general situation is typical of major inventory initiatives. Such initiatives are funded from the policy realm and have to respond to changing policy priorities. They are costly and therefore have to bow to the dictates of policy. From a technical point of view, the net result is that although the names of the land cover maps announce their similarity (Land Cover Map of Great Britain and Land Cover Map 2000), even if both were perfectly correct for the whole country, any comparison between them to establish landscape change is fraught with uncertainty because of the changed class definitions (ontologies) and methodology.

The endemic problem indicated by the case of LCM2000 is not overwhelming: we believe that it can be overcome by applying methods developed around the notion of establishing formal relations between datasets (either semantic or statistical) and then using formalisms (probabilistic and qualitative) to model the uncertainty at the level of the polygon. These are being tackled within the REVIGIS project. We would urge land cover data commissioners and producers to follow the example of the CEH and LCM2000 to include some level of process history and indications of thematic heterogeneity as part of their standard products.

Acknowledgements

This paper would not have been possible without the co-operation of a number of those involved in commissioning and executing CS2000, including Robin Fuller, Geoff Smith, Andrew Stott, Neil Veitch and Roy Haines-Young. Others have helped too, but to preserve anonymity we are not listing them by name. This paper describes work done within the REVIGIS project funded by the European Commission, Project Number IST-1999-14189. We thank our partners in the project, especially Andrew Frank, Robert Jeansoulin, Arko Lucieer, Alfred Stein, Nic Wilson, Mike Worboys and Barry Wyatt.

References

Barr CJ, Bunce RGH, Clarke RT, Fuller RM, Furze MT, Gillespie MK, Groom GB, Hallam CJ, Hornung M, Howard DC, Ness MJ (1993) Countryside Survey 1990: main report. Countryside 1990 Series: Volume 2. Department of the Environment, London

Bunce RGH, Barr CJ, Gillespie MK, Howard DC, Scott WA, Smart SM, van de Poll HM, Watkins JW (1999) ECOFACT 1 - Vegetation of the British Countryside. London, HMSO

CEC (1992) Council Directive 92/43/EEC of 21 May 1992 on the conservation of natural habitats and of wild fauna and flora. Official Journal of the European Communities. L206

Cherrill AJ, McClean C, Fuller RM (1995) A comparison of land cover types recognised in an ecological field survey of Northern England and in the first remotely sensed Land Cover Map of Great Britain. Biological Conservation 71: 313-323

Coleman A, Maggs KRA (1968) Land Use Survey handbook: an explanation of the Second Land Use Survey of Britain on the scale of 1: 25,000. Second Land Use Survey. London

Fisher PF (1999) Models of Uncertainty in Spatial Data. In P Longley, M Goodchild, D Maguire, and D Rhind (eds) Geographical Information Systems: Principles, Techniques, Management and Applications, Vol 1. Wiley and Sons, New York, pp 191-205

Fuller RM, Groom GB, Jones AR (1994) The Land Cover Map of Great Britain: an automated classification of Landsat Thematic Mapper data. Photogrammetric Engineering and Remote Sensing 60: 553-562

Fuller RM, Smith GM, Sanderson JM, Hill RA, Thomson, AG ('No Date') Land Cover Map 2000: construction of a parcel-based vector map from satellite images. (forthcoming)

Fuller RM, Wyatt BK, Barr CJ (1998) Countryside Survey from ground and space: different perspectives, complementary results. Journal of Environmental Management 54: 101-126

Groom GB, Fuller RM, (1996) Contextual correction: techniques for improving land cover mapping from remotely sensed images. International Journal of Remote Sensing 17: 69-89

Haines-Young RH, Barr CJ, Black HIJ, Briggs DJ, Bunce RGH, Clarke RT, Cooper A, Dawson FH, Firbank LG, Fuller RM, Furse MT, Gillespie MK, Hill R, Hornung M, Howard DC, McCann T, Morecroft MD, Petit S, Sier ARJ, Smart SM, Smith GM, Stott AP, Stuart RC, Watkins JW (2000) Accounting for nature: assessing habitats in the UK countryside. DETR, London

Haines-Young RH, Swanwick C (1996) Scoping study for Countryside Survey 2000 (Part A) First Interim Report to Department of the Environment. University of Nottingham, Nottingham

Haines-Young RH, Swanwick C (2000) Countryside Survey 2000: Scoping Study. University of Nottingham, Nottingham

Haines-Young, R.H., Bunce, R.G.H. & Parr, T.W (1994) Countryside Information System: An Information System for Environmental Policy Development and Appraisal. Geographical Systems 1: 329-345

Jackson DL (2000) Guidance on the interpretation of the biodiversity Broad Habitat Classification (terrestrial and freshwater types): definitions and the relationship with other habitats. JNCC Report, no. 307

Secretariat on the Convention on Biological Diversity (2000) Sustaining Life on Earth: How the Convention on Biological Diversity promotes nature and human well-being [online]. Available from: http://www.biodiv.org/doc/publications/cbd-guide-en.pdf [Accessed 13/10/01]

Smith GM, Fuller RM (2001) An integrated approach to land cover mapping: the land cover of Jersey. International Journal of Remote Sensing 22: 3123-3142

Smith GM, Fuller RM ('No date') Land Cover Map 2000 and meta data at the land parcel level. Proceedings of the RSPS meeting Uncertainty in Remote sensing and GIS. 3-4 July, 2001, University of Southampton (in press)

Stamp DL (1939) The Land Utilisation Survey of Britain. London

UK Biodiversity Steering Group (1994) Biodiversity: the UK action plan. London, HMSO

Wyatt BK, Greatorex-Davies NG, Bunce RGH Fuller RM, Hill MO (1993) The comparison of land cover definitions. Countryside 1990 Series: Volume 3. Department of the Environment, London

A Methodology for Updating Geographic Databases using Map Versions

Ally Peerbocus[1], Geneviève Jomier[1], Thierry Badard[2]

LAMSADE, Université Paris Dauphine, Pl. Mal de Lattre, 75775 Paris Cedex 16, France. Email: (jomier,peerbocus)@lamsade.dauphine.fr[1]
Institut Géographique National, COGIT, 2 à 4 Avenue Pasteur, 94165 Saint Mandé Cedex, France. Email: Thierry.Badard@ign.fr[2]

Abstract

This paper addresses issues concerning the exchange and integration of geographic data between producers and users. Once a producer has delivered a geographic database to a user, who then uses it as a reference for specific applications, the database may be updated on both sides. Consequently, the integration of future updates - delivered by the producer -in the user's geographic database is a complex operation due to possible conflicts between updates performed by both parties. The resulting database may become inconsistent and the user's added information may be lost. Users therefore need mechanisms to help them in the process of update integration. This paper provides a methodological framework for the updating of geographic databases. It relies on a multi-version GIS, allowing an automatic detection of conflicting updates between two map versions.
Keywords: geographic database, multi-version GIS, updates, map version

1 Introduction

Geographic Information Systems (GIS) are increasingly used in a large spectrum of applications. Since, implementing such systems is complex, users, generally, purchase reference geographic data from producers in order to set up their GIS. For instance, a transportation company purchases from a producer a geographic database representing the road network of a given region for its transport planning application. For the user, the database delivered by a producer serves as a *reference map,* on which to develop the application. Geographic data producers are responsible for producing and maintaining up-to-date databases, delivered to users on a regular basis. Meanwhile, users may need to add information on the reference map or update the map to take into account real world changes, or information they are interested in; for instance bus lines and bus stops (see Fig. 1).

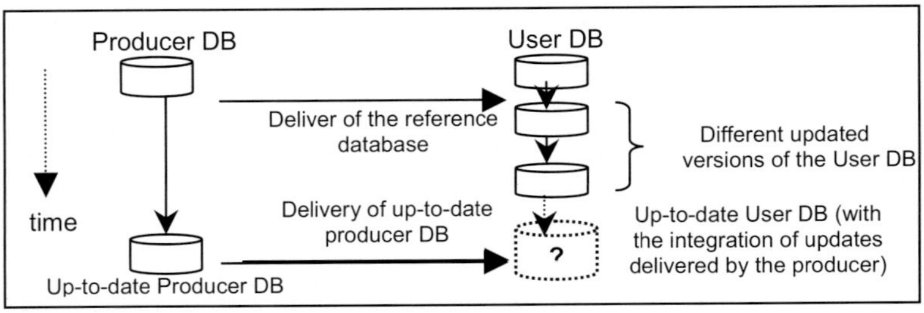

Fig. 1. Integration of updates

Consequently, the integration of the producer's updates in the user's database may result in conflicts with those already performed by the user as described in (Badard 1998). For instance, if the producer changes the location of a road, to increase information accuracy, the bus line and bus stops along the road must be changed too, otherwise the user's database is in an inconsistent state.

The first step for a proper integration of these updates requires the identification of the updated objects in the user's database as well as those in the producer's database in order to detect possible conflicts. At present, producers generally deliver a whole up-to-date database to the user. Even if the percentage of change is small between two updates, current GIS do not provide any mechanism for the extraction of updates between two versions of a database representing the same area at two different times (Raynal 1996). Several techniques have been proposed to achieve this purpose. They are based on the exhaustive comparison of all the objects in the two versions of the database (Badard 1998), relying on geographic data matching algorithms (Lemarié *et al* 1996). Such an approach, well adapted in a general context where no hypothesis on the data model is assumed, is based on complex algorithms and needs considerable effort to be implemented.

This paper proposes a mechanism for an automatic detection of conflicting updates performed in two different versions of a database. It is based on the version approach proposed in (Gançarski *et al* 1994). This paper is organised as follows: section 2 describes the context through an example and presents an overview of related work; section 3 details our approach and the way it is implemented; section 4 concludes the paper.

2 Context and Related Work

Before presenting an overview of related work, this section describes an example of exchange of geographic data between a producer and a user, illustrating conflicts between producer's and user's updates.

2.1 Context

The context used to illustrate the producer/user scenario is based on a road network application relying on Georoute®, a database produced by the IGN (the French National Geographic Institute) and dedicated to car navigation services. Fig. 2 depicts part of the producer's map, identified as $prod_0$, which is delivered in a geographic database to the user. The map shows the state of the modelled road network, identified as R_1 to R_5, and land parcels, identified as P_1 to P_6. Fig. 2 also represents the new up-to-date producer's map, identified as $prod_1$, reflecting the new state of the road network after:

1. the construction of a new road R_6, splitting parcel P_2 into P_{2a} and P_{2b};
2. the deletion of R_2;
3. the construction of a new roundabout, identified by P_7, at the junction of roads R_1, R_3, R_4, R_5 and R_6 implying the update of all these road sections.

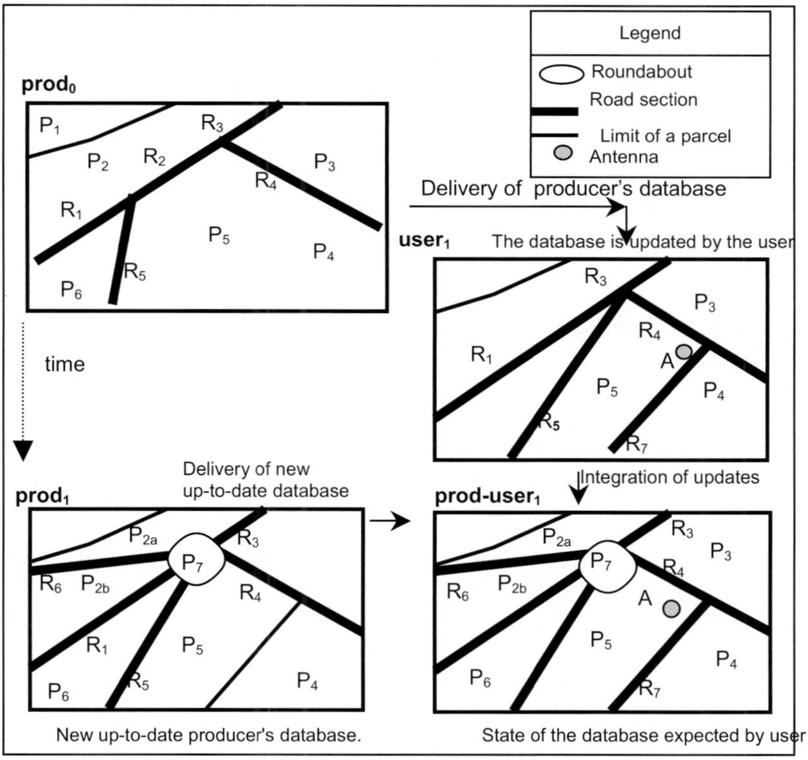

Fig. 2. The different states of the producer and user maps

On the user side, the initial geographic database has been updated to obtain a new version of the map, identified as *user*$_1$, different from *prod*$_1$. The updates performed in *user*$_1$ are illustrated in Fig. 2:

1. the update of roads R_1 and R_5;
2. the deletion of R_2;
3. the creation of a road section R_7 at the limit between P_4 and P_5;
4. the creation of an antenna, *A,* representing the user mobile phone company.

Finally, the map identified as *prod-user*$_1$ corresponds to what the user would like to obtain after the integration of updates from *prod*$_1$ and *user*$_1$.

In general, updates in a geographic database can be performed on both the schema and the objects. Updates on an object represent its evolution - i.e., creation, deletion, and thematic and/or geometric changes. Schema updates are required when new kinds of objects appear as bus-stops, mobile phone antennae which are specific objects for the user and whose representation is not provided in the producer's schema of the database. When the producer delivers the new up-to-date database to the user, (*prod*$_1$ in Fig. 2), the user must then consider this new information for an update to his current database (*user*$_1$ in Fig. 2). He cannot just replace the old reference database with the new one since the resulting database may be in an inconsistent state; for instance, if a road section is enlarged in *prod*$_1$, he should displace any antenna on this road. Besides, specific information added by the user may be lost and some of the updates may be in conflict with the producer's updates, like road R_7, which exists only for the user.

2.2 Sources of Conflicting Updates

Several situations may result in conflicts between updates separately performed by the producer and the user. First, conflicts may be due to the update of the same object by both actors, defined as a 1-to-1 update, such as parcels P_5, P_6, and roads R_1, R_5, updated by the producer and the user. Secondly, conflicts are very likely to occur in case of group updates like:

- 1-to-N update: one object is deleted and is replaced by several objects; e.g., the splitting of parcel P_2 into P_{2a} and P_{2b} in map *prod*$_1$,
- N-to-1 update: several objects are deleted and replaced by one object; e.g., the merge of R_1 and R_2 in *prod*$_0$ resulting in R_1 in *prod*$_1$,
- M-to-N update: several objects are deleted and in their place several other objects are created; e.g., the creation of the new roundabout P_7 by the producer in *prod*$_1$.

A complete taxonomy of conflicts hindering the updating of geographic database is described in (Badard 1998).

2.3 Related Work

Several techniques concerning the detection of update differences between two geographic databases, modelling the same region, have been proposed. They are generally based on the comparison of the geometry (Devogèle 1998. (Badard *et al* 1999) proposes to isolate these differences by using the "geographic data-matching method", which goes through every database object in a region and computes the correspondence relationships between objects, from their geometry stored in the two versions. The resulting relationships can be classified, considering their cardinality: a) 1-to-0 or 0-to-1: an object of one database does not match with any object of the other one; b) 1-to-n or n-to-1, with n>0: an object of a database matches with one or several objects of the other one; c) n-to-m, with n>1 and m>1: several objects of a database match with several objects of the other one.

The correspondence relationships are then analysed and updated objects are classified according to the evolution they have undergone - the typology is defined in (Badard 1998b) -, and which can either concern the object level only, or both schema and object levels. Furthermore, new delivery modes dedicated to the exchange of updating information have been proposed (IHO96, Poupart-Lavoie 1997, Badard 1998b, Badard *et al* 2001) to help the integration process in databases.

Together with these methods for the detection of updates in geographic databases, propagation mechanisms of these effects have been proposed in a multi-scale database context (Badard 2000, Kilpelaïnen 1997, Uitermark *et al* 1998). In all these papers, no hypothesis is made about the data model used and only a general solution is provided. It's clear that detecting changes in the whole database requires tremendous efforts and sophisticated algorithms.

A proper updating of a user's database implies the preservation of the integrity of the map delivered by the producer. This means that users must perform their updates on versions of the reference map, and the comparison of the different map versions should be possible in order to detect changes between two map versions. However, as far as we know, in the geographic context only one technical paper of *SmallWorld* GIS (Easterfield *et al*) has focused on the management of version in GIS. Few implementation details, however, have been provided.

We propose a methodology for the updating of geographic databases called *Updating by Map Versions* (*UMV*). It is based on the use of a multi-version geographic database as described in (Bauzer *et al* 1993), which supports the management of map versions. The detection of conflicting updates is based on the automatic identification of all the database objects, and not on comparisons performed on the geometry of geographic objects as proposed in (Badard 2000). The next section deals with the main features of the UMV methodology and describes how it is implemented.

3 Updating by Map Version Methodology

From here on, the producer database will be addressed followed by the user database as *producer-DB* and finally, the *user-DB*. Initially the *producer-DB* contains one version of the map representing the modelled geographic area. This version is identified as $prod_0$. The UMV methodology, comprised of four steps, is illustrated in Fig. 3 and detailed in the next sub-sections.

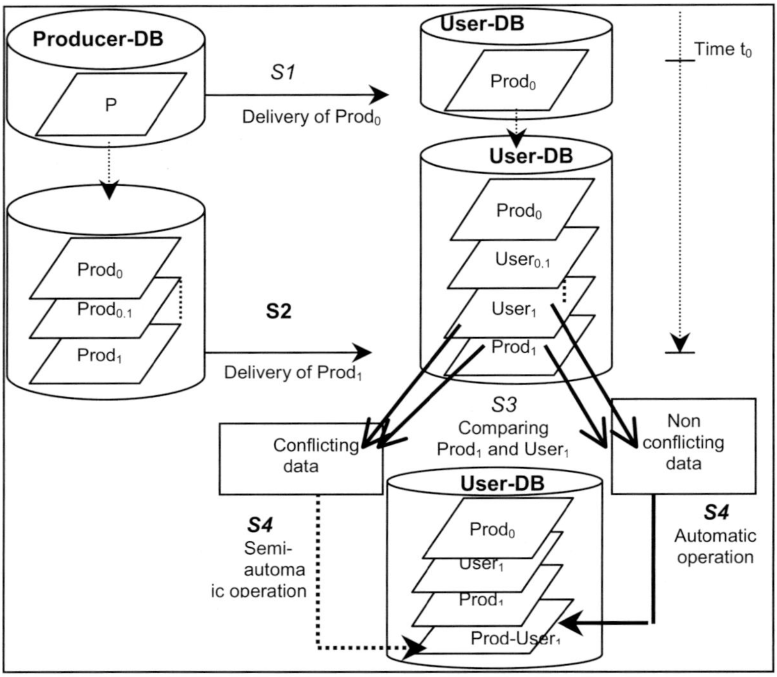

Fig. 3. The steps comprising the UMV methodology

These steps are:

S1. At time t_0, the producer delivers to the user the initial reference map version, $prod_0$. This map version is inserted in *user-DB*, and identified as $prod_0$. It serves as the reference map for the user. $Prod_0$ is preserved in the delivered state, frozen in both *user-DB* and *producer-DB*. Updates are performed on successive map versions generated from $prod_0$ on both sides. The newly generated map versions are identified as $prod_{0,i}$ in *producer-DB* and $user_{0,j}$ in *user-DB*.

S2. Then, at time t_1, an up-to-date database version is delivered by the producer to the user, identified as $prod_1$. The user's map version at this moment is identified as $user_1$. The delivered map version, $prod_1$, is inserted in *user-DB*.

S3. The map versions $user_1$ and $prod_1$ in *user-DB* are compared to detect conflicting (from an updating point of view) and non-conflicting data.
S4. Finally, using the strategy, discussed in section 3.3, a new map version is created, *user-DB*, to include part or all of the user and producer's updates. An automatic integration of data can be used for non-conflicting updates. A semi-automatic operation is needed to integrate the conflicting data and the consequences of the conflicts in the final user map version, identified as $prod\text{-}user_1$. For instance, the bus line of the user present in $user_1$ is moved to follow the new location of a road recorded in $prod_1$.

The UMV methodology is supported by a multi-version database (Gançarski *et al* 1994), which is introduced in the next sub-section. We assume that both the producer and the user have a multi-version geographic database. The underlying version mechanism is now detailed.

3.1 The Multi-version Geographic Database

Two levels are distinguished: the user level and the database level. At the user level, for external users, the multi-version database appears as a set of independent map versions, representing the same area, which coexist within the same storage space. This means that each map version can be managed (read and updated) separately and independently from the other map versions of the same region. A new map version is always generated or *derived* as a copy of an existing map version.

At the database level, however, one important feature of the multi-version database mechanism is that it automatically allows keeping track of all database objects that compose a consistent map version. Thus, several versions modelling the same real world object may coexist in the database. This gives origin to the concept of *multi-version object*, which is a "repository" of all versions of a given object – i.e., multi-version object O encapsulates the mapping between all different states of an object and the corresponding map versions. One of the main advantages of the mechanism is that it minimises storage occupancy and avoids redundancy while storing multiple map versions. The derivation relationships among the distinct map versions are recorded in a structure called *map version tree*. Updates to objects in one map version are handled without side effects on other map versions, due to an appropriate management of internal version identifiers. To obtain the value of an object in a map version, the system applies a rule stating that it has the same value as that in the map version from which it is derived except if another value is explicitly specified. This rule is recursive and called *implicit sharing rule*.

When an object is deleted in a map version, its value in the database is set to $\perp$, meaning it does not exist. When an object O in a map version v is involved in a group operation - splitting (e.g. the splitting of parcel P_2 into P_{2a} and P_{2b} in map version $prod_1$) or merging operation -, the link between O and the resulting object(s) is stored in a *genealogy graph* (Sperry *et al* 1999). A special value "#"

for *O* in map version *v* is used to denote a group operation. When geographic objects are created from one or several other geographic objects - i.e. the object has one or several ancestors, the resulting geographic objects are initialised with a special value "*" in the map version parent of the map version in which the operation has been performed. Thus, the genealogy graph represents 1-to-N, N-to-1 and N-to-M evolution of geographic objects.

The system uses the internal identifier of objects to follow the evolution of objects through time. Internal identifiers are managed only by the system, conversely to external identifiers, which are managed by users. For further details on this approach, the reader is referred to (Cellary *et al* 1990, Bauzer-Medeiros *et al* 1993, Gançarski *et al* 1994, Cellary *et al* 2000).

3.2 Illustration of the Multi-version Geographic Database

The top of Fig. 4 shows a part of the producer multi-version database corresponding to map versions $prod_0$ and $prod_1$ described in Fig. 2. For the sake of clarity, we ignore the intermediate map versions between $user_0$ and $user_1$ in *user-DB*, considering that the database is composed only of $user_0$ and $user_1$. Each multi-version object in the figure is represented by a table with two columns: *MV-id* (for multi-version identifier) and *Value*. Parcel P_6 has a different value for each of the two map versions: *valp6* in $prod_0$ and *valp6a* in $prod_1$. The Parcel P_2 in $prod_1$ has been split and replaced by parcels P_{2a} and P_{2b} as illustrated by the genealogy tree. Parcel P_1 has only one value, represented by *valp1*, corresponding to map version $prod_0$. According to the implicit sharing rule and the producer's map version tree, *valp1* is also the value of P_1 in map version $prod_1$.

The bottom part of Fig. 4 shows a part of the user's multi-version database. Road section R_7 and the antenna *A* have only one value, *valr7* and *valA* respectively, corresponding to map version $user_1$, meaning that they have been created in $user_1$. Parcel P_1 has only one value, $valp_1$, for map version $prod_0$, thus its value in map version $user_1$ is implicitly shared with that in $prod_0$. Parcel P_6 has two values, *valp6* in map version $prod_0$ and *valp6x* in map version $user_1$, because it has been updated in map version $user_1$.

Notice that the identifier of new objects created in $prod_1$ or $user_1$ must not be in conflict. This can be the case if the same identifier is used in $prod_1$ and $user_1$ to represent two different real world entities. To prevent this, the identifier of new objects is prefixed with the name of the database in which it is created. For example, the identifiers of P_{2a} and P_{2b} in Fig. 4 are in fact *prod-DB*.P_{2a} and *prod-DB*.P_{2b}. For simplicity sake, this does not appear on the figure.

This sub-section has described the main principles of the multi-version approach. In reality, the geometry of some geographic objects may be represented by complex objects. As such, updates are carried out on the elementary objects that compose the geometry (Peerbocus *et al* 2001).

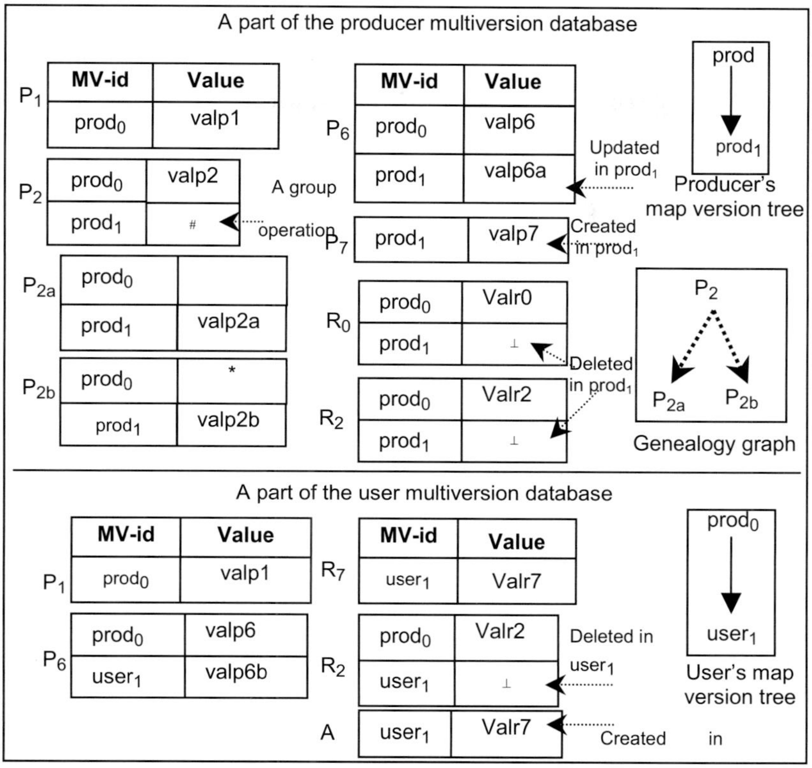

Fig. 4. Part of the multi-version databases

3.3 Delivery of Producer's Updates

On delivery, the producer's updated map version $prod_1$ is inserted in *user-DB* to enable the detection of conflicts between the producer's and the user's updates. This insertion operation is performed automatically as follows:

1. The system first modifies the map version tree to include map version $prod_1$ as derived from $prod_0$; $prod_1$ and $user_1$ become alternative map versions, both derived from $prod_0$, as illustrated in Fig. 5.
2. The system then verifies for each object in the delivered map version $prod_1$ whether the object has been updated or created by the producer - i.e., it has a value explicitly associated with $prod_1$. If so, the system inserts the value corresponding to $prod_1$ in the corresponding multi-version object in *user-DB*.

Finally, the multi-version database is composed of multi-version objects having values corresponding only to $prod_0$ and/or $prod_1$ and/or $user_1$ (see Fig. 5). For instance, parcel P_6 has three distinct values corresponding respectively to map versions $prod_0$, $prod_1$ and $user_1$, parcel P_1 has only one value for $prod_0$, shared implicitly by $user_1$ and $prod_1$, and so on.

The next step consists in comparing, in *user-DB*, the user's and the producer's map versions, $user_1$ and $prod_1$, for the detection of possible conflicting and non-conflicting updates.

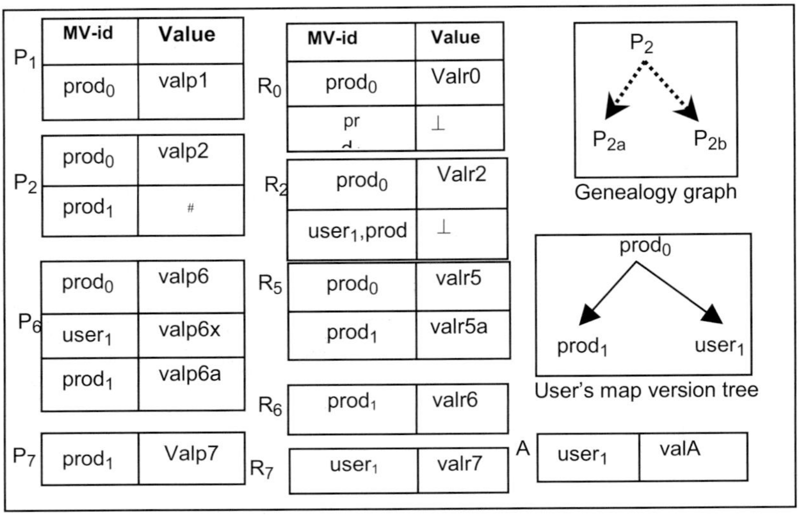

Fig. 5. Part of the user's database after the insertion of $prod_1$

3.4 Comparison of Updates and Detection of Conflicts

When the system compares the values associated with $prod_1$ and $user_1$ for the different multi-version objects in the database, the two following situations are possible:

Case 1. An object has the same value in both map versions. This can occur because the object has not been updated - i.e., the multi-version object contains only one value for $prod_0$, shared implicitly by $prod_1$ and $user_1$. Alternatively this may happen when the object has been updated or created in both the *prod-DB* and *user-DB*, and the values are equal - e.g., road section R_2 which has been deleted in map versions $user_1$ and $prod_1$. In these cases, there is no conflict.

Case 2. The object has different values in the two map versions, $user_1$ and $prod_1$. This situation is possible in the following cases:

a) the object has been updated or created in both $user_1$ and $prod_1$ and the two values are different; e.g., parcel P_6 has value *valp6* in $prod_0$, *valp6x* in $user_1$ and *valp6a* in $prod_1$, and *valp6x* and *val6a* are different. Here, the two updates or creations are in conflict.

b) the object has been updated either in *prod*$_1$ only or in *user*$_1$ only. The update corresponds to one of these operations:

- a creation: roundabout P_7 and road R_6 have been created in map version *prod*$_1$, and antenna *A* in *user*$_1$.
- a deletion: road R_0 is deleted in *prod*$_1$ and still exists in *user*$_1$.
- an update of its value: road R_5 has value *valr5a* in *prod*$_1$ and value *valr5* in *user*$_1$ (implicit sharing with prod$_0$).
- a group operation: the value of parcel P_2 in *user*$_1$ is *valp2*, implicitly shared with *prod*$_0$. Its value in *prod*$_1$ is denoted by #, meaning a group operation which is explained by the genealogy graph of P_2: it has been split into two new parcels P_{2a} and P_{2b}. These two parcels have only one value in *prod*$_1$ (corresponding to their creation). Conflicts exist in these cases and, for each object, its value in the new map version to be created in the *user-DB* depends on the user's decision.

These different situations can be visualised on the map by using special colouring of the object, revealing non-conflicting and conflicting updates and the types of conflicts.

This section has focused on updates relating to objects only. For schema updates a similar procedure is adopted [BCJ98]; e.g., antenna if it exists only in *user-DB*.

3.5 Proposed Strategy for Updates Propagation

The previous steps of the UMV methodology help the user in the visual detection of conflicts both at schema and object levels. Moreover, it supplies information concerning the types of evolution underlying the different updates. Now, the remaining step concerns the propagation of the detected updates in the user database.

This step needs an appropriate strategy, which may depend on many factors of the application concerned - e.g., knowledge about the underlying topology of the spatial objects (Egenhofer *et al* 1994, Badard *et al* 1999). For instance, the user may decide or not to favour his update in place of the producer's one in case of conflict. This choice may affect the user's added information, which may need to be readjusted. It is, therefore, wiser to use already proposed strategies such as (Badard *et al* 1999, Badard 2000, Kilpelaïnen 1997, Uitermark *et al* 1998), where the propagation problem has been thoroughly studied. The final map version of the user after the propagation of updates may contain the updates of the producer as well as those of the user. Suppose that the final user's map version is *prod-user*$_1$ as illustrated in Fig. 2.

To obtain the map version *prod-user*$_1$, the user first derives a map version identified as *prod-user*$_1$ from *prod*$_1$ (see Fig. 6), since *prod*$_1$ contains a large part of updates that the user needs to represent in his map version. Thus, all objects in the new map version are shared implicitly with *prod*$_1$; e.g., parcels P_1, P_6 and road sections R_5, R_6. Next, he includes in *prod-user*$_1$ his specific updates: the road section R_7 and the antenna *A*. Fig. 6 shows part of the state of *user-DB* after the creation of the final map version *prod-user*$_1$, resulting from the merging of *user*$_1$ and *prod*$_1$. In the user's database, the value, *valr7*, of the road section R_7 in *prod-user*$_1$ is the value coming from *user*$_1$ and it is shared explicitly with *user*$_1$ as illustrated in Fig. 6. The situation is the same for antenna A that the user preserves in *prod-user*$_1$. In case, an object in *prod-user*$_1$ has a value different from that in *prod*$_1$ and *user*$_1$, for instance road section R_4 in Fig. 6, the system creates a new entry for this value, *valr4b*, which is associated with *prod-user*$_1$.

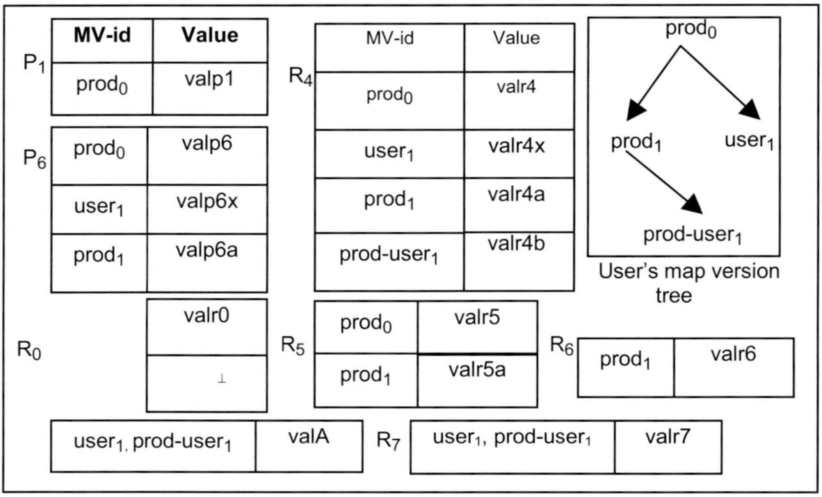

Fig. 6. Part of *user-DB* after the integration of data from *prod*$_1$

Finally, from *user-DB* of comprising of four map versions, the system reads the value of an object *O* in *prod-user*$_1$, as follows: if *prod-user*$_1$ appears in the multi-version object, then the value of *O* is the one corresponding to *prod-user*$_1$ (see Fig. 6). Otherwise, if *prod-user*$_1$ does not appear in the multi-version object, then the value of *O* is shared implicitly with the value of the nearest ancestor, obtained from the user's map version tree, which is either *prod*$_1$ (e.g., P_6, R_6), or *prod*$_0$ (e.g., P_1); finally, if only *user*$_1$ appears in the multi-version object, then *O* does not exist in *prod-user*$_1$.

After this propagation step, the user will work on new versions of *prod-user*$_1$ for further updates, whereas the producer uses new versions of *prod*$_1$. Thus at time t_2, $t_2 > t_1$, a new operation of integration of updates may take place: the new up-to-date map version, *prod*$_2$, is inserted in the user's database and compared to the current user's map version, *user*$_2$, which has been created by derivation and updates from *prod-user*$_1$. A new map version *prod-user*$_2$ is created, integrating updates coming from *prod*$_2$ and *user*$_2$. To help users in understanding updates performed in the different map versions for integration purposes, the updates should be documented as depicted in (Peerbocus *et al* 2001).

4 Conclusions

The focus of this paper was on how to help the exchange of updating information between a geographic data producer and a user. The main advantage of the UMV methodology is that it allows the automatic detection of updates whereas existing techniques require an exhaustive retrieval within the different versions of the database. The UMV methodology responds as well when updates are delivered on a given frequency as for real time updates. The UMV methodology can also be applied in a general context where there is a need of exchanging geographic data between any two users or between a user and a producer.

A prototype of a multi-version geographic database has been developed using MapInfo® in the LAMSADE Laboratory, University of Paris Dauphine. It requires the implementation of the version mechanism in the geographic database, which must be managed by a version manager. It allows the representation of the different states of geographic objects. All changes are documented. The prototype allows the retrieval of updated geographic objects between any two map versions of the multi-version geographic database and provides the user with the associated change documentation (Hedjar 2001).

The integration of the updates and propagation of their effects in geographic databases requires handling all the spatial relationships between entities in an effort to preserve consistency or added information. Several research works have set up tools for the retrieval of these relationships necessary to the updating of geographic databases. Ongoing researchers (IGN) investigating the development of a formalism and a model for the design of geographic databases, which are easier to maintain. The UMV methodology thus appears as a key element of a global methodology for the design of easy-to-update GIS.

References

Badard T (1998a) Towards a generic updating tool for geographic databases. In: GIS/LIS'98, Fort Worth, Texas, pp 352-363

Badard T (1998b) Extraction des mises à jour dans les Base de Données Géographiques. Revue Int. de Géomatique 8(1-2):121-147

Badard T, Lemarié C (1999) Propagating updates between geographic databases with different scales. In: Innov. in GIS VII:GeoComputation, London

Badard T (2000) Propagation des mises à jour dans les bases de données géographiques multi-représentation. Ph.D. thesis, Univ. Marne-la-Vallée, France

Badard T, Richard D (2001) Using XML for the exchange of updating information between GIS. In: CEUS 25. Elsevier, Oxford, pp 17-31

Bellosta MJ, Cellary W, Jomier G (1998) Consistent Versioning of OODB Schema and its Extension. 14-èmes Journées BDA, Hammamet, Tunisia

Bauzer-Medeiros C, Jomier G (1993) Managing Alternatives and Data Evolution in GIS. In: ACM Workshop on Advances in GIS. Arlington,Virginia

Cellary W, Jomier G (1990) Consistency of versions in object-oriented databases. In: VLDB. Brisbane, pp 432-441

Cellary W, Jomier G. (2000) The Database Version Approach. Networking and Information Systems Journal 3(1): 177-214

Devogèle T (1998) Le processus d'intégration et d'appariement des BD géographiques. Ph.D. thesis, Univ. Versailles-Saint Quentin, France

Egenhofer MJ, Clementini E, Di Felice P (1994) Evaluating inconsistencies among multiple representations. In: 6th SDH. Edinburgh, UK, pp 901-920

Easterfield ME, Newell RG, Theriault DG ('No Date') Version Management in GIS-Applications and Techniques. Smallworld technical paper no. 4

Gançarski S, Jomier G (1994) Managing Entity Versions within their Context: a Formal Approach. In: DEXA'94. Athens, LCNS no. 856, pp 400-409

Hedjar M (2001) A prototype for documenting spatiotemporal evolution. Research report, DEA 127, LAMSADE, Univ. Paris-Dauphine, France

International Hydrographic Organisation (1996) IHO transfer standard for digital hydrographic data. Publication S-57, Edition 3.0

Kilpelaïnen T (1997) Multiple representation and generalisation of geo-databases for topographic maps. Finnish Geodetic Inst., 124, 51-711-212-4

Lemarié C, Raynal L (1996) Geographic data matching: First investigations for a generic tool. In: GIS/LIS'96. Denver, Colorado, pp 405-420

Peerbocus MA, Bauzer Medeiros C, Jomier G, Voisard A (2001) Documenting Changes in a Spatiotemporal DB. In: XVI BSDB. Rio

Poupart-Lavoie G (1997) Développement d'une méthode de transfert des mises à jour de données à réf. spatiale. M.Sc. , Univ. Laval, Québec

Raynal L (1996) Some elements for modelling updates in topographic database. In: GIS/LIS'96. Denver, Colorado, pp 405-420

Sperry L, Claramunt C, Libourel T (1999) A Lineage Metadata Model for the Temporal Management of a Cadastre Application. In: Int. Workshop on Spatio-Temporal Models and Languages. Firenze, Italy

Uitermark H *et al* (1998) Propagating updates: Finding Corresponding objects in a multi-source environment. In: 8th SDH. Vancouver, pp 580-591

Revealing Uncertainty in Maps of Glacial Lake Algonquin

Scott Drzyzga, Ashton Shortridge and Randall Schaetzl

Michigan State University, Department of Geography, 314 Natural Science Building, East Lansing, Michigan 48824 USA

Abstract

This paper examines the effects that uncertainty in digital elevation data has on results obtained from a spatial model. Ongoing analysis of glacial Lake Algonquin in northern Michigan has used a spatial model to interpolate unknown shoreline locations between known shoreline positions. Extant shoreline features were surveyed with global positioning technology and ancient shorelines were reconstructed using a DEM and a statistical model of isostatic rebound. However, shoreline data contain relatively small measurement errors and relatively large errors are associated with the DEM. These errors, when propagated through a series of GIS operations, may render uncertain results. This research recognises and attempts to assess these errors in order to produce a new map of Lake Algonquin shorelines - one that illustrates shorelines and areas of positional uncertainty. Results indicate that even small errors in input data can contribute uncertainty to model output. Understanding these uncertainties can be valuable to further research concerning these and other ancient shorelines.
Keywords: error propagation, spatial modelling, simulation, Lake Algonquin

1 Introduction

According to Heuvelink (1998, p.3), most work on spatial modelling "has been concentrated on the business of deriving computational models that operate on spatial data, on the building of large spatial databases, and on linking computational models with GIS". However, some serious attention has also been paid to the issue of spatial data quality. Goodchild and Gopal (1989) concluded that virtually all spatial data stored in a GIS are, to some extent, contaminated by error. Subsequently, Heuvelink (1998) has argued that a potential danger exists in GIS because such errors can be propagated through GIS operations, corrupt results and render interpretations meaningless.

In work about mapping glacial shorelines, Schaetzl *et al.*, (in press), reported results obtained from a spatial model. These results contain an unknown quantity of uncertainty resulting from errors propagated from input data and through the GIS operations used. Therefore, the purpose of this research is to re-evaluate the spatial model employed by Schaetzl *et al.*, and account for the effects that input data errors had on results.

1.1 Mapping Lake Algonquin

Lake Algonquin was a prominent proglacial lake in the Great Lakes basin (Fig. 1). It maintained a relative high stage from approximately 11,200 to 10,400 BP and developed conspicuous shoreline features (Larson and Schaetzl, 2001). As glaciers receded from the area lower outlets were uncovered, the land surface rebounded as the weight of ice was lifted, and lake levels fell in stages. Locations in northern Michigan have since rebounded at a faster rate than locations in southern Michigan (Futyma, 1981). This crustal uplift process is called isostatic rebound; Fig. 2 illustrates the differential effects this process has had on the highest known Lake Algonquin shoreline.

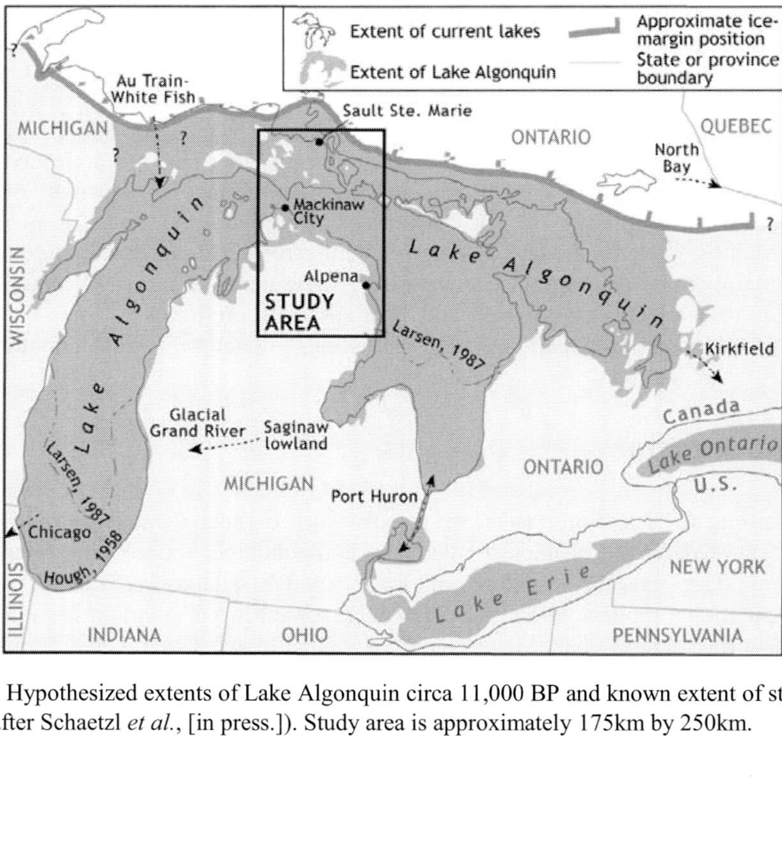

Fig. 1. Hypothesized extents of Lake Algonquin circa 11,000 BP and known extent of study area (after Schaetzl *et al.*, [in press.]). Study area is approximately 175km by 250km.

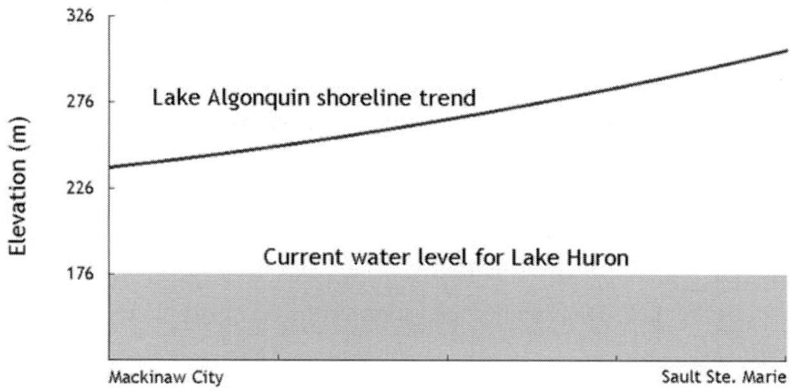

Fig. 2. Differential effects of isostatic rebound on the Algonquin shoreline in Michigan. Trend elevations estimated along an 85km transect from Mackinaw City to Sault Ste. Marie. Elevation values are in meters above sea level.

Much has been written about how Lake Algonquin was formed (Spencer, 1891; Gilbert, 1898; Goldthwait, 1908; Leverett and Taylor, 1915; and Stanley, 1936), but little agreement exists about the exact spatial extent of the water surface (Hough, 1958; Futyma, 1981; Larsen, 1987, see Fig. 1). Little agreement exists because the terrain and some shoreline features have been altered over the past 10,000 years by natural or human processes, and other evidence has been buried or reworked by later lake transgressions. Uncertainty about the extent of this lake prompted an extensive survey by Schaetzl *et al.*, (in press). The field element of the survey employed global positioning technology to measure positions {x,y,z} along the highest known Lake Algonquin shoreline. In the laboratory, a GIS was used to conflate and analyse the shoreline data and a digital elevation model (DEM) of the region in order to interpolate unknown shoreline locations and generate a map of results. In sum, a GIS was used to evaluate a computational model - a simple representation of a complex, crustal deformation process.

1.2 Statement of Problem

The certainty with which one can assume a DEM represents true surface form is a function of the quality of the elevation data provided. Kyriakidis *et al.*, (1999, p. 678 after Bolstad and Stowe, 1994) maintain, "that DEM source data and production methods are not perfect, and consequently mismatch errors occur between actual and DEM-reported elevation values." GIS operations can propagate error in the DEM through processing and analysis, thereby producing uncertain results. Therefore, a need exists to quantify and represent uncertainty in results obtained from the spatial model used to interpolate Lake Algonquin shorelines.

The objectives of this research are to quantify and model errors in Schaetzl's shoreline data and DEM (a mosaic of USGS 3 arc second DEMs resampled to a

90 meter ground resolution), re-evaluate the shoreline interpolation model, and represent uncertainties in a map of results. Six related questions were posed in an effort to achieve the research objectives:

1. How much elevation error is contained within the shoreline data set?
2. How much elevation error is contained within the DEM?
3. What influence do elevation errors in shoreline data have on interpolated shoreline locations?
4. What influence do elevation errors in DEM data have on interpolated shoreline locations?
5. What influence do combined elevation errors have on interpolated shoreline locations?
6. Where, if anywhere, do uncertainties exist in a map of results?

2 Methods used for Summarising Error

The area selected for study is located in Michigan's northern Lower and Upper Peninsulas (Fig. 1). This region contains many conspicuous Lake Algonquin bluffs. Local relief varies within the region, which includes moraines, steep bluffs and relatively flat outwash plains.

2.1 Measurement Errors in the Shoreline Data

While Schaetzl *et al.*, (in press) collected position data, efforts were also made to assess the accuracy of GPS-measured elevation data. Positions were measured for eight different benchmarks of known elevation (one was measured twice, but during different trips to the field). Measured benchmark elevations were compared to actual elevation values published by the National Geodetic Survey for the same monuments (NGS Information Services Branch, 1999). A root mean squared error (RMSE) of 1.20 meters represents the average magnitude of difference between measured and published values. In addition, RMSE values for eastings and northings were not greater than 0.8 meters (not shown). Because the methods used to gather and process shoreline data were the same as those used to gather and process benchmark data, the average vertical error in the shoreline data was inferred to be similarly small. Summary statistics were calculated to describe the distribution of error within the benchmark data (Table 1). Although the sample size is small (n=9), the statistical evidence in Table 1 does not indicate that the error distribution is not normal. Hence, we assumed the distribution of error to be normal. Further assessment indicated that error did not appear to be a function of space, position, or some form of landscape covariation process. Therefore, we believe these errors are measurement errors and error magnitudes are associated with the limits of equipment precision. We assume this error process is random and spatially independent.

Table 1. Summary statistics for elevation error distributions

Statistic	Shoreline data	DEM
RMSE	1.20	5.09
Min	-1.63	-21.20
Max	1.75	22.10
Mean	0.00	0.01
Variance	1.61	25.94
Skewness	0.25	0.24
Kurtosis	-1.64	2.23

2.2 Errors in the DEM

During their survey, Schaetzl *et al.* collected additional position data for shorelines that existed later and lower than the Main Algonquin shoreline. These ancillary position data and National Geodetic Survey benchmark data (NGS, 2001) were used to assess the accuracy of DEM elevations at 1353 locations. All data sets were referenced to the same coordinate system, and horizontal and vertical control networks. A root mean squared difference of 5.09 meters (see Table 1) indicates a larger average magnitude of error exists in the DEM than in the shoreline data set. Therefore, it is reasonable to suspect that larger errors in the DEM contribute more uncertainty to results than smaller errors in the shoreline data.

Summary statistics were calculated to describe the distribution of error within the DEM (Table 1). The Kolmogorov-Smirnov test was used to test the cumulative frequency distribution for non-normality. Significance test results did not provide evidence to reject the standard null hypothesis; *the sample distribution fits an expected normal distribution*, using a 90 percent confidence interval. Hence, we assumed the distribution of error in the DEM to be normal.

Although we assumed errors to be normally distributed, we did not assume errors to be spatially independent. Bolstad and Stowe (1994), Brown and Bara (1994) and Kyriakidis *et al.* (1999) have demonstrated reasons for why errors in USGS DEM products exhibit spatial autocorrelation and not spatial independence structures. Isotropic and anisotropic semivariograms were calculated and analysed to characterise the distribution of errors; each indicated an isotropic and positively autocorrelated covariance structure exists. However, unlike Bolstad and Stowe (1994), further analysis did not reveal any linear relationship between error values (signed or unsigned) and position values (x or y or z). Hence, we treated the covariance structure associated with DEM errors as a spatially dependent random process of unknown origin.

2.3 The Spatial Model

The spatial model used by Schaetzl *et al.* (Fig. 3a) is relatively simple, but it simulates a complex, crustal deformation process. In essence, the model reconstructs the ancient shoreline by adjusting a DEM of modern terrain to account for isostatic rebound. First, former water surface elevations are modelled as a function of location using the shoreline data and a second-order polynomial trend surface with parameters estimated via ordinary least squares. This trend surface represents the upwarped nature of the ancient landscape sans local relief (e.g., Fig. 2).

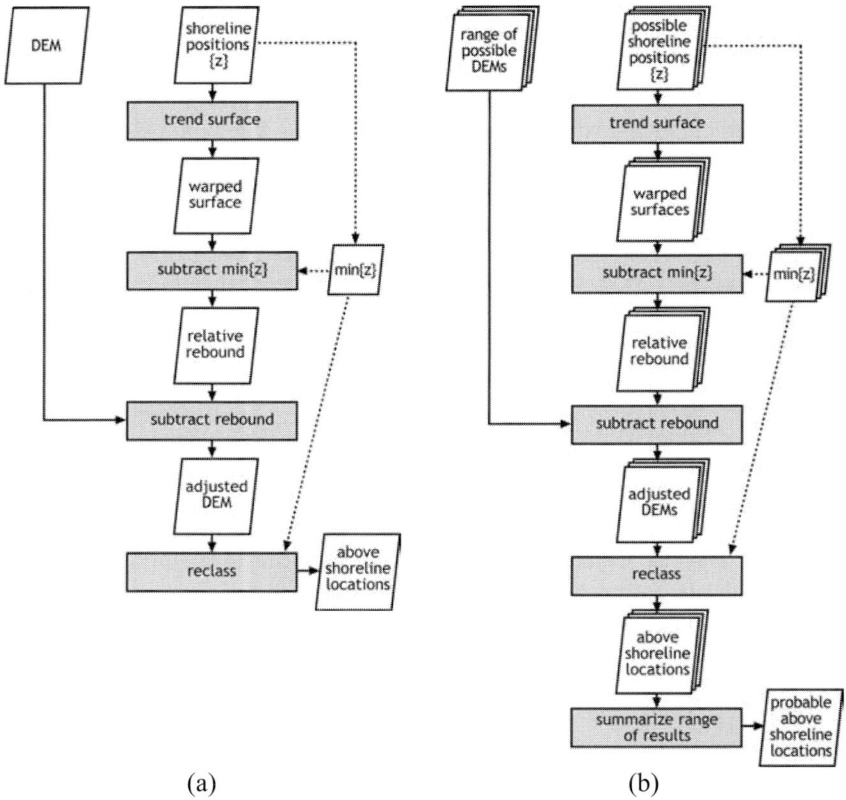

Fig. 3. The spatial models used by **(a)** Schaetzl *et al.*, and **(b)** this research

Second, the minimum elevation value in the shoreline data set is identified then subtracted from all elevation values in the warped surface. The resulting surface represents isostatic rebound relative to the lowest known shoreline elevation. This step is necessary because much disagreement exists in the literature regarding the actual lake level and controlling outlet elevations (e.g., Hough, 1958 and Larsen,

1987). Therefore, the isostatic rebound surface is calculated relative to the lowest known elevation and not a controlling outlet.

Third, the isostatic rebound surface is subtracted from a surface model of modern elevations, a.k.a. the DEM, thereby producing an isostatically depressed elevation model with local relief. Finally, classifying all elevation values with respect to the minimum shoreline elevation value cartographically floods the adjusted elevation model. Nodes with elevation values greater than the minimum value are classified as *above shoreline*, while the balance are classified as *not above shoreline.*

3 Methods Used for Modelling Uncertainty

Monte Carlo simulation is a computer intensive technique used to reveal the contributions of error to results from input data, as well as differences across a geographic area (Burrough and McDonnell, 1998). In essence, a typical simulation consists of evaluating a spatial model many times, each time using inputs that have been modified. Individual input values are modified randomly via some model that specifies the error distribution of the input and makes assumptions about the behaviour of error. Parameters for the model are derived with respect to known or assumed information about errors in the input data. After a large number of realisations are generated, the set of spatial model results is summarised to determine the range of possible outcomes.

The goal of a Monte Carlo simulation is not to determine a single correct answer, but rather to define probable limits associated with a large number of possible outcomes given random errors in data. The Monte Carlo simulation technique is appropriate for this analysis because it can reveal contributions of error to results from input data, which is the purpose of this research.

The error model employed during this work adopts four assumptions. First, the distribution of error associated with each input value in each data set is normal. Second, because each error may be treated as a random variable, every measured value comprises one out of an infinite set of possible yet, equally valid representations. The third assumption is employed with respect to the shoreline data set only; the error associated with any one input value is independent, random, and cannot be systematically corrected. Last, the fourth assumption is employed with respect to the DEM only; the errors associated with input values are spatially structured, random, and can be simulated via the sequential Gaussian simulation technique described below (see Section 3.2 and Goovaerts, 1998).

3.1 Simulation One: Uncertainty from Shoreline Data

During the first simulation, 100 realisations of shoreline elevations were generated. Each realisation consisted of original shoreline elevation values that were perturbed independently by error values drawn randomly from the normal probability distribution function with mean = 0 and variance = 1.61 (see Table 1). The adjusted spatial model (Fig. 3b) was run 100 times, each time using one of the perturbed data sets, rather than the original, as input. The DEM values were held constant. This simulation produced a set of 100 binary grids, each containing cells that represented *above shoreline* (1) and *not above shoreline* (0) locations. Using grid algebra, the set of 100 binary grids were summed and divided by the scalar 100 to produce a single and final grid. Final grid cells with a zero value indicate locations that were classified *above shoreline* zero percent of the time. Final grid cells with a unity value indicate locations that were classified *above shoreline* 100 percent of the time. Values ranging between zero and one indicate the proportion of results that were above the minimum shoreline elevation value.

3.2 Simulation Two: Uncertainty from the DEM

During the second simulation, 100 realisations of the DEM were generated. Each realisation consisted of DEM values that were perturbed by random and spatially structured errors, which were made conditional to ancillary shoreline and NGS benchmark data.

Realisations were generated using a two-step procedure. First, the spatial structure of DEM error was estimated by employing the geostatistical software program Gstat (Pebesma and Wesseling,1998) to fit a compound exponential variogram model [Eq. 1] to an empirical variogram of measured DEM errors. And second, we employed sequential Gaussian simulation (sGs and succinctly described by Goovaerts, 1998:4-5) to account for the spatially dependent error process, and condition the DEM to higher accuracy spot data. According to Goovaerts (1998:4),

> "sequential simulation amounts to modelling the conditional cumulative distribution function (ccdf)... ...then sampling it at each of the grid nodes visited along a random sequence. To ensure reproduction of the *z*-semivariogram model [Eq. 1 - authors inclusion], each ccdf is made conditional not only to the original *n* data but also to all values simulated at previously visited locations."

In sum, we used sGs to develop 100 models of spatially structured DEM errors that were used to perturb the original DEM.

$$\gamma(h) = 1 + 8(1 - e^{-3h/270}) + 13(1 - e^{-3h/3800}) \tag{1}$$

where: nugget = 1,
$sill_1 = 9$,
$range_1 = 270$ m,
$sill_2 = 22$, and
$range_2 = 3800$ m.

The adjusted spatial model (Fig. 3b) was run 100 times, each time using one of the perturbed data sets, rather than the original, as input. Original shoreline data values were held constant. This simulation produced a second set of 100 binary grids, which were summarised in the same manner described in Section 3.1. It was expected that this simulation would produce a wider range of results, and therefore more uncertainty, than did the first simulation because the average magnitude of error associated with the DEM is greater than that associated with the shoreline data set.

3.3 Simulation Three: Uncertainty from Error Interaction

The interaction between input errors can take many forms. For example, errors in one data set may cancel errors in the other - a best-case scenario. A worst-case scenario would reveal a compounding effect. Moreover, the degree of interaction may vary over space, thereby perturbing results in some locations differently than others. The third simulation, which incorporated 100 realisations each of shoreline data and the DEM, was used to reveal influences that the interaction of errors in both data sets had on results. Accordingly, the adjusted spatial model (Fig. 3b) was run 100 times to produce a third set of 100 binary grids. Again, binary grids were summarised in the same manner as describe in Section 3.1.

4 Mapping the Results

Maps were prepared to illustrate the spatial distributions of shoreline locations associated with the three final grids. For display purposes, grid cells with values less than 0.05 were reclassified as *below shoreline* and illustrated with white. Grid cells with values greater than 0.95 were reclassified as *above shoreline* and illustrated with a gray tone. Values including and between 0.05 and 0.95 were reclassified as *uncertain* and highlighted with black.

4.1 Influence of Elevation Errors in the Shoreline Data Set

Relatively small measurement errors inherent to shoreline data contributed some uncertainties to results. Fig. 4a shows the probable distribution of shoreline locations and areas over which the position of the Lake Algonquin shorelines remains in doubt. Much of the land and water interface is distinct but some sections are peppered with relatively small areas of uncertainty. Several larger areas of uncertainty exist at locations apparently reworked by fluvial activity (e.g., the Thunder Bay River system west of Alpena) and spaces possibly associated with emergent wetlands (e.g., islands in the Upper Peninsula). However, some of these uncertainties may also be products of data distribution. Much of the shoreline survey data was collected in the Lower Peninsula. Many but fewer positions were measured in the Upper Peninsula and none were gathered in Canada. Given what is commonly known about the behaviour of trend surfaces calculated beyond the extent of input data (i.e., for locations in Canada), we hypothesise that some of the uncertainties illustrated in Fig. 4a were contributed by trend surface flutter in the adjusted spatial model. However, that is a hypothesis to be entertained during future research.

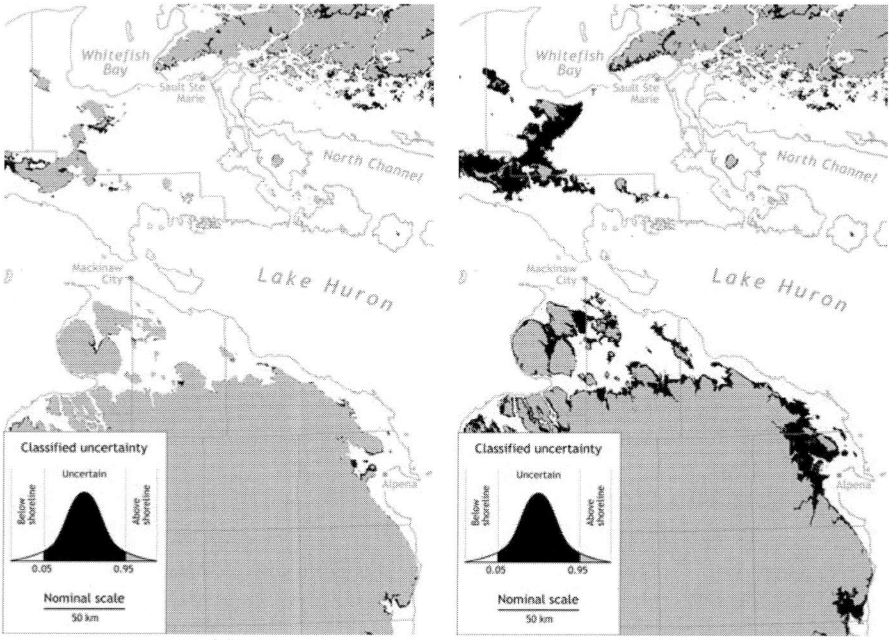

(a) (b)

Fig. 4. Shoreline uncertainties propagated from, **(a)** measurement errors in shoreline elevation data, and **(b)** spatially structured errors in the DEM. Modern water body names, county boundaries, and places mentioned in the text were added as references.

4.2 Influence of Elevation Errors in the DEM

Fig. 4b shows the distribution of above shoreline locations and areas over which the position of the Lake Algonquin shoreline remains uncertain. Although the larger errors in the DEM were suspected to contribute more uncertainty to results, distributions in the output are surprisingly large. While similarities exist between the spatial distributions of uncertainty in Figs. 4a and 4b, some striking differences are present. In some areas, the width of the band of uncertainty is greater than 10 km. For example, the peninsula located north of Alpena (Fig. 4a) was cartographically reduced to two islands (Fig. 4b). Much of the land area in the Upper Peninsula (Fig. 4a) has also been reduced by uncertainty (Fig. 4b). However, modelling DEM elevation errors has served to reveal some new information. For example, results from simulation two show a massive spit complex located approximately halfway between Alpena and Mackinaw City. This complex was not apparent in Schaetzl's original findings or after simulation one. Large errors in the DEM may have propagated into broad uncertainties, but when rigorously modelled, they may also contribute to discovery.

4.3 Influence of Interacting Errors

The interaction between errors during simulation three manifested into uncertainty distributions nearly identical to those produced during simulation two. Hence, a separate figure is not included here. Shadows of uncertainty caused by errors in the shoreline survey were masked by shadows caused by errors in the DEM. No compounding effect was evident. Most of the slight differences occurred in the island and spit complex located just south of Mackinaw City and the island complex located north of the North Channel. As hypothesized above, however, the observable differences in Canada may have been induced by trend surface flutter in the adjusted spatial model.

5 Discussion

The shoreline survey data used in this research consist of position measurements made along relict, and thus static, lacustrine features formed by Lake Algonquin. Yet, active coastal environments are dynamic and their associated shorelines can change quickly via the combined influences of erosion and deposition processes. For that reason, the exact position of any shoreline, whether modern or ancient, cannot be known with absolute certainty. However, based on arguments put forth by Schaetzl *et al.* (in press), we accept the shoreline survey data set as the set of best proxy indicators of Lake Algonquin shoreline locations.

Burrough and McDonnell (1998:243) consider the question, "if a new attribute U is defined as a function of inputs $A_1, A_2,...A_n$, we want to know what is the error associated with U, and what are the contributions from each A_n to that error?" In this research, the new attribute U represents Lake Algonquin shoreline position, and we want to know the contribution of error in shoreline position from the shoreline survey data and the DEM, A_1 and A_2 respectively.

Results from this research illustrate the significant impact that even small errors in spatial data can have on model results. Recall that elevations in the shoreline data set contain an average error magnitude of only 1.2 meters. These small errors, after being propagated through several GIS operations, manifested into hundreds of meters of positional uncertainty in some locations.

The DEM contained elevation values with an average error magnitude greater than 5 meters. This error range is larger than the range associated with the shoreline data, and subsequently, was propagated into larger uncertainties.

5.1 Uncertainty Distributions in Maps of Results

Obvious spatial patterns are evident in results when errors in the DEM are propagated. Review of Fig. 4 reveals several locations that are sensitive to the larger errors inherent to the DEM. A re-examination of field notes, USGS topographic maps, and the original DEM provided grist for an explanation. Relatively flat areas, like those formed by alluvial outwash processes, contain little local relief and subtle changes in elevation. Should the interpolated position of a shoreline vary along the vertical dimension of these relatively flat locations, even by only a few meters, then the intersection of the water surface and the land surface can vary widely over the horizontal plane. Conversely, those locations with relatively steep gradients and significant local relief, like those associated with wave-cut bluffs (e.g. like those located just south of Alpena), will tend to be less sensitive to errors in interpolated shoreline positions. Although differences in shoreline elevation will change the position of the shoreline along the vertical axis, less change would occur along the horizontal plane. In sum, those areas most sensitive to errors tended to exhibit shallow gradients.

6 Conclusions

Lake Algonquin shorelines in Michigan cannot be mapped with the same degree of precision at all locations, at least not with the data sets, spatial model or terrain data currently available. This finding in no way discredits those data sets or the work done by Schaetzl *et al.*, because they have delineated the boundaries of Lake Algonquin much more extensively and precisely than any previous work. However, it seems appropriate to suggest limitations on how their results could be used. For example, their results would be quite useful for reference purposes when illustrated at smaller cartographic scales, like the one used for Fig. 1, because the width of the *band of uncertainty* may be concealed within the width of a pen stroke. At larger nominal cartographic scales, like the one used for Fig. 4, their results would not be as useful for determining shoreline position because the exact extent of the former lake boundary remains uncertain in many locations. Yet, future efforts to map the extent of Lake Algonquin and other paleolakes will benefit from the research presented here as uncertainty maps can be used to focus subsequent fieldwork and surveying efforts. An iterative process of planning, fieldwork, spatial modelling, and uncertainty mapping, with refinements made along the way, may be what is required to finally map Lake Algonquin.

Acknowledgements

This work was supported by a NSF (Geography and Regional Science) grant (BCS-9819148) to RJS. We acknowledge the many contributions of students from GEO 809 (Fall 1999) to this project: B. Weisenborn, K. Kincare, X.

Cordoba, K. Shein, C. Dowd, and J. Linker. Also, we thank Beth Weisenborn for sharing with us her template of Lake Algonquin, which we used to create Fig. 1.

References

Bolstad PV, Stowe T (1994) An evaluation of DEM accuracy: elevation, slope, and aspect. Photogrammetric Engineering & Remote Sensing 60(11):1327-1332

Brown DG, Bara TJ (1994) Recognition and reduction of systematic error in elevation and derivative surfaces from 7-1/2 minute DEMs. Photogrammetric Engineering and Remote Sensing 60(2): 189-194

Burrough PA, McDonnell RA (1998) Principles of Geographic Information Systems. New York: Oxford University Press Inc.

Futyma RP (1981) The northern limits of glacial Lake Algonquin in upper Michigan. Quaternary Research 15: 291-310

Gilbert GK (1898) Recent earth movements in the Great Lakes region. U.S. Geological Survey 18th Annual Report, Part 2. Washington D.C.

Goldthwait JW (1908) A reconstruction of water planes of the extinct glacial lakes in the Lake Michigan basin. Journal of Geology 16: 459-476

Goodchild MF, Gopal S (1989) Accuracy of Spatial Databases. London: Taylor & Francis.

Goovaerts P (1998) Impact of the simulation algorithm, magnitude of ergodic fluctuations and number of realizations on the spaces of uncertainty of flow predictions. Stanford Center for Reservoir Forecasting, Stanford University, Unpub. annual report No 11.

Heuvelink GBM (1998) Error propagation in environmental modelling with GIS. Bristol, PA: Taylor & Francis

Hough JL (1958) "Geology of the Great Lakes." Urbana: University of Illinois Press

Kyriakidis PC, Shortridge AM, Goodchild MF (1999) Geostatistics for conflation and accuracy assessment of digital elevation models. International Journal of Geographical Information Science 13(7):677-707

Larsen CE (1987) Geological history of Glacial Lake Algonquin and the Upper Great Lakes. U.S. Geological Survey Bulletin 1801

Larson G, Schaetzl RJ (2001) Origin and evolution of the Great Lakes. Journal of Great Lakes Research 27(4):518-546

Leverett F, Taylor FB (1915) Pleistocene of Michigan and Indiana and the history of the Great Lakes. U.S. Geological Survey Monograph 53

National Geodetic Survey, Information Services Branch. (2001) NGS Datasheets [online]. Available from: http://www.ngs.noaa.gov/datasheet.html (Viewed on March, 2001)

Pebesma EJ, Wesseling CG (1998) Gstat, a program for geostatistical modelling, prediction and simulation. Computers and Geosciences 24(1):17-31

Schaetzl RJ, Drzyzga SA, Weisenborn BN, Kincare KA, Cordoba XD, Shein KA, Dowd CM, Linker J (in press) Correlation and mapping of Lake Algonquin shorelines in Michigan. Annals of the Association of American Geographers.

Spencer JW (1891) Deformation of the Algonquin beach, and birth of Lake Huron. American Journal of Science 4:12-21

Stanley GM (1936) Lower Algonquin beaches of Penetanguishene Peninsula. Bulletin of the Geological Society of America 47: 1933-1960

Spatio-Temporal Methods to Reduce Data Uncertainty in Restricted Movement on a Road Network

Nirvana Meratnia, Wolfgang Kainz, and Rolf de By

University of Twente, Department of Computer Science, Enschede, The Netherlands, [Meratnia, Kainz]@cs.utwente.nl, deBy@itc.nl

Abstract

Much of the success of location-based applications comes from using spatial analysis in query processing and graphic visualisation. Recently, with the arrival of miniature and inexpensive GPS receivers and cellular phones, new horizons have been opened for these applications. The mobile telephone and Internet technology within the GIS environment has fostered new applications with higher degrees of user satisfaction. The advance this technology affords for data acquisition, facilitates the provision of real-time location-based services.

Recently, considerable work has been undertaken in feasibility assessments and in identifying the type of services and applications that can be offered to mobile persons in an environment equipped with mobile telephones, Internet technology and GIS. No equivalent attention has been given, however, to the data that will be used for running such services. The acquired data have some degree of uncertainty mainly due to system error associated with acquisition devices and limitations in representing continuous data temporally.

By focusing on restricted movement on a road network, this paper proposes a spatial registration technique to reduce the error associated with location data; it also proposes consistency checks and time registration in an effort to compensate for times when no observation is available. The processed data will also reduce the ambiguity in determining the current location of an object within the network.

Keywords: restricted movement, mobile person, spatial/temporal registration, data uncertainty, consistency checks

1 Introduction

The situation in which the whereabouts of objects[1] are constantly monitored and stored are an important class of problem that present-day database/GIS users will find hard to tackle with their current systems. The problem arises as a result of inappropriate data models, data structures and operations, indexing methods and query processing techniques, all of which are necessary to handle continuous changes in moving objects along with the volume of data that is generated. For example, existing query languages are mainly non-temporal and limited to accessing a single (current) database state, not past and future ones (Sistla et al, 1997).

Much of the success of a GIS project depends on the quality of the data that is entered into the system (de By, 2000). Of course, the data uncertainty has various implications on data modelling, querying and indexing (Wolfson et al, 1999). To handle the uncertainty, different methods should be used. As (Márkus, 1997) emphasised, uncertainty in a GIS environment may occur in three phases: data acquisition, data processing and in using information. In general, the uncertainty in data acquisition may occur due to instrument errors, inappropriate frequency of data acquisition, data transmission errors, and poor data acquisition environments. With respect to a poor environment, it is clear that certain data acquisition devices cannot work properly in some situations, such as poor weather conditions. On the other hand, factors such as inappropriate sampling methods, data storage, data representation methods and interpolation techniques also affect uncertainty during the processing phase. Although errors occur as a result of data processing and data use, most GIS errors result from improper or inadequate data input (Demers, 1998).

One of the sources of uncertainty for a moving point object, comes not only from the error associated with the data acquisition devices, but also from the fact that data is acquired in a discrete way. Since computer systems cannot directly represent continuous phenomena of an arbitrary nature, such phenomena must be approximated using a finite structure. The uncertainty of the representation of an object's movement is affected by the frequency with which position samples are taken, i.e., the sample rate. The sample rate may be defined by considering the speed of an object and the desired maximum distance between consecutive samples (Pfoser et al, 1999). It is also necessary to think about how to handle cases when the connection between the object and data acquisition device is lost and data cannot be updated (Wolfson et al, 1999). Uncertainty in the data can accumulate from different sources. Research on uncertainty in geo-spatial information is usually concerned with sources of uncertainty such as the correctness and completeness in the measurement, and the analysis and interpretation of digitally represented, geo-referenced phenomena (Pfoser et al, 1999). As error in input data can be high and may be propagated in further

[1] Of all physical objects, i.e., objects with clear boundaries that follow the Newtonian laws in their movement, cars are covered in this study as an example of objects with restricted movement on a road network.

processes, this paper focuses on reducing error in the object locality data while ignoring other types of error.

It is also important to notice that uncertainty has cost, which means higher uncertainty leads to less retrieved information (Wolfson et al, 1998) and consequently results in partially or misleading outcomes.

Another critical issue concerns the sufficiency of the number of observations. Too few observations give ambiguous results in interpolation techniques. Although it is possible that an application adds data, it should be done on the basis of rules describing object movement. On the other hand, too many observations consume storage space and higher data acquisition frequency causes more data traffic in wireless environments and therefore may be expensive (Chamberlain, 1999).

Obtaining a "faithful representation" of moving objects with a sufficient number of discrete (possibly erroneous) data points is the goal of this paper. Such representations can open the door for recognition of movement patterns and object classification. These may consequently introduce information about the environment where the objects move and assist the query processing. For example, in locating the shortest path between two points in a road network or in finding the closest feature of interest, not only the static spatial aspects should be considered, but also the characteristics of the road network (such as average speed for each road, delay times at intersections, among other aspects). Such information can be obtained by studying patterns of movement. Thus, unambiguous and optimal representation of moving objects is crucial.

This paper proposes spatial and temporal registration for the reliable representation of moving objects by minimising possible errors in the data. Section 2 contains the description of the methods, addressing the issue of reducing spatial error in the data, followed by explaining the consistency check approach to provide more reliable data for further processing. Time registration is also introduced to handle situations in which no observations are available. Results of experiments and description of data used in these experiments are presented in section 3. Finally, the paper concludes with results of some experiences and recommendations for future work.

2 Methods

Computations about moving objects require the use of positional interpolation techniques, in which one of the fundamental issues is the "medium" via which the objects are travelling. The medium may or may not impose restrictions on where objects go. Among all three possible types of movement, namely unrestricted movement in 2D or 3D, restricted movement in 2D or 3D and restricted movement in a 2D or 3D network, this study focuses on the latter. The observed object can be a car equipped with a GPS receiver or a person travelling in a car carrying a cellular phone that enables position determination.

It is assumed that there is a central base station that receives the location and time data in addition to the queries (if any) from the moving person/car and sends the query results back. The central base station can process two types of query. The first type aims to provide statistical information about a group of objects, i.e. information about the road network status such as average speed for a certain road, or waiting time at a certain intersection. Such information can be obtained based on the historical data from previously monitored objects and can be computed at different times a day, different days a week and so on. The second type of query is initiated by a travelling person and concerns the position of the individual with respect to other persons and phenomena. In this case, the history is collected in addition to the information about the road network (that is provided by the statistical information) and can be used to answer the query. For example, the current and estimated future speed and acceleration of the car, together with information about the estimated waiting time and average speed along the road segments are used to answer the query about the minimum estimated time it takes to reach a destination.

An essential assumption that we have made about data is that there is no error associated with time measurements, while the location data may have error. This can be a fair assumption since at each time (whether correct or incorrect) there is a location, while at a certain location (if it is wrong) there is no time that the person/car passes through. Since there is uncertainty about the location of the various cars at any point in time, we may not be able to answer the queries with absolute certainty; therefore, a degree of probability may be used (Sistla et al, 1999).

Our objects of study include *moving objects,* such as cars or persons, and the *road network*, which consists of nodes with geometry and links with topology. Regardless of the data acquisition method, the main part of the moving object data is in the form of <*O_id, X, Y, T*> and we call it *census point.* For a census point, *X* and *Y* represent the position of the moving object, with identification *O_id,* at time *T.* While data can be acquired in equal or random time intervals, the size of time intervals should be chosen in a way that does not lead to too few or too many observations. For the sake of simplicity, <*Road_id, From, To, Speed*> are assumed to be the main attributes of a road network segment. *From* and *To* indicate start and end nodes, *Speed* is the maximum allowed speed for a road segment with identification *Road_id.* We assume road network segments are one way. Other attributes can be added. The geometry is stored in the nodes.

2.1 Spatial Registration

As mentioned before, data coming from the data acquisition phase may be erroneous and before any further processing, it is important to correct the data to the best of our knowledge. To do so, the following procedure is proposed, in which restricted movement on the road network plays an important role:

- For each census point, the nearest road segment(s) will be identified. To do so, the minimum distance between each census point and road segment is computed.
- The *Road_id* corresponding to the minimum distance will be assigned to the census point, indicating the most probable road that the object was on. Fig. 1 shows the details. In our figures, road segments are illustrated as *R1, R2, ...* and *R15* and census points are illustrated as i, i+1, i+2, i+3, ….
- Next, the consistency of the assigned *Road_ids* will be checked. We will examine three distinct consecutive *Road_ids* at a time and the key to accept a sequence is the connectivity of the middle road to the next and previous one.

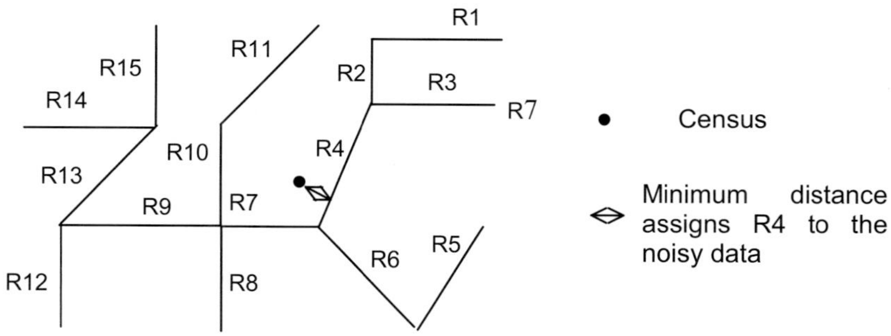

Fig. 1. The selected road segment

Fig. 2 shows acceptable and unacceptable sequences. For the census point sequence shown in A, the logical road sequence will be R1, R2, R3 and R4. Having a data in R'2 obviously indicates an error. For such a census point sequence, case B is not acceptable while case C is an acceptable road order.

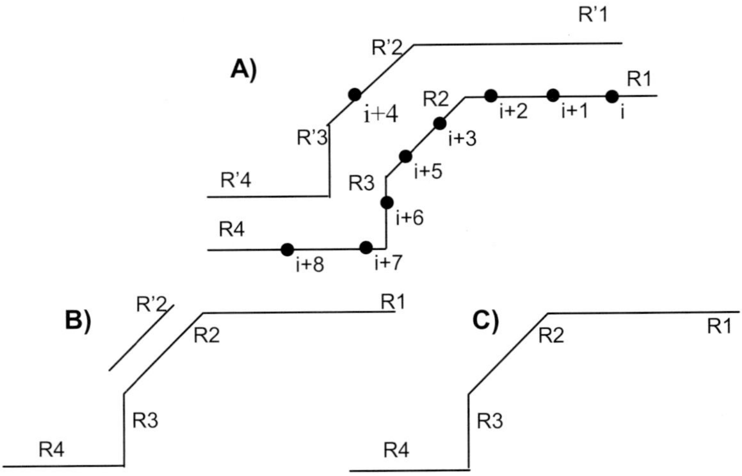

Fig. 2. (Un) Acceptable sequence in consistency check

- Following the consistency check, the census point corresponding to the unacceptable road segment is re-projected to the previous or next road, whichever is closer in order to maintain the consistency.
  The projection of the raw data into the correct or the most probable road leads to the identification of the error associated with the data.
- The length of the projection (perpendicular to the road) describes the error in one dimension; X, Y, or inclined direction depends on the direction of the road. It seems logical to assume an homogeneous error, in which the error is considered to be circular; this implies the error in X is equal to the error in Y or in other words, the error along the road equals the error perpendicular to the road on average.

This method is used to reduce uncertainty in the data. Since the census point is projected spatially to the most probable road segments, this procedure is called "spatial registration".

An important question remains, namely "when should data be modified?" The first option is real-time modification, which leads to storing the modified data. Although it seems to be a time consuming operation, data will be ready for further processes and any posed queries can be answered quickly later on. The second option is storing the noisy data and modifying them when it is necessary, meaning when a query is posed. This option speeds up the data acquisition and storage phases but slows down the query-processing phase. Choosing any of the options definitely depends on the application goals.

2.2 Time Registration

As data is acquired and represented in a discrete way, an appropriate technique is needed to provide the data when no observation is available to determine the historical path and future locations. In such cases, positional interpolation is used. As the method of interpolation is entirely dependent on the characteristic of the moving object, i.e. its maximum speed, maximum acceleration, maximum angular speed, and so on, an important question is which interpolation technique should be chosen (Meratnia, 2000). On the other hand, concerning our example other factors, such as traffic jams, amount of fuel and human decisions have an important influence in predictions concerning location and time.

Since we do not measure the positions in-between two consecutive position samples, the best we can do is to limit the possibilities of where the moving object could have been (Pfoser et al, 1999). Our assumption about restricted movement on a road network necessitates that the location coincides with the road network and helps to narrow the possible area that a car has visited. The procedure to obtain the data in-between two consecutive position samples consists of three main steps:

- Define a general function that describes the car movement at any point in time. It has to be mentioned that the path of the car following the road segments has been determined in the consistency check and is restricted to the road network. The time is known, however, only for those points where there is data. Therefore, the time registration function is chosen as "D(t)" or distance along the network as a function of time. Different interpolation techniques can be used (and compared) to provide better understanding of the car movement. In this study, two interpolation methods are used, namely spline and linear interpolation.
- Constrain the function "D(t)" to the location and time data obtained from the spatial registration step. This means that to estimate the parameters involved in the distance function, in addition to the corresponding time values, the distances along the road network between the census points are used. Fig. 3 shows how the distance between points is restricted to the road network. In this figure, the object is moving from road segment R1 to road segment R3, according to the spatial registration of the points and the consistency check. Distance is measured along the road network. Therefore, the distance D_{i+2} is being forced to pass through the intersection node between R1 and R3, instead of passing through a straight line connecting data points i+1 and i+2. This information would not be available if the spatial registration and the consistency check had not been performed.

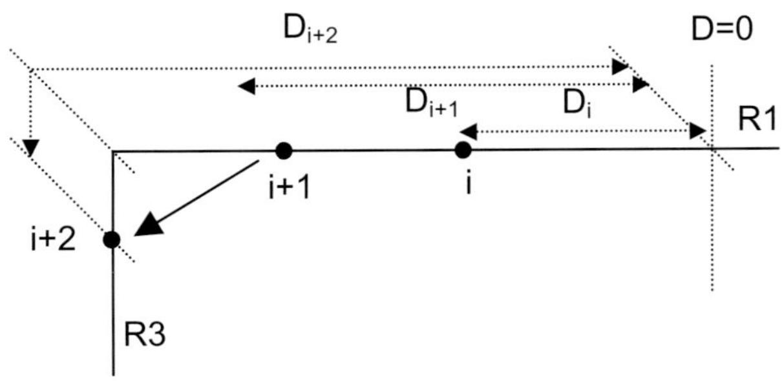

Fig. 3. Constraining distances to the road network

- As a rule of thumb, whenever the number of equations using the available data exceeds the number of parameters to be estimated, a least-square adjustment is used to minimise the possible errors and introduce a confidence measure for the estimated parameters. Examples can be found in (Meratnia et al, 2001).

3 Experiments

The proposed methods were tested using simulation techniques. The evaluation procedure starts by simulating so-called true data for both the object and the road network. In the next two subsections, the data description and the analysis of the results are discussed.

3.1 Data Description

The data has two parts; the road network data and the moving object (car/person) data. The road network consists of road segments (or links) and intersections (or nodes). The structure of the road network has been simulated and includes topology and geometry. The topology is associated with the links as each link has a start and end node. The geometry is associated with nodes as each node has a location. We assume that the road network data is more accurate than the car data. No error is assumed associated with the road network, however, errors can be incorporated in the registration analysis.

Other data that we have simulated are the speed limits and the direction of each road segment. The intersections have been designed to have green and red traffic lights to regulate the incoming and outgoing cars. The green and red traffic light time intervals were simulated independent of the cars and were defined as a global function of time. Therefore, at any point in time, it is determined which car is allowed to pass and which car has to wait and for how long.

The simulation had the cars follow the above road network rules. Each car starts from a randomly selected road segment and continues by maintaining the connectivity between the consecutive links. The average speed of a car in a road segment was simulated to have a random value, which does not exceed the permitted speed in that road segment. Acceleration (at starting points as well as intersection points) the same as deceleration (when the car approaches an intersection as well as when it is going to stop) were also simulated.

The path determined for cars was temporally discrete in two ways; every second (data was provided for accuracy determination), and every minute (data were introduced to test the distributed errors, for mean, and standard deviation to both X and Y direction of the data).

3.2 Results

Comparing true and noisy data, and computing the RMSE indicates a 2.4m error on average in the data. The RMSE was computed as the average of the spatial errors in the data. Although the directions of the error vectors may differ, this value gives an indication about the average size of the error buffers. The procedure continues by applying the proposed correction method on the noisy data by projecting them on the road network and performing the consistency check. Comparing the noisy and the corrected data, the estimated error results in 1.8m, measured perpendicular to the link directions. It is clear that the average error in data has been reduced by 25%. We will assume that the error is homogeneous and therefore the error component in the direction of the links equals that in the perpendicular directions. The computed RMSE for the corrected data compared to the true data (which we only have in the simulation) is 1.5m, which is very close to our original estimate. Therefore, if we introduce this value as uncertainty along the link we can achieve more accurate results in an adjustment procedure. Fig. 4 shows the difference between the true, noisy (raw), and the corrected data.

In time registration evaluation, spline and linear interpolation are used separately to provide the complete travelled path with distance as a parametric function of time.

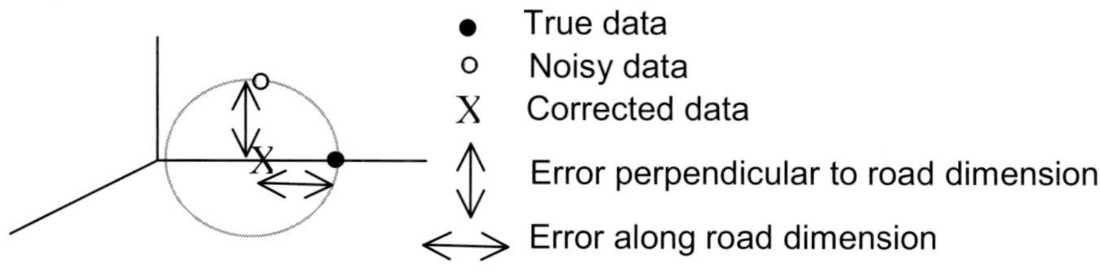

Fig. 4. Homogeneous error

The computed spatial RMSE between the interpolated data points using spline and true data is 33.8m, while spatial RMSE between the interpolated data points using linear interpolation and true data is 27.6 km. At first, one can consider it a very large error compared to the errors associated with the discrete raw data points. However, they are acceptable since an object is travelling this distance, (with a speed of 50 km/h) in just 2 seconds. It is noticeable that in using both Spline and linear interpolation techniques, errors mainly occur at the intersection points. It is an expected result, since the behaviour near intersections is not smooth and is highly influenced by human decision and abrupt changes. Fig. 5 and (6) show the distribution of the errors in the direction of the links for the spline and linear time registration, respectively. It should be mentioned that the errors have been exaggerated and plotted normal to the links for better visualisation.

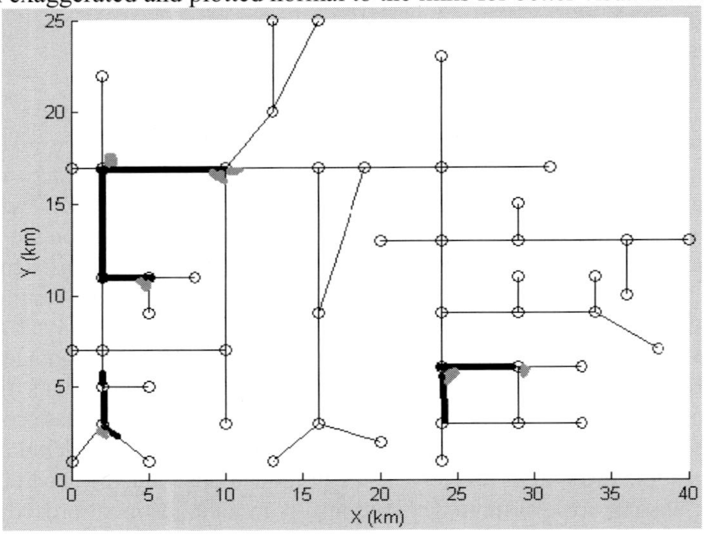

Fig. 5. The spatial error along the road network associated with spline time registration

It is also noticeable that although the linear time registration has less RMSE than the one of the spline time registration, the maximum errors are larger and their distribution is less homogeneous. This can be explained by the fact that more parameters of the Spline (near the intersections) model variations of the movement better, while the same number of parameters may not be required far from the intersections.

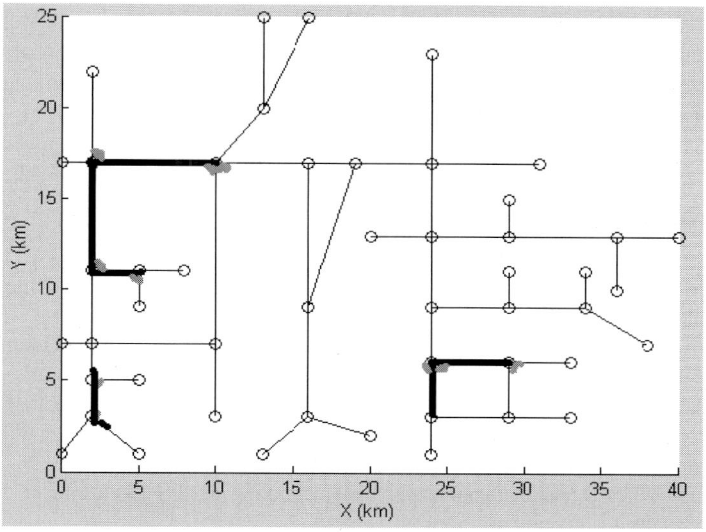

Fig. 6. The spatial error along the road network associated with linear time registration

4 Conclusions and Future Work

In this paper, methods for handling error in restricted movement on a road network have been proposed. They play an effective role to reduce error associated with the data. A consistency check has been suggested to reduce ambiguity in the data. Although effectiveness of any interpolation technique depends on the type of application using it, including the estimated error components, can be of help in identifying the appropriate function for each application type. More parametric time registrations are recommended, including automatic detection of the location and time, when abrupt changes occur in the movement. We assumed that error exists only in location of census point, while we recommend studying the case when time also is uncertain. For the time being, error in the GIS itself is ignored, while it needs to be taken into account in future work. Performing these methods on real data is useful in trying to overcome complex problems that have not been fully addressed. More error analysis is recommended including estimation of confidence intervals for the spatio-temporal results.

Issues such as defining the maximum effective time interval (i.e. the minimum data acceptable for economic reasons) to clearly represent the object within tolerance values of space and time, modelling intersections, and defining a cost model (time) are areas to be addressed in future work.

Moving objects can be used to provide valuable information about the environment through which they pass. Obtaining an accurate representation of cars opens the door to studying patterns and classes of cars/persons, which can assist in obtaining spatio-temporal information about the road network in addition to information about traffic conditions. Accessing such information helps to answer many spatio-temporal queries about the history of persons/cars. Additionally, the information assists in predicting the future location of monitored persons/cars or providing spatio-temporal information for persons looking for closest (in terms of time and/or space) features of interest.

References

Chamberlain S, Wolfson O, Dao S, Jiang L, Medsez G (1998) Cost and Imprecision in Modeling the Position of Moving Objects. In: Proceedings of the 14th International Conference on Data Engineering (ICDE98). IEEE Computer Society, Orlando, Florida, pp 588-596

De By RA (2000) Principles of Geographic Information Systems. ITC. The Netherlands

Demers NM (1998) Fundamentals of Geographic Information System. John Wiley & Sons, INC

Márkus B (1997) Decision Support and Error Handling in GIS Environment. Habilitation presentation. University of Sopron, Sopron

Meratnia N (2000) Implementation of Data Structures and Operations for Moving Objects Modeling. M.Sc. thesis, ITC, The Netherlands

Meratnia N, De By RA (2001) Moving objects data modeling. In: Proceedings of ASPRS 2001 Annual Conference. St. Louis

Pfoser D, Jensen C (1999) Capturing the Uncertainty of Moving-Object Representations. In: Proceedings of the 6th International Symposium (SSD99). Springer-Verlag series LNCS 1651

Sistla P, Wolfson O, Chamberlain S, Yesha Y (1999) Updating and Querying Databases that Track Mobile Units. Invited paper in special issue of the Distributed and Parallel Databases. Journal on Mobile Data Management and Applications, pp 257-287

Sistla P, Wolfson O, Chamberlain S, Dao S (1997) Modeling and Querying Moving Objects. In: Proceedings of the 13th International Conference on Data Engineering (ICDE97). IEEE Computer Society, pp 422-432

Wolfson O, Sistla P, Xu B, Zhou J, Chamberlain S (1999) Tracking Moving Objects Using Database Technology in DOMINO. In: Proceedings of the 4th Workshop on Next Generation Information Technologies and Systems (NGITS). Zikhron-Yaakov, Israel. Lecture Notes in Computer Science, no. 1649, Springer-Verlag, pp 112-119

Wolfson O, Xu B, Chamberlain S, Jiang L (1998) Moving Objects Databases: Issues and Solutions. In: Proceedings of the 10th International Conference on Scientific and Statistical Database Management (SSDBM98). IEEE Computer society, pp 111-122

Quantifying Uncertainty of Digital Elevation Models Derived from Topographic Maps

Qihao Weng

Department of Geography, Geology, and Anthropology, Indiana State University, Terre Haute, IN 47809, USA

Abstract

This paper explores a methodology for quantifying the uncertainty of DEMs created by digitising topographic maps. The origins of uncertainty in DEM production were identified and examined. The uncertainty of DEM data was quantified by computing a vector total of Root Mean Square Error (RMSE) from the source map, sampling and measurement errors, and the interpolation process. Distributional measures including accuracy surfaces, spatial autocorrelation indices, and variograms were also employed to quantify the magnitude and spatial pattern of the uncertainty. The test for this methodology utilises a portion of a 1:24 000 topographic map centred on Stone Mountain in northeastern Georgia, USA. Five DEMs, constructed with different interpolation algorithms, are found to have the total RMSE ranging from 4.39 to 9.82 meters, and a highly concentrated pattern of uncertainty in rugged terrain. This study suggests that the RMSE provides only a general indicator of DEM uncertainty. Detailed studies should use distributional measures to understand how the uncertainty varies over a surface.
Keywords: DEM, uncertainty, origins; distributional measures

1 Introduction

Two methods are frequently employed for obtaining digital elevation model (DEM) data: cartographic digitising and photogrammetric (analytical or digital) methods. The cartographic digitising method is widely used since topographic maps are usually available. The basic procedure involves the transformation of contour lines on existing maps into digital coordinate data using manual or automatic digitizers. A surface is then fitted to these point observations in interpolation of the elevation at every grid point. Alternatively, an automatic raster scanner is used with the vectorization technique. This method is straightforward, and the coverage can be very large. However, the general quality of the derived DEM is poor unless great efforts are made to extract terrain characteristic features and break lines as well (Ackermann, 1994).

The quality of a DEM is dependent upon a number of interrelated factors, including the methods of data acquisition, the nature of the input data, and the methods employed in generating the DEMs (Shearer, 1990). Of all these factors, data acquisition is the most critical one. Previous studies on DEM data acquisition have focused either on examination of generation method(s), or on case studies of accuracy testing (Ackermann, 1978; Ebner and Reiss, 1984; Torlegard *et al.* 1987). These studies are not adequate, however, for the purpose of understanding uncertainty (an indicator used to approximate the discrepancy between geographic data and the geographic reality that these data intend to represent) associated with DEM data and the propagation of this uncertainty through GIS based analyses. The development of strategies for identifying, quantifying, tracking, reducing, visualising, and reporting uncertainty in DEM data are called for by the GIS community (Zhu, 1997; Fisher, 1999; Heuvelink, 1999; Veregin, 1999).

The objectives of this study are: (1) to understand the sources and reasons for uncertainty in DEMs produced by cartographic digitising; and (2) to develop methods for quantifying the uncertainty of DEMs using distributional measures.

2 Measurement of DEM Uncertainty

Measurement of errors in DEMs is often impossible because the true value for every geographic feature or phenomenon represented in a geographic data set is rarely determinable (Goodchild *et al.* 1994; Hunter *et al.* 1995). Uncertainty, instead of error, should be used to describe the quality of a DEM. Quantifying uncertainty in DEMs requires comparison of the original elevations (e.g., elevations read from topographic maps) with the elevations in a DEM surface. Such a comparison results in height differences (or residuals) at the tested points. To analyse the pattern of deviation between two sets of elevation data, conventional ways are to yield statistical expressions of the accuracy, such as the root mean square error, standard deviation, and mean. In fact, all statistical measures that are effective for describing a frequency distribution, including central tendency and dispersion measures, may be used, as long as various assumptions for specific methods are satisfied.

The most widely used measure is the Root Mean Square Error (RMSE). It measures the dispersion of the frequency distribution of deviations between the original elevation data and the DEM data, mathematically expressed as:

$$RMSE_z = \sqrt{\frac{1}{n}\sum_{i=1}^{n}(z_{di} - z_{ri})^2} \tag{1}$$

Where: Z_{di} is the ith elevation value measured on the DEM surface;
Z_{ri} is the corresponding original elevation;
n is the number of elevation points checked.

The larger the value of the RMSE, the greater the discrepancy between the two data sets. When the true values such as ground truth are used as the reference data, the "uncertainty" becomes "error". Accuracy is the reverse measurement of error. The accuracy of a DEM can be defined as the average vertical error of all potential points interpolated within the DEM grid (Ackermann, 1996). In other words, it is the vertical root-mean-square accuracy of all points (infinitely many) interpolated in the DEM grid.

The main attraction of the RMSE lies in its easy computation and straightforward concept. However, this index is essentially a single global measure of deviations, thus incapable of accounting for spatial variation of errors over the interpolated surface (Wood, 1996). This index fails, besides, to reveal any information about the mean deviation between the two measures of elevation, neither the form of frequency distribution such as balance between small and large deviations and skewness of the distribution (Wood, 1996). Most interpretations of the RMSE value assume a zero mean deviation, which is often invalid (Li, 1993a; Li, 1993b; Monckton, 1994). Moreover, the magnitude of the RMSE value depends on the variance of a true elevation distribution, and is subject to the influence of relative relief and scale of measurements (Wood, 1996). To correctly report and compare the RMSE values for areas with different relative relief values a composite index such as RMSE-to-contour density ratio should be developed and utilised in order to minimise the influence of this "natural" variance (Gao, 1997).

3 Uncertainty from Elevation Measurements for Topographic Maps

The vertical accuracy of topographic maps with which terrain elevations are photogrammetrically measured depends on the base-height ratio, the relationship of the ground distance between successive exposures of photographs to the flying height. The larger the ratio, the higher the heighten accuracy. For a particular flying height, camera focal length and angular coverage are the two factors determining the base-height ratio. Therefore, theoretically speaking, super-wide-angle photography is better than wide- and narrow-angle photography for height measurements. In practice, however, wide-angle photography is most frequently used, since a compromise must be made between heighten accuracy and the requirements for good scale and resolution (Petrie, 1990).

If a particular base-height ratio is set, the accuracy of height measurements is then largely limited by flying height. Petrie (1990) reports that the expected accuracy of spot heights using wide-angle photography will be between 1/5,000 and 1/15,000 of the flying height, depending on the type of stereo-plotter used. For example, from flying heights in the range of 1 to 15 km, the heighten accuracy will range from 0.1 to 3.0 meters (RMSE).

There are four data sampling patterns generally employed in obtaining elevation measurements: systematic, random, composite sampling, and contouring. Systematic sampling is a grid-based measurement of spot heights in a regular geometric pattern, square, rectangular, or triangular. In contrast, random sampling measures heights at significant points selected by the photogrammetrist, e.g., on hilltops, along break lines and streams. Composite sampling combines elements of both of the above approaches. The final method is contouring, which systematically measures contours over the whole area of the stereo model supplemented occasionally by the measurement of spot heights along terrain break lines. This mode of measurement consistently yields significantly lower accuracy than the first three methods (Petrie, 1990; Shearer, 1990). The minimum possible contour interval will lie between 1/1,000 and 1/2,000 of the flying height. The expected or specified accuracy of such measurements is that 90 percent of all the points shall be within half of the contour interval. This photogrammetric criterion has been applied to the United States National Map Accuracy Standards. Thus, if the error in the vertical dimension is assumed to be a normal distribution, then the vertical RMSE can be computed as follows:

$$1.645\text{RMSE} = \text{contour interval} / 2 \tag{2}$$

Solving for the RMSE, then:

$$\text{RMSE} = 0.304 * \text{contour interval} \tag{3}$$

Table 1 summarises the relationship between flying height, minimum possible contour interval and the heighten accuracy within one standard deviation (assume that the mean is the true value). If we assume that the camera used to take the source aerial photos has a focal length of 15 cm (wide-angle, most commonly used), then the relationships between photo scale, contour interval, and vertical accuracy can be established.

Table 1. Flying height, minimum possible contour interval, and heighting accuracy

Flying height (m.)	Minimum possible contour interval (m.)	RMSE (90%)
1 000	0.5~1.0	0.25~0.5
5 000	2.5~5.0	1.25~2.5
10 000	5.0~10.0	2.50~5.0
15 000	7.5~15.0	3.75~7.5

Source: Petrie, 1990. Compiled by the Author.

4 Uncertainty from Spatial Interpolation

4.1 Methods for Assessing the Uncertainty

It has been demonstrated that DEM accuracy can vary to a certain degree with different interpolation algorithms and interpolation parameters (Eklundh and Martensson, 1995; Weng, 1998). Five commonly used spatial interpolation algorithms are examined and compared in this study. They include inverse distance squared (Davis, 1986), minimum curvature (Briggs, 1974), modified Shepard's method (Shepard, 1968; Wang, 1990), radial basis function (Hardy, 1990), and triangulation with linear interpolation (Wang, 1990). These interpolators convert randomly spaced data points into regularly gridded data points, which can then be used to generate DEM surfaces. These interpolators can be classified as either an exact or approximate algorithm, if the original data points are preserved on the interpolated surface (Wren, 1975; Lam, 1983).

The test site is selected from a 7.5-minute (1:24,000) USGS topographic map that has a contour interval of 20 feet. It is located in Stone Mountain, Georgia, a hill of moderate relief composed of granite bedrock, and covers a rectangular area of 13,000 by 16,000 feet (Fig. 1). The maximum and minimum elevation of this site is 1683 and 800 feet respectively, resulting in a relative relief of 883 feet. The mean elevation is 976.79 feet, with the standard deviation of 205.93 feet. These statistical indicators should be compared when a new study area is selected, because the accuracy of interpolation is, to a certain degree, subject to the influence of terrain complexity (Ackermann, 1996; Gao, 1997).

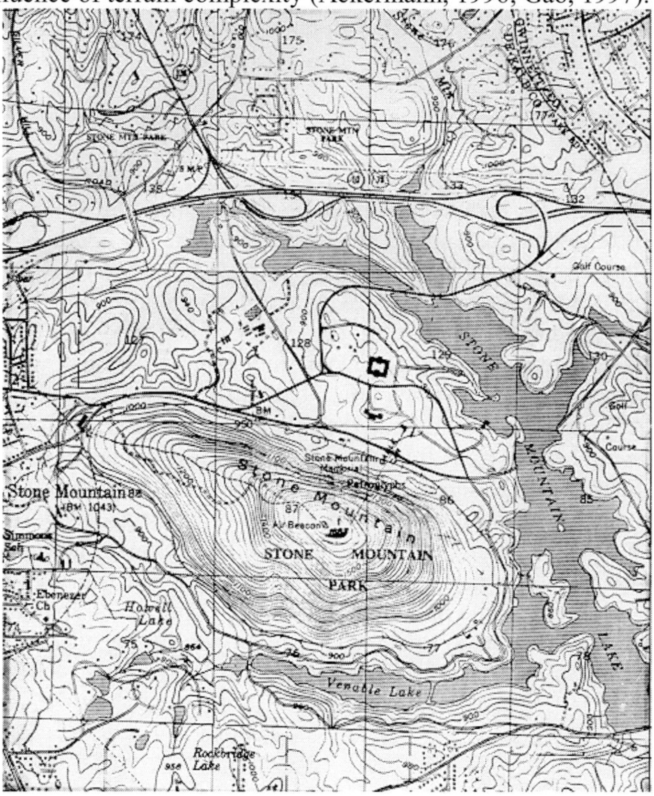

Fig. 1. Test site: Stone Mountain of the northeastern Georgia

The data capture process involves using the CAPTURE (R-WEL, Inc.) and entering point elevations into the database by keyboard. The rule for the selection of points is to distribute them evenly throughout the area examined. Ideally, the points should be those with exact locations identified, such as benchmarks, spot heights, or road intersections. Alternatively, the points where a geographic feature intersects with a contour line are used. The initial data set includes a total of 50 points. With an increment of 50 randomly distributed points overtime, each subsequent data set is generated. At the end, eight data sets with the density of 50, 100, 150, 200, 250, 300, 350, and 400 points become ready for gridding.

The interpolation package used in this study is SURFER. The following three questions are taken into account in evaluating the accuracy of interpolation of elevation data: (1) Among the interpolation algorithms examined, which will produce the most accurate result, statistically and spatially? (2) How does grid resolution influence the accuracy of interpolation? (3) Does data point density have an impact on interpolation uncertainty? If so, what is the relationship between data point density and interpolation uncertainty? To answer these questions, the RMSE of the residuals is computed to reveal the closeness of an interpolated surface to the reality. Check points were picked up along the contour lines of the topographic map and compared with the corresponding elevations on an interpolated surface.

4.2 Test Results

In assessing the accuracy of various interpolation algorithms, the same grid resolution, i.e., 50 meters, is used. Table 2 shows the result of the interpolation. Among the interpolators, the radial basis function (RMSE=6.6628) generates nearly identical result. Triangulation with linear interpolation algorithm provides the worst interpolation, producing the largest total deviation (RMSE=9.5403). The contour map created has distinct triangular faces, indicating that too few data points have been used for the interpolation. The inverse distance squared algorithm generates a reasonably high statistical accuracy (RMSE=6.8404), comparable to kriging and radial basis function. If the RMSE is the only concern, then the modified Shepard's method is the best interpolator (RMSE=3.9448). However, both the modified Shepard's method and the inverse distance algorithm tend to produce a "bull's eye" pattern. Minimum curvature produces acceptable smooth interpolated surfaces, but its statistical accuracy is low (RMSE=8.6937). The selection of parameters for interpolation is listed in Table 3. It should be noted that the changes in interpolation parameters might improve or worsen the statistical performance of an interpolator (Weng, 2001).

Table 2. Effect of different input densities on the quality of interpolation

Input data points	RMSE (meters)				
	Inverse distance squared	Minimum curvature	Modified Shepard's method	Radial basis function	Triangula-tion
50	7.4302	8.5602	3.6579	7.8448	7.7563
100	6.8536	8.4038	3.5165	7.2330	7.6556
150	6.7879	8.3581	3.2428	6.7818	7.4228
200	6.7663	8.2080	2.7764	6.7567	7.2442
250	6.6793	8.1543	2.1534	6.0605	7.1802
300	6.3638	8.0112	2.1210	5.7965	7.0263
350	5.8988	7.9794	2.7277	6.1366	8.4747
400	6.8404	8.6937	3.9448	6.6632	9.5403

Table 3. Interpolation parameters used

Algorithms	Interpolation parameters
Kriging with linear variogram	C = 292; A = 5.7 Nugget effect: Error variance = 0; Micro variance = 0 Drift type: No Drift
Inverse distance squared	ß = 3; δ = 0
Minimum curvature	Max residual = 0.08; Max iteration = 10,000
Shepard's method	Smoothing = 0
Radial basis functions	$R^2 = 0.01$
Triangulation with linear interpolation	N.A.

5 Spatial Distribution of DEM Uncertainty

5.1 Measurement of Spatial Uncertainty of DEMs

It is necessary to note that DEM uncertainty reported by the RMSE assumes a uniform error value for an entire DEM surface. This assumption is often not true. Many authors have suggested that the distribution of errors in DEMs will show some forms of spatial pattern (Guth, 1992; Li, 1993a; Wood and Fisher, 1993;

Monckton, 1994). The best way to observe and analyze the spatial pattern of uncertainty is to have a graphical representation - creating an accuracy surface. This representation has the advantage of clearly indicating where serious, and perhaps anomalous errors occur. Comparison of such a surface, for example, with a plot of the original input contours, can be extremely informative with respect to the occurrence and magnitude of errors in relation to such factors as the terrain slopes and distribution of input data (Shearer, 1990).

Furthermore, a spatial autocorrelation index can be computed to measure the extent of error clustering. When interval/ratio data are of concern, then either the Moran's I or the Geary's C can be used to measure spatial autocorrelation. The former is more commonly used in mapping accuracy. Both are inversely related, and often imply the same results (Griffith and Amrhein, 1991). Moran's I statistic is computed as follows:

$$I = \frac{n\sum_{u=1}^{n}\sum_{v=1}^{n} w_{uv}(dz_u - \overline{dz})(dz_v - \overline{dz})}{\sum_{u=1}^{n}(dz_u - \overline{dz})^2 \sum_{u=1}^{n}\sum_{v=1}^{n} w_{uv}} \tag{4}$$

Where: dz_u is the deviation between the two models for each cell;
dz_v is the deviation between the two models for some neighbouring cells;
w_{uv} is the weighting given to neighboring cells; and
$\overline{dz}$ is the average deviation between the two models.

The value of this statistic usually ranges from -1 to 1, with the value 1 indicating similar value clustering, 0 random pattern, and -1 dissimilar value clustering.

An accuracy surface was created to show the spatial pattern of uncertainty resulted from the interpolation (Fig. 2), in which the magnitude of uncertainty is represented by means of contours. It is clear from these maps that uncertainty tends to cluster in rugged areas of the test site where elevation changes rapidly, especially around the crest of the mountain. All surfaces have a value of the Moran's index larger than 0.8, indicating a reasonably high degree of clustering. However, a close look at these diagrams reveals that some interpolators do a better job than others in terms of revealing the systematic errors that result from under-representation of rugged areas. The modified Shepard's method produces the least degree of clustering. The minimum curvature algorithm, on the other hand, generates a great deal of clusters around the Stone Mountain. In between them are kriging, radial basis function, inverse distance, and triangulation with linear interpolation. A correlation between the RMSE and the Moran's index gives a coefficient of 0.87, indicating that statistically better-interpolated surfaces may result in less error clustering.

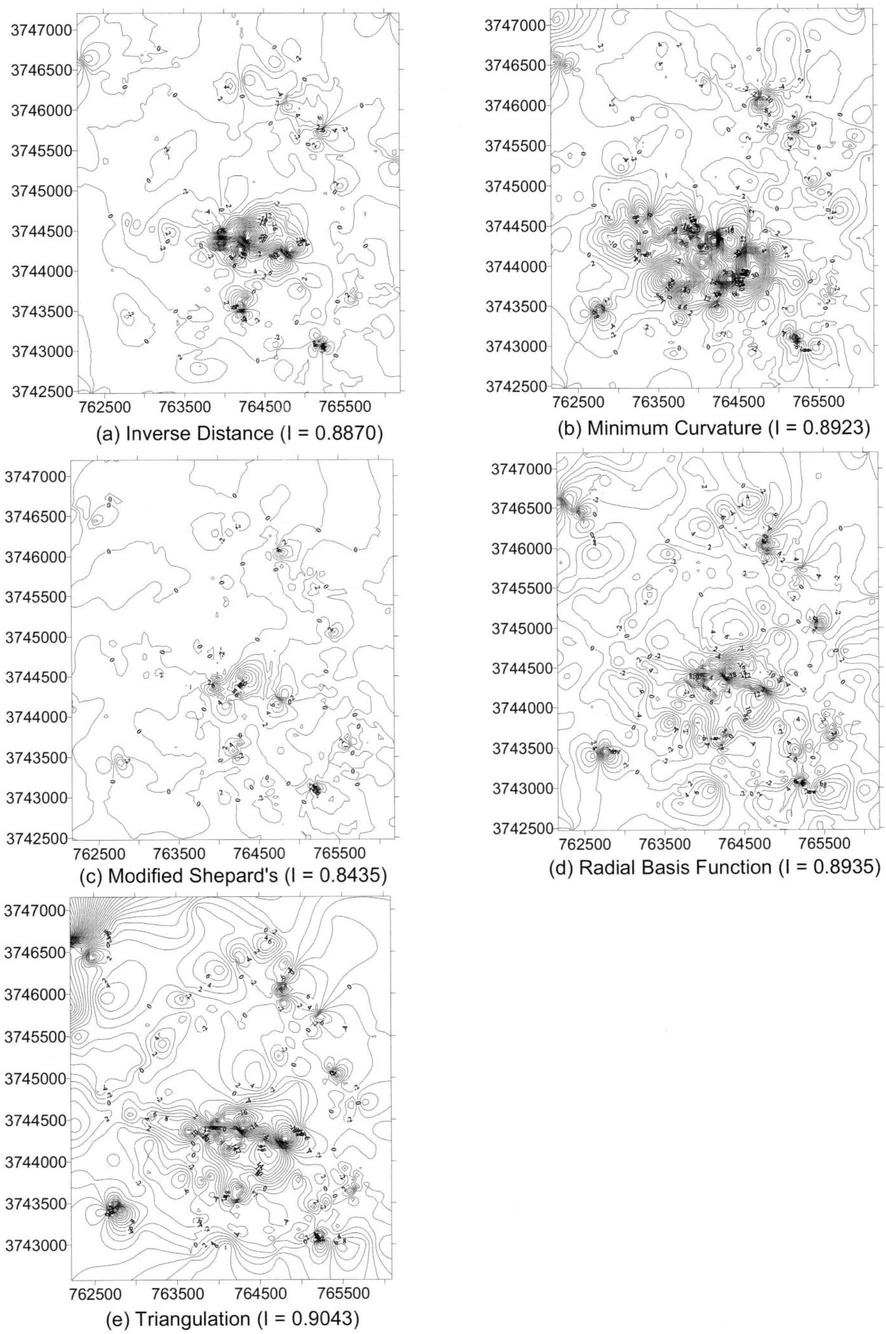

Fig. 2. Accuracy surfaces of the DEMs of Stone Mountain

5.2 Variogram Modelling Spatial Variability of the Residuals

The apparent pattern of clustering in the elevation residuals suggests that there exists an underlying spatial trend in the residual values in each accuracy surface. A common tool to investigate this spatial trend is the empirical variogram (Houlding, 1994). Variogram modelling allows for describing and measuring how difference in residual value changes with distance and direction. The classical variogram function is essentially a measure of dissimilarity between two observations at a separation distance (the lag) h apart, and can be defined as (Carr, 1995):

$$\gamma(h) = \frac{1}{2n(h)} \sum_{i=1}^{n} (z_i - z_{i+h})^2 \tag{5}$$

where z_i and z_{i+h} are the values of residual at locations i and $i+h$, and n is the number of pairs considered. Applying this equation to the residual data, six empirical variograms are constructed (of which five are shown in Fig. 3). These variograms show the spatial variance of the residual values at every 200 feet. In order to evaluate the spatial variance at any separation distance, a continuous function (curve) must be fitted to these point observations. Table 4 shows the results of model fitting and parameters estimated.

The exponential model gives us the best quality of fit for the accuracy surfaces of kriging, modified Shepard's method, and radial basis function; while the spherical model gives the best fit for the accuracy surfaces of inverse distance squared, minimum curvature, and triangulation with linear interpolation. The estimated range for the models, over which the influence of spatial dependency of the residual values is in effect, differ from 320 to 1300 feet. This is equivalent to 0.16 to 0.65 inch in the map. The highest range goes to triangulation with linear interpolation (range = 1300 ft), because this algorithm basically does not account for the underlying trend in the data. The remaining five interpolation algorithms have an average range of 508 feet (i.e., a quarter inch in the map). Beyond these ranges, the increase in separation distance will no longer cause a corresponding systematic increase in the average squared difference between pairs of residual values. This parameter relates closely to the terrain complexity that a variogram model attempts to represent. The information listed in Table 4 therefore provides also a guide for determining input data sampling interval and density.

The estimated nugget takes into account both sampling and measurement error and microscale variation (the spatial variation occurring at distance closer than the sample spacing). This study finds that microscale variation in all of the models equals to zero. Since the minimum separation distance in all the six models is small (Fig. 3), it is safe to conclude that sampling and measurement error are primarily related to the nugget effect, which has a range of 3.43 to 23 feet. Based on the value of nugget effect, the standard deviation of the sampling and measurement error (RMSE) can be calculated. It is estimated to be from 1.85 to 4.79 feet, or from 0.56 to 1.46 meters.

Table 4. Estimated variogram models for the interpolation residuals

Interpolators	Estimated variogram model	Anisotropy (ratio; angle)
Kriging with linear variogram	$\gamma(h)=\begin{cases} 11 & h=0 \\ 11+29*[1-\exp(-\frac{h}{570})] & h>0 \end{cases}$	2; 81.88
Inverse distance squared	$\gamma(h)=\begin{cases} 7 & h=0 \\ 7+11*[\frac{1.5h}{450}-0.5*(\frac{h}{450})^3] & 0<h<450 \\ 18 & h\geq 450 \end{cases}$	1; 0
Minimum curvature	$\gamma(h)=\begin{cases} 23 & h=0 \\ 23+68*[\frac{1.5h}{700}-0.5*(\frac{h}{700})^3] & 0<h<700 \\ 91 & h\geq 700 \end{cases}$	2; 169.1
Modified Shepard's method	$\gamma(h)=\begin{cases} 3.43 & h=0 \\ 3.43+2.5*[1-\exp(-\frac{h}{320})] & h>0 \end{cases}$	2; 51.38
Radial basis functions	$\gamma(h)=\begin{cases} 13 & h=0 \\ 13+24*[1-\exp(-\frac{h}{500})] & h>0 \end{cases}$	2; 76.59
Triangulation with linear interpolation	$\gamma(h)=\begin{cases} 22 & h=0 \\ 22+37*[\frac{1.5h}{1300}-0.5*(\frac{h}{1300})^3] & 0<h<1300 \\ 59 & h\geq 1300 \end{cases}$	2; 117.6

6 Discussion

The uncertainty of digital elevation models resulting from cartographic digitising is related to three aspects: the elevation measurements for the topographic map, sampling and measurement error, and the interpolation process. Assuming the uncertainty from each source is an independent vector, the total DEM uncertainty can be computed as follows:

$$RMSE_{total} = \sqrt{[(RMSE_m)^2 + (RMSE_s)^2 + (RMSE_i)^2]} \quad (6)$$

where $RMSE_m$, $RMSE_s$, and $RMSE_i$ are the uncertainty from the source map, from the sampling and measurement error, and from the interpolation process respectively. With the Eq. 6, the overall quality of the DEMs for the Stone Mountain with different spatial interpolation algorithms can be assessed. More generally, an error budget table can be created to give an estimate of the uncertainty from various sources (Table 5). This table provides the reader with some guidelines for creating DEMs with the cartographic digitising method. However, this table and the findings of this paper should be applied with caution before the following points have been fully considered:

(1) Correlation of uncertainty. The uncertainty in the source map and the uncertainty caused by the interpolation are often correlated. High uncertainty tends to concentrate in rugged terrain areas in both sources. The Eq. 6 needs to be adjusted to account for this correlation in any future research effort.

(2) Point density and distribution. The accuracy of spatial interpolation of elevations is subject to input data point density and distribution, among many other factors. The analysis should be extended into the impact of different data collection patterns (e.g., random vs. systematic; significant points vs. contouring). Equally important is the problem of how to take the complexity of terrain into account to determine point density and distribution. Comparative studies may be conducted by using different input data density for different degrees of roughness in a surface or by taking break lines.

(3) Scale. Scale issue comes into terrain analysis in many ways. The scale of a source map has a direct impact on the quality of DEM generated. The accuracy of terrain parameters and features derived from DEMs exhibits scale dependencies (Hutchinson and Gallant, 1999). Moreover, the degree of clustering of DEM uncertainty may vary at different scales (Monckton, 1994). There are also problems of how the terrain characteristics of a multi-scale can be mathematically described. The analysis should be extended into more complicated types of terrain in different geomorphologic settings at different scales.

(4) Variation in interpolation parameters. The variation in interpolation parameters may significantly improve or worsen the DEM accuracy. Some exemplary questions of further investigation include: How does the RMSE of residual values change as the smoothing parameter of the radial basis function becomes larger, and how will it change with a larger smoothing parameter in the modified Shepard's method, and how will it decrease as the weighting power of inverse distance algorithm becomes larger? Are these effects related to terrain complexity and scale?

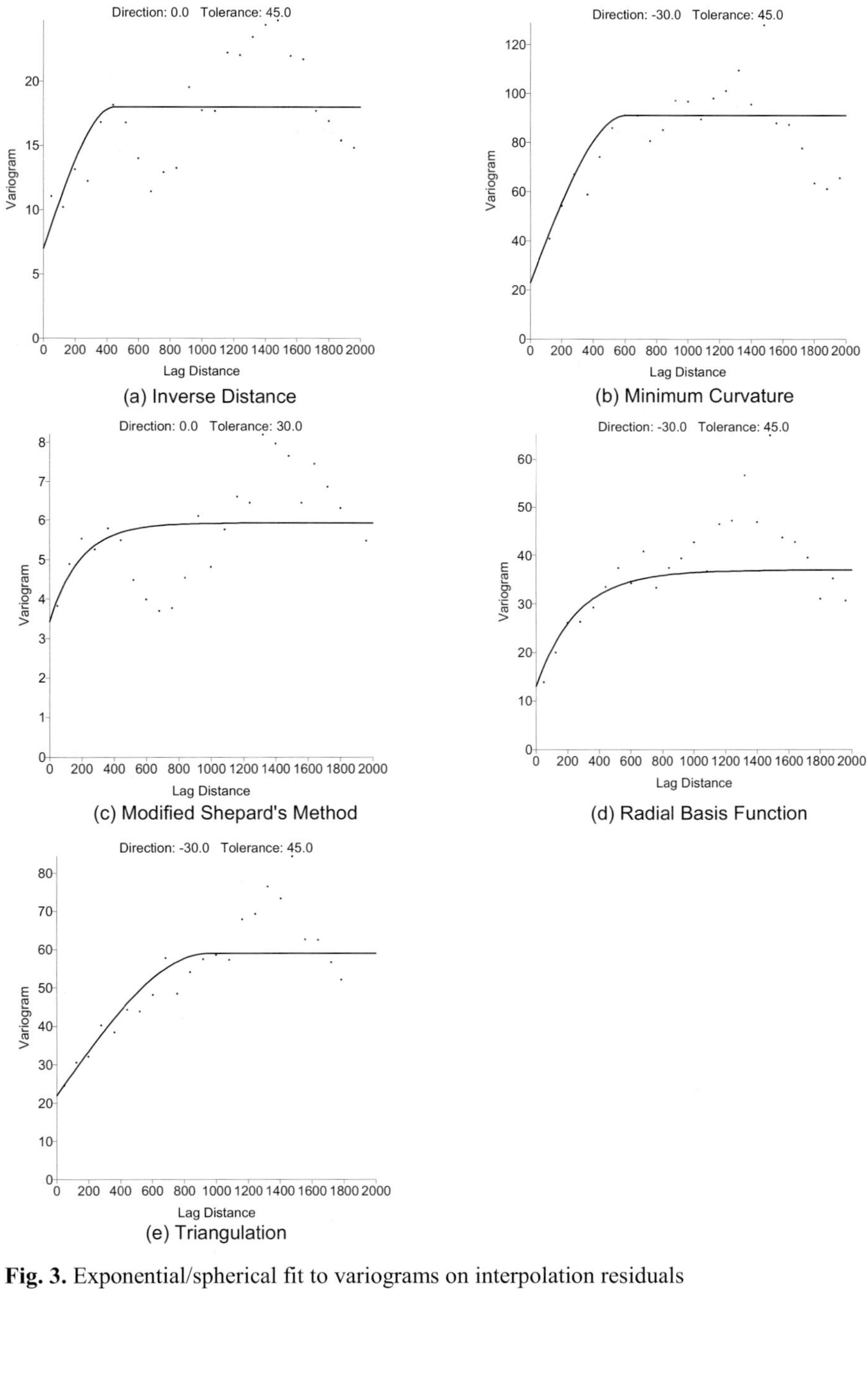

Fig. 3. Exponential/spherical fit to variograms on interpolation residuals

7 Conclusions

Table 5. Error budget for the DEMs based on 1:24 000 topographic maps

Error sources	Range of RMSE (meters)
Source map	1.85
Interpolation	
-Variation in interpolation algorithm	3.9448 ~ 9.5403
-Variation in input data density (50 ~ 400 points)	5.7960 ~ 7.8322
-Variation in grid resolution (10 ~100 meters)	2.2675 ~ 9.8154
Sampling and measurement	0.5645 ~ 1.4618
Total	4.3935 ~ 9.8226

This paper has addressed issues related to the uncertainty of DEMs based on the cartographic digitising method. It is suggested that the uncertainty of a DEM may be quantified by calculating a vector total of the RMSE from the source map, from the sampling and measurement error, and from the interpolation.

The RMSE of the source map may be computed as RMSE = 0.304 * contour interval, assuming a 90 percent rule for 1/2 contour interval errors. Both the positional and heighten accuracy of a topographic map are specified based on photogrammetric and surveying criteria, and they, in turn, define the contour interval of the map.

The determination of the uncertainty from the interpolation requires computing the value of residuals between the interpolated and the original elevations at tested points. The magnitude of uncertainty from the spatial interpolation is subject to many factors, but primarily to interpolation algorithm parameters and input data point density and distribution at a given grid resolution.

The RMSE alone is not sufficient for quantifying DEM uncertainty, because this measure rarely addresses the issue of distributional accuracy. To fully understand and quantify the DEM uncertainty, spatial accuracy measures, such as accuracy surfaces, indices for spatial autocorrelation, and variograms, should be used. The uncertainty is found concentrated in rugged terrain areas, although the five interpolators produce various spatial patterns of clustering. The degree of clustering is substantially high in all surfaces, as indicated by the Moran's index. Variogram modelling allows for quantitatively assessing the spatial variability of the residual data, and for calculating the RMSE that resulted from the sampling and measurement errors.

References

Ackermann F (1978) Experimental investigation into the accuracy of contouring through DTM. In: Proceedings of Digital Terrain Modelling Symposium. St. Louis, pp 165-192

Ackermann F (1994) Digital elevation models: techniques and applications, quality standards, development. In: ISPRS (ed) Proceedings of the Symposium on Mapping and Geographic Information Systems. University of Georgia, Athens, Georgia, pp 421-432

Ackermann F (1996) Techniques and strategies for DEM generation. In: Greve C (ed) Digital Photogrammetry: An Addendum to the Manual of Photogrammetry. American Society of Photogrammetry and Remote Sensing, Falls Church, VA, pp 135-141

Briggs IC (1974) Machine contouring using minimum curvature. Geophysics 39(1): 39-48

Carr JR (1995) Numerical Analysis for the Geological Sciences. Prentice Hall, Englewood Cliffs, NJ

Davis JC (1986) Statistics and Data Analysis in Geology (Second Edition). John Wiley and Sons, New York

Ebner H, Reiss P (1984) Experience with height interpolation by finite elements. Photogrammetric Engineering and Remote Sensing 50(2): 177-182

Eklundh L, Martensson U (1995) Rapid generation of digital elevation models from topographic maps. International Journal of Geographic Information Systems 9(3): 329-340

Fisher P (1999) Models of uncertainty in spatial data. In: Longley PA, Goodchild MF, Maguire DJ, Rhind DW (eds) Geographical Information Systems (Volume 1): Principles and Technical Issues (Second Edition). John Wiley and Sons, New York, pp 191-205

Gao J (1997) Resolution and accuracy of terrain representation by grid DEMs at a micro-scale. International Journal of Geographic Information Science 11(2): 199-212

Goodchild MF, Buttenfield BP, Wood J (1994) Introduction to visualizing data quality. In: Hearshaw HM, Unwin DJ (eds) Visualization in Geographic Information Systems. John Wiley and Sons, New York, pp 141-149

Griffith DA, Armhein CG (1991) Statistical Methods for Geographers. Prentice-Hall, Englewood Cliffs, NJ

Guth P (1992) Spatial analysis of DEM error. In: Proceedings of ASPRS/ACSM Annual Meeting. American Society of Photogrammetry and Remote Sensing, Washington DC, pp, 187-196

Hardy RL (1990) Theory and applications of the multiquadratic-biharmonic method. Computers and Mathematics with Applications 19: 163-208

Heuvelink G (1999) Propagation of error in spatial modeling with GIS. In: Longley PA, Goodchild MF, Maguire DJ, Rhind DW (eds) Geographical Information Systems (Volume 1): Principles and Technical Issues (Second Edition). John Wiley and Sons, New York, pp 207-217

Houlding SW (1994) 3D Geoscience Modeling: Computer Techniques for Geological Characterization. Springer-Verlag, New York

Hunter GJ, Caetano M, Goodchild, MF (1995) A methodology for reporting uncertainty in spatial database products. Journal of the Urban and Regional Information Systems Association 7: 11-21

Hutchinson MF, Gallant JC (1999) Representation of terrain. In: Longley PA, Goodchild MF, Maguire DJ, Rhind DW (eds) Geographical Information Systems (Volume 1): Principles and Technical Issues (Second Edition). John Wiley and Sons, New York, pp 105-124

Lam SN (1983) Spatial interpolation methods: a review. The American Cartographer 10 (2): 129-49

Li Z (1993a) Theoretical models of the accuracy of digital terrain models: An evaluation and some observations. Photogrammetric Record 14(82): 651-659

Li Z (1993b) Mathematical models of the accuracy of digital terrain model surfaces linearly constructed from square gridded data. Photogrammetric Record 14(82): 661-673

Li Z (1994) A comparative study of the accuracy of digital terrain models (DTMs) based on various data models. ISPRS Journal of Photogrammetry and Remote Sensing 49(1): 2-11

Monckton C (1994) An investigation into the spatial structure of error in digital elevation data. In: Innovations in GIS 1. Taylor and Francis, London, pp 201-211

Petrie G (1990) Photogrammetric methods of data acquisition for terrain modelling. In: Petrie G, Kennie TJM (eds) Terrain Modeling in Surveying and Engineering. Whittles Publishing Services, Caithness, pp 26-48

Shearer, JW (1990) The accuracy of digital terrain models. In: Petrie G, Kennie TJM (eds) Terrain Modeling in Surveying and Engineering. Whittles Publishing Services, Caithness, pp 315-336

Shepard D (1968) A two dimensional interpolation function for irregularly spaced data. In: Proceeding 23rd National Conference ACM. Brandon/Systems Press, Princeton, pp 517-523

Torlegard K, Ostman A, Lindgren R (1987) A comparative test of photogrammetrically sampled digital elevation models. In: Transactions of the Royal Institute of Technology, Photogrammetric Reports Nr 53, Sweden

Veregin H (1999) Data quality parameters. In: Longley PA, Goodchild MF, Maguire DJ, Rhind DW (eds) Geographical Information Systems (Volume 1): Principles and Technical Issues (Second Edition). John Wiley and Sons, New York, pp 177-189

Wang L (1990) Comparative Studies of Spatial Interpolation Accuracy. Master thesis, Department of Geography, University of Georgia

Weng Q (1998) Comparative assessment of spatial interpolation accuracy of elevation data. In: Proceedings of 1998 ACSM Annual Convention and Exhibition. Baltimore, Maryland

Weng Q (2001) A methodology for conceptualizing and quantifying uncertainty of Digital Elevation Models. In: Forer, P, Yeh A, He J (eds) Advances in GIS Research III: Towards Holistic Spatial Data Handling. Springer-Verlag, Berlin

Wood J (1996) The Geomorphological Charaterisation of Digital Elevation Models. Ph.D. thesis, Department of Geography, University of Leicester, Leicester, UK

Wood J, Fisher P (1993) Assessing interpolation accuracy in elevation models. IEEE Computer Graphics and Applications 13(2): 48-56

Wren AE (1975) Contouring and the contour map: a new perspective. Geographical Prospecting 23: 1-17

Detecting Building Alignments for Generalisation Purposes

Sidonie Christophe and Anne Ruas

Cogit Laboratory - IGN-France, 2 avenue Pasteur 94160 Saint Mandé France, anne.ruas@ign.fr

Abstract

Generalisation is well recognised as a complex process that should trigger specific algorithms, on different types of objects in some logical or appropriate order. To guide the process (where and how to generalise) one solution is to distinguish characterisation from the generalisation process. Characterisation aims at finding and describing relevant 'working areas' that can be a part of an object or a set of objects. As a result, the choice of an appropriate algorithm(s) becomes easier and can be constrained by the detected properties of this new entity. This paper presents a method to both detect and characterise building alignments in an effort to improve the use of generalisation algorithms namely typification and displacement. The first step consists of the identification of building alignments from straight-line templates. The second step characterises these alignments to retain only those that are perceptually regular. The characterisation is based on an analysis of the spatial location of buildings as well as on the properties of the buildings that belong to the alignment in question. To evaluate the regularity of the distribution, estimators are proposed for each property.. At the end a global quality estimator of the perceptual alignment- based on the aggregation of the estimators - is proposed. This global estimator is used to retain the best building alignments that will then be carefully generalised. The results presented have been implemented in the Lamps2 GIS software.

Keywords: pattern recognition, spatial Analysis, typification, generalisation

1 Introduction

1.1 Why Detect Building Patterns?

There are a number of approaches for automation of the generalisation process. For example one approach followed by the GIUZ laboratory of Zurich uses constrained algorithms in which internal conditions transform information while at the same time preserve some properties. Another approach which separates the task of identification/characterisation from the task of transformation has been investigated at the COGIT laboratory for a number of years. Detecting building patterns is a well recognised characterisation issue. The intent of this paper is to report on some of the findings related to the detection and characterisation of selected groups of buildings, , and then present some specific algorithms that effectively generalise these building groups. –

The generalisation of buildings concerns both the generalisation of each building - shape simplification, size dilation and local enlargement – and involves the contextual generalisation of a set of buildings - selection, typification, aggregation, displacement. Contextual generalisation usually occurs at a medium scale, whenever the main information to preserve no longer relates to each building but rather, illustrates that some buildings exist. The challenge is to preserve the appropriate building characteristics while at the same time reduce the quantity. The detection of building patterns involves finding a relevant group of buildings that have some particular spatial location of interest and then in identifying the properties that should be preserved during the process. The operation of 'reducing the number of buildings while preserving the properties of the group, is referred to as typification.

1.2 What Makes a Pattern?

When viewing a map, most people are able to conceptually group entities together. The action of grouping is usually based on some similarity in the criteria. Things that 'look the same' can be grouped together. This definition of perceptual groups comes from the Gestalt theory [Wertheimer 12]. A perceptible group consists of entities that have similar characteristics and do not exhibit too many differences. The separation between two groups generally occurs whenever there are sufficient inconsistencies in pattern regularity. For example, some are considered 'strong' group members when they have an important regularity of some criteria, or are considered not strong when irregularities or inconsistencies occur. Others can be considered as 'light' either because there are some important irregularities or because the regularities are not 'excellent'. All these qualitative and fuzzy definitions are intuitive to humans but are complex for computing purposes. Moreover, working on a classification rapidly illustrates that there is no universal partitioning of entities in the mathematical sense or in any consistent computational process. The grouping process is fuzzy in its definition and fuzzy in its borders.

The aim of this research has been to detect a particular case of building groups that can be referred to as building perceptual alignments. The geometry of the buildings used for testing is 2D polygons, with 1 meter of geometric resolution. A building pattern is a set of close buildings that have a 'well organised' spatial distribution and that have some geometric similarities (shape, size, and orientation). Consequently not all groups of buildings are appropriate for testing; preferable are those that are spatially organised and share perceptual similarities. Amongst the possible spatial distributions, this study focuses on alignment (aligned along a straight line) as follows;

1. It is a frequent pattern
2. It can be easily perceived and it structures the space
3. It needs to be explicitly preserved during generalisation.

This latter point is relevant since the number of operations tend to remove contextual regularity (such as displacement and object removal). Other objects such as roads, have not been integrated, as the detection of patterns was to be based on a single object category. Further, the approach was to study the relationship of buildings with roads after the identification of perceptual alignments. Roads were then subsequently considered as an object resource with which to reinforce the process. Thus, in summary, the proposed mechanism is to look for buildings that are close to one another and that are aligned, in a regular manner i.e. their distance is regular or consistent along a line. Additionally they should exhibit similarities in terms of shape, size and orientation.
A group of aligned buildings (i.e. building alignment) is therefore considered a prerequisite condition but is not all encompassing vis-à-vis 'building perceptual alignment'. The term 'perceptual' implies the regularity of the alignment. According to these definitions, however, it is clear that a perfect perceptual alignment does not exist. As a result, it is necessary to allow some flexibility in the definition of regularity and to try to quantify it as much as possible. The difficulty of such an exercise involves aspects about resolving the thresholds (i.e. what is the limit between nearly regular and not regular) and determining the aggregation of independent criteria (e.g. size and shape) that characterise the global 'quality' of the building perceptual alignment.

1.3 Previous Work in the Generalisation Field

Amongst previous related research, work has been undertaken in both the area of generalisation and pattern recognition. In these areas, works that compliment this particular research endeavour can be categorised into two sets and are as follows

- Research on detection of building structures for building typification;
  - [Hangouët 98] detects buildings aligned along roads,
  - [Regnauld 98] detects close buildings organised along a graph using Minimum Spanning tree (MST),

- [Boffet and Rocca-Serra 01] detect and aggregate triplets of buildings to constitute building alignment using a bottom-up approach,

- Research on building typification;
  - [Sester 00] randomly removes some buildings and relocates the remaining ones as regularly as possible by means of attraction forces computed in an iterative way using neural networks,
  - [Ruas 99] groups buildings by means of a network (Delaunay triangulation) and progressively removes the 'worst building' by means of a cost function based on size, density and directional proximity.

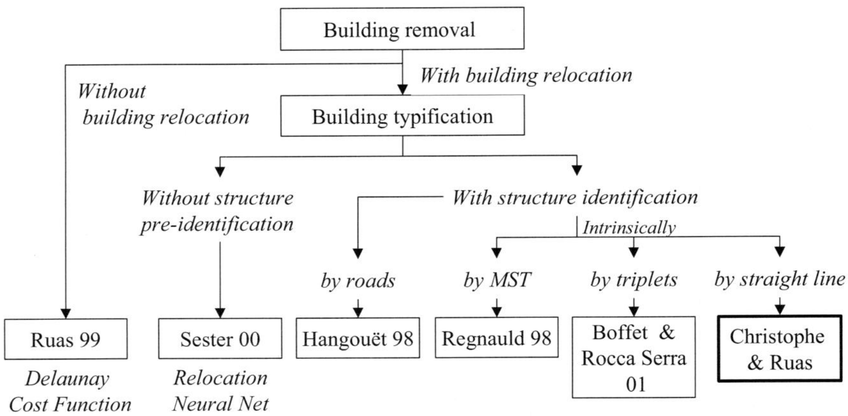

Fig. 1. Some existing methods for building removal

The method proposed here builds on a selection of work, see [Regnauld 98] and [Boffet and Rocca-Serra 01].. The challenge is to propose a method less complex than [Regnauld 98] (i.e. without requiring the computation of MST) but more robust than [Boffet and Rocca-Serra 01].

1.4 How to Detect Building Perceptual Alignments?

To detect a straight line alignment pattern, one can either group objects by means of a spatial structure such as MST, Delaunay or Voronoï, and determine if straight alignments exist within these linked objects, or one can use the straight line to directly group the objects. This latter method has been chosen as it provides hypothetical straight line alignments without requiring the development of complex spatial structure analysis tools.

The proposed method is divided into two steps:

1. The first step (section 2) aims at identifying aligned buildings within an urban block (a close cycle of streets). The result is a set of aligned buildings that are potential candidates for becoming building perceptual alignment. At this stage, the distance between buildings may or may not be regular.

2. The second step (section 3) qualifies the potential candidates according to their various regularities; eliminates the ones that are clearly not regular and, quantifies the quality of the retained candidates.

Section 2 and 3 below address each of these steps. The implementation has been achieved using the GIS Lamps2 of Laser Scan.

2 The Detection of Straight Line Building Alignment

This section presents the methodology used and identifies straight line alignments of buildings. A top-down method has been developed. Straight lines are predefined patterns used to detect the alignment. Starting from a geometric description of alignment, we determine which buildings are concerned by their structure (2.1). Further, the structures identified are assessed and the incomplete ones are recomposed (2.2). The intent is to use an exhaustive search of all alignment possibilities by means of a one-line template that takes 'all' orientations from a single anchor point, located on the xmin and ymin of the urban block.

2.1 The Detection of Potential Alignments Based on Projection

It is worth noting that the orthogonal projection of aligned points gives close projected-points on a straight line of projection (Fig. 2). As such, the centroid of all buildings of the urban block is projected on an 'exhaustive' set of straight lines. Then on each line, groups of very close projected-points are searched. These groups of points represent the potential alignments. The relation between a projected point and its building is maintained throughout the process: thus, a projected point is linked to one single building.

Questions raised by this choice are:

- Which straight lines are to be used for this projection?
- How should the distribution of projected-points be assessed?
- How should the point clusters be filtered to detect the aligned buildings?

2.1.1 Straight Lines Choice

To find all the potential candidates, each urban block is assessed using a set of straight lines defined by an angle α and an anchor-point. The angle α varies from 0 to 180 degrees with a 1 degree step. The complete scan highlights all alignments.

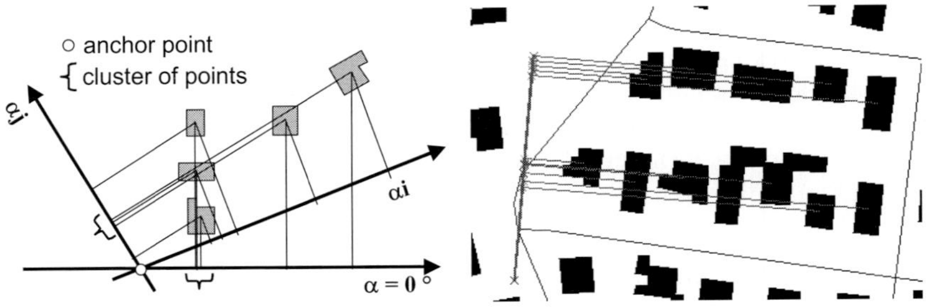

Fig. 2. Straight line assessment of an urban area

2.1.2 Study of Distribution of Projected-Points on a Straight Line α

Each straight line holds a set of projected points amongst which some clusters may exist. To detect the clusters a simple method has been chosen that consists of gathering the projected-points together in equal size packs from the anchor-point and assessing the proximity within each pack. The greatest distance between two successive points in each pack - called *δmax* - represents the proximity inside the pack. Other methods could also have been used for this step such as an ascending hierarchical classification method.

The size of the packs refers to the number of successive points per pack. An alignment should likely be composed of at least 4 buildings. As alignments often involve a limited number of buildings, we have experimentally chosen to study structures of 4, 5 and 6 points. Consequently, for a single straight line, three collections of packs are generated corresponding to size 4, 5 and 6.

2.1.3 Selection of the Best Packs

At this stage, there are 180 * 3 collections of packs to assess ($line_\alpha$ = (coll4 {$pack_i$, δmax_i}, coll5 {$pack_j$, δmax_j}, coll6 {$pack_k$, δmax_k})). Most of these do not present any alignment. To filter the best lines, the minimum value of δmax - *δmax min* - per collection and per line is computed. This value represents the "efficiency" of the line to identify the clusters. It is used to order the line by efficiency. Experimentally, it is apparent that all lines and collections that have a *δmax min* bigger than 3 meters can be rejected.

The same alignment, however, can still occur in different packs:

- An alignment of 6 buildings will be described for the same α, 3 times in a collection of 4 points, 2 times in the collection of 5 points and 1 time in the collection of 6 points
- An alignment may be described with successive α values.

Based on point's identifier, a first filter is done for each line to retain only the longest pack representing the same points. A second filtering - based on the buildings identifier- is done between lines to reject redundant alignments. We will see later on that the alignments are not perfect and at this stage are considered only a 'potential alignment'. To view alignments, a set of segments between the centroid of the consecutive buildings is created (Fig. 3, left). To find a better geometry of the straight-line alignment, the adjustment of the alignment is measured using a classical regression between centroids. A set of alignments of different sizes is obtained (Fig. 3, right).

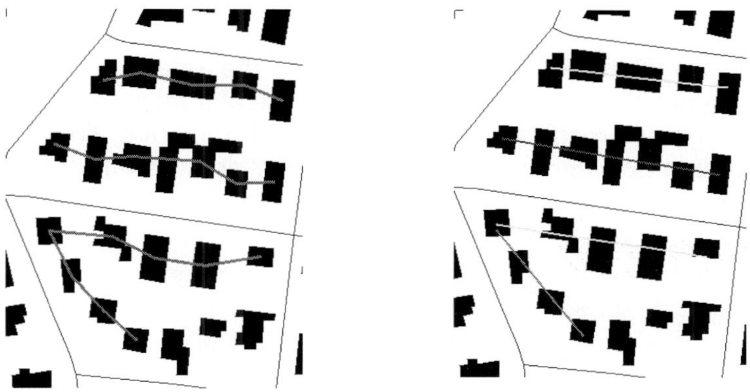

Fig. 3. Adjustment of alignments with regression (black :6, white:5, gray:4)

At the end of this step some overlapping or incompleteness still exists and should be corrected.

2.2 The Filtering and Composition of Alignments

The first step determined that a number of potential alignments are incomplete. As such the potential alignments and filter overlapping must be extended.

Only small alignments (maximum 6 buildings) are detected but larger ones have to be detected in their entire size (case 1). Alignments are extended to catch buildings using an iterative buffer at the extremity (of 10m) (Fig. 4). Moreover some buildings fall 'under' alignment, but are not actually a component of it (case 2). This can result either from the packs filtering or from a certain deviation of the centroid. Those buildings *are* caught by a buffer of 10m on both sides of the alignment (Fig. 4).

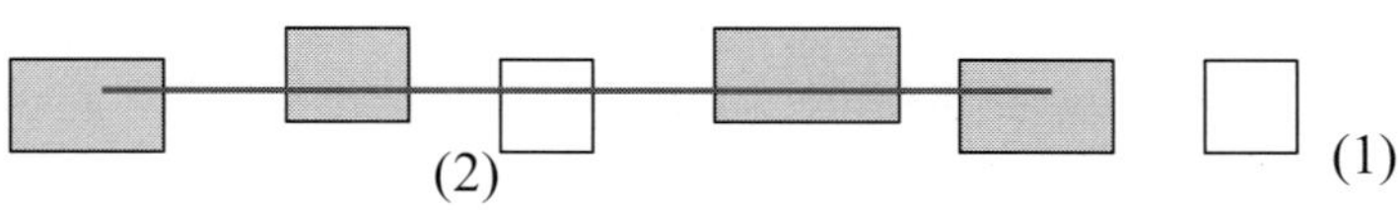

Fig. 4. Situation of alignments: recuperation and extension

After the extension of alignments some alignments could not be merged (Fig. 5): if they have the same general orientation, they are merged and the adjustment of the alignment geometry is re-computed.

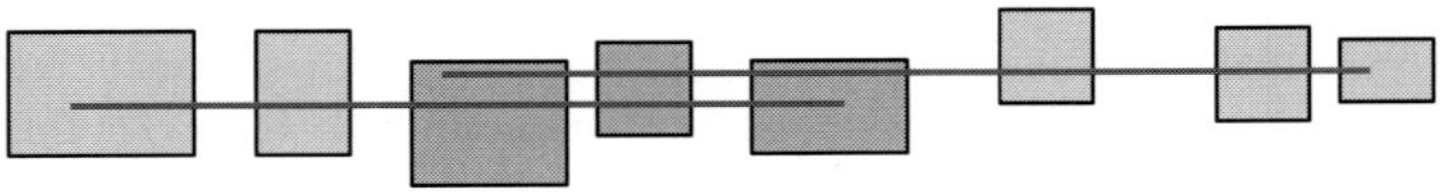

Fig. 5. Situation of alignments: merging

All the identified alignments are added to the database. The "alignment" class is created with two particular attributes: the slope of the regression and the list of component-buildings. The association between one object 'alignment' and the object 'buildings' is a composition relationship (cardinality 1:n).

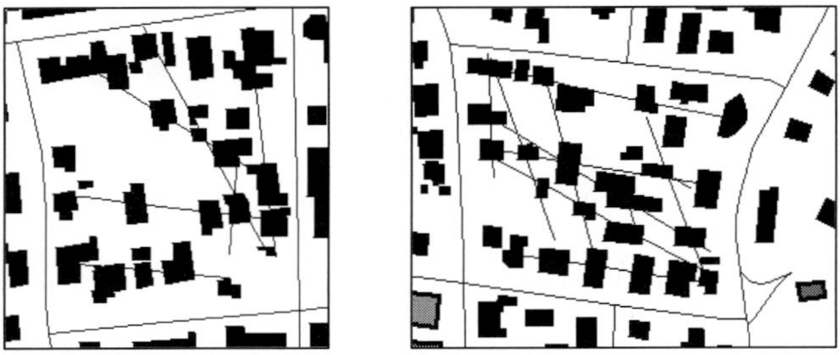

Fig. 6. Examples of results on two different urban blocks (in 0-90° only)

In Fig. 6 it's apparent that an alignment is not always a perceptual structure since the inter-distances are large and irregular and/or because the buildings that belong to the alignments are different one from another. The alignments now have to be characterised to become building perceptual alignments.

3 From Alignments to Perceptual Alignments

The detected alignments must be qualified and ranked according to their importance. Some alignments will be rejected while others will be preserved and qualified for the typification process. Some visual criteria, based on principles of perception, are defined to describe potential alignments (3.1). The *selection of the "best" alignments* is based on regularity of all perceptual criteria. Results and required improvements are proposed in section 4.

3.1 The Criteria of Characterisation

Perceptual criteria are used to select and to describe the alignments. Two kinds of criteria were selected:

- The position of objects (which represent the quality of the pattern); in other words, the proximity and the alignment of their sides and centre.
- The similarity between objects (which represent the quality of the component of the pattern) and refers to the shape, the size and the orientation of buildings.

At this stage measures are not aggregated, they are simply listed.

3.1.1 The Position of Objects

Proximity: The proximity which is the projection (minimal distance) between the boundary of two consecutive buildings is computed (Fig. 7, left).

Arrangement of objects: Qualities of the sides of alignment and alignment of centres are computed by using angle variation.

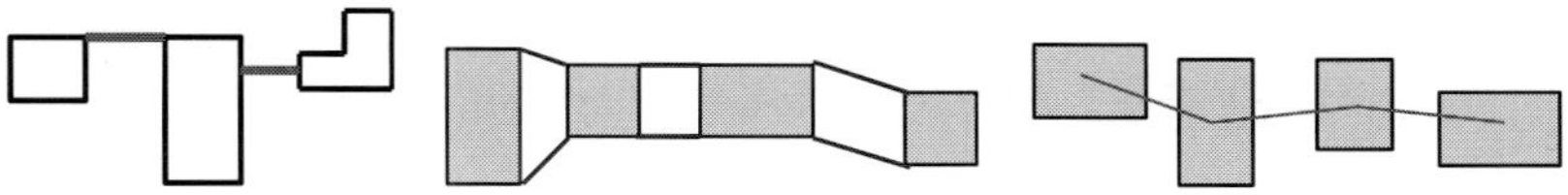

Fig. 7. Measures of proximity,alignment of sides, and, alignment of centres

3.1.2 The Similarity of Objects

Size: The size is the area of a building.

Shape: Refers to the concavity of buildings, using the comparison of the area and the area of the convex hull (concavity = 1 - area/area (convex-hull)). Other measures could be added such as elongation ratio.

Orientation: Various definitions have been described in [Regnauld 98] such as the general orientation and the orientation of walls. The general orientation is measured by the average between the two longest segments between the points of the border (Fig. 8 left). This method gives approximate results as follows: first for squared buildings, the mean is proves insignificant second, for very irregular buildings, the results are invalid.

As a result, wall orientation, which is the mean of the orientation of the wall in [0, pi/2], weighted by the length of each wall has been chosen. Another orientation of interest, is the perpendicular orientation (Fig. 8 right).

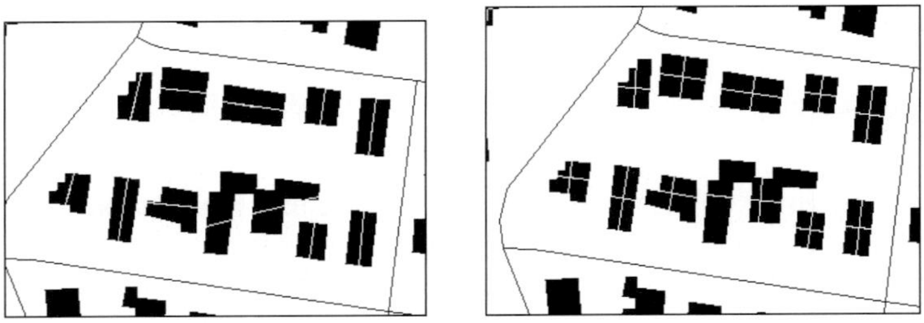

Fig. 8. Measure of general orientation and orientation of walls

The orientation of walls supports a more effective comparison of buildings and is more suitable than a general orientation.

3.2 Aggregation of Criteria Necessary to Define Pattern Quality

The detected alignments must now be qualified according to perceptual criteria. Each pattern is characterised by a distribution of values concerning proximity, arrangements, size, shape and orientation of the buildings. Each criterion is perceptually 'good' if the values are close one to another. The perception is thus based on the regularity of values.

An estimator of homogeneity needs to be developed for each criterion and then aggregated to a global one, for each significant pattern. Assuming that a scale of comparable values is required, a normalisation must be done. The classical normalisation of values prior to computing the standard deviation ($X_i = (\max(x_j) - x_i) / \max(x_j)$) did not yield appropriate results.

3.2.1 Estimator of Homogeneity for Each Criterion

The measure of regularity is based on the standard deviation. The mean of the values has not been used since there are only a few objects and thus, the influence of any exceptions would be too high. Consequently, in the standard deviation, the mean has been substituted with the median, which in this case, is more appropriate. Several estimators E, were then developed, one per criterion, $\sigma' = \sqrt{(1/n * \Sigma_i (x_i - med)^2)}$:

When the value of x_i is larger than 1 (area, distance), we divide the standard deviation by the median. The estimator values, however, should be comparable. If 0.5 is perceived as medium for one property, it should be perceived as medium for the others. Consequently, each estimator has to be adapted to reflect the homogeneity of each criterion:

- Arrangement, orientation and shape, $E = \sigma'$
- *Proximity*, $E_p = \sigma' / median$
- Area, $E_a = \sigma'' / median$, with $\sigma'' = \sqrt{(1/n * \Sigma_i (x_i - med))}$. The quadratic value placed too much emphasis on large areas.

The standard deviation only computes the regularity of the distance and not the distance itself. Consequently, if the median for proximity is superior to 10m, some values are added to the test value to penalise it (E_p can be higher than 1).

We have a set of estimators E_i defining each criterion of the structure. The quantitative values given by each estimator are experimentally tested and validated. To clarify the results quantitative values are transformed into qualitative values. If the values of E_i are superior to 1 it is “very bad”, between 1 and 0,7 it is "bad", between 0,7 and 0,5 it is "medium", between 0,5 and 0,25 it is "good", under 0,25 it is "very good". (see Fig. 9)

Finally, a global estimator GE is proposed that defines the global quality of the pattern, measured by $1/n*\Sigma_i E_i^2$. The use of the square enforces the influence of small values (i.e. regular) and high values; a value > 1, i.e. not regular at all will correspond to what is required. Experimentally, we saw that if GE < 0.25, the alignment of buildings was perceptually good. Consequently, among the alignments defined in section 2, only those having a GE value under this threshold are considered as being a 'building perceptual alignment'. Therefore, $0{,}25= 0{,}5^2$, corresponds to the limit between good and medium values.

3.2.2 Accounting for the Number of Buildings

These preliminary experiments were undertaken with samples of between 4 and 15 buildings. The more elongated the pattern, the less regular it might be. The number of objects could be included in the final cost function but this requires further assessment.

3.3 Results and Comments

A methodology has been presented to detect and qualify building perceptual alignments. Some comments on the limitations and necessary improvements are briefly outlined.

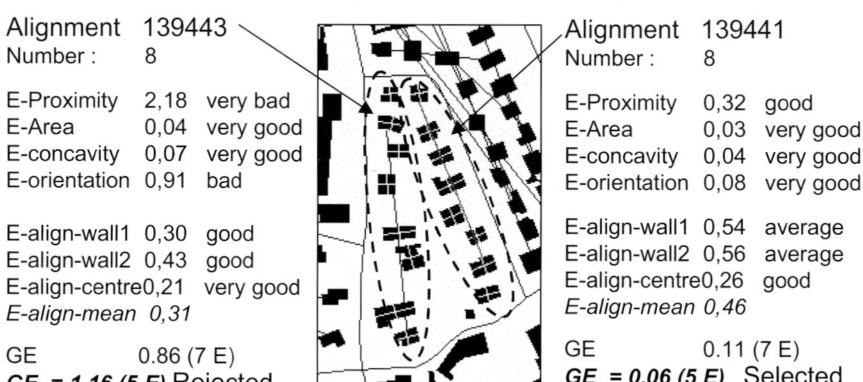

Fig. 9. Visual results : example 1

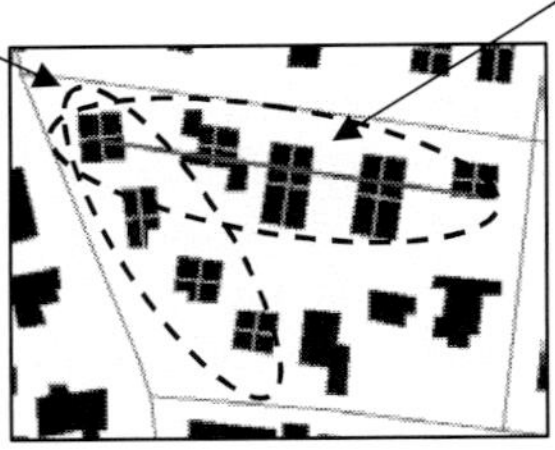

Fig. 10. Visual results example 2

Without having addressed optimisation concerns, we observe that some estimators E_i, are still inadequate. For example,

- The Area and Concavity estimator values are always too small. Thus, the normalisation should be reconsidered.
- The estimator of alignment (E-align) based on the variation of angles that connects close buildings is not as strict as a visual inspection might require, (e.g. Fig. 9, alignment 139443). Indeed the visual perception of alignment depends not only on angles but also on distances. A solution to improve E-align lies in replacing α by sinus α. Moreover; a combination of the three indicators should be developed. Taking the three indicators (7E) puts too much weight on alignment, while taking the average (5E) may be not be the best solution.

Even if the visual results are acceptable, some more tests are needed to tune these functions. The difficulty is partly related to the fact that statistical measurements have been used for only a limited number of objects.

The comparison between the orientation of the buildings and the orientation of the alignment, as well as the comparison between the orientation of the alignment and the orientation of the closest road would be useful information to add.

The aggregation of estimators for the global qualification of the perceptual alignment also needs improvements. In the above assessment, all estimators are equally weighted and an average is computed. The only exception is the distance estimator, which combines the regularity and the value of the distance. The aggregation of two concepts (regularity and value) is certainly not optimal. An improvement might be achieved using the following rule: the smallest the object size, the smallest the acceptable distance.

The detection of alignment (section 2) could include some buildings that may decrease the quality of the global structure. Therefore, these structures might be rejected during the selection step (section 3). Whenever a large structure is removed, consideration could be given to the effect it has on decreasing the global quality.

4 The Use of Perceptual Alignments for Generalisation

The aim of this research was to detect building perceptual alignments for generalisation purposes. This generalisation process has not yet been implemented but an explanation of how these structures will be used is provided below:

- For building typification, minimum size and minimum interdistance, are used to find the number for required buildings. The appropriate number of buildings should be removed (either the less regular or the redundant ones) and the final alignment should be rearranged while maintaining the global characteristics of the structure.
- In case of an excessive building density, if the building block is too thin and if there are two parallel building alignment structures (which is often the case as there are two road parallel roads), we could preserve the best one and remove the buildings of the other alignment..
- Last but not least, perceptual alignments should be useful for the building displacement process. New algorithms such as [Bader 01] perform displacements based on various strengths between objects. In case of the displacement of buildings that belong to a perceptual alignment, special strengths should be introduced to preserve the alignments as much as possible.

5 Conclusion

In this paper, we present research on detecting and qualifying building perceptual alignments. These types of patterns have been chosen as they are frequently found in urban areas and are necessary elements in cartographic presentation and generalisation activities, such as typification and displacement. The idea is to detect the structures and to propose new and simpler algorithms that would better maintain geographical patterns. The proposed method is a top down detection approach as it starts with predefined straight-line templates and seeks buildings that participate in these structures. The detection phase gives acceptable results. The characterisation and selection phase should still be improved (see 3.3) even if the results are already reasonable sound. The concept of estimator functions requires further work and fine-tuning. The next step is to build new generalisation algorithms that will, as a result, be easier to develop as it is anticipated that the pattern derivation will resolve a number of the problem issues.

References

Bader M (2001) Energy Minimization Methods for Feature Displacement in Map generalization. PhD thesis, University of Zurich, Suisse

Boffet A, Rocca Serra S (2001) Identification of spatial structures within urban blocks for town qualification. In: Proceedings XX International Cartographic Conference, Vol 3. 6-10 August, 2001, Beijing, China, pp 1974-1983

Hangouët JF (1998) Approche et méthodes pour l'automatisation de la généralisation cartographique ; application en bord de ville. Ph.D. thesis, University of Marne-La-Vallée France[1]

Regnauld N (1998) Généralisation du bâti: Structure spatiale de type graphe et représentation cartographique. Ph.D. thesis, University of Provence – Aix-Marseille 1 [1]

Ruas A (1999) Modèle de généralisation de données géographiques à base de contraintes et d'autonomie. Ph.D. thesis, University of Marne-La-Vallée France[1]

Sester M (2000) Generalization based of least squared adjustments. In: Proceedings International Archives of Photogrammetry and Remote Sensing, 16-23 July 2000, Amsterdam, The Netherlands, 23: 931-938

Wertheimer M (1912) Über das Denken der Naturvölker: I. Zahlen und Zahlgebilde. Zeitschrift für Psychologie, 60:321-378. Reprinted in Drei Abhandlungen zur Gestalttheorie, 1925

[1] See ftp://ftp.ign.fr/ign/COGIT/THESES/

Temporal GIS and Statistical Modelling of Personal Lifelines

Marius Thériault[1], Christophe Claramunt[2], Anne-Marie Séguin[3] and Paul Villeneuve[1]

Centre de recherche en aménagement et développement, Université Laval, Québec, CANADA G1K 7P4, tel. (418) 656-7558, fax (418) 656-2018, {Marius.Theriault, Paul.Villeneuve}@crad.ulaval.ca[1]
Institut de recherche de l'École navale, BP 600, 29240 Brest Naval, FRANCE, Claramunt@ecole-navale.fr[2]
INRS Urbanisation, Culture et Société, Université du Québec, 3465 rue Durocher, Montréal, CANADA H2X 2C6 ; tel. (514) 499-4047, Anne-Marie_Seguin@inrs-urb.uquebec.ca[3]

Abstract

Spatio-temporal data models for describing complex lifelines and trajectories of persons, as well as events that affect their evolution, are prerequisites for the statistical analysis of their relationships. Such analyses are useful to develop a better understanding of urban dynamics and social transformations. This paper develops a spatio-temporal database model for handling personal trajectories along a time line (many complementary lifelines) allowing for the statistical analysis of any pre-defined event. It combines survival analysis, Cox regression and temporal GIS. The combination of these aspects support an assessment of the likelihood of any event to occur in the life of persons at risk, after a given time delay and under some specific conditions. Our model was implemented and tested using a geo-relational approach that also supports spatial and temporal reasoning at complementary levels of abstraction. It allows the cross-analysis of several multi-dimensional lifelines to form individual trajectories. The application example is based on an historical survey of personal biographies (spatially located) of 418 professional workers living in the Quebec Metropolitan Area in 1995-96.

Keywords: temporal GIS, multi-dimensional lifelines, event history analysis, individual trajectories

1 Urban Dynamics, Temporal GIS and Event History Analysis

Modelling the evolution of urban systems implies the analysis of human activities at different levels of abstraction along their temporal and spatial dimensions. Understanding persons' behaviour in a dynamic perspective and the constraints of their environment can provide clues about processes responsible for the evolution of cities. Such analysis, however, clearly implies integration of several multi-dimensional episodes describing the behaviour (biography) of individuals. Within urban studies, space-time representations of individual trajectories have been widely studied using time-geography modelling and analysis concepts following Hägerstrand's seminal work (1967). Other investigations have explored time-geography principles to analyse the multi-dimensional structure of spatial behaviour of people involved in specific activities (Taylor and Parkes, 1975; Goodchild and Janelle, 1984; Miller, 1991; Odland and Shumway, 1993; Whiters, 1997; Janelle *et al.*, 1998; Johnson, 2001).

During the last decade, important progress was made in integrating the time dimension within GIS (Langran, 1989; Peuquet, 1994; Frank, 1994; Peuquet and Quian, 1996; Böehlen *et al.*, 1999; Hornsby and Egenhofer, 2000). The statistical analysis of spatio-temporal processes, however, is still lagging far behind the development of GIS technology. Some of the temporal GIS models identified so far are based on an explicit description of spatial evolution with respect to change in status for identifiable entities (Claramunt and Theriault, 1996; Hornsby and Egenhofer, 1998 and 2000). These approaches aim at the modelling of transition patterns among events in an effort to identify recurring spatio-temporal processes. They provide an explicit description of changes in the geographic phenomena modelled. Analysis of basic transition primitives and their combination to record geographical change can reveal patterns that could further our understanding of evolution processes. Such an approach, requires exploratory analysis of successive events. To the best of our knowledge, database models are not currently linked to existing statistical analysis of successive events, survival analysis, and modelling of choice processes and therefore not ideally suited to theses types of assessments. In fact, because of the complexity of their data structures, they only allow computation of some basic descriptive statistics on transitions among successive states (E.g. mean values and frequencies).

On the other hand, the literature is replete with existing statistical methods for modelling discrete choices and for performing temporal analysis of state transitions (logistic regression, log-linear models, survival analysis using Kaplan-Meier techniques, Cox regression). These statistical approaches are linked to probability theory. Making assumptions on the error distribution of observations, they provide hypothesis tests based on the comparison of actual events to their theoretical distribution, in space and time. Thus, among a set of events, they could identify those which are unlikely to appear at random in the actual spatio-temporal configuration. Adding the power of inferential statistics to GIS applications would clearly enhance their usefulness for urban studies. It will allow scientists to assess the effect of data sampling procedures on the accuracy of simulation results. As well, it will provide strong guidelines to distinguish patterns of events which are significant from those which are not (Pötter and Blossfeld, 2001).

Event history modelling is a specific type of longitudinal statistical analysis that focuses on the survival rate of a given status and considers various attributes of the observed individuals. It is used to estimate the probability of an event to occur, considering the time elapsed after some condition is met (Blossfeld and Rohwer, 1995; Blossfeld, 1996; Le Bourdais and Marcil-Gratton, 1998). For example, event history analysis (Cox, 1972) can compute the likelihood of a woman to give birth to a child, after she gets married (or she starts a cohabiting union), taking into consideration her age and her income. This likelihood is the complement of her survival function for not having a child under the same conditions.

Survival analysis uses Kaplan-Meier (product-limit) technique to estimate the length of time elapsed before occurrence of an event after a change in personal status enables it to occur. Using observed duration of elapsed time, event, survival analysis computes the base line hazard function of an event to occur. However, at survey time, the event would not have occurred for some enabled individuals. These are then called censored. Because censored cases were at risk, they must be considered when estimating the proportion of individuals experiencing the event. Therefore, one cannot estimate hazard using only those objects experiencing a given event. The stochastic part of the phenomenon must also be considered explicitly to estimate the odds ratio (number of individuals experiencing the event divided by those who do not). It must be accumulated over time to model hazard evolution. When the event occurs for a person, he is generally no longer at risk. Thus, the number of persons at risk changes over time while hazard decreases or increases, depending on the nature of the event (E.g., death and marriage). Moreover, the chance that an event occurs depends partly on the attributes of each individual and the context in which he/she lives. For example, risk of death increases with age of the person, someone living in dangerous conditions is also more at risk. The marginal effect of the local context on hazard assessment (decrease in survival) is modelled using time regression based on exponential mathematical models: $h(t) = e^{(\beta_0 + \alpha \ln t)} e^{(\beta_1 X_1 + \beta_2 X_2 + \ldots + \beta_n X_n)}$. In this equation, hazard is modelled using two terms. The first part estimates the base hazard using time elapsed after enabling occurred (natural logarithm of t). The second one is specified using any number of independent variables (X) providing values specific to each observation. Coefficients are computed for observed data using maximum likelihood techniques. This defines the basis of Cox regression techniques (Cox, 1972) available in most statistical software, like SPSS.

Unfortunately, SPSS, like all other statistical software, can manipulate only flat files. It allocates columns to variables and rows to observed cases. For most applications in time regression (and survival analysis), the task of restructuring information to build this flat file is time-consuming and prone to errors. Moreover, computation of many attributes depends on the choice of reference events (enabling conditions and modelled event). While it is certainly feasible to integrate spatial context in survival analysis (even time-varying context), to the best of our knowledge, there is no existing example of such an application. This requires integration of data models that defines events, their ordering and semantics within a GIS environment. In addition, a data manipulation language that allows for retrieval of transition sequences among events is required. We believe that such an improvement would open new frontiers for the application of GIS in urban studies.

The objective of this paper is to present a GIS modelling strategy to efficiently generate event history tables compatible with most statistical software. The procedure is illustrated and tested using an historical survey of 418 professionals living in the Quebec Metropolitan Area during 1995-96. The purpose of this survey was to collect detailed information about all significant events that occurred during their adult life (residential, familial and professional trajectories), with specific references to space and time. The following sections will discuss some lifeline modelling principles needed for this project (Section 2) and the linkage to temporal GIS (Section 3). Section 4 presents a geo-relational extension of our model. Section 5 discusses the application of the survey and discusses some preliminary results. Finally, the conclusion assesses the overall efficiency of the procedure and outlines further work.

2 Lifeline Modelling Principles

Modelling individual trajectories within GIS might offer a better data support for the development of urban land-use and transportation models. An important achievement of recent urban modelling is the integration of the decision-making behaviours of urban actors. These include activity location and travel decisions which are intricately linked with household structures and professional profiles of persons (Hunt and Simmonds, 1992). Information about these contextual attributes is needed for temporal regression analysis (Ben-Akiva and Lerman, 1985) or computer-intensive micro-simulation. Temporal regression is also required for modelling complex systems, which integrate decision rules of many interacting individuals.

During the last decades, individual trajectories generated patterns of events of increasing complexity. This is linked to many factors such as an escalation of divorces, extension of contractual short-term employment and increasing geographical mobility. This trend is highly related to the economic restructuring occurring in most countries since the mid seventies (Rose and Villeneuve, 1993; Séguin, 1994). Within cities, individual trajectories aggregate to yield demographic, professional and residential patterns that can be observed using census data. However, the processes by which personal biographies aggregate to form macro patterns cannot be derived from censuses. The latter give only the barest snapshot reports on complex situations (Thériault *et al.*, 1999).

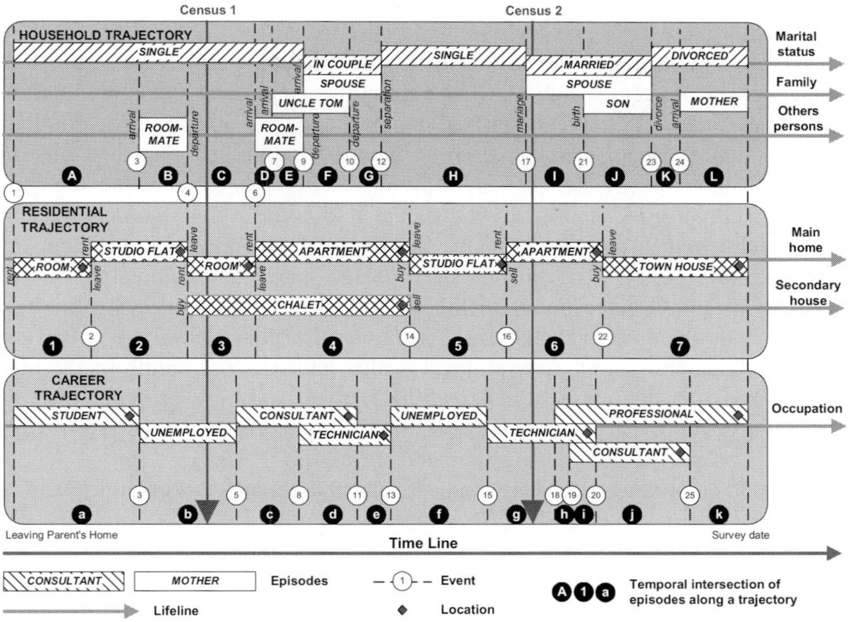

Fig. 1. Schematic example of an individual history

In our database schema, an individual history (Fig. 1) is formed by a set of complementary lifelines defining three trajectories (household, residential and professional career). An individual's history is altered when an event occurs modifying one of his lifelines, or in other words, the episodes of one of his lifelines. Let us remark that such an event may alter - simultaneously or afterwards - several episodes of many lifelines, leading to interdependent episodes. For example, finding employment could be a preliminary condition for renting an apartment. Other events can be anticipated and lead to prior adjustment: a household's move may anticipate a new child birth. Therefore, trajectories show interlocked evolution based on personal behavioural strategies. They are formed of partially or totally ordered episodes. Moreover, each of these trajectories is somewhat different in structure. A household composition changes with arrival or departure of any member. The residential trajectory depicts the succession of living places. The career pathway may describe successive mixes of potentially simultaneous occupational activities (Thériault *et al.*, 1999).

An individual's lifeline is defined as a set of time-stamped episodes along one of his life dimensions (E.g. marital status, having children). These lifelines are described using episodes and events. The global status of an individual for an instant of time can be derived from an aggregation of his different lifelines. A lifeline has a form that is either linear or multi-linear. For example, some individuals may hold several jobs or residential locations simultaneously. Lifeline episodes constitute logical sequences related to life cycles. Studying these evolution (or decision) patterns is certainly more relevant for urban planning than knowing the exact timing of events for each individual.

Fig. 1 illustrates a complex mix of real world phenomena (persons, trajectories, lifelines, episodes and events). All these notions, including the individuals and trajectories themselves, are integrated in our model. The first basic modelling concept is the episode, which is used to describe individuals and their historical properties. An episode denotes an individual's homogeneous value along one modelling dimension, that is, a lifeline. An individual's history is composed of many lifelines that include partially ordered time-stamped (or not) episodes, using temporal periods. An episode shows some status that endures over a given period of time and corresponds to the individuals lifeline. Following Snodgrass (1999), we define a period as an anchored interval of time delimited by two time-stamps (begins and ends). A time interval also indicates the length of the period (its duration). In the sense of Allen (1984), an episode - a property in Allen's terminology - is divisible. Therefore, if an episode holds over a period of time *i*, then all its properties hold during any subinterval of *i*.

A second important modelling concept relates to events. An event models one-to-many changes of episodes from one-to-many individuals' lifelines. Each episode is bounded by two events (begins and ends), potentially building multidimensional networks of ordered successive episodes or events. An event is time-stamped by either an instant or a period. An event using a period of time is indivisible. For example, if a marriage occurred in June 1999 (time granularity of a month), we cannot derive that this marriage happened on the 15th of June 1999.

For prototyping purposes, our modelling concepts are specified using a relational database approach. However, these modelling concepts are general enough to be applied to other database models such as object-oriented models. Two relations constitute the main components of this database: events and episodes (Fig. 2). They are described as follows:

An **episode** is defined as a tuple Episode (EpisodeId, EpisodeName, LifeLineDimension, SetOfIndividualId, SetOfRelatedIndividualId, EpisodeTime, QuestTime, EpisodeSpatial). EpisodeId uniquely identifies each episode. EpisodeName represents and names the class of episodes represented along a lifeline (E.g. single or married). LifeLineDimension indicates the dimension modelled by this lifeline (E.g. individual's life, marital status, and occupation). SetOfIndividualId keeps the list of individuals to whom this episode belongs (E.g. the person answering the questionnaire, and his/her relatives actively involved in this episode). SetOfRelatedIndividualId lists the other persons linked to this episode (E.g. members of an individual's household playing only a passive role). EpisodeTime models the temporal period over which this episode is valid. EpisodeSpatial indicates the spatial location of this episode. QuestTime keeps the date when the questionnaire was completed (important for surveys including persons answering at different times).

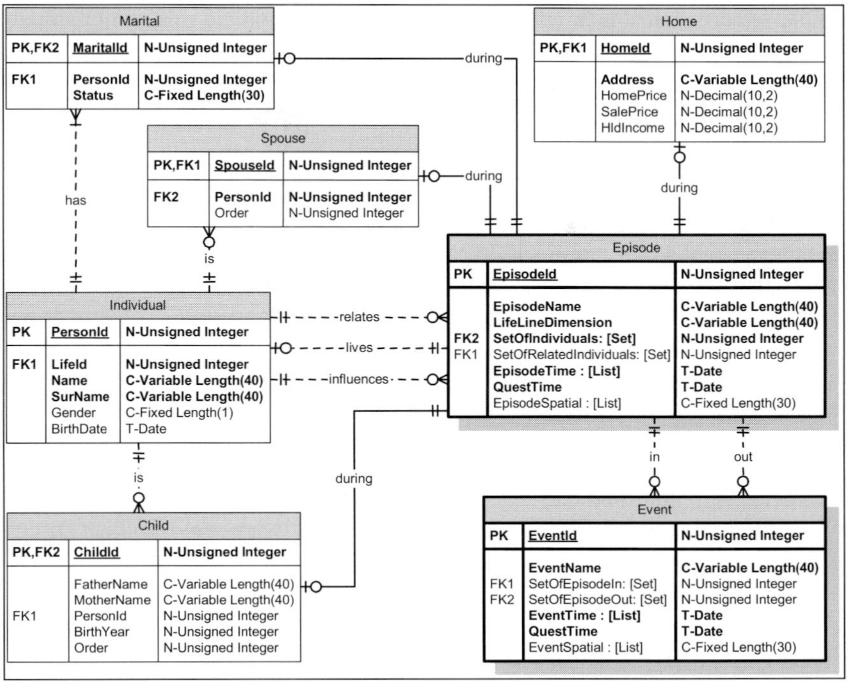

Fig. 2. Relational schema for describing household and residential lifelines

An **event** is defined as a tuple *Event* (*<u>EventId</u>*, *EventName*, *SetOfEpisodeInId*, *SetOfEpisodeOutId*, *EventTime*, *QuestTime*, *EventSpatial*). *EventId* uniquely identifies each event. *EventName* represents the class of events represented (E.g. marriage, birth). *SetOfEpisodeInId* gives the set of episodes which are terminated by this event. *SetOfEpisodeOutId* gives the set of episodes which are initiated by this event. As previously mentioned, an event can be linked to several episodes. Conversely an episode might be either terminated or generated by the conjunction of several events. These links provide means of selecting the previous and the following episodes, using their explicit sequence ordering. This could work even when time stamps are missing or when these episodes are non-immediate successors or predecessors. *EventTime* models the time when this event happens (using either temporal instants or periods). *QuestTime* indicates the date when the information was collected. *EventSpatial* keeps the spatial location of this event.

Temporal and spatial attributes use multi-level data representations: a calendar of type (*Year: Month*) for temporal attributes *EventTime, EpisodeTime* and *QuestTime*; and, a spatial hierarchy of type (*MunCode: Municipality: NeighbourhoodType: PostalCode: Longitude: Latitude*) for spatial attributes *EventSpatial* and *EpisodeSpatial*. Such representations support reasoning at various levels of abstraction as introduced in Claramunt and Jiang (2000). As necessary, this primary structure can be extended using additional relations to handle specific attributes (thematic, temporal and spatial) that complement the semantics of the application (E.g. Child, Spouse, Home as illustrated in Fig. 2).

3 Temporal GIS Modelling

Lifeline histories are modelled at the basic individual level using events and episodes. Thanks to a graph structure linking successive episodes and events, this model also provides means to handle chronological time and historical sequences (ordered events) in the same structure. For example, *SetofEpisodeIn* points to every episode ended by the current event, while *SetofEpisodeOut* lists all those following it (Fig. 2). Additionally, because individuals move within the city, the same database schema can be used to analyse spatio-temporal patterns.

In Fig. 2, the relational schema uses sets to model groups of individuals related through an episode (E.g. a newborn child and his parents). Lists are used to bound time periods (begin- and end-date) and to record multi-level spatial location. In an effort to map this relational schema to the software environment used for prototyping, (combination of MapInfo and MS Access) some minor adaptations were required (Fig. 3). Sets were replaced by intermediate tables (*EpisodeIn* and *EpisodeOut*), thus-allow for many-to-many relationships among episodes and events. Lists of individuals were mapped to tables (*ActingIndividuals* and *RelatedIndividuals*) and linked to episodes. *EpisodeTime* and *EventTime* were disaggregated in two fields giving their beginning and ending dates. Spatial lists (holding spatial locations) are defined using a *Spatial* table that maintains relationships with both episodes and events. Finally, the resulting mapping is monitored by an administration table (*MapInfo_MapCatalog*) that provides instructions for generating map symbols when specified tables and views are opened within MapInfo, using ODBC services.

Events and episodes are stored into two tables using the fields *EpisodeName*, *LifeLineDimension* and *EventName* to distinguish their type, and to identify the lifeline to which the episode belongs. All geographical locations are recorded in a unique table, named *Spatial*. Therefore, several events and episodes can be related to a single location tuple as required. Time management is operated in two ways by four interrelated tables: *Episode*, *Event*, *EpisodeIn*, *EpisodeOut*. Date fields, in both *Event* and *Episode* tables, keep track of chronological time, using T-Date types. Links to *EpisodeIn* and *EpisodeOut* define multi-dimensional ordering of related episodes and events. All spatio-temporal features managed by these five general-purpose tables, greatly eases the implementation of spatio-temporal views and the formulation of spatio-temporal queries.

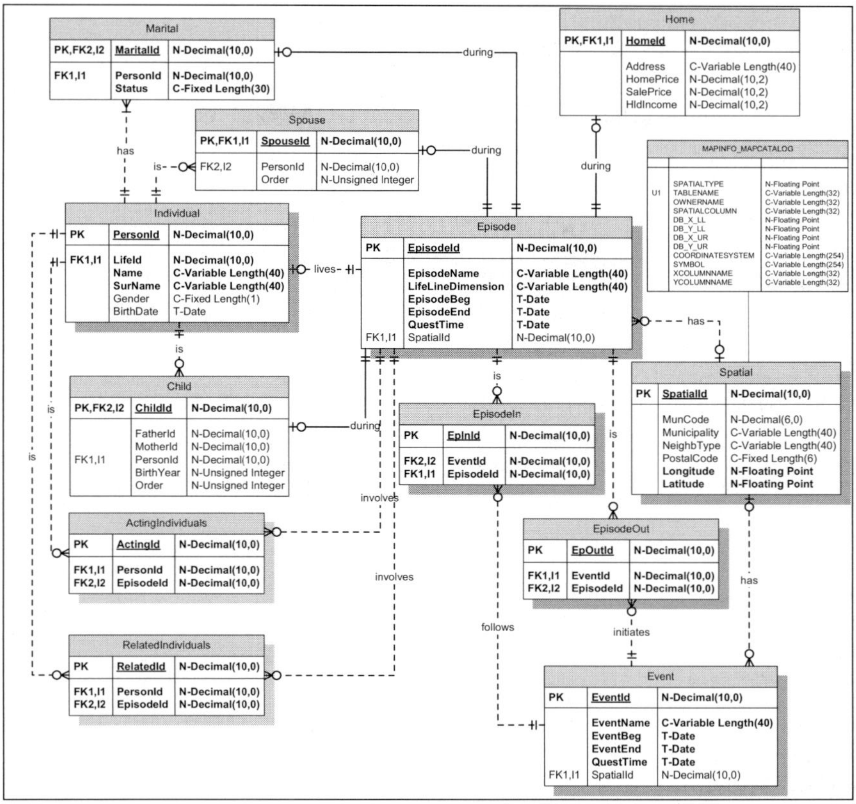

Fig. 3. Access-MapInfo expanded schema for household and residential lifelines

4 Spatio-Temporal Views

Manipulating several tables using a given query is not a straightforward and efficient task. Complex queries can be predefined using conventional relational views embedded in the database schema. Fig. 4 presents some examples of views integrating various components of respondents' events and episodes with data about the respondent itself. Using views, one can model some specific schemas showing any type of episodes or events, providing an efficient link with the secondary tables of Fig. 3 (E.g. Home, Marital). Moreover, these views can integrate any spatial and temporal information in simple structures. Depending on the DBMS capabilities, these views can later be joined to build more complex views, thereby defining a comprehensive system to solve queries.

Table 1 details a list of tables and views involved in defining views for our application, indicating join relationships used. *Respondent_Episodes* and *Respondent_Events* views (Fig. 4) give, for each respondent of our survey, the list of his/her reported episodes and events. Using database triggers, these views automatically add the respondent's *Age* (*EpisodeBeg-BirthDate*) and the *Duration* of the episode (in months) to each tuple. They also indicate if, at survey time, the

Table 1. Relational links defining spatio-temporal views of individual lifelines

Resulting View	From Tables / Views	Relational / Sorting Conditions
RespEpis	[ActingIndividuals] As [c1] [Episode] As [c2] [Individual] As [c3] [Spatial] As [c4]	[c1].[EpisodeId] = [c2].[EpisodeId] ^ [c2].[SpatialId] = [c4].[SpatialId] ^ [c1].[PersonId] = [c3].[PersonId]
IndivEvent	[RespEpis] As [c1] [Event] As [c2] [EpisodeOut] AS [c3] [Spatial] As [c4]	[c1].[EpisodeId] = [c3].[EpisodeId] ^ [c2].[EventId] = [c3].[EventId] ^ [c2].[SpatialId] = [c4].[SpatialId]
Couples	[Spouse] As [c1] [Episode] As [c2] [RespEpis] As [c3]	[c1].[SpouseId] = [c2].[EpisodeId] ^ [c2].[EpisodeId] = [c3].[EpisodeId] ^ [c3].[Respondent]
Respondent_Episodes	[RespEpis] As [c1] [Episode] As [c2] [Individual] As [c3]	[c1].[EpisodeId] = [c2].[EpisodeId] ^ [c1].[Respondent] ^ [c1].[PersonId] = [c3].[PersonId] Order By [c1].[PersonId], [c1].[EpisodeBeg]
Respondent_Events	[IndivEvent] As [c1] [Event] As [c2] [Individual] As [c3]	[c1].[EventId] = [c2].[EventId] ^ [c1].[PersonId] = [c3].[PersonId] ^ [c1].[Respondent] Order By [c1].[PersonId], [c1].[EventBeg]
Episodes_Before_Event	[RespEpis] As [c1] [Respondent_Events] As [c2]	[c1].[PersonId] = [c2].[PersonId] ^ [c1].[EpisodeEnd] <= [c2].[EventBeg] Order By [c1].[PersonId], [c2].[EventBeg], [c1].[EpisodeBeg]
Episodes_After_Event	[RespEpis] As [c1] [Respondent_Events] As [c2]	[c1].[PersonId] = [c2].[PersonId] ^ [c1].[EpisodeBeg] >= [c2].[EventEnd] Order By [c1].[PersonId], [c2].[EventEnd], [c1].[EpisodeBeg]

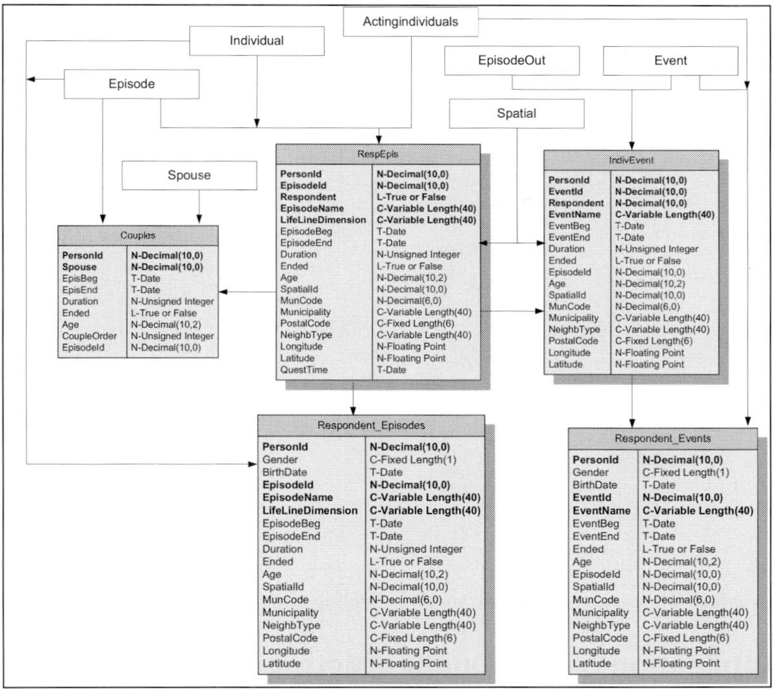

Fig. 4. Spatio-temporal integrated views of respondents' episodes and events

episode was *Ended*, or still happening. Using time-stamps to order each respondent's episodes and events, the next step defines multi-dimensional links among lifelines (Fig. 5; Table 1). Two multi-dimensional views (*Episodes_Before_Event* and *Episodes_After_Event*) show respectively, for each respondent, the list of all episodes that started before any event occurred during his entire life and all those that started after the event. In the view *Episodes_After_Event*, the field *TransTime* indicates elapsed time (in years) between the event and the episode (*[c1].[EpisodeBeg]-[c2].[EventEnd]*). If an episode was not ended at survey time (*[c1].[EpisodeBeg]>=[c1].[QuestTime]*), it is called *Censored.*

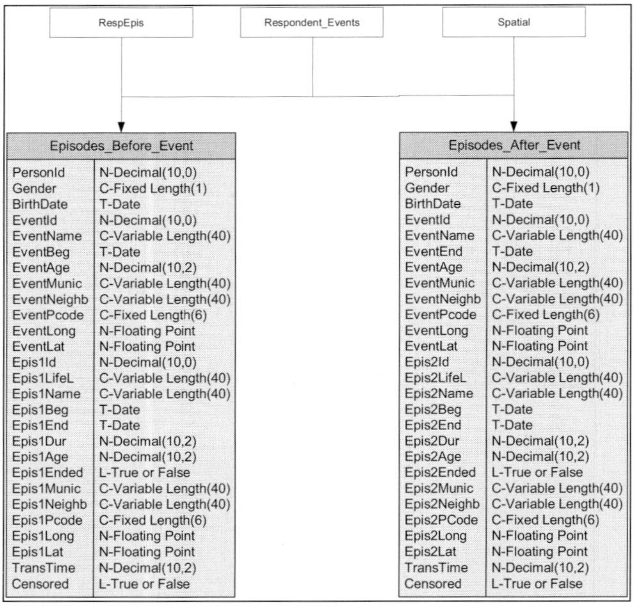

Fig. 5. Multi-dimensional lifeline transitions among episodes through each event

5 Application of the Quebec Metropolitan Area

In our application example, 418 respondents reported 5,541 events and episodes. Episodes are distributed into four lifelines (Family, Individual, Marital and Residential), 10 episode types and 12 event types (Table 2). For each respondent, an episode and an event were assigned to the interview because its timing is needed to distinguish censored events and ongoing periods. Building multi-dimensional sequences of periods preceding and following each event during the life of each individual, generated 22,676 tuples in view *Episodes_Before_Event*, and 44,839 tuples in view *Episodes_After_Event*. This last view provides relevant data and logical structure to generate the flat files needed for time regression in SPSS.

Table 3 presents a relational-like query selecting sequences of episodes and events according to the following criteria: Among people who were tenants when their first child was born, distinguish those who decided to buy a house afterwards from those who did not; then compute how long they postponed their decision and to which distance they moved.

Table 2. Typology of episodes and events

	Episodes		Events	
Lifeline	EpisodeName	Count	EventName	Count
Family	Child	750	Alone	234
Family	Child Departed	102	Birth	418

Individual	Interview	418	Buy	646
Individual	Life	418	Child Adoption	3
Individual	Self	418	Child Arrival	64
Marital	Couple	533	Child Birth	683
Marital	Separated	174	Child Departure	102
Marital	Single	234	Inhabit	169
Residential	Owner	646	Interview	418
Residential	Roomate	169	Leaving Parent	418
Residential	Tenant	1679	Rent	1679
			Separation	174
			Union	533

Table 3. Relational-like query to generate the event history flat file for SPSS

Query command	Notes
Select *,	Computing moving distance
Distance([c1].[Longitude],[c1].[Latitude],[c2].[Epis2Long],[c2].[Epis2Lat],"Km") "DIST"	Using Spatio-temporal views
	Restricting to each person's life
From [Respondent_Episodes] As [c1], [Episodes_After_Event] As [c2]	Buying home after child's birth
Where [c1].[PersonId]=[c2].[PersonId]	Censored: still tenants at survey time
And (([c2].[EventName] = "Child Birth" and [c2].[Epis2Name] = "Owner")	Respondent was tenant at that time
	Respondent was living in Quebec CMA
or ([c2].EventName = "Child Birth" and [c2].[Epis2Name] = "Interview"))	Birth date is during residential episode
and ([c1].[EpisodeName]="Tenant")	Retaining only one child birth by person
and ([c1].[NeighbType] in ("City Core","Old Suburbs","New Suburbs","Urban Fringe"))	Ordering to retain the first set (child-buy)
and {[c2].[EventEnd] .TDuring. [c1].[EpisodeBeg;EpisodeEnd]}	
Group By [c1].[PersonId], [c1].[EpisodeId]	Generating a line between locations
Order By [c1].[PersonId], [c1].[EpisodeId], [c2].[Epis2Beg]	Generating the event history table
Object [Line, [c1].[Longitude]:[c1].[Latitude];[c2].[Epis2Long]:[c2].[Epis2Lat]]	
Into Event_History_Table	

In Table 3, the relational-like query uses a join between the view *Respondent_Episodes* and the table *Episodes*. The view *Respondent_Episodes* gives the home location of some tenants when the first child was born. This describes the neighbourhood where the 151 respondents who were tenants when their first child was born were living at that time (City core, Old Suburbs, New suburbs, Urban fringe). This gives an indication of their relative location within the Quebec Metropolitan Area. Knowing which ones did not change their housing tenure afterwards (*Censored*), and which ones did, we can test the hypothesis of a relationship linking neighbourhood types where young parents live and their decision to buy a home less than five years later (Table 4). The chi-square test indicates that no significant relationship was found between these two facts in our sample of tenants (probability = 0.566). Therefore, willingness to purchase a house after the first child's birth seems invariant across the city.

Table 4. Cross-tabulation of home location of tenants and their decision to buy a home

Decision to buy a home within 5 years after the first child is born	Type of living neighbourhood when the first child was born								
	City Core		Old suburbs		New suburbs		Urban fringe		
	Count	Expected	Count	Expected	Count	Expected	Count	Expected	Sum
Did buy a home	33	34.8	64	60.5	7	8.3	1	1.4	105
Did not buy a home	17	15.2	23	26.5	5	3.7	1	0.6	46
Sum	50	50	87	87	12	12	2	2	151

$\chi^2 = 2.032; df = 3; probability = 0.566$

Table 5 shows the result of a Cox regression estimating attributes influencing the decision of tenants to buy a home after the first child is born. According to the overall chi-square test, the relationship is highly significant. Three factors were found to have a significant influence (Wald's statistics) on the willingness to buy a home: decade of birth of the first child (*PeriodB*: odds ratio increasing with time), distance at which people are ready to move in order to access home ownership (*Dist*: odds ratio increases with distance), and duration of their stay in the new home (*Epis2Dur*: odds ratio increases with expected duration of stay).

Table 5. Attributes influencing tenants for buying a new home after the first child is born

	B	SE	Wald	df	Sig.	Odds e^B
EPIS2DUR : Duration of stay at destination home (Years)	0.137	0.019	50.650	1	0.000	1.147
DIST : Distance from child birthplace to new home location (Km)	0.007	0.002	12.545	1	0.000	1.007
PERIODB : Decade of birth of the first child			26.687	3	0.000	
1960-69	-2.115	0.734	8.311	1	0.004	0.121
1970-79	-1.574	0.485	10.530	1	0.001	0.207
1980-89	-0.267	0.426	0.393	1	0.530	0.765
Reference 1990-95	0.000					1.000

$\chi^2 = 61.314; df = 5; probability < 0.0001$

Table 5 can be used to build a mathematical expression for the likelihood of tenants to buy a home (*Censored=0*) at time *TransTime* after their first child is born (*EVENTNAME="Child Birth"*):

$$h(t) = e^{(\beta_0 + \alpha \ln TRANSTIME)} e^{(0.137\,EPIS2DUR + 0.007\,DIST - 2.115\,Period60 - 1.574\,Period70 - 0.267\,Period80)}$$

, using respondent's spatio-temporal attributes. This function has two parts:

$e^{(\beta_0 + \alpha \ln TRANSTIME)}$, the base line hazard function related to elapsed time after the first child is born (see Fig. 6),

and $e^{(0.137\,EPIS2DUR + 0.007\,DIST - 2.115\,Period60 - 1.574\,Period70 - 0.267\,Period80)}$, the marginal effect of independent variables on respondent's decision to buy a home. Moving distances come from measurement of moving paths from the location of previous home to that of the newer (Fig. 6).

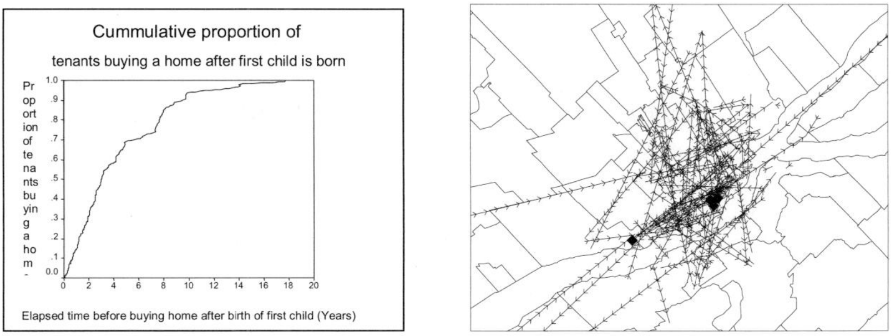

Fig. 6. Likelihood function of buying a home and moving paths of house places

6 Conclusion

This paper introduces a GIS database approach illustrated by an application example that clearly demonstrates the interest of analysing event histories at the individual level using temporal statistical methods. Moreover, integrating events and episodes management within GIS provides efficient means for integrating geographical criteria into the modelling of people's decision. This goes far beyond the mere mapping of individual evolution path over space. Because, they isolate space-time attributes in specialised tables, the geo-relational schemas of this application can be readily used in many other temporal GIS applications needing to relate successive events occurring along lifelines.

Further work is needed to extend the concept of view towards spatial and temporal operators. Integrity constraints have to be enforced to check the convergence of chronological and ordered temporal information handled in the database. A user-friendly query interface has to be developed for better manipulation of data. Finally, adding dynamics would improve mapping of results.

Acknowledgements

This project was funded by the Quebec Province's FCAR, the Canadian SSHRC, the Canadian NSERC, the Québec Province's Ministère de l'Éducation and the Canadian Network of Centre of Excellence in Geomatics (GEOIDE).

References

Allen JF (1984) Toward a general theory of actions and time. Artificial Intelligence 23:123-154

Ben-Akiva M, Lerman SR (1985) Discrete choice analysis: Theory and application to travel demand. MIT Press, Cambridge, Mass

Blossfeld HP (1996) Macro-sociology, rational choice theory, and time: A theoretical perspective on the empirical analysis of social processes. European Sociological Review 12(2):181-206

Blossfeld HP, Rohwer G (1995) Techniques of event history modelling. In: New approaches to causal analysis. Lawrence Erlbaum Associates, Mahwah, New Jersey

Böhlen M, Jensen C, Scholl M (1999) Spatio-temporal database management. In: Lecture Notes in Computer Sciences #1678, Springer-Verlag, Berlin

Claramunt C, Jiang B (2000) Hierarchical reasoning in space and time. In: Forer P, Yeh AGO and He J (eds) Proceedings of the 9th International Symposium on Spatial Data Handling. Beijing, China, 3a, pp 41-51

Claramunt C, Thériault M (1996) Toward semantics for modelling spatio-temporal processes within GIS. Proceedings of the 7th International Symposium on Spatial Data Handling. Delft, Holland, 47-63

Cox DR (1972) Regression models and life tables. Journal of the Royal Statistical Society B-34:187-220

Egenhofer M (1991) Reasoning about binary topological relations. In: O. Günther and H.-J. Schek (eds) Advances in Spatial Databases. Springer-Verlag, Berlin, pp 143-160

Frank AU (1994) Qualitative temporal reasoning in GIS – Ordered time scales. Proceedings of the 6th International Symposium on Spatial Data Handling. Zurich, Switzerland, pp 410-430

Goodchild MF, Janelle DG (1984) The city around the clock: Space-time patterns of urban ecological structure. Environment Planning A 16:807-820

Hägerstrand T (1967) Innovation Diffusion as a Spatial Process. University of Chicago Press, Chicago, Illinois

Hornsby K, Egenhofer M (1998) Identity-based change operations for composite objects. Proceedings of the 8th International Symposium on Spatial Data Handling. Vancouver, Canada, pp 202-213

Hornsby K, Egenhofer M (2000) Identity-based change: A foundation for spatio-temporal knowledge representation. International Journal of Geographical Information Systems 14(3):207-224

Hunt JD, Simmonds DC (1992) Theory and application of an integrated land-use and transport modelling framework. Environment and Planning B: Planning and Design 20:221-244

Janelle DG, Klinkenberg B, Goodchild MF (1998) The temporal ordering of urban space and daily activity patterns for population role groups. Geographical Systems 5(1-2):117-137

Johnson MP (2001) A spatial decision support system prototype for housing mobility program planning. Geographical Systems 8(1):49-67

Langran, G. (1989) A review of temporal database research and its use in GIS applications. International Journal of Geographical Information Systems 3(3):215-232

Le Bourdais C, Marcil-Gratton N (1998) The impact of family disruption in childhood on demographic outcomes in young adulthood. In M. Corak (ed) Labour Markets, Social Institutions and the Future of Canada's Children. Statistics Canada, Ottawa, pp 91-105

Miller HJ (1991) Modelling accessibility using space-time prism concepts within geographical information systems. International Journal of Geographical Information Systems 5(3):287-301

Odland J, Shumway JM (1993) Interdependencies in the timing of migration and mobility events. Papers in Regional Science 72(3):221-237

Peuquet DJ (1994) It's about time: A conceptual framework for the representation of temporal dynamics in geographical information systems. Annals of the Association of American Geographers 84(3):441-461

Peuquet DJ, Quian L (1996) An integrated database design for temporal GIS. Proceedings of the 7th International Symposium on Spatial Data Handling. Delft, Holland, pp 1-11

Pötter U, Blossfeld HP (2001) Causal inference from series of events. European Sociological Review 17(1):21-32

Rose D, Villeneuve P (1993) Work, labour markets and households in transition. In Bourne L and Ley D (eds) The Social Geography of Canadian Cities. McGill-Queens Press, Montreal, pp 153-174

Séguin AM (1994) Transformations sociales et carences de l'habitation. In: Dumont F, Langlois S, Martin Y (eds) Traité des problèmes sociaux. Institut québécois de recherche sur la culture, Québec, pp 65-79

Snodgrass RT (1999) Developing Time-Oriented Database Applications in SQL. Morgan Kaufmann, New York

Taylor PJ, Parkes DN (1975) A Kantian view of the city: A factorial-ecology experiment in space and time. Environment and Planning A 7:671-688

Thériault M, Séguin AM, Aubé Y, Villeneuve P (1999) A spatio-temporal data model for analysing personal biographies. Proceedings of the 10th International Workshop on Database and Expert Systems Applications. IEEE Computer Society, Florence, Italy, pp 410-418

Whiters SD (1997) Methodological considerations in the analysis of residential mobility: A test of duration, state dependence, and associated events. Geographical Analysis 29(4):354-374

Urban Simulation Using Neural Networks and Cellular Automata for Land Use Planning

Anthony Gar-On Yeh[1] and Xia Li[2]

Centre of Urban Planning and Environmental Management, The University of Hong Kong, Pokfulam Road, Hong Kong SAR, P.R. China (hdxugoy@hkucc.hku.hk, lixia@graduate.hku.hk)[1]
Guangzhou Institute of Geography, Guangzhou 510070, P.R. China, xlib@gis.sti.gd.cn [2]

Abstract

The paper presents a method for integrating neural networks, GIS and Cellular Automata (CA) that can be used in land use planning for simulating alternative development patterns according to different planning objectives. Neural networks are used to simplify model structures and facilitate the determination of parameter values. Unlike traditional CA models, the proposed model does not require users to provide transition rules, which may vary for different applications. Historical remote sensing data are used as the training data to calibrate the neural network. The training is robust because it is based on the well-defined back-propagation algorithm. Moreover, original training data are assessed and modified according to planning objectives to generate alternative development patterns.
Keywords: neural networks, cellular automata, GIS, urban simulation, urban planning

1 Introduction

Urban Cellular Automata (CA) have been developing rapidly for the simulation of complex urban systems since the late 80s. A number of interesting investigations have been documented (Batty and Xie, 1994; White and Engelen, 1993; Clarke and Gaydos, 1998). Urban systems involve spatial and sectoral interactions, which cannot be easily adapted to the functionality of current GIS software (Batty et al., 1999). CA-based approaches have applications in the study of urban and regional spatial structure and evolution. Modelling cities with Cellular Automata is a relatively new approach although it has its roots in geography and relates to the work of Hägerstrand (1965) and Tobler (1979) (Clarke and Gaydos, 1998).

Three main types of urban CA models can be considered. The first type uses CA models to generate results that can be explained by urban theories. These models are used to test ideas and assumptions for hypothetical cities. For example, Webster and Wu (1999) present an interesting CA model to implement urban theories concerning developers' profit seeking and communities' welfare-seeking behaviours. In so doing, they explore the mediating effects of alternative systems of land-use rights. The second type is to apply CA models for the simulation of real cities. Clarke and Gaydos (1998) apply CA models to simulate and predict urban development in the San Francisco Bay region in California and the Washington/Baltimore corridor in the Eastern United States. White et al. (1997) provides a realistic simulation of the land-use pattern of Cincinnati, Ohio. A constrained CA model is developed to simulate urban expansion and agricultural land use loss of Dongguan in the Pearl River Delta, a rapidly growing area in southern China (Li and Yeh, 2000). The third type is to use CA to develop normative planning models to simulate different urban forms based on planning objectives. Yeh and Li (2001) use CA to generate different urban forms, ranging from monocentric to polycentric urban development. These urban forms can be assessed to meet selected criteria for sustainable development through minimising agricultural land use in an effort to achieve compact development. Ward et al. (2000) also develop a constrained CA model which has been applied to an area in the Gold Coast, a rapidly urbanising region of eastern Australia. They demonstrate that CA models can simulate planned development as well as realistic development by incorporating sustainable criteria in the simulation.

In the simulation of real cities, model calibration is needed to find suitable values of simulation parameters that can best fit actual development. Unfortunately, there is no universally applicable method of calibration due to the complexity of nature. Another reason is that appropriate methods have not been well developed. There are only very limited studies in addressing the calibration issues in CA simulation. Wu and Webster (1998) use multi-criteria evaluation (MCE) to heuristically define the values of parameters for CA simulation. Clarke et al. (1997) consider that visual tests are useful to establish parameter ranges and to make rough estimates of parameter settings. The impact of each parameter is assessed by changing its value while holding other parameters constant. Clarke and Gaydos (1998) suggest that calibration can be done by statistically testing the observed criteria against the expected. These methods are very time-consuming because they need to compare all possible combinations of parameters. Another problem is that the combinations are infinite and a sound search procedure is difficult to design.

This paper presents a CA model using neural networks to simulate potential or alternative urban development patterns based on different planning objectives. The simulation of the conversion of non-urban land use to urban land use should be very useful to urban planning. The structure of neural networks is simple and the calibration is easy. Neural network can be used to replace the transition rules used by conventional CA models in a simple and effective way. The following sections will discuss how to develop neural networks to simulate different urban forms based on different planning objectives.

2 Neural Networks for Urban CA Simulation

Neural networks have the capacity to recognise and classify patterns through training or learning processes. They have been used in urban studies, such as journal-to-work flows and airline and telecommunication traffic (Fischer and Gopal, 1994; Openshaw, 1993). These studies indicate that neural networks provide superior levels of performance to those of conventional statistical models because they can well handle the uncertainties of spatial data.

The proposed model consists of two separate parts – training and simulation (Fig. 1). Training is based on the back-propagation procedure, which can generate optimal weights from a set of training data. In this study, remote sensing and GIS are used to provide the historical empirical data to reveal the relationships between site attributes and urban development. There are two ways in using the empirical data for training. The original data can be directly used to generate realistic simulation, which assumes that urban growth proceeds according to historical trends. However, the original data can also be modified according to some criteria, which are related to planning objectives.

In creating new sets of training data, some criteria should be provided to evaluate the past development 'points' in the original data set. It is assumed that some better and worse development 'points' can be identified according to their costs and benefits, which are measured by the criteria. A simple way is to remove some 'bad points' based on the evaluation. The modified data set can allow the training to ignore 'false' information. This is important for the model to generate ideal or optimal simulation results. A number of modified data sets can be obtained by using different sets of criteria, which correspond to planning objectives.

The new sets of modified training data can lead the training procedure to obtain new sets of parameter values. These new sets of parameter values will then be used by the neural network to generative alternative or ideal urban patterns. This can help to correct the historical land use problems and search for more appropriate urban growth in the future. The whole procedure is simple and robust because the parameter values are not arbitrarily defined.

Site attributes determine development probability. It is very convenient to obtain site attributes when GIS analysis functions are employed. The site attributes for CA simulation usually includes the development level in the neighbourhood, as well as various types of proximity attractiveness (Batty and Xie, 1994; Wu and Webster, 1998). Common GIS analysis functions, such as buffer and overlay, can be carried out to obtain these spatial variables.

The CA simulation is based on the algorithm of neural networks. At each iteration, the neural networks will determine the development probability, which is subject to the input of site attributes and weights. A cell may have *n* site attributes (variables):

$$(S_1, S_2, S_3, S_4, S_5, S_6, \ldots, S_n) \tag{1}$$

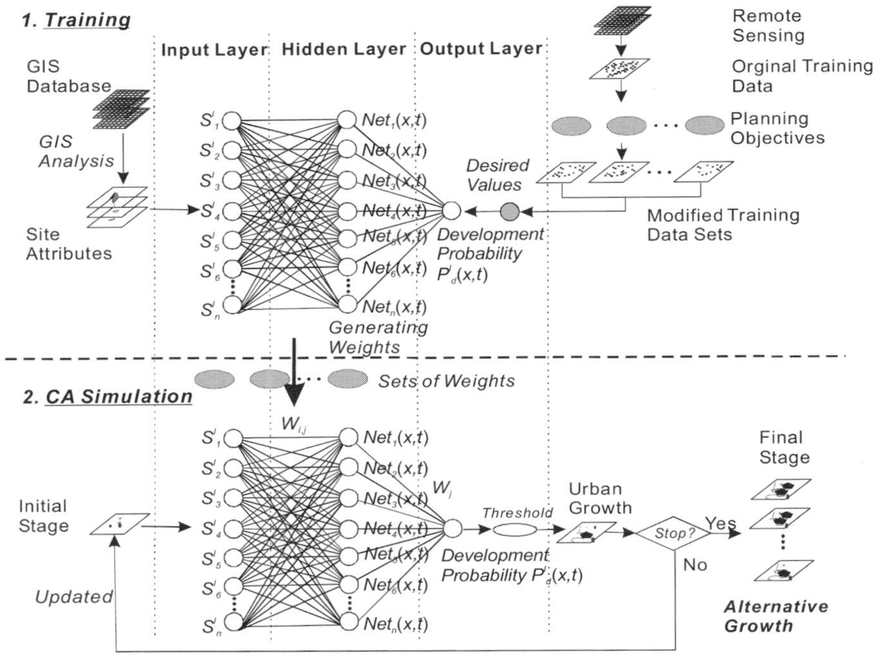

Fig. 1. Simulation of urban forms using the neural-network-based CA model

A neural network can be designed to estimate development probability at each iteration of the CA simulation. The neural network has three layers, one input layer, one hidden layer and one output layer. The input layer has *n* neurons with regard to these site attributes. The hidden layer may also have *n* neurons. The output layer has only one neuron, which calculates the development probability. The site attributes of a cell will be input into the first layer and the neural network will determine its development probability at the output layer at each iteration.

The original data are usually scaled into the range of [0, 1] before they are input to neural networks (Gong, 1996). Scaling each variable treats them as equally important inputs to neural networks and makes them compatible with a sigmoid activation function that produces a value between 0.0 and 1.0. The following linear transformation is used:

$$S_i' = (S_i - \text{minimum}) \,/\, (\text{maximum} - \text{minimum}) \tag{2}$$

The algorithm for the CA model is based on a simple three-layer network. In the neural network, the signal received by neuron j of the hidden layer from the first input layer for cell x is calculated by:

$$net_j(x,t) = \sum_i W_{i,j} S_i'(x,t) \tag{3}$$

where x is a cell, $net_j(x,t)$ is the signal received for neuron j of cell x at time t, $W_{i,j}$ is the weight of the input from neuron i to neuron j and $S'_i(x,t)$ is the site attributes for variable (neuron) i.

The activation of the hidden layer for the signal is:

$$\frac{1}{1+e^{-net_j(x,t)}} \tag{4}$$

The development probability (P_d) for cell x is then calculated by:

$$P_d(x,t) = \sum_j W_j \frac{1}{1+e^{-net_j(x,t)}} \tag{5}$$

This simulation is loop-based. Development probability is calculated according to site attributes using the neural network at each iteration. The development probability decides whether a cell is converted or not for development. A stochastic disturbance term can be added to represent unknown errors during the simulation. This can allow the generated patterns to be more close to reality. The error term (RA) is given by (White and Engelen, 1993):

$$RA = 1 + (-\ln \gamma)^{\alpha} \tag{6}$$

where γ is a uniform random variable within the range {0, 1}, and α is a parameter to control the size of the stochastic perturbation. In this case, α can be used as a dispersion factor in the simulation. The development probability is revised as:

$$\begin{aligned} P_d'(x,t) &= RA \times \sum_j W_j \frac{1}{1+e^{-net_j(x,t)}} \\ &= (1 + (-\ln \gamma)^{\alpha}) \times \sum_j W_j \frac{1}{1+e^{-net_j(x,t)}} \end{aligned} \tag{7}$$

A cell with a high value of development probability will be likely urbanised during the simulation. A predefined threshold value should be used to decide whether a cell is developed or not at each iteration. If a cell has a probability greater than the threshold value, it will be converted for development. The number of cells in the neighbourhood is recalculated and the site attributes are updated at the end of each iteration. The simulation will continue until the total number of converted cells is equal to the required land consumption.

3 Applications

3.1 Study Area and Site Attributes

A real city, Dongguan, was selected for the experiment to demonstrate the advantages of the neural-network-based CA model. The model was used to simulate the conversion from non-urban land use to urban land use. Located in the Pearl River Delta region, China, it has a total land area of 2,465 km^2. It is a fast growing region with a tremendous amount of land use changes and urban sprawl in recent years (Yeh and Li, 1997; Yeh and Li, 1999). In this study, seven spatial variables are defined to represent the site attributes of each cell for the simulation of urban development. These variables include:

1. Distance to the major (city proper) urban areas S_1;
2. Distances to sub-urban (town) areas S_2;
3. Distance to the nearest road S_3;
4. Distance to the nearest expressway S_4;
5. Distance to the nearest railways S_5;
6. Neighbourhood development level (the window of 7 × 7 cells) S_6;
7. Agricultural suitability S_7.

The reason to choose these variables is that they are important factors to decide development probability (Wu and Webster, 1998). These site attributes were obtained by employing GIS analyses and remote sensing classification. The distance variables were calculated using the *Eucdistance* function of *ARC/INFO GRID*. These distance variables were dynamically updated during the simulation. The neighbourhood development level was measured based on the number of developed cells in the 7 × 7 cells adjacent to the central cell. It was calculated using the *Focal* function of *ARC/INFO GRID*. This variable was also dynamically updated during the simulation. The initial neighbourhood development level was calculated from the 1988 binary image.

3.2 Training

Training data for urban growth were obtained by the classification of satellite TM images of 1988 and 1993 (Fig. 2). The classification results were imported to *ARC/INFO GRID* in the grid format. These grids were used as the empirical data for the calibration of the CA model. Although the original TM images had a ground resolution of 30 × 30 m, the cell size was reduced to 50 × 50 m by re-sampling for faster simulation. Binary values were used to represent developed and non-developed areas in 1988 and 1993. The value of 1 represents developed (converted) cells and 0 represents non-developed ones. The urban areas of 1988 were used as the starting point of the simulation. The urban areas of 1993 were also obtained for testing the simulation by comparing the actual development with the result of the simulated development. The CA simulations were carried out from 1988 to 1993 so that different urban forms generated by the model can be compared and evaluated against the actual development in 1993.

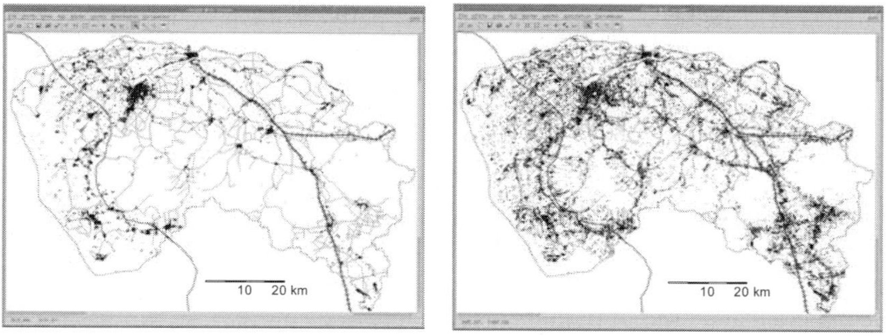

1988 1993

Fig. 2. Urban areas of Dongguan in 1988 and 1993 classified from satellite TM images

Training was carried out by using a neural network package referred to as THINKS PRO[1] . An essential task is to design the network structure for the CA simulation. The design of the network structure is relaxed since the numbers of layers and neurons in each layer can be subjectively determined. There are, however, some principles that can be used to guide the determination of the network structure. The increase of the numbers of layers and neurons will drastically increase the computation time for the loop-based CA model. It is practical to use the numbers of layers and neurons as few as possible without severely compromising the model accuracy. Kolmogorov's theorem suggests that any continuous function $\phi\colon X^n \rightarrow R^c$ can be implemented by a three-layer neural network which has n neurons in the input layer, (2n+1) neurons in the single hidden layer, and c nodes in the output layer (Wang, 1994). De Villiers et al. (1992) also suggest that a neural network with one hidden layer may be more preferable than one with two hidden layers in terms of learning speed and performance. A three-layer network is most suitable for CA models, which are based on many iterations. Practically, (2n+1) neurons in the single hidden layer may seem to be too much for actual applications. Experiments also indicate that a network of (2n / 3) neurons in the hidden layer can generate results of almost the same accuracy level but requires much less time to train than that of (2n + 1) neurons (Wang 1994).

Therefore, it is appropriate to use three layers of the neural network for urban simulation. The input layer has seven neurons corresponding to the seven variables of the site attributes. The hidden layer also has seven neurons. The output layer has only one neuron to output the development probability. There are $7 \times 7 = 49$ weights to be determined for the links between the input layer and the hidden layer, and 7 weights between the hidden layer and the output layer. A total of 56 parameters were used for the neural-network-based CA model.

Original training data were obtained by the overlay of historical urban growth (1988-93) from remote sensing and site attributes from GIS. The training data can be used to calibrate the network to produce the realistic simulation of the study area. It is inappropriate to use the whole data set for training because the size is too large and the data may have spatial correlation. A random sampling procedure was carried out to reduce the data volume and data redundancy.

The random stratified sampling points were generated by the *ERDAS IMAGINE*[1] package. Their co-ordinates were then imported to *ARC/INFO GRID* for the retrieval of the site attributes that were associated with these sampling points using the *Sample* function. 1,000 random sampling points were obtained and used to train the neural network.

3.3 Simulation Results

The simulation was totally based on the neural network. After appropriate weights had been obtained by the training, they were imported into the neural-network-based CA model for the simulation. The model was implemented in a GIS platform by the integration of neural network and GIS. GIS facilitates access to spatial data, which are used as site attributes for the simulation. The model was developed in *ARC/INFO*[2] *GRID* using the Arc Macro Language (AML). The GIS package also provides powerful spatial handling functions that are useful for CA simulation.

The output neuron of the network generated the development probability for each cell at each iteration. The cells with development probability greater than the threshold of 0.85 were converted into developed cells. The neighbourhood development level was then recalculated again to update the site attributes. The parameter α of the random disturbance was set to 1 so that only a small amount of uncertainty was presented in the simulation. The simulation time was automatically determined to ensure that the amount of land conversion was finally equal to that of actual development in 1993.

The model can be conveniently used to predict urban development based on past trends. However, an import task of our CA simulation is to generate alternative development patterns based on different pre-defined planning objectives. A way to generate alternative development patterns is to integrate MCE with CA models (Wu and Webster, 1998). Parameter values can be adjusted corresponding to various planning objectives. However, there are uncertainties about the method because the determination of parameter values is quite relaxed.

---

[1] *ERDAS IMAGINE* is a trademark of ERDAS, Inc.

[2] *ARC/INFO* is a trademark of Environmental Systems Research Institute, Inc.

A way to remove the uncertainties is to obtain the parameter values based on training. A number of different training data sets can be defined based on quantitative evaluation. The sample points from the original data sets were evaluated according to some criteria related to a planning objective. Unsuitable development cells were identified according to these criteria. For example, the sites with the distance greater than 30 km from town centres are considered as unsuitable for development of the objective of town-centre-based development. If a development point in the original sampling data set has a distance greater than the threshold, its desired value (development or not) will be adjusted from 1 to 0. For the objective of agricultural-conservation, all development points with the suitability score greater than 0.8 will be considered as unsuitable for development. Their desired values will be changed to 0 accordingly so that the neural network can be trained to avoid the encroachment on good agricultural land in the simulation of urban growth.

The modification of the training data set can help the neural network to remember the 'failure' and prevent the problem in the simulation. It is easy to create a couple of alternative training data sets from the original data set based on the evaluation of past development. The modification is related to planning objectives. We only use four typical planning objectives for urban growth because other options can be easily defined by the same method (Table 1).

Table 1. Rules for creating alternative training data sets

Planning Objectives	Modification Rules
1. Continuation of past development trend	Use original data without any modification
2. Promotion of mono-centric development	Change the desired values to 0 for all the development sites with $S_1 > 200$
3. Promotion of poly-centric development	Change the desired value to 0 for all the development sites with $S_2 > 30$
4. Promotion of agricultural-conservation development	Change the desired value to 0 for all the development sites with $S_7 > 0.8$

After these training data sets had been created based on the above modification rules, the *THINK PRO* package was used to obtain the corresponding sets of parameters or weights by the training procedure. Four different sets of weights were obtained for the simulation of different urban forms. Normal weights were obtained using the original data (planning objective 1), but adjusted weights were obtained using the modified data (planning objectives 2, 3 and 4). Different sets of weights were then input to the CA model to generate different development patterns. The model provides a useful tool making it possible to directly link the training network to the formation of distinct urban development patterns.

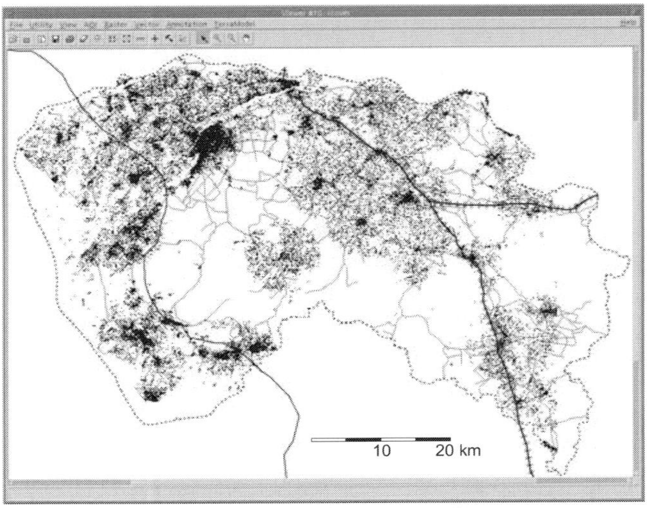

Fig. 3. Simulation of development in 1988-1993 using normal weights (planning objective 1 - continuation of past development trend)

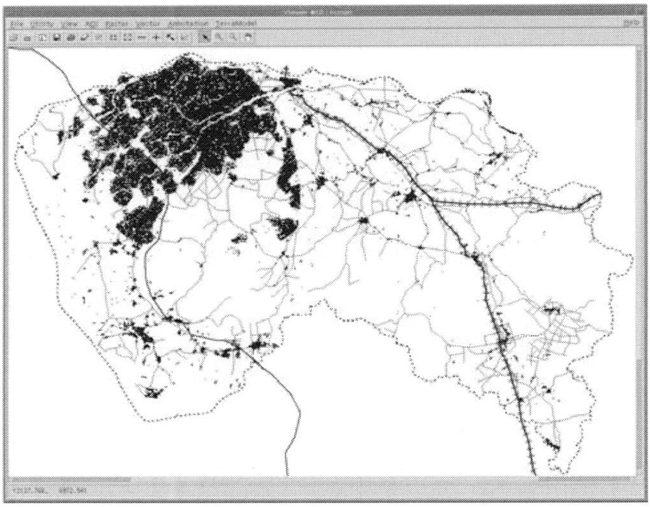

Fig. 4. Simulation of mono-centric development (planning objective 2) using adjusted weights

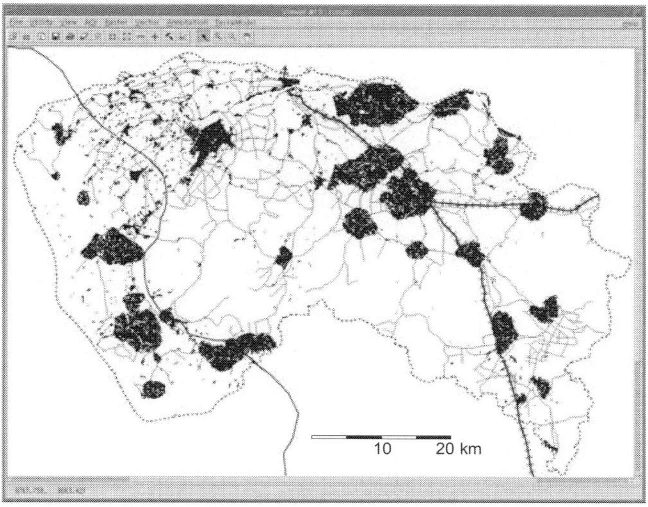

Fig. 5. Simulation of poly-centric development (planning objective 3) using adjusted weights

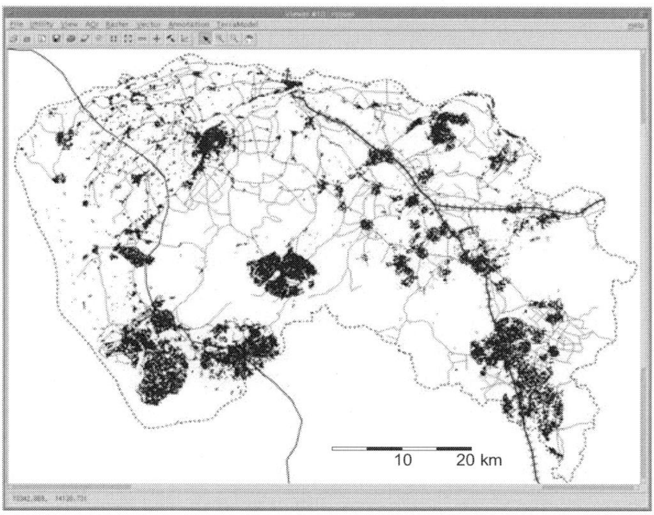

Fig. 6. Simulation of urban development with strict protection of agricultural land (planning objective 4) using adjusted weights

Fig. 3 shows the simulation output by using the normal weights obtained from the original training data set which is based on the past growth trend without any modification. As a result, the simulation is quite similar to the actual development (Fig. 2). Alternative urban forms can be derived by proper training of the network and through using different planning objectives as shown in Table 1. Fig. 4 shows the effects of promoting mono-centric development around the existing city proper by using the adjusted weights obtained from planning objective 2. Fig. 5 shows the simulation of promoting poly-centric development around the existing 29 town centres. In this case the adjusted weights obtained from planning objective 3 are used. Finally, the strict control on the conversion of the best agricultural land can be implemented by applying the fourth set of adjusted weights according to planning objective 4 (Fig. 6). It can be found that land development is well controlled in the alluvial plain in the northwestern part of the region.

4 Conclusion

It is very tedious to define model structure, transition rules and parameter values for conventional CA models. The proposed CA model can simplify these jobs considerably. Although multi-criteria evaluation (MCE) techniques can be used to define various parameter values, the method substantially relies on expert knowledge and experiences. There are also uncertainties because parameter values are usually defined in a relaxed way. The neural network-based model can significantly reduce the requirements for explicit knowledge for identifying relevant criteria, for assigning scores, and for determining criteria preferences. The model can effectively map the non-linear features of urban systems because it uses neural networks.

In this model, the original training data set can be evaluated to identify the 'good' or 'bad' performance of developed cells according to some criteria, which are related to planning objectives. New training data sets can be formed to train the network to accommodate interventions in the simulation process. The model can be used in two situations: simulation of urban development based on the existing development trend, and generation of alternative development patterns based on different planning objectives. Distinctive development patterns can be easily simulated based on the training of neural networks using different adjusted weights to reflect different planning objectives. Remote sensing and GIS data can be used to prepare the training data sets for a simulation that is more realistic. Based on planning objectives and development evaluation, original training data sets can be modified to obtain different sets of adjusted weights through the training procedure of neural networks. These adjusted weights can be applied to the CA model in generating preferable patterns.

The neural-network-based model is simple and convenient to use, but it can generate very complex features of urban systems. However, an inherent problem with neural network models is that they are black box in nature. The meanings of the parameter values are difficult to explain because the relationships among neurons are quite complex. Moreover, the determination of the network structure is also subject to user's preferences. It seems that there is no way to determine what is the best network structure.

The simulation also assumes that transport systems (railways and roads) would not change during the simulation period. It is assumed they were just upgraded and should therefore remain stable for the short period of time during which the simulation takes place. Future studies should accommodate the possible changes in the transport system for testing in the model. There are substantial uncertainties in simulating future changes of road systems since they are subject to relatively frequent changes and are often affected by urban transportation policies. It is more reasonable to tackle this issue by considering future transport development as an exogenous factor, which can be addressed by GIS for subsequent input to CA models.

Acknowledgements

This research is supported by the funding from the Croucher Foundation, Hong Kong.

References

Batty M, Xie Y (1994) From cells to cities. Environment and Planning B: Planning and Design 21: 531-548

Batty M, Xie Y, and Sun ZL (1999) Modeling urban dynamics through GIS-based cellular automata. Computers, Environment and Urban Systems 23: 205-233

Clarke KC, Gaydos LJ (1998) Loose-coupling a cellular automata model and GIS: long-term urban growth prediction for San Francisco and Washington /Baltimore. International Journal of Geographical Information Science 12(7): 699-714

Clarke KC, Gaydos L, Hoppen S (1997) A self-modifying cellular automaton model of historical urbanization in the San Francisco Bay area. Environment and Planning B: Planning and Design 24: 247-261

Fischer MM, Gopal S (1994) Artificial neural networks: a new approach to modelling interregional telecommunication flows. Journal of Regional Science 34: 503-527

Gong P (1996) Integrated analysis of spatial data from multiple sources: using evidential reasoning and artificial neural network techniques for geological mapping. Photogrammetric Engineering & Remote Sensing 62: 513-523

Hägerstrand T (1965) A Monte-Carlo approach to diffusion. European Journal of Sociology VI: 43-67

Li X, Yeh AGO (2000) Modelling sustainable urban development by the integration of constrained cellular automata and GIS. International Journal of Geographical Information Science 14(2): 131-152

Openshaw S (1993) Modelling spatial interaction using a neural net. In: Fischer MM, Nijkamp P (eds) GIS, Spatial modelling and policy. Springer, Berlin, pp 147-164

Tobler WR (1979) Cellular geography. In: Gale S, Olssen G (eds) Philosophy in Geography. Reidel, Dordrecht, pp 379-386

Ward DP, Murray AT, Phinn SR (2000) A stochastically constrained cellular model of urban growth. Computers, Environment and Urban Systems 24: 539-558

Wang F (1994) The use of artificial neural networks in a geographical information system for agricultural land-suitability assessment. Environment and Planning A 26: 265-284

Webster CJ, Wu F (1999) Regulation, land-use mix, and urban performance. Part 2: simulation. Environment and Planning A 31: 1529-1545

White R, Engelen G (1993) Cellular automata and fractal urban form: a cellular modelling approach to the evolution of urban land-use patterns. Environment and Planning A 25: 1175-1199

White R, Engelen G, Uijee I (1997) The use of constrained cellular automata for high-resolution modelling of urban land-use dynamics. Environment and Planning B: Planning and Design 24: 323-343

Wu F, Webster CJ (1998) Simulation of land development through the integration of cellular automata and multicriteria evaluation. Environment and Planning B: Planning and Design 25: 103-126

Yeh AGO, Li X (1997) An integrated remote sensing and GIS approach in the monitoring and evaluation of rapid urban growth for sustainable development in the Pearl Rive Delta, China. International Planning Studies 2: 195-222.

Yeh AGO, Li X (1999) Economic development and agricultural land loss in the Pearl River Delta, China. Habitat International 23: 373-390

Yeh AGO, Li X (2001) A constrained CA model for the simulation and planning of sustainable urban forms using GIS. Environment and Planning B: Planning and Design 28: 733-753

Creation of the Anamorphoses-Based 3D-Pyramidal Block-Diagrams

Nikita Bogomolov, Ilya Rylskiy, and Vladimir Tikunov

Moscow State University, Moscow 119899 Russia, Phone 7-095-936-14-98, Fax 7-095-932-88-36, Tikunov@geogr.msu.su

Abstract

This paper reports on work undertaken to create a 3D image, which displays as many indices as possible (thematic attributes) on a map. A new method has been developed for displaying the information. It resembles the classical method of the block-diagrams (prism maps), but it has more advantages and the efficiency is much greater. In addiltion, this method can be used simultaneously with anamorphosing. The technique of 3D anamorphoses (cartograms) compilation is outlined. Accordingly, algorithms for dynamic anamorphoses compilation for multi-dimensional images and multi-dynamic animations are discussed.
Keywords: 3D images, anamorphoses, multifaceted pyramids, multi-dynamic animations, algorithms

1 Anamorphoses

The main objective of the work reported here was to create 3D images that display as many indices as possible (thematic attributes) on the map. As is generally recognised, classical cartography offers a vast array of methods to display data, such as using colour, hatching, symbols and so on. For example, in creating a map of administrative units, there are a number of indices that should be shown. The usage of widespread methods - like column diagrams - allows us to show a number of indices - perhaps as many as 30-40 different ones. The cartographic legibility of such a map decreases greatly, and the quality of the information presentation becomes in some ways, almost equal to the presentation of the data table, on which the map is based. As a result, different methods of presentation are important if the map message is to be effectively communicated. The efficiency of the method is also very important. The colour used in a map presentation is one of the more effective methods, while point symbols, hatching, diagrams, and others can also be extremely effective. Anamorphosing is also very efficient, but varies from the other methods greatly.

Anamorphoses could be defined as graphic images that are derived from traditional maps, with varying scales according to the values of parameters taken from the original map. The English-speaking world uses such terms as *transformed maps, pseudo-cartograms, topological cartograms* and other terms for anamorphoses. We prefer to use the term 'anamorphoses'. The process of their compilation should be referred to as anamorphosing (from the Greek "anamorphoo"), which means to display the essence of transformation of the image proportions. In addition, it should be noted that these terms are widely used in a number of countries, especially in Eastern Europe. Anamorphoses could be classified into linear, areal and volumetric characteristics. In this paper only volumetric anamorphoses will be addressed.

In this context, a new method of displaying map information has been created and described in the following sections. It resembles the classical method of the block-diagrams (prism maps), but it has more advantages and its efficiency is much greater. Using this method simultaneously to anamorphosing, is entirely possible.

2 3D Anamorphoses

The main purpose in presenting this method is as follows. We have two indices (thematic attributes) to be mapped: for example, the first one is the real GNP of the country; while the second one is the total population of the country. Both values are from 1996 and cover the whole world.

Real GNP is the gross national product converted to 'international' dollars using purchasing power parity (PPP) rates. An 'international dollar' has the same purchasing power over GNP as the U.S. dollar in the United States. Real GNPs of developing countries tend to substantially exceed their conventional GNPs and reflect more accurately the average welfare level of their population. On the contrary the interaction potential of the PPP measured GNP continues to be determined by official exchange rates, and therefore in developing countries it is usually smaller than the values of the real GNP.

First of all, the anamorphoses is produced according to the total population and real GNP of the countries (Gusein-Zade, Tikunov, 1992, 1993). The results are 2D images as shown in Figs. 1 and 2.

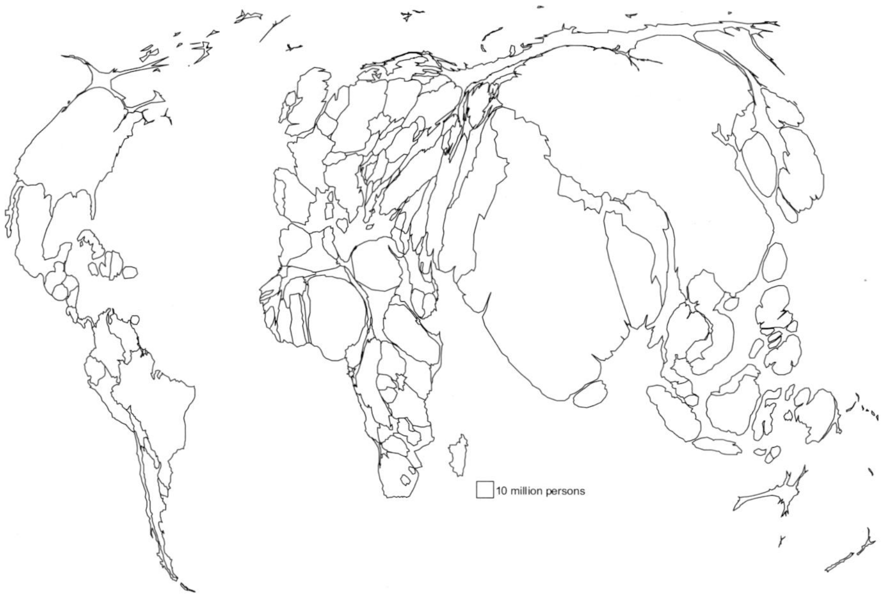

Fig. 1. Population, 1996 midyear estimates

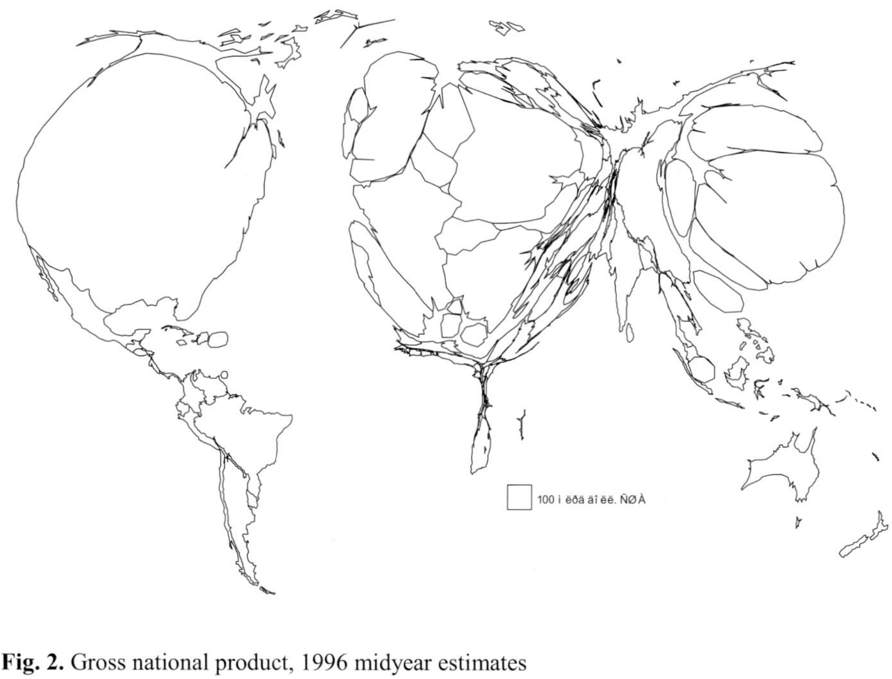

Fig. 2. Gross national product, 1996 midyear estimates

Following this presentation, the centre point for each country is used. In case a country is divided into two or more polygons – e.g., Indonesia, Philippines, etc. – the centre point of each polygon is used. The centre points are chosen "on-sight", but the main condition for these points is to be inside the contour of the country. We define centre points of the polygons in the Cartesian coordinate system (XY), according to the anamorphoted image of the population of the countries. All of the points of the country boundaries are given the Z value equal to 0 (zero), and all of the centre points of the countries – equal to the value of the second index expected to show (GNP per person). If we build the 3D surface according to these points (using the linear interpolation of Z value), then we have a number of the multifaceted pyramids. Their bases will have just the same form as the country itself, and the number of faces will be equal to the number of the node points in the polygon of this country.

The anamorphoses of the world according to the total population have a good property. The area of the country here is proportional to the population. The height of the pyramid we built is proportional to the GNP. As is generally known, the volume of the pyramid is: $V=S*H/3$, where S is the square of the basement, H is the height. Consequently the volume of pyramids we compiled is proportional to the GNP of the countries.

It is necessary to mention that comparison of the GNP values of different countries is difficult because of the different shapes of the countries.

The resulting image is very spectacular, and this is one of the most positive features of this method. The comparison of the values of GNP per person and total population becomes very easy. Here two indices are shown, however, the more common methods of displaying the data were not used and were kept in reserve. All information about these two indices is shown in the described manner.

Of course, it is possible to build the block diagrams (prism maps) based on the anamorphoses, but the result will be somewhat worse since the columns cover one another. In addition, it would not be able to see the country with a low value of GNP per person, if neighbours with greater values surround it - it looks like a hole. The zero level is invisible, and it makes the comparison very difficult. The 3D-pyramidal block-diagrams are free from these bad features. They ensure showing the zero level for each country, and provide a good overview (Fig. 3).

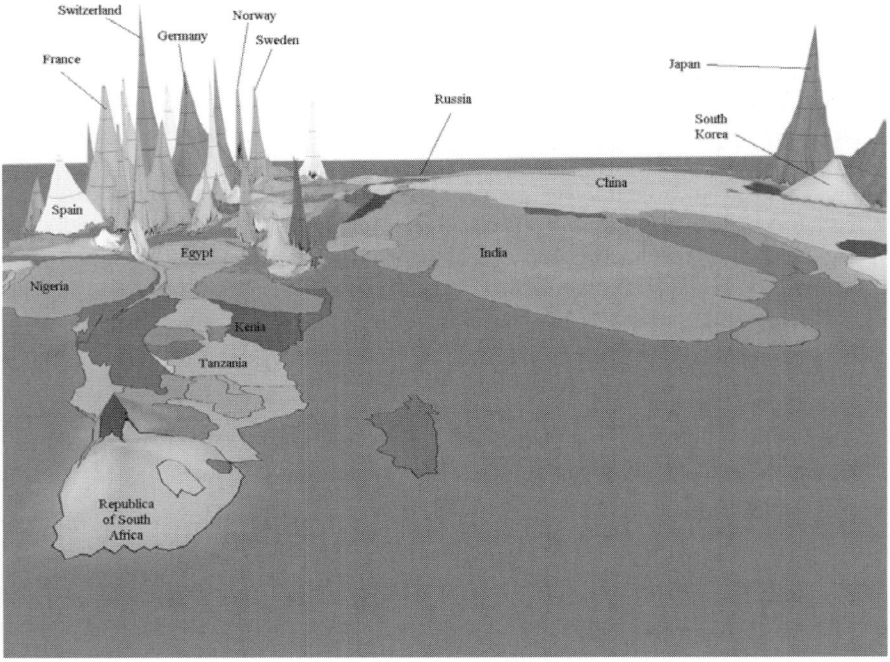

Fig. 3. 3D-pyramidal block-diagram of GNP per person based on population anomorphose

2.1 Technology and Building Methods

Let us consider the technology of producing the images in detail. The base data consist of the political map of the world, the data about the total population of the countries and about the real GNP in each country in 1996.

1. The initial world map was transformed into the anamorphose using the program and mathematical transformation that were developed specifically for generating such images. The resulting data (the files with the XY-coordinates of nodes and the object's topology) was converted into the ArcView 3.2 data format (*.SHP).
2. Centroids were extracted from polygons using X-TOOLS (ArcView 3.2 additional program module) into single point theme. After extracting the position of each centroid, they were checked and in case of an incorrect location, the centroid was manually removed. The new field for the value of GNP per person has been added, and values were given for each centroid. The polygon theme has been converted into a line theme (also using X-TOOLS), and the Z value equal to 0 was given to all lines.
3. Two sets of geometries (centroids and lines) were converted into ArcInfo coverages using ArcInfo 7.5. Following this, a grid surface was built from the

coverages using the ArcInfo TOPOGRID tool (using line point, and additional polygon data for borders of the resulting grid).

4. Additional thematic data has been prepared in ArcView Layout and exported into *.JPG (raster) file in high resolution.
5. Additional thematic raster data has been draped over the grid (surface) using the GEOMETRIC CORRECTION tool from Erdas Imagine 8.3.1 (we used polynomial approximation).
6. Units and projection information has been applied to the grid (surface) and thematic raster. In addition, an annotation file in Erdas Imagine (centers of annotations were linked to centroids of countries).
7. Using VIRTUAL GIS (an Erdas Imagine module) we combined the grid (surface) with thematic raster, and annotations were added. The resulting scene has been saved in Virtual GIS project and rendered from different points of view. The interactive virtual flight through the scene is also possible.

The resulting surface is smooth, though the algorithm intended to produce triangular faces. There are two reasons for this:

- the surface is a regular grid, and raster cells cannot form pure triangles,
- the method of 3D-scene rendering in ERDAS IMAGINE Virtual GIS smoothes all surfaces for “soft imaging”. It seems to be impossible to disable this function.

These images we produced for visually analysing the results. Visual analysis is not very precise, and some deviations are allowed. As far as we calculated, the smoothed surfaces differ from the triangular ones only about 0.2-2.5 percent.

3 Dynamic Anamorphoses

Undoubtedly, the anamorphoses are one of the most impressive types of images. They have, however, some negative features: in case of great disparity of the country's area and the value of the displaying index, the contour of the polygon can be deformed greatly (just squashed), and the quality of perception of the information decreases rapidly. The identification of the countries in greatly deformed regions can be made easier by using the dynamic anamorphoses.

In addition, the animation of the anamorphoses must be very interesting, showy and impressive, especially in case of displaying the dynamics of the same values (for example, the growth of the population in the world during the last 10 years).

We produced the program module, which allows us to build the dynamic anamorphoses interactively. Building the intermediate phases requires the interpolation of X,Y coordinates of nodes. We use the linear interpolation. The user can generate the intermediate phases between two images: it can be phases between the original map and anamorphose, or between two anamorphoses. It is possible to build them in real-time mode (25 frames per second, fps) and to see the animation in both directions.

The user can carry out following operations:

1. He can see the generated maps during the building of the anamophosis, scale it, resize it, and save it into raster (*.BMP) or vector (*.SHP, for ArcView GIS 3.0) file.
2. He can name two files with maps with the same topology, and build all of the phases of morphing from one image to another. These two images can be anamorphosed.
3. The user can watch the intermediate phases automatically (25 fps animation) or interactively (by mouse dragging). If automatically play is selected, then the duration of the animation can be defined.
4. The results of the automatic interpolation can be saved in .BMP frames (the number of frames can be set) or ArcView GIS shape files (*.SHP). In case of choosing raster export, the size of the raster can be set. It is also possible to resize and scale any part of the image.

Using this program module for the dynamic visualization of the anamorphoses, makes it easier to identify the contours through the animation. In addition, generating an animation between two anamorphoses seems to be very useful, since identification without this method is difficult.

4 Multidimensional and Multidynamic Anamorphoses

We are using cartographic images very often. We also use 3D images – like surfaces and block-diagrams, and others. After the introduction of multimedia, when it became possible for personal computers to show the full-screen video, the possibility of the displaying objects in their dynamics (thereby increasing its dimensions by 1). For example, animated 2D maps are 3D in fact (here time is the third dimension), and animated surfaces are 4D objects (Berlyant, 1996)..

In economic geography, we frequently use a number of dynamic rows of indices that are interesting to show on the map. Therefore, we can produce a series of dynamic 2D images – one animation according to the dynamics of one index. It will be easy to analyse each of these films. It will be very difficult, however, to make any correlation between them and to fully understand the significance. As we can see, there is a problem of displaying two or more dynamic indices in one image simultaneously, without building dynamic maps of correlation.

Let us solve this task for three dynamic indices, namely, the total population of the country, the GNP per person, and the expected average duration of life. We have these three values for each country in the world for every year from a period of the last 30 years.

1. Displaying the population. For every year, an anamorphose can be produced to illustrate population characteristics of all countries. All of the images must be raster in order to drape them over the grid surface. In case of creating the animation using these images, we would be able to see 2D deformations and

distortions of the country's shapes. Because the animated anamorphose is so expressive, one can readily see the object dynamics throughout their changing form.

Following this process a central point for each country's polygon must be selected. Let us define their X,Y coordinates. All of these points can be chosen "on sight", or by using any criteria – it does not matter. In case the country consists of several polygons, we must choose a central point for each polygon. This point must be immovable, and it must stay inside of the polygon during all period of animation.

2. Expected average duration of life. Its dynamics can be shown by using colours. Let's make multi-grade colour scale (for example, 100 different grades) to display each level of the showing index by separate colour. Therefore, if we paint each country polygon corresponding to its expected average duration of life colour, and if we do it 30 times – one time per one year, then we see crudely changing colours of the countries.
3. GNP per person. For this characteristic, the method of 3D-pyramidal block-diagrams, which were mentioned earlier, will be used. Here the 2D anamorphose, as mentioned in section 1 are used. All points of the country's contours have their X,Y coordinates, while the Z coordinate is equal to 0. The centres of the polygons, (as defined by the authors), also have their X,Y coordinates, however, their Z coordinate must be equal to the GNP per person for each country in one particular year. Thus an array of points is produced with their XYZ coordinates, and it is now possible to build a surface using them as reference.

Finally, if all of these methods are combined, a surface with "hills" and "planes" can be built, where the height of the hill is proportional to the average GNP per person in this country in one year; the form of its base and the base's area are equal to the population of this country in the same year; and the colour of this hill corresponds with their expected average duration of life (also in the same year). We can build one object like this for every year for the period or interest. Further, if they are animated the dynamic surface characteristics can be seen, where the hills are changing their shape, size, height, and colour.

It is also possible to show horizontals on these "hills", and so these images become metric and quite comprehensible.

This animated image has many dimensions: it includes the 3D surface, and here we have shown 3 dynamic processes.

5 Conclusion

The development of computer technologies enables new methods for displaying data of different kinds in one map using such methods like: 3D-anamorphosing, dynamic 3D-anamorphosing, multidimensional and multidynamic anamorphosing. All these methods allow us to show the data using the shape of the object (in addition to colour, hatching, and other traditional methods). The animated shape of the object can clearly illustrate the geographic statistics in a dynamic and effective way. Finally, these methods allow map users to visualise several different static or dynamic characteristics (attributes) in one image simultaneously.

References

Berlyant AM (1996) Geoiconics. Astreya Press, Moscow, (in Russian).

Gusein-Zade SM, Tikunov VS (1992) New methods of compilation of anamorphoses and their using. In: Proceedings 5th International Symposium on Spatial Data Handling, vol.1. August 3-7, 1992, Charleston, South Carolina, USA, pp 51-62

Gusein-Zade SM, Tikunov VS (1993) A New Technique for Constructing Continuous Cartograms. Cartography and Geographic Information Systems 20(3):167-173

Schematic Networks: An Algorithm and its Implementation [1]

Sergio Cabello and Marc van Kreveld

Institute of Information & Computing Sciences, University Utrecht, P.O. Box 80.089, 3508TB Utrecht, The Netherlands, {sergio, marc}@cs.uu.nl

Abstract

This paper addresses research issues and describes a prototype implementation for the computation of schematic maps. The generated layout mainly depends on two parameters, namely the shape of the schematic connections and the distance between connections. The quality of the results as well as different issues regarding its construction are discussed. Additionally several extensions that can be implemented leading to an improvement of the final layout of the schematic maps and related applications are suggested.
Keywords: network, schema, map, visualisation

1 Introduction

Most underground systems in cities and most railroad companies show their maps with highly schematised connections between the major stations. The positions of these stations are usually preserved (approximately), but the shape of each connecting route or connection has been reduced to a polygonal line with only a few links (straight line segments) that are either horizontal, vertical, or diagonal (45° or 135°). The main advantage of such maps is that they provide a quick overview of the layout of the network, without showing unnecessary information like the precise shapes of the connections.

Several earlier papers discussed the automated construction of schematic maps (Elroi 1988a, Elroi 1988b, Elroi 1991), but the topic has reappeared lately in more papers (Avelar & Müller 2000, Barbowsky et al. 2000). All of these papers make use of some local operator, such that after applying it several times, the map converges to a schematic one. Different criteria based on different measures are used to decide in which order, and to which parts of the map the operator should

[1] This research is supported by the dr.ir. Cornelis Lely Stichting.

be applied. Iterative approaches like these can have prohibitive costs in time in interactive situations, and in those papers, no theoretical time bounds, nor actual computation times are given. Even convergence has not been shown formally.

A related topic that is relevant to schematisation is the rendering of particular routes under queries (Avelar & Huber 2001, Agrawala & Stolte 2001). Although the topic has its own characteristics, there are several common aspects.

In a recent paper (Cabello et al. 2001), a relatively simple and efficient algorithm to produce schematised maps was presented. Its efficiency comes from the fact that it is combinatorial, in the sense that it does not apply methods to make the map converge slowly to a schematised one. Instead, it does some pre-computation and then places the connections immediately at their final positions. The algorithm is set up to obey a given set of requirements that are needed for the algorithm to work. There are also some choices the user can make for different visualisation styles. Then the algorithm produces a schematised map, if one exists, with the given requirements and choices. These are:

- Junctions of connections are not displaced.
- The allowed types of schematic connections can be chosen beforehand, but are fixed in the algorithm (schematisation models).
- Shared departure of different connections from the same junctions may be allowed or disallowed.
- A minimum separation distance between two connections may be chosen (not taking into account the first link of connections from the same junction).

The theoretical and algorithmic aspects of the algorithm were studied and proved in the aforementioned paper. This paper describes the implementation, results of experiments, and several practical aspects that were not previously addressed. The research questions of this paper are:

- Given the requirements above, and a schematisation model, how many connections (what percentage) can be placed for different data sets?
- How fast is the algorithm in practice?
- If not all connections can be placed, how should the algorithm deal with non-placeable connections?
- Can the algorithm be adapted to incorporate through-going connections at junctions, that is, to enforce that some connection is the prolongation of some other connection at a junction?
- How should the algorithm deal with very short connections?
- How should the minimum separation distance be implemented?
- Can obstacles (other map features) be incorporated in the algorithm?
- Can the output of the algorithm be improved aesthetically, without sacrificing the number of placed connections?
- Can the minimum separation distance be maximised, without placing fewer connections?

These questions are addressed by developing methods that extend the algorithm of Cabello et al. (2001), and provide a prototype implementation in Java which can be tested at the web page http://www.cs.uu.nl/archive/sw/schematic-map/. Three different data sets with railways and roads of countries were chosen as test data. The most important conclusion is that the algorithm is fast in practice, places nearly all connections, and gives topologically correct, aesthetic maps as in the output for Fig. 1.

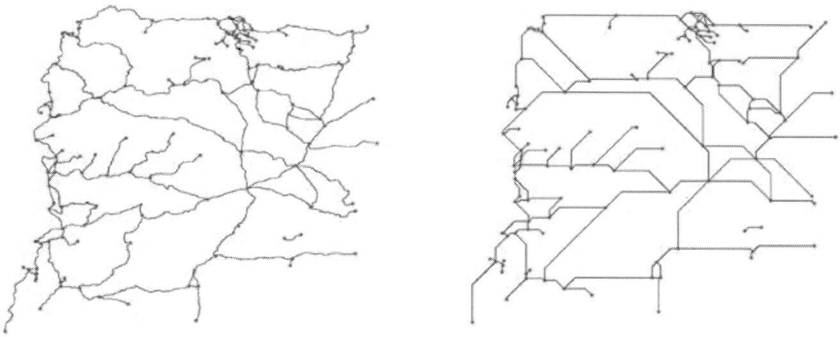

(a) (b)

Fig. 1. Northwest of Iberic Peninsula, **(a)** original map after editing with the prototype, **(b)** the schematised version made by the prototype

This paper begins with a global description of the algorithm of Cabello et al. (2001), which is needed to understand its features and limitations. The implementation prototype and its possibilities are then described. Section 4 gives the results obtained from experiments on different data sets, with different settings of the implementation. Further visualisation possibilities of the method are presented followed by the conclusions.

2 The Algorithm and its Parameters

In this section, an explanation is provided concerning the fundamental ideas about the algorithm of Cabello et al. (2001). Assume a network or map is presented, such as a railway map, and the desired type of schematic connections to be used, such as the 3-link connection type are specified. In brief , the algorithm is as follows: first a top-to-bottom ordering of the connections of the original map is computed, and then the schematic connections are placed, one by one, following this order. As much freedom as possible is left for connections to be placed later. The concepts and ideas are explained in the following sections..

Imagine a physical model of the original map where each connection is made of an elastic cord with fixed endpoints. One natural approach to construct the schematic map could be to continuously deform some connections in order to make them more schematic and simple whilst keeping the topology of the whole network. This process can be continued until some schematisation level is reached. A similar idea is used in Avelar & Müller (2000), where the positions of the junctions and important points are changed iteratively to create a schematisation. With this type of approach, the schematic map keeps the topological relations from the original map, as well as some spatial relations. For example, the cyclic order of the roads leaving one city is as in the original map.

Some observations on this approach imply the need for a combinatorial algorithm. As a rule, when a connection is deformed, as described, it is useful to provide as much freedom as possible for the next deformation. If, for example, one connection in the centre of the map is deformed, its schematised placement will limit the possibilities of placement of other connections both above and below it. Hence, it is not possible to decide which is the best possible deformation. If, however, the connection at the top of the map is deformed and it is deformed such that it leaves as much space as possible below it, will then influence later deformations in a limited manner. This process is therefore continued in which the top-most connections are deformed among the ones not yet schematised.

When deforming one connection to leave as much space as possible for later, all connections below it play no role (that is, do not influence its placement). All connections already placed above it, however, have to be observed, otherwise the topology of the network could be changed. This brings us to the final structure of the algorithm:

- Extract all connections from the original map and order them from top-to-bottom in a list.
- Remove all connections from the original map, but keep the endpoints where they are.
- Place, one by one and leaving as much space as possible for later, each connection from the ordered list in the map. Each newly placed connection must be of the desired shape and below the previously placed connections.

To order the connections a decision must be reached for each pair of connections, with respect to their above-below relation. Observe that this order relation is computed in the original map and will be preserved in the schematic map, so it is an invariant relation under continuous deformations of the connections. For this reason, one point P is considered to be below a connection c if the upper vertical half-line from P intersects any continuous deformation of the connection c that does not pass over P. To decide if P is above is done in the same way but using the lower half-line. It can also be the case that any deformation of c intersects the half-line above P and the half-line below P, in which case P is both above and below c. Some cases are shown in Fig. 2.

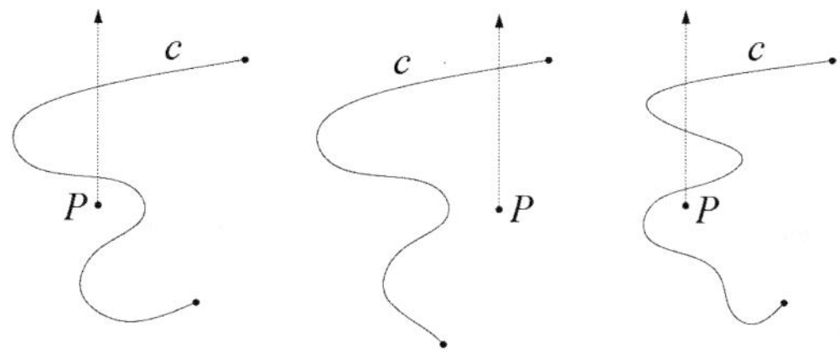

Fig. 2. Left: P is neither above nor below c. Center: P is below c. Right: P is below and above c

To determine the above-below relation of two connections the endpoints of one connection are compared with the other connection, and vice versa. Details of the efficient computation of the ordering are given in Cabello et al. (2001).

In the algorithm, for placing the next connection, some considerations have to be taken into account. For example, to facilitate the legibility, the schematic connections must retain some distance between them. Observe that the distance between two connections depends on whether they are sharing an endpoint or not (see Fig. 3). When two connections do not share any endpoint, the distance is the minimum over all the vertical distances from any link (edge) to the other connection. If two connections share an endpoint, the minimum must be taken, but without taking into account the distances from the first link to the other connection. The reason to use only the vertical distance is due to the fact that an ordering of the connections from top to bottom is used

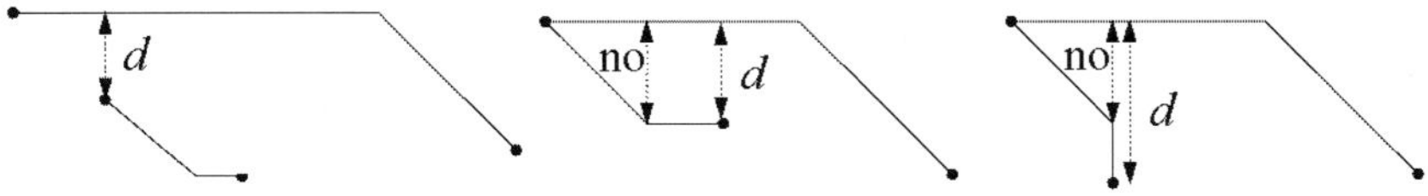

Fig. 3. The distance between connections depends on whether they share endpoints or not. The distances labelled "no" are not considered

The algorithm can fail in two different ways. When computing the order, it can happen that the connections cannot be ordered. For example, the left part of Fig. 4 gives an example where no connection is topmost, since each one has another connection above it. The algorithm can also fail because the next place of the connection cannot be placed. For example, to the right in Fig. 4, use is restricted to the 2-link orthogonal (axis-aligned) schematic connections. As a result the dashed connection cannot be placed without intersecting the other one.

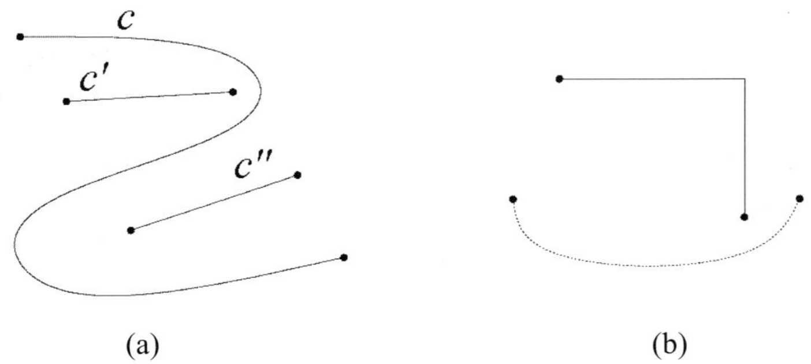

Fig. 4(a) the connections cannot be ordered: *c* above *c'*, *c'* above *c''*, and *c''* above *c*, **(b)** the schematised version of the dashed connection cannot be placed if the orthogonal 2-link connections are restricted

Although the idea of the algorithm can be applied to any type of schematic connection, certain statements can be proved in the following cases:

- 2-link orthogonal;
- 3-link: orthogonal, diagonal, orthogonal;
- 3-link: diagonal, orthogonal, diagonal;

where all of them have to be shortest in their class.

Theorem: For the above types of schematic connections, there is an algorithm that, in $O(n \log n)$ time, schematises any map with n segments. Furthermore, if the algorithm fails to report a schematic map, then no schematic map with the desired connection type exists.

3 The Prototype

An experimental prototype using the ideas above has been implemented as a Java applet. Although some techniques to speed up the algorithm that were explained in Cabello et al. (2001), are not used, the processing time does not appear to be problematic.

Editing tools were included in the prototype in order to make it more flexible and interactive for the user. Observe that the algorithm described above does not move the endpoints of connections, but some relocation of them may be desired, so this tool can be important. For example, it could be useful to identify and merge the endpoints of connections when they are very close in the schematic map. Observe that when two connections have a slight overlap in the vertical direction and they are closer in the vertical direction than allowed, that this algorithm would not schematise the lower one, see Fig. 5. In addition, the presence of intersections close to endpoints plays an important role, since problems arise with ordering the connections. Sometimes such problems can be resolved by editing, see Fig. 5.

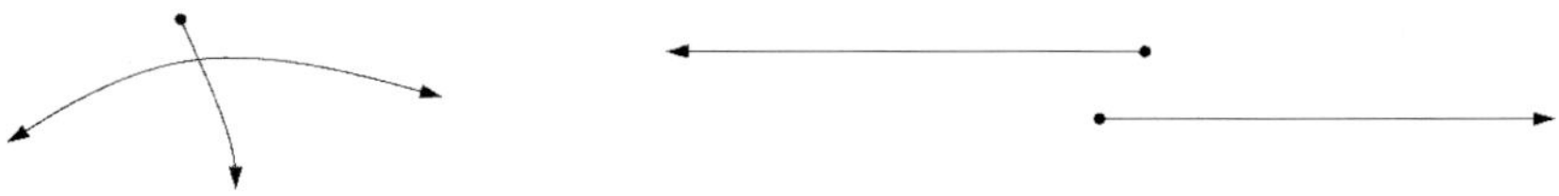

(a) (b)

Fig. 5. Some data may produce problems if it is not edited, **(a)** intersections may produce a problem to decide which connection is above. Thus, it must be topolgically correct. **(b)** If two endpoints are closer than the minimum allowed distance, and they are not identified, the lower connection cannot be schematised.

The prototype has four different screens, each one referring to a different map (see Fig. 6).

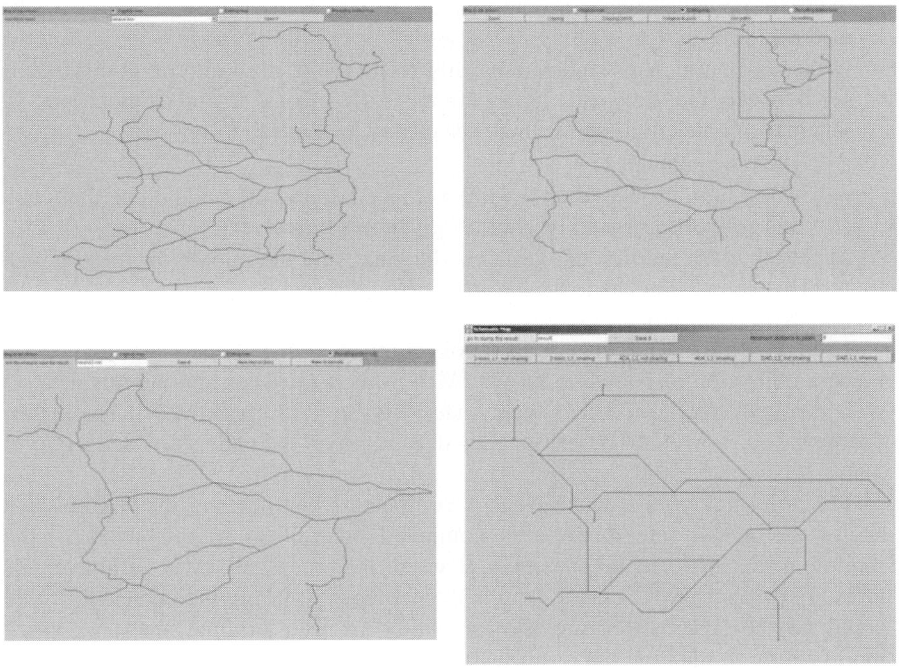

Fig. 6. Four screenshots of the prototype

- The first screen (original map) is used for opening the file with the data. This map cannot be changed and it remains the same all the time.
- The second screen (editing map) allows to select a rectangular region from the map and perform some operation on that region:
  - Zoom on that region;
  - Clip: selects all connections partially in the rectangle;
  - Clip (strict): selects all connections lying completely in the rectangle;

  - Collapse to point: moves all endpoints and junctions in the rectangle to the centre of it;
  - Join connections: if there are only two endpoints in the rectangle, both incident to only one connection, it joins the corresponding connections to one new connection (the two endpoints merge and become an intermediate vertex of the connection). If there are several connections inside the rectangle, it merges any two connections incident to some endpoint by turning it into an intermediate vertex;
  - Do nothing: do not make any change.
- The third screen is used to visualise the editing results, which includes buttons to compute the intersections of the connections. As this is an expensive operation for maps with many connections, it is recommended - after using it - to save the map such that it does not have to be repeated. Pressing the button "Make schematic" opens a new window in which the schematic map will be made.
- In the schematic map window (fourth screen), the desired minimum distance is set between the schematic connections. As a result, the different types of schematic connections can then be selected, which are as follows;
  - 2-link orthogonal;
  - 3-link, with the first and last link orthogonal, and the middle one diagonal. This connection has to be shortest given this type.
  - 3-link, with the first and last link diagonal, and the middle one orthogonal. This connection has to be shortest given this type.

The input data is organised as a collection of connections, each connection being a collection of points. A format (MIN) was created for reading this data, as well as a converter from ESRI shape (.shp) files to the format (MIN). The results can be saved as MIN or Postscript files depending on the extension used (.min or .ps).

The order among the connections is computed when opening the window that shows the fourth screen. So it is computed only once. In this window, the minimum distance can be specified. Because it is a visualisation parameter, its units are pixels. The schematic map is then computed, where the connections that could not be schematised are shown in blue in their original shape. A small modification to improve the result was introduced in the algorithm> it computes a schematic map *M* using the specified minimum distance *d*; *d is then increased* to find the biggest value that produces a schematic map with the same number of schematised connections as *M*. The schematised map with those settings is the output map. The search for the best value of *d* is done by a binary search, so that not all values for *d* have to be tried incrementally.

4 Experimental Results

The prototype was used with maps from Canada, Ireland and Spain, the number of connections per map are given in Table 1. In all of them, the time of computation is within 3 seconds. However, the most time consuming part is the binary search to find the optimal distance, as explained at the end of the previous section. In all examples and results in this section, the smallest allowed vertical distance between two connections was set to 10 pixels. Some of the results can be seen in Figs. 7-9.

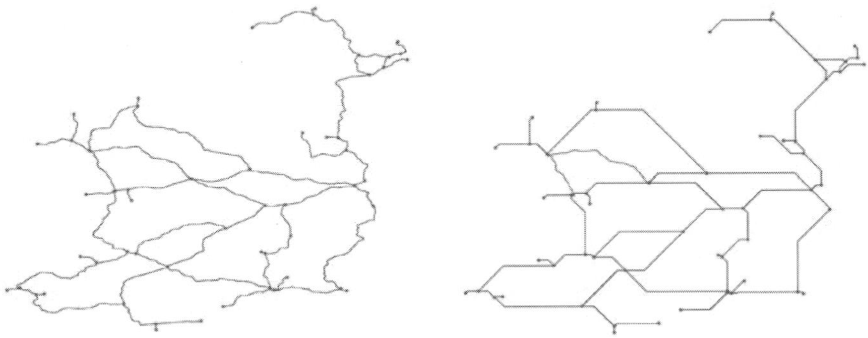

(a) (b)

Fig. 7. Ireland, **(a)** original railway map, **(b)** schematic map with (orthogonal, diagonal, orthogonal) connections and without sharing departure

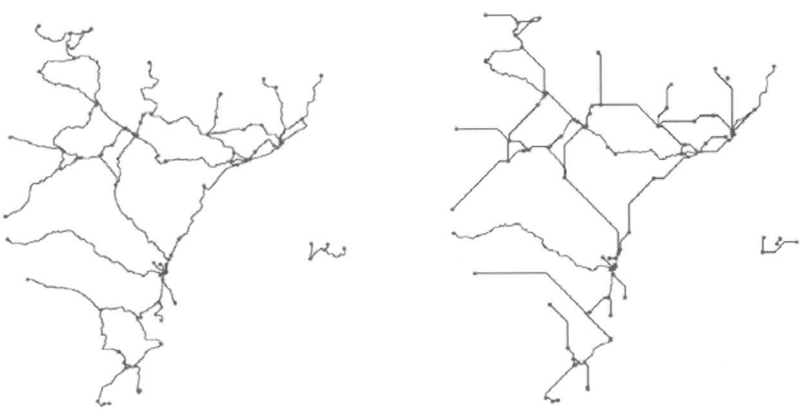

(a) (b)

Fig. 8. East of Spain, **(a)** original railway map, **(b)** schematic map with (orthogonal, diagonal, orthogonal) connections and without sharing departure

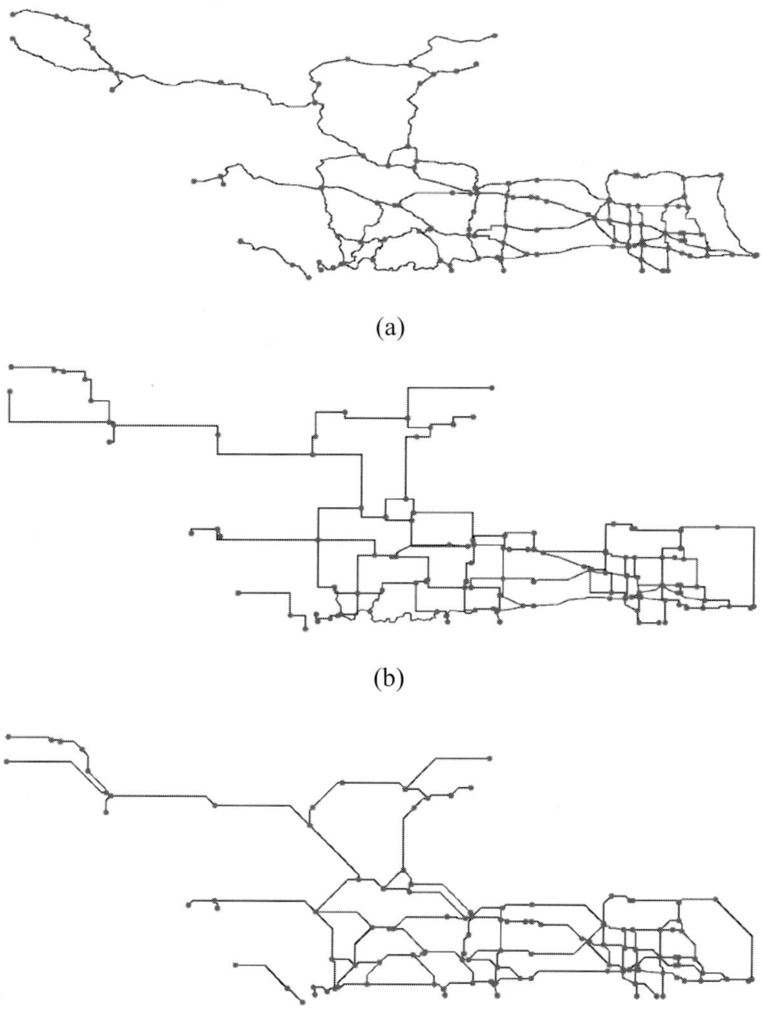

(a)

(b)

(c)

Fig. 9. West of Canada, **(a)** original road map, **(b)** schematic map with two links per connection without sharing departure, **(c)** schematic map with (orthogonal, diagonal, orthogonal) connections without sharing departure

For the maps of Ireland and Canada, the top-to-bottom placement order was computed without any problem. Problems were encountered when computing among the connections for the map of Spain. This was due to inaccuracies in the data, which is topologically not correct. In particular, in the part shown in Fig. 8, there were intersections between different connections. After computing and resolving those intersections, the placement order was computable.

As can be seen in the pictures, some connections cannot be placed because they would intersect some connection already placed, they would give sharing departure, or they would not keep the proper distance. Table 1 shows the connections that were present in the original data, and the percentage of them can be placed without and with sharing departure. As can be seen, the improvement is only small when sharing departure is allowed. After analysing the pictures in the prototype, which allows zooming in the original map, one can see that the biggest source of problems is that the endpoints close together were not identified and merged. So, most of the non-schematised connections would have been schematised after some editing of the map beforehand.

Table 1. Percentage of placed connections for the maps shown in this article

	Number of connections	Not sharing departure	Sharing departure
Fig. 7 right	65	93.8 %	93.8 %
Fig. 8 right	100	76 %	79 %
Fig. 9 centre	155	74.1 %	86.4 %
Fig. 9 bottom	155	89.6 %	89.6 %

5 Other Visualisation Issues

Some features can be listed that improve the visualised results. Most of them could be included in the prototype in a straightforward manner, however, this has not been done yet. These extensions are:

- Obstacles can be managed when the connections keep the same spatial relation with them (passing above or below). For example, when the obstacles are points, they can be treated as connections of length zero in the algorithm.
- The same algorithm can be applied to boundaries of countries or administrative maps as well (for instance, for choropleth maps). For this purpose, the boundaries should be split into a few pieces.
- Usually, sharp corners do not look nice, and the visualisation can be improved by rounding them.
- Sharing departure from one endpoint can be avoided by representing the endpoints by rectangles. The connections can then leave the rectangle as a bundle of parallel links. With a few adaptations, it is possible to produce schematic maps that are commonly used for subway lines in cities.
- When some endpoint represents the junction of two or more major transportation lines, it is desirable to have continuity of the connections of the same line, where one is the prolongation of the other one on opposite sides of the junction. In some of the cases, this can be handled with a bottom-up post processing in the schematic map.
- After computing the order among the connections, other ways of placing the connections, such as alternating between the topmost and the bottommost

connection, may give better results. Here, the bottommost connections should be placed in the bottommost position. This improvement avoids the risk of a cushioning effect at the bottom of the map. In the examples, however, such effects are not apparent even without this adaptation.

6 Conclusions and Further Research

In this paper the algorithmic approach of Cabello et al. (2001), has been presented and discussed through the review of a prototype implementation. Several extensions to that paper, which improve the visualisation, have been incorporated, and also editing tools to obtain topological consistency of the input map. Experiments show that most problems can be resolved with some editing of the data, especially using identification of junctions and the merge operator The prototype presented in this paper can be accessed and used via Internet at http://www.cs.uu.nl/archive/sw/schematic-map/.

The algorithm presented (Cabello et al. 2001) is the first method to take a combinatorial approach to schematic map construction, and as such is very efficient. This paper shows that the method is fast, and the approach is versatile. In addition, various adaptations and extensions allow the visualisation of other styles of schematic maps, including subway maps and schematised choropleth maps.

References

Agrawala M, Stolte C (2001) Rendering effective route maps: Improving usability through generalisation. In: Proc. Siggraph 2001. pp 241-250

Avelar S, Huber R (2001) Modeling a public transport network for generation of schematic maps and location queries. In: Proc. 20th Int. Cartographic Conference. pp 1472–1480

Avelar S, Müller M (2000) Generating topologically correct schematic maps. In: Proc. 9th Int. Symp. on Spatial Data Handling. pp 4a.28–4a.35

Barbowsky T, Latecki L J, Richter K-F (2000) Schematizing maps: Simplification of geographic shape by discrete curve evolution. In: Spatial Cognition II, LNAI 1849. pp 41–48

Cabello S, de Berg M, van Dijk S, van Kreveld M, Strijk T (2001) Schematization of road networks. In: Proc. 17th Annu. ACM Sympos. Comput. Geom. pp 33–39

Elroi D (1988a) Designing a network line-map schematization software-enhancement package. In: Proc. 8th Ann. ESRI User Conference [online]. Available from: http://www.elroi.com/fr2_publications.html

Elroi D (1988b) GIS and schematic maps: A new symbiotic relationship [online]. In: Proc. GIS/LIS'88. Available from: http://www.elroi.com/fr2_publications.html

Elroi D (1991) Schematic views of networks: Why not have it all. In: Proc. of the 1991 GIS for Transportation Symposium. pp 59–76, Available from: http://www.elroi.com/fr2publications.html

Developing Lightweight, Data-Driven Exploratory Geo-Visualization Tools for the Web

Erik Steiner, Alan MacEachren, and Diansheng Guo

GeoVISTA Center, Department of Geography, 302 Walker, Penn State University, University Park, PA, USA, Phone: 814-865-7491; Fax: 814-863-7943
Correspondence to: maceachren@psu.edu; www.geovista.psu.edu

Abstract

This paper details efforts to development a set of prototype, web-based geo-visualisation tools. The focus is on design and implementation of lightweight tools that can run in standard web browsers and access data, stored in a remote database. The web geo-visualisation tools incorporate several standard exploratory spatial data analysis methods, including linked brushing and interactive animation. These tools are constructed using Macromedia Flash, a commercial software application that produces content for the web using a publicly available file specification called SWF. The advantages and disadvantages of Flash for geo-visualisation are discussed. One advantage is that it supports links to remote databases. Accordingly, the use of XML as the communication syntax for the geo-visualisation tool-to-database link is discussed. Two solutions to dynamic database access have been implemented, one using a stand-alone Java server and the other using Java servlet technology; the advantages and disadvantages of each approach are discussed.
Keywords: interactive maps, visualisation, exploratory spatial data analysis, remote database access, web mapping

1 Introduction

Dramatic advances in 'mapping' have occurred during the past decade. Among the most fundamental are the change from maps that communicate a single message, statically, to maps that support exploration of multiple perspectives, dynamically. Although dynamic maps have been appearing on the WWW (world wide web – hereafter, simply web) for several years, most of the innovation in geo-visualisation during this time has been directed to expert users working on desktop (or more powerful) computers. Technological and societal changes are making it possible to extend from this base toward development of highly interactive geo-visualisation tools designed to support non-expert needs as well as the web-delivery of such tools.

In this paper, a set of lightweight, data-driven geo-visualisation tools for the web that are being developed as part of a 'Digital Government' (DG) research initiative in the U.S are both described and demonstrated. The focus of our contributions to the wider DG effort is on dynamic maps and graphics to support visual delivery and analysis of federal statistical summaries (compilations of numerical data, such as those from the national Census, that are published in aggregate form).

The initial section below provides some background for both the kinds of exploratory geo-visualisation tools being developed and for the DG project they are a part of. Next, the implementation of a web-based prototype using Macromedia Flash as the development platform is briefly presented. The prototype includes dynamic and linked exploratory geospatial data analysis methods (methods that are now relatively standard for systems targeted at expert analysts but, in the implementation presented here, are designed for use by non-expert users). The subsequent section outlines the mechanisms constructed to support data access from a commercial database (Oracle) through the Flash (map-based) interface. We close with a brief discussion of possible extensions to these tools, some of the limitations encountered, and plans for application of an iterative, human-centred design approach to tool development and assessment.

2 Background

Interactive maps for the web have become quite commonplace, and substantial investments in both geospatial research and applications are being directed toward serving of geospatial data via the web; see (MacEachren, 1998) for a recent review. There have also been several successful web-implementations of exploratory geo-visualisation methods; see (Andrienko & Andrienko, 1999; Dykes, 1997; Harrower, MacEachren, & Griffin, 2000). Still, it remains a challenge to design web-based interactive geo-visualisation tools that meet professional cartographic design standards while being flexible enough to achieve the core geo-visualisation goal of supporting a multi-perspective approach to data exploration. This challenge generally requires that developers (and often users) master formal computer programming languages; and even the best of current tools impose serious limitations from the perspective of graphic design – limitations that are likely to impede the transitioning of multi-perspective geo-visualisation methods from expert to non-expert user.

The goal of our particular DG project is to develop 'quality graphics' that support the understanding and flexible analysis of statistical summaries produced by eight federal government agencies in the U.S. The research has a dual focus on provision of relatively sophisticated exploratory data analysis tools to support expert work in generating these statistical summaries (e.g., to support population projections by the Bureau of the Census) and to support access to derived information by the public (with the target audience here ranging from the ordinary citizen interested in health-environment issues to the university researcher conducting a demographic or economic analysis).

The Flash-enabled tools described below are intended to support this dual focus. First, Flash is being used to provide both an environment for formal cognitive experimentation and for rapid prototyping. The rapid prototyping allows dynamic data exploration methods to be quickly implemented and assessed, before committing resources to final implementation as part of a suite of Java-based tools being constructed within GeoVISTA *Studio* (MacEachren et al., 2001) and/or Illumitek's nViZn. Second, we are experimenting with Flash as a final web-delivery mechanism that supports user-friendly but flexible exploratory geo-visualisation tools that can be used effectively by non-experts.

In sections 3 and 4 below, we describe prototype, Flash-based, web geo-visualisation tools we have implemented using a client-server approach. The first section focuses on design and implementation of interactive exploratory maps that run in standard web browsers and that support dynamic linking between visual representation forms and user construction of temporal and non-temporal animated sequences. Then, the next section details the process of linking these interactive maps dynamically to a remote database.

3 Design of Dynamic Maps

We have created several dynamic maps using Flash in an effort to determine capabilities and limitations of this. Here we illustrate, and describe briefly, three of these prototypes.

3.1 Linked Geographic Brushing

In our initial prototype, we implemented a choropleth map and scatterplot display using state-level data of the United States. The interface was designed in an earlier version of Flash (4.0) that supported data access only through loading URL Encoded text files. Later prototypes (discussed in sections 3.2 and 3.3) implement more sophisticated database support. With our initial tool, once the data are loaded, they are linked to their geographic entities (states) and to a scatterplot representation. This functionality is illustrated in Fig. 1 (below). The figure shows a screen capture from a custom dynamic application built using Flash's ActionScript (a scripting language that is similar, conceptually, to JavaScript while emphasising graphic applications). Flash's graphic design tools allow the designer to construct the visual elements from which the interface is composed, while ActionScript supports control of the global characteristics of the display objects (colour, size, location, transparency, etc.). In the case of the choropleth map shown, state elements were given a lightness value that corresponded to their data value, and in the case of the scatterplot, individual points were given positions according to their values on two separate attributes.

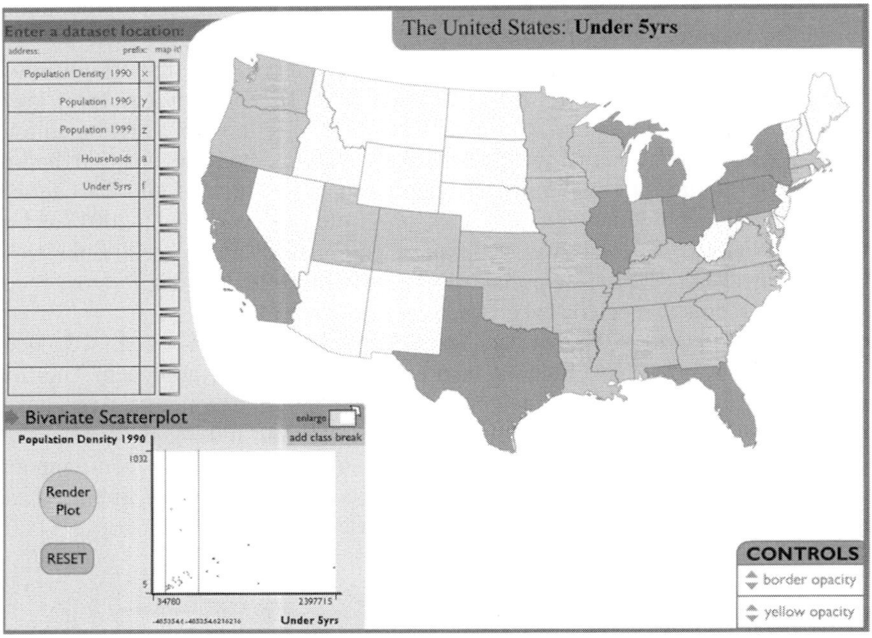

Fig. 1. Prototype 1: Linked geographic brushing interface

The prototype demonstrates the concept of linking a scatterplot and choropleth map dynamically, so that manipulation of one has an impact on the other. For previous research on the topic of linked geographic brushing; see (Dykes, 1997; Haug et al., 1997; Monmonier, 1989). Here, we implemented a one-way communication from the scatterplot to the map in which user interactions within the scatterplot are propagated to the map. Specifically, users can define class breaks on the x-axis of the scatterplot (the axis depicting the variable mapped) by dragging a vertical line that represents each break point. These breaks are then immediately propagated to the map display, with the colour assigned to each state altered to reflect its new class assignment. Thus users can define classes based on the characteristics of one variable's relationship to another and evaluate the impact of those class break choices on the spatial pattern of that initial variable. An interactive version of the interface may be accessed at the following address: http://www.geovista.psu.edu/publications/Beijing01/SteinerICA01/loadusa/loadusa5.html.

Our success with the Flash 4.0 environment demonstrated the feasibility of developing linked displays that include lightweight and dynamic statistical maps (the example above involves a 93KB download to the web browser). One drawback that we encountered in our initial testing was a lack of flexibility related to loading new data into the interface. Flash 4.0 included very limited support for robust data structures. The highly customised nature of the individual components of this first prototype (scatterplot, choropleth map) also constrained the future application of these representation forms to new interfaces.

3.2 User-Controlled Animation Sequences

A second prototype interface we designed combines the linked brushing concept with user-controlled animation sequences and more sophisticated data access. The interface is designed to allow users to load and explore county-based thematic data attributes through a cartographic display. Variables are loaded using eXtensible Markup Language (XML) syntax (supported in Flash 5.0 and higher). XML is rapidly becoming a standard information exchange language; it is supported by commercial databases, with efforts to develop a geospatial variant (GML) well under way (Open GIS Consortium, 2000; Zaslavsky, 2000). Support for XML makes it possible for Flash to access a database efficiently. In our case, an Oracle database server stores the relevant datasets (see section 4 for details on database access issues).

The dynamic environment illustrated in Fig. 2 is designed to interpret the structure of an incoming XML file and, subsequently, dynamically construct

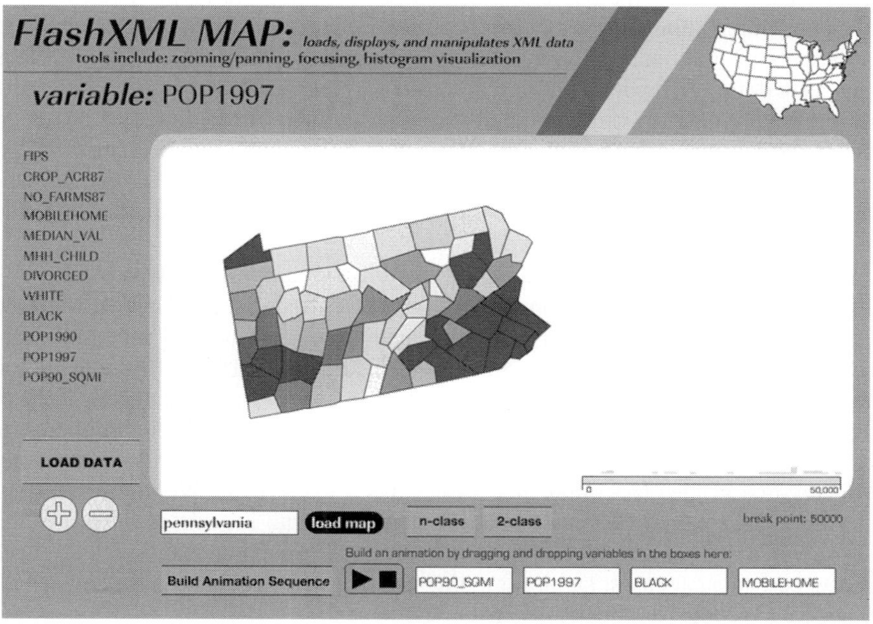

Fig. 2. Prototype 2: XML-based map incorporating animation sequences

visual and data elements from this code. In this prototype, once the data are processed by the Flash Player, they are stored locally and may be explored and visualised without further communication with the server.

In prototype 2, three visual representation forms are generated upon data input: a choropleth map, a histogram display, and roll-over buttons for each variable. The number and names of the variables are defined based on the XML metadata. The prototype includes standard zooming and panning functions (for the map) and the three elements of the interface are dynamically linked. For example, if a user clicks or does a 'mouse over' on a variable button, the attribute selection is immediately propagated to the map and histogram; if the user interacts with the histogram, the changes are immediately propagated to the map.

Flash is an object-oriented environment in which each display entity (e.g., county) controls aspects of its own display. Labels for the elements in the variable list (left column of the interface) are generated from metadata within the XML import file. On import, the data associated with each variable are distributed to the geographic entities (e.g., counties). When the user chooses a new variable (by mousing-over its label), each geographic unit now 'knows' to refer to the selected locally stored attribute. Thus, each map object accesses data simultaneously, without referring to the original dataset or making a new call to the server. Similarly, the histogram object recognises the shift to a new variable and will immediately update its form based on the user's selection.

The histogram supports linked geographic brushing in the form of a moveable class break similar to the scatterplot described in the first prototype. The user may click and drag along the base of the histogram display to create a two-class representation of counties above and below a certain class break; see Fig. 3. Users can also define a break point for a single variable and then jump among the different variables to compare the spatial distributions of the variables based on this cut-off criterion. The dynamic interaction between a statistical distribution representation (histogram) and a cartographic representation (choropleth map) may allow users to explore geographic datasets in greater depth with the possibility of generating novel hypotheses.

Users can visualise each individual variable alone or may choose to rapidly shift between variables to compare attribute values and distributions, an example of non-temporal animation. For more information on non-temporal animation, see: (DiBiase, *et al.*, 1992; Egbert & Slocum, 1992; Peterson, 1993). Building such a sequence is accomplished by dragging variable labels from the list in the left margin to the slots in the "build animation sequence" row. As the animation is run, users can interact with the histogram display to produce a dynamic brushing effect with a single class break that is transmitted to each variable in the animated sequence. The same tools will work for temporal data sequences, although the drag-and-drop method for building sequences is not practical for long time sequences. In such cases, supporting drag-and drop for sequence end points (with other data accessed automatically) may be practical.

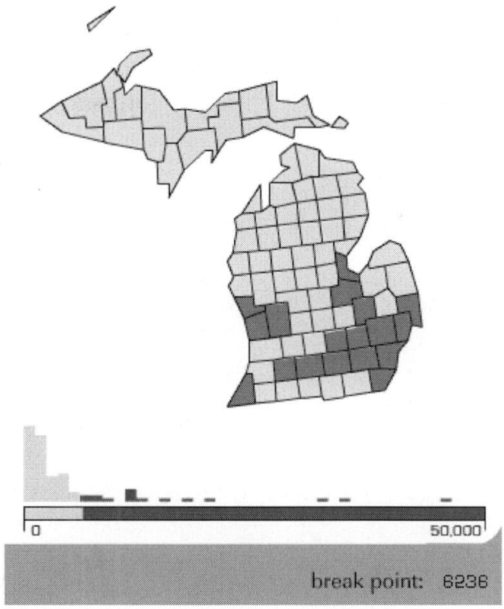

Fig. 3. Prototype 2: Histogram-Map link with draggable class break

This prototype demonstrates the feasibility of using dynamically-loaded XML files to create dynamic visual representation forms using Flash. While the interface remained lightweight (114KB), the data loading process proved to be a performance bottleneck (taking perhaps 5 seconds to load Pennsylvania). Once the data were loaded, however, the system response was immediate and satisfactory.

3.3 U.S. by County - Interacting with Detailed Maps

The first two prototypes described, dealt with both a limited number of data variables and limited number of geographic entities. To further explore the capabilities and limitations of Flash as an environment for web-based geo-visualisation (particularly in support of the missions of our Digital Government partner agencies), we created a new application for county-level data of the entire United States.

Generating a base map of the U.S. to accept the XML input was a labour-intensive process. As explained in more detail below, it involves manual labelling of each geographic entity (more than 3000 here). Once completed, of course, the map is reusable.

While we streamlined the previous ActionScript code to perform only absolutely necessary operations on the dataset, there was a substantial improvement of system performance with this county-based map. An XML file containing "fips" codes (a standardised coding scheme used in the U.S.) with a single attribute took about one minute to load and the system response once loaded was unsatisfactory for linked brushing and animation. This is a recent prototype and we are currently evaluating the application in an effort to speed up the processing.

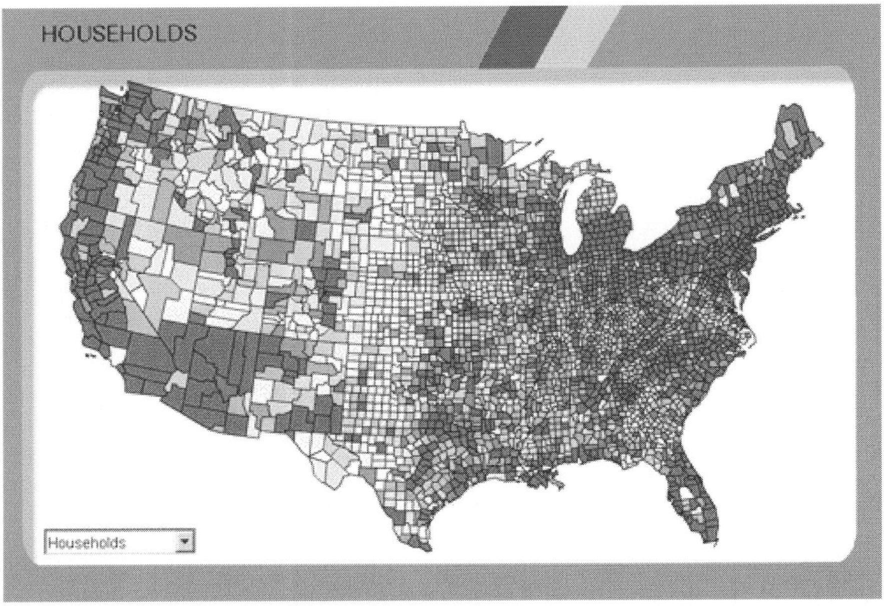

Fig. 4. Prototype 3: U.S. by county, loaded through large XML file

Flash provides support for interpreting and constructing XML files to allow efficient and robust client-server communication with small datasets. Flash's basic XML parser allows the designer to build interpreters for the expected XML file in order to store attribute names, values, and metadata. The following section details the database side of the prototype designs we have implemented.

4 Database Access

To support access to and visual exploration of a wide range of data using the tools described above (and their extensions), we are building a database of demographic, health, and related statistics on a dedicated Oracle database server. The initial implementation of each example above relied on pre-processed data stored as XML files, with data embedded in the client application at download time. Creating a dynamic link between the web-based geo-visualisation client and the database server, in contrast, offers several advantages. Among them: 1) the user can freely select and focus on any interesting subset of data for exploration; 2) with a centralised database, we eliminate many problems regarding consistency checks, maintenance, and updates of data; 3) through a standard interface (here we adopt the XML format as the communication syntax), the visualisation modules and database management are separated, which can ease the evolution of both components (e.g., allowing us to easily replace the Flash-based geo-visualisation modules with ones written in Java). Due to the limited ability for Flash applications to process data (particularly if we are to achieve the desired lightweight client that is sufficiently responsive for real-time interaction), we also need the server to do most of the data query and processing tasks and then serve the client with a small ready-to-visualise dataset.

We evaluated two technical solutions to this dynamic connection. One is through Java Servlet technology. The other is to develop a dedicated Java server. While there are other ways to achieve the dynamic connection between Flash and a database (namely through ColdFusion MX, Macromedia newest server technology) we chose to test two available solutions in our lab. Both are described and compared below.

4.1 Connection through Java Servlets

A servlet is a Java class used to extend the capabilities of web servers. Servlets provide a component-based, server- and platform-independent method for building web-based applications, without the performance limitations of CGI programs. So, a servlet can be thought of as an 'applet' (a Java class for web browsers) that runs on the server side—without a 'face'. It works as follows (Fig. 5).

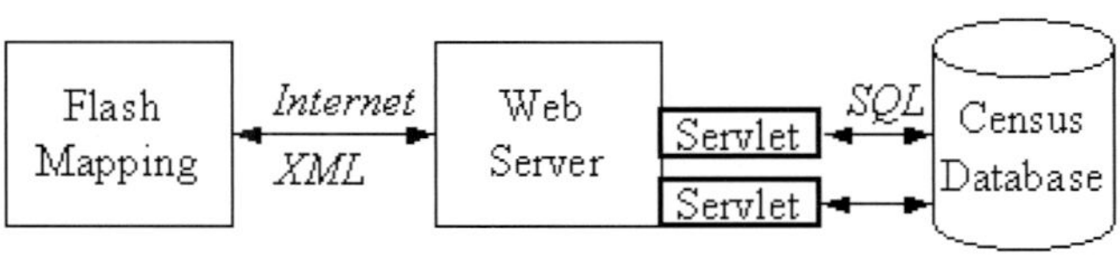

Fig. 5. Connection through Java Servlet

The Flash client can send a request, such as:
http://rangoon.geog.psu.edu/servlet/OracleConn?tablename=counties&statename=Pennsylvania&attribute=pop1990,white,black

Here, http://rangoon.geog.psu.edu/ is the HTTP address of the web server, OracleConn is the name of the servlet, tablename=counties &statename= Pennsylvania &attribute= pop1990, white, black is the parameter string for this request. The OracleConn servlet can parse these parameters and compose an SQL query:

Select POP1990, WHITE, BLACK from COUNTIES where statename= 'Pennsylvania'.

The servlet will then connect to the database through JDBC (Java Database Connectivity) APIs, execute the above query, and return the data. In our implementation, in addition to retrieving data, the servlet also pre-processes the data (e.g. calculates the mean value, median value, and other information) as required by the Flash client. Then the retrieved data and the derived information will be packed in XML format and passed to the client. From the Flash client's perspective, this request is exactly the same as downloading an XML file from the web server—it does not know or care how the file is generated. To be aware of which data are available in the database, the Flash client needs to retrieve associated metadata from the database before issuing any query requests. The metadata is also passed in XML format.

4.2 Connection through a Dedicated Java Server

Another choice for implementing the connection between the web geo-visualisation tools and the remote database is to build a dedicated Java server. Adopting this strategy, we built a mini server that does not require embedding the service in a web server (as we did with the Java servlet). The communication protocol is still XML. With our own server, we have the flexibility and freedom to design a more complex protocol than the parameter string used by Java servlets. For this implementation, the Flash client will first build a socket connection with the server. Then all the communication (including query, metadata, and data) between the Flash client and the Data server will be achieved through XML messages sent back and forth through this socket.

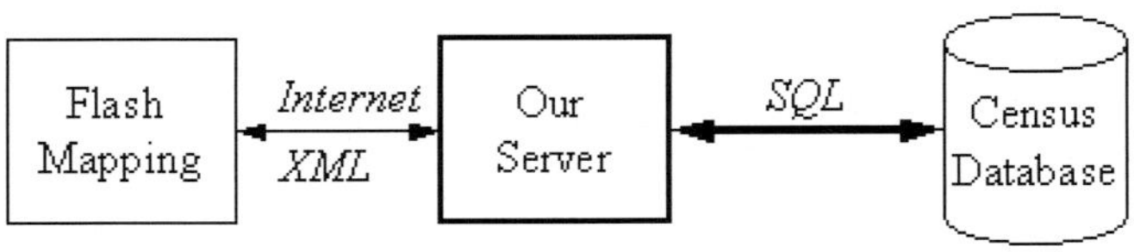

Fig. 6. Connection through a dedicated server.

For example, our server runs on the machine rangoon.geog.psu.edu at port 2001. The Flash client first builds a socket connection to this server. Then it can issue a query such as:

```
<select>pop1997, white, black</select> <from>COUNTIES</from>

<where> <statename>Pennsylvania</statename> </where> <order> pop1997 </order>

<derive> mean, median </derive>
```

The communication protocol can be more complex than this; this is dependent upon the agreement between the client and the server. The server then parses the above request into a SQL statement:

Select pop1997, white, black from COUNTIES where statename = 'pennsylvania' ORDER BY pop1997

Once parsed, the server will calculate the mean and median values of the data (as <derive> tag pair required) and pass it to the client in XML format.

4.3 Comparison of Two Techniques

There are both advantages and disadvantages for the two alternative forms of client-server connection described above. These are outlined for each below.

With a Java servlet, it is easy to implement and deploy a new service. However, the request-response communication is limited to a parameter string. Thus, it is hard to express a complex request. The servlet methods seem suitable for usability experiments, each of which can be simple and often need quick implementation. Another advantage of a Java servlet is that it is embedded in web servers and can be accessed easily through standard http protocol and the standard port 80, which is accessible through almost any firewall. As security issues become increasingly important in universities and government agencies, this advantage may become an overriding one.

A stand-alone Java server, while being no more difficult to build than a servlet, makes flexible and complex services possible. One example where this is important would be a collaborative environment in which distributed users could interact and observe interactions on the same dataset. A drawback of this approach is that many firewalls are configured to limit access to a small number of standard ports (ftp, http, etc.), which means the server used in the example above (port 2001) cannot be accessed outside of the local firewall.

5 Discussion

The DG project reported here is focused on improving the quality of maps and graphics available in geo-visualisation tools for the web and on addressing the usability of these tools through implementing and assessing a series of interface prototypes. Our broad goals are to support sophisticated exploratory graphics for experts and comparable statistical summaries in a form that is accessible to, and usable by, non-expert public users through web-based interfaces. We have now implemented several prototype interfaces in Flash (the three above are representative). We are now moving on to the tasks of assessing the effectiveness of specific geo-visualisation concepts and to determining the extent to which Flash can support both rapid prototyping and cognitive experimentation.

Developing flexible and robust geo-visualisation tools for the web is a challenge. Meeting this challenge often requires trade-offs between tool functionality and ease of construction. Developing effective map-based interactive graphics 'from scratch' using a formal programming language such as Java or C++ requires considerable programming skill. Such development is often hampered (even for those with such skill) by the limitations of these languages when faced with implementing subtle display variations needed to support good cartographic design. Furthermore, public access to these tools is limited by web browser versions and the presence of special plug-ins on the end-users' machine.

The web increasingly supports vector-based interactive display formats in SVG (Scaleable Vector Graphics) and SWF. SVG is an XML-based open standard for displaying vector graphics (not currently accompanied by high-level authoring tools that incorporate interactivity). An open XML standard is promising for the flexible display of vector graphics, but at this point, the feasibility of rapid-prototype SVG maps and devices is questionable. SWF and its accompanying authoring environment (Macromedia Flash) offer a lightweight tool (~15% of SVG file size) who's plug-in is supported on the majority of computers (estimates suggest that 95% of current users have at least the 4.0 plug-in installed in their web browser; most computers now come pre-installed with the plug-in).

Designing in Flash holds some distinct advantages over existing development techniques in both prototyping and producing functional geo-visualisation interfaces, not the least of which is design time. Flash designers with limited programming knowledge can produce functional and graphically rich interfaces very quickly for prototyping and for testing conceptual extensions. The ActionScript programming interface is intuitive enough for inexperienced programmers, while flexible enough to support complex object-oriented development by expert designers. These qualities make Flash well suited to graphics prototype development and may support sophisticated geospatial products in the future. Flash also dramatically improves the graphical flexibility of the designer to apply existing cartographic theory to interactive displays. Furthermore, Flash applications are lightweight and run on the web through the small plug-in that has achieved very wide distribution. The transition from development environment to the web is as simple for designers as choosing 'Publish' from the Flash-authoring interface. Finally, our examples illustrate the ability of Flash to support a database connection through a dedicated server or a Java servlet. Once a connection has been made to the database, the data may be stored locally and thus support multi-perspective interactions without further database queries.

Although Flash has many features that make it attractive for web-based geo-visualisation applications, we have also encountered some serious limitations. The main drawback of designing geo-visualisation applications in Flash is the inability of the software to recognise any existing spatial data formats. The geographic base for the maps produced in these three prototypes required a time-consuming manual step—assigning a label to each display object (county). Now that the geographic boundaries for U.S. states and counties have been transformed for use with Flash, we are able to prototype county- and state-based data. However, each new ‘geography’ requires tedious manual encoding. While the format for Flash SWF export files used in web browsers is public, the Flash software and its

ActionScript language is not; and it does not support automated import of objects and conversion to movie clips (the format required in order to program subsequent behaviours). We also observed a considerable processing slow-down for loading and displaying large datasets. This limitation may become increasingly restrictive as we begin to evaluate our county-based US map (3000+ entities) while adding further interactive controls and animation. The most recent release of Flash (FlashMX) and its associated plug-in appear to dramatically improve XML parsing speed. Unfortunately, these tools were not available to test at the time of our experimentation.

Despite the disadvantages listed here, Flash-based geo-visualisation is proving useful for our cognitive and usability experiments. We are currently running a cognitive map animation experiment and a usability assessment of conditioned choropleth maps with two additional prototypes. Designing and testing rapid prototypes in an environment such as Flash has the potential to speed up development and improve the graphical integrity of exploratory geo-visualisation tools generally. The rapid prototypes allow us to explore the viability of geo-visualisation concepts and assess implementation feasibility before committing scarce resources to their development in a more robust and extensible Java environment.

Acknowledgements and Disclaimer

This paper is based upon work supported by the National Science Foundation under Grant No. 9983451. Any opinions, findings, and conclusions or recommendations expressed in this material are those of the author(s) and do not necessarily reflect the views of the National Science Foundation. Thanks go to Luis Mejias for his considerable efforts in building the geographic database of the U.S.

References

Andrienko GL, Andrienko NV (1999) Interactive maps for visual data exploration. International Journal of Geographic Information Science 13(4):355-374

DiBiase D, MacEachren AM, Krygier JB, Reeves C (1992) Animation and the role of map design in scientific visualisation. Cartography and Geographic Information Systems 19(4):201-214, 265-266

Dykes JA (1997) Exploring spatial data representation with dynamic graphics. Computers & Geosciences 23(4):345-370

Egbert SL, Slocum TA (1992) EXPLOREMAP: An exploration system for choropleth maps. Annals of the Association of American Geographers 82(2):275-288

Harrower M, MacEachren AM, Griffin A (2000) Design, implementation, and assessment of geographic visualisation tools to support earth science education. Cartography & Geographic Information Systems 27(4):279-293

Haug D, MacEachren AM, Boscoe F, Brown D, Marrara M, Polsky C, Beedasy J (1997) Implementing exploratory spatial data analysis methods for multivariate health statistics. In: Proceedings of GIS/LIS '97, Cincinnati, OH, Oct. 28-30, 1997

MacEachren AM (1998) Cartography, GIS and the world wide web. Progress in Human Geography 22(4):575-585

MacEachren AM, Hardisty F, Gahegan M, Wheeler M, Dai X, Guo D, Takatsuka M (2001) Supporting visual integration and analysis of geospatially-referenced statistics through web-deployable, cross-platform tools. Paper presented at the Proceeding, dg.o.2001, National Conference for Digital Government Research, Los Angeles, CA, May 21-23

Monmonier, M (1989) Geographic brushing: Enhancing exploratory analysis of the scatterplot matrix. Geographical Analysis, 21(1):81-84

Open GIS Consortium (2000) OpenGIS Web Map Server Interface Implementation Specification

Peterson, MP (1993) Interactive cartographic animation. Cartography and Geographic Information Systems 20(1):40-44

Zaslavsky, I (2000) A New Technology for Interactive Online Mapping with Vector Markup and XML. Cartographic Perspectives 37:12-24

GAP-Tree Extensions Based on Skeletons

Tinghua Ai[1] and Peter van Oosterom[2]

Department of Cartography and GIS, Wuhan University, P. R. China,
aith@wuhan.cngb.com[1]
Section GIS technology, Department of Geodesy, Delft University of Technology,
the Netherlands, oosterom@geo.tudelft.nl[2]

Abstract

The Generalized Area Partitioning tree (GAP-tree) is a model that supports on-the-fly generalisation of planar partitions of polygon objects. This paper focuses on solving some limitations of the existing GAP-tree and explores two extensions. The first extension enables the handling of a disjoint polygonal cluster (within the partition) and allows aggregation operations to be performed. The skeleton partitioning model, which is based on the constrained Delaunay triangulation for the polygonal cluster, is used to define the bridge areas between aggregated polygons. The second extension involves the improvement of removing insignificant objects by separating them into parts around the adjusted skeleton and assigning these parts to different neighbours. The adjusted skeleton is defined by the compatibility between the removed object and its neighbours, which considers not only topological relationships but also importance and semantic similarity. This process again uses the Delaunay triangulation. The algorithm is given to construct the extended GAP-tree.
Keywords: on-the-fly generalisation, spatial partitioning, and neighbourhood analysis, GAP-tree

1 Introduction

Multi-scale representation of spatial data is the intent of seamless data navigation, progressive web transfer, self-adaptable visualisation and other applications. Recently it has gained considerable significance in such fields as map generalisation and on-demand mapping. To obtain geographic data representation suitable for multiple resolutions, one has to build hierarchical spatial partitioning and clustering. The Generalized Area Partitioning tree (abbreviated as GAP-tree) developed by Van Oosterom (1991, 1994, 1995) is a spatial data model for representing multiple detail levels of area objects within a planar partition, such as topography, land use and soil type. It supports display of increased detail during continuous zoom-in rather than just exaggeration of graphic symbols. In the GAP-tree structure, a node in the tree represents a generalised polygon region, which may contain unimportant lower level polygons, and the trace from leaves to root corresponds to polygon selection and simplification in map generalisation. The GAP-tree construction is based on answers to two questions: "Which object is the least important one?" and "Which is the most compatible neighbour of the unimportant polygon?" The first question is resolved based on the object classification and on its attribute and geometric properties such as size. The latter is based on the computation of neighbours and semantic similarity. The neighbourhood in the original GAP-tree does not consider disjoint polygons having an approximate neighbour relationship (with close distance). The original GAP-tree construction work is applicable when regions are connected to each other, which means that the polygon distribution has neither overlaps nor hole areas (or empty space).

This paper extends the GAP-tree data structure for on-the-fly generalisation (continuous zoom-in/out after pre-processing the data) in two ways. First, the neighbourhood is extended to include also polygons at a certain distance. This is used when an unimportant object has to be removed and the neighbour with the highest compatibility value (most compatible neighbour) has to be found. The techniques applied will be spatial searching (using a spatial index) with a buffer surrounding the object and using interpolated constrained Delaunay triangulation (Preparata and Shamos, 1985) to compute distance measures between the two objects. Furthermore, the constrained Delaunay triangulation is used to compute a polygon between the two disjoint neighbours, which will be called the 'neighbour region' of the two objects. This neighbour region is influenced by the importance and the (in)compatibility of the object types belonging to the triangles, which might form the neighbour region. The unimportant object will be merged into the most compatible neighbour and this will include the neighbour region between the two objects.

Second, a different extension has been considered and instead of assigning an unimportant object completely to the neighbour with the highest compatibility value, the object is subdivided in parts and neighbours get assigned pieces based on their compatibility value. This is achieved through an adjusted skeleton of the object that has to be removed. The higher the compatibility value of a neighbour, the more it pushes away the skeleton (and the larger piece of the object will be assigned to this neighbour). An algorithm will be described, that will not produce the 'equal' parts skeleton lines, but rather will give a compatibility value based on weighting. These two developments are an addition to the original GAP-tree and are improvements for features such as (parallel lines) described in earlier papers (Van Putten and Van Oosterom, 1998).

The remaining part of the paper is arranged as follows. In section 2 we present the refined neighbourhood analysis. The aggregation of 'disjoint polygonal cluster' is presented in section 3. Section 4 discusses the extension of another operation amalgamation based on the adjusted skeleton. Section 5 gives the progressive refinement of the extended GAP-tree construction and provides the example application in land use data generalisation. Finally, some future work is outlined in the conclusion, section 6.

2 Neighbourhood Analysis

The GAP-tree is based on neighbourhood analysis. In spatial cognition, neighbourhood is a general concept influenced by topological, geometric, semantic, and Gestalt nature. If two objects touch and share a common boundary, they are neighbours to each other in an adjoining relationship. Two objects with disjoint topological relationship, but in close proximity, can also be regarded as having an approximate neighbourhood relationship. In this case, the geometric distance plays a major role. In a spatial cluster, in which entities have similar distance to each other, the (near) equal attribute values and the similar distribution properties in Gestalt nature consider the corresponding set of objects to be in the neighbourhood of each other. In the natural world, the distribution of geographic phenomena generally has an uncertainty boundary rather than a sharp one. An example would be soil types or vegetation classes in which there is usually some form of gradation in the real world phenomena as opposed to a finite demarcation often depicted in map form. Although in data representation, we can draw a sharp boundary to separate different categories of phenomena, in relationship judgments we should not rely too much on it. In the original GAP-tree, the neighbourhood judgement is simply based on a topological touch relationship.

Each object in a GIS has a semantic description (at least a classification or feature category) and a spatial description. Selecting all objects from a single feature category results in a sub-set of objects within the planar partition. One polygon object, on the one hand, has adjoining neighbours with objects possibly from other categories. On the other hand, it can also have approximate neighbours with the same classification. As far as the adjacency degree is concerned, the former is spatially close but may be semantically far, the latter is semantically close but spatially far. What is the most compatible object in the neighbourhood? In earlier research (Van Oosterom 1991, 1994), the compatibility formula has been introduced to resolve this question, but only for adjoining neighbours. The impact of distance should also be included in the compatibility formula to allow approximate neighbours to participate.

Approximate neighbourhood representation assigns those objects with distance to each other less than the tolerance (buffer zone) into a set called approximate neighbourhood set. Objects with touch relationship can be thought of as special cases in which objects have zero distance to each other. In this sense, the adjoining and the approximate neighbourhood can be unified. For the semantic contribution in neighbourhood identification, we define the following principle: within a neighbourhood cluster, the closer the semantic relationship between object elements, the closer the neighbourhood relationship to each other is. In Fig. 1, an example of a neighbourhood cluster representation is given: object *a* has neighbours *d, c, f, b, h*. When comparing two candidate objects *b* and *f*, we find that object *b* has the same category as object *a,* so it has closer neighbourhood than object *f* to object *a,* although object *f* touches object *a*. This is just an intuitive judgement and the strict judgement depends on compatibility computation.

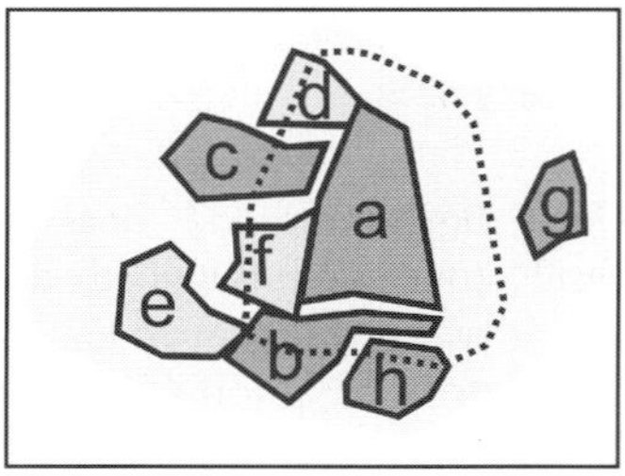

Fig. 1. Neighbourhood cluster of polygon a.

In neighbourhood definition for disjoint objects, another consideration is the context influence (Hernandez and Clementini, 1995). For a large neighbourhood tolerance distance, two objects having less distance are regarded as strong neighbours. However, when the third object appears between them, it is considered that they no longer have a neighbourhood relationship. This implies we cannot simply use the distance metric to find a neighbour, such through application of a buffer operation in commercial GIS software. An additional test has to be designed. The Voronoi diagram of the polygonal cluster is computed and each polygon is extended with its influence region. The result is a partition of space based on the objects with a certain classification. The approximate neighbourhood is based on the topological touch relationship between the partitioning polygons in the Voronoi diagram.

Based on the above analysis, objects are referred to as having neighbourhood relationship if they satisfy one of the following conditions:

1. having topological touch relationship to each other;
2. among objects with a similar classification, their Voronoi cell polygon shares a boundary and the distance between objects is less than some tolerance.

Furthermore, the semantic adjacency and the length of a shared boundary decide the neighbourhood strength. The impact of these parameters has been used in the computation of the most compatible neighbours in the original GAP-tree construction. With the new neighbourhood definition, the shared boundary includes not only boundaries between objects themselves, but also between the touching partitioning polygons in the Voronoi diagram in case of disjoint objects.

According to the definition, in Fig. 1, object *a* has neighbours *b, c, f, d* and this order also represents the neighbourhood strength. Object *g* has the same category but is far away from object *a*, so it cannot act as the neighbor of *a.* Although object *h* has the same category as object *a* and is located within the buffer region of object *a*, the appearance of object *b* makes their corresponding Voronoi diagram partitioning polygons not touch. Therefore *h* is not the neighbour of object *a*.

3 Aggregation Based on the Skeleton Partitioning Model

In spatial data handling, we can distinguish two combination operations: the **aggregation** for polygon objects with homogeneous semantics, and the **amalgamation** for polygon objects with heterogeneous semantics. The results of aggregation will inherit the attribute of the original ones,, while the result of amalgamation will change the classification of some of the original objects. We will discuss the amalgamation operation in section 4, while in this section will focus on aggregation.

Aggregation is an important operation in the generalisation of data sets such as land use, urban building clusters among other applications. The original GAP-tree focuses on the selection and combining of objects. However, no distinction is made between amalgamation and aggregation. In the GAP-tree structure, the merge operation just happens based on the compatibility with the least important object and the result is a ‘parent and child’ edge in the GAP-tree. When considering the aggregation of non-adjoining neighbours, we find that the operation implies that a third object (with another classification) is affected.

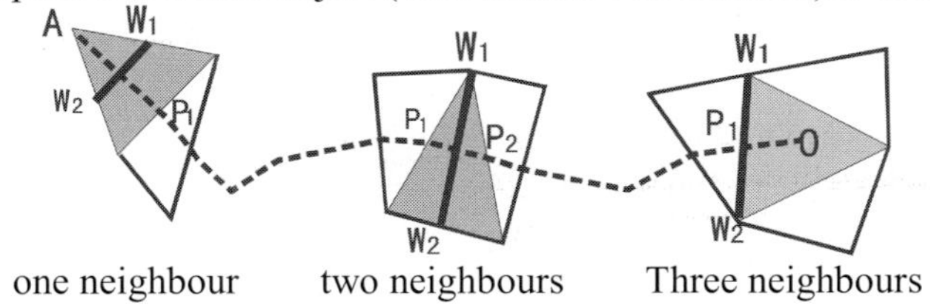

Fig. 2. W_1, W_2 local distance for three triangle types

For a set of objects, from a similar category, we construct the constrained Delaunay triangulation. The triangles connecting different polygons are considered which implies that those triangles within and outside the polygons but with three vertices located in the same polygon will be removed from the triangulation. For the remaining triangles, according to the number of neighbour triangles, the dashed skeleton segments are linked as depicted in Fig. 2 (where P_i is the middle point of the corresponding triangle edge, and O is the weight centre of the triangle). This is done to get the skeleton partitioning model. The result is illustrated in Fig. 3, in which every object polygon is surrounded by one partitioning polygon, and the partitioning skeleton separates the ‘empty’ space equally between two or more object polygons. If two partitioning polygons touch, the corresponding contained object polygons are possible neighbours, depending on the distance. Next we compute distance measures (minimum, average) between object polygons based on the skeleton edges.

The aggregation of approximate neighbour polygons behaves as filling a certain area between them and changing its classification (a little bit of amalgamation).

We call this area the *neighbour region*. Define a neighbourhood tolerance distance *d*, and then select those triangles whose local distance between the connected object polygons is shorter than *d*. Group these selected triangles into triangle clusters. Each triangle cluster goes across a part of the partitioning skeleton edges and plays the bridge role connecting two or more neighbour objects, see Fig. 3.

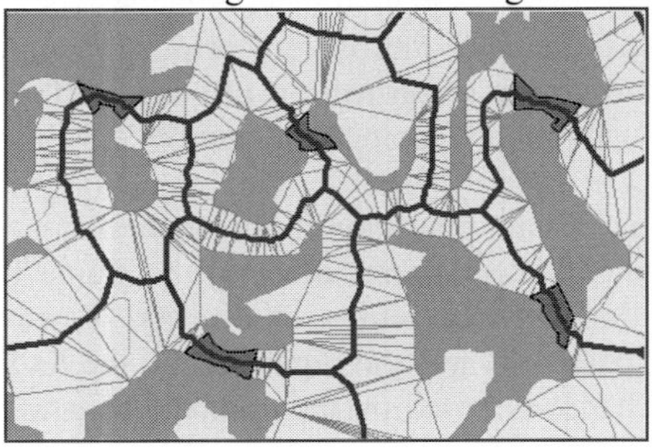

Fig. 3. One category of land use object distribution and *neighbour region* extraction

4 Amalgamation Based on the Adjusted Skeleton

In the iterative process of GAP-tree construction, each step will remove the most insignificant polygon object among the current remaining objects. Therefore, the most compatible neighbour will fill the space. This process can also be understood as the amalgamation operation in the case of two semantically different objects. In the original GAP-tree, this operation is based on topological structure and semantic compatibility computation, and the insignificant object will completely be assigned to one of the neighbours. We cannot always find an obviously strong neighbour to completely replace the insignificant object. From the point of view of error adjustment, separating the insignificant polygon around its skeleton and assigning different parts into different neighbours is better than assigning the whole polygon into one of the neighbours. Bader and Weibel (1997) presented an idea to separate an insignificant object based on the constrained Delaunay triangulation and the skeleton. However, if the insignificant object does not have an obvious lineal stretch distribution, the method cannot guarantee that the terminal point of the skeleton exactly meets one of the intersection points between two neighbours. Adjustment of the skeleton is required. Next we will provide the detailed method of this amalgamation operation from the geometric algorithm.

The basic idea of this method is to generate the skeleton within the insignificant polygon and then create the polygon parts composed of some skeleton edges and one boundary edge. The generated polygon parts are assigned to the neighbour polygon on the other side of the boundary edge. Here the Delaunay triangulation is used again, but this time within the studied polygon. To seamlessly partition the insignificant polygon, we expect that the terminal point of each skeleton edge meets in a node of the boundary edge. This condition includes two aspects: one is that each terminal point of the skeleton edge must exactly be located in one node of boundary edge, another is that each node of the boundary edge must exactly have one linking skeleton edge. Note that a node is defined as the location where at least three edges meet. Obviously, directly extracting the skeleton based on all triangles within one polygon cannot guarantee the above conditions, see Fig. 4b. We use the following two steps to adjust the skeleton edges.

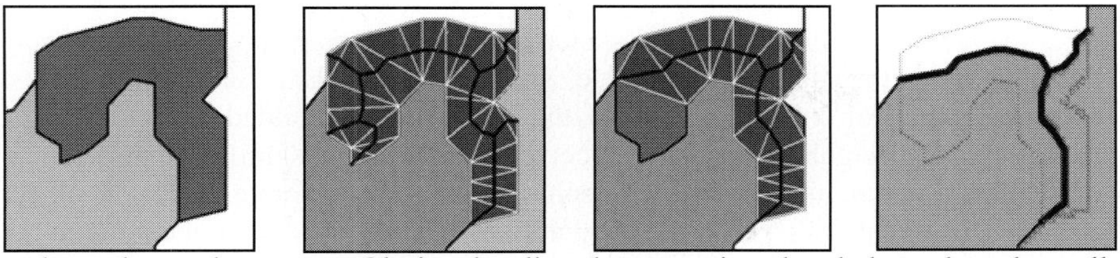

(a) (b) (c) (d)

Fig. 4. The generation of skeleton edges within one insignificant polygon, **(a)** original scene, **(b)** original skeleton, **(c)** adjusted skeleton, **(d)** partitioning result

First, all triangles linking the same boundary edge (that is, with all three points on the same boundary of one neighbour polygon) are removed and will be assigned to the corresponding polygon, as shown in Fig. 4c. Based on the remaining set of triangles within the polygon, the new skeleton edge is constructed. We now can guarantee that each terminal point exactly meets in one node, and is not located in the middle of a boundary edge. As illustrated in Fig. 4c, from the point of view of neighbour, the area in which triangles are removed serves as the concave part. Since this part is strongly surrounded by one neighbour polygon, in the partitioning it should not act as a competition region for different neighbours to separate. Second, a skeleton edge begins from a triangle with only one neighbour triangle, which appears in some places of the boundary, which is not smooth.

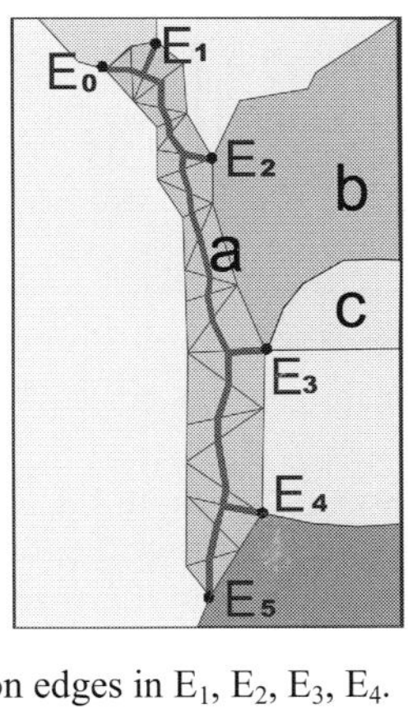

Fig. 5. Adding additional skeleton edges in E_1, E_2, E_3, E_4.

As described above, the skeleton begin point will always be a node. However, in some situations there is no skeleton edge connected to a node (a relative smooth part), see E_1, E_2, E_3, and E_4 in Fig. 5. In Fig. 5, we can just get one skeleton edge from E_0 to E_5. From nodes without a skeleton edge, search the closest point in the existing skeleton edge along a triangle edge. Some nodes relate to just one triangle edge intersecting with the existing skeleton edge, such as E_1 and E_3. Others may relate to more than one triangle edge, such as E_2 and E_4. Select the shortest connection (half of one triangle edge) and let it act as the added 'skeleton' edge. Note that, if a neighbour wants to receive a part of an insignificant polygon, it must share a common boundary with some length. For example, in Fig. 5, object c shares just one point E_3 with the insignificant object a, it will not merge with the part of object a.

Through these processes for matching boundary nodes and skeleton terminal points, the conditions discussed earlier are fulfilled. Now, select all the boundary edges of the insignificant polygon and all its skeleton edges to organise the polygons. Each boundary edge will make up one polygon together with some skeleton edges. The generated partitioning polygon can now be assigned to the other side neighbour of the boundary edge. This amalgamation maintains the topological consistency, not resulting in gaps or overlapping regions. Fig. 6 shows an experiment example of this amalgamation. In the left, the objects surrounded by the wide dark boundaries are unimportant (small size) and will be removed. The right graphic is the result of subdividing the insignificant objects.

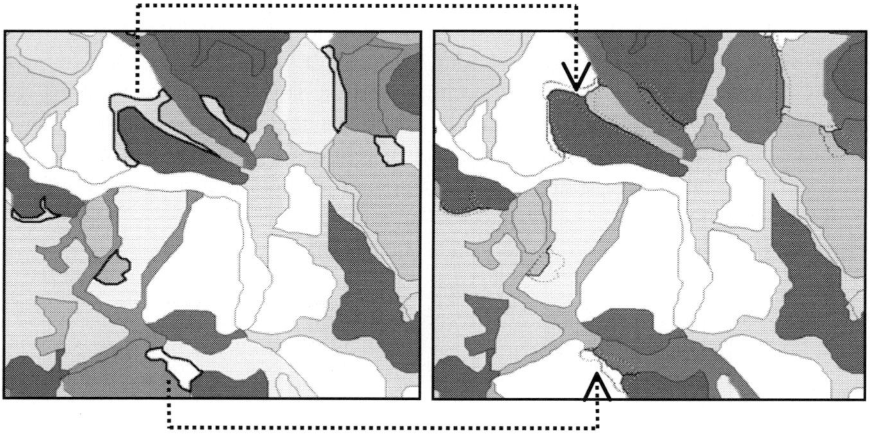

Fig. 6. An experiment of removing insignificant objects based on adjusted skeletons.

The skeleton generation above is based on the assumption that the neighbours have the same influence to divide the insignificant object and the skeleton edge partitions left/right region in an equal way. Further improvement, in future work, can be obtained by considering geometric relationship and semantic similarity. We use function compatibility(a,b) to compute the compatibility between object a (to be removed) and its neighbour object b, taking into consideration:

- The compatibility of the types: *type_comapt(a,b)*;
- The length of the shared edge: *length(a,b)*;
- The importance of *b: Imp(b)*.

The function could be: compatibility(a,b)=type_compat(a,b)*length(a,b)*imp(b).

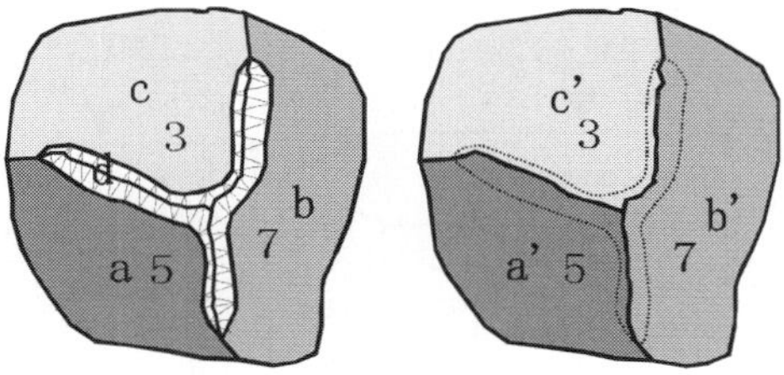

Fig. 7. The generation of adjusted skeleton.

We call this kind of skeleton, the adjusted skeleton, since it considers the compatibility difference between polygons competing for extending their area.
In the geometric algorithm, the adjusted skeleton can be an adjusted version of the normal skeleton. In Fig. 2, the skeleton points P_i need to be computed according to two segment ratios based on the compatibility values, rather than the middle position unless the two compatibility values are the same. For example, in Fig. 7, object *c* has compatibility value 3 and *b* has 7. For the region between object *c* and *b,* 30 percent will be assigned to object *c* and 70 percent will be assigned to object *b*.

5 Progressive Refinement of the GAP-Tree Construction

The aggregation of adjoining neighbours and the improvement of the amalgamation, changes the GAP-tree construction. The basic idea is that once an unimportant object is determined, it has the possibility either to be aggregated with its approximate neighbour or to be amalgamated into adjoining neighbours. The algorithm of the GAP-tree construction, which is an iterative process, is described in Fig. 8.

As usual, an edge in the GAP-tree represents the parent-child relationship between the unimportant polygon (child) and the polygon(s) to which it has been assigned (parent). Fig. 9 shows an example with the iterative process of the GAP-tree construction using both aggregation (P_1 steps) and amalgamation (P_2 steps).

It is possible that one 'neighbour region' lies on top of more than one polygon; see for example the upper-right image and the one below. This is because the 'neighbour region' is computed using a Delaunay triangulation of only the selected polygons, which have similar classification and are within tolerance distance of the least important polygon.

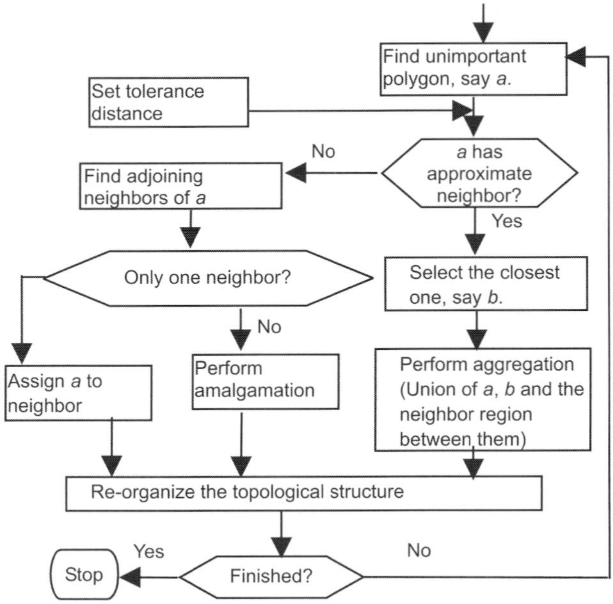

Fig. 8. The progressive construction of the extended GAP-tree.

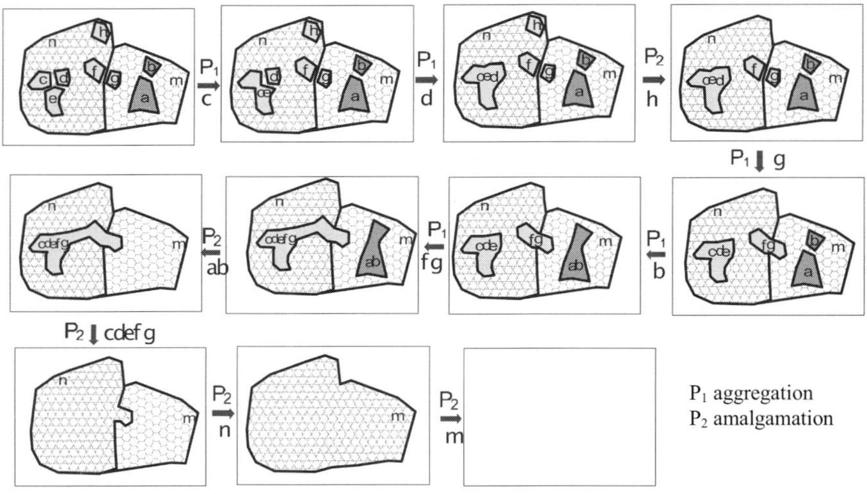

Fig. 9. The scene change during the extended GAP-tree construction.

Using nodes and edges to record the result of two kinds of operation, as shown in Fig. 10, we get a new data structure. As in the amalgamation operation, an object can be split into parts and each is assigned to a different parent, this will destroy the real tree structure of the GAP-'tree'. It has now become a 'directed acyclic graph', because one child may have several parents (GAP-DAG). In the parent-child edges, it must be stored how big the part of the child object assigned to this parent is (*child-part,* with values between 0 and 1). The original area formula has to be changed into: *area*(object)=*Area*(object-GAP-node)-*sum*(*child-part* * *Area*(child-node))

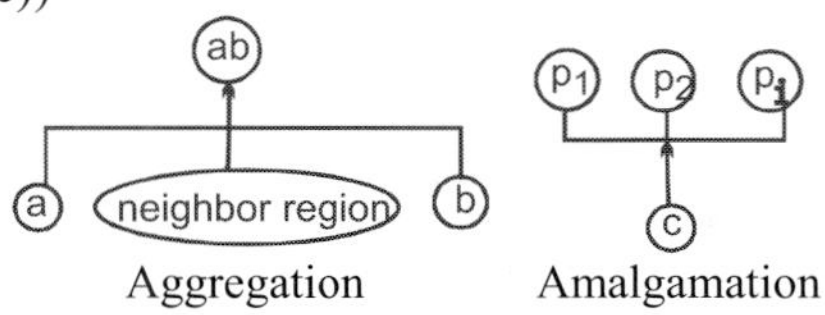

Fig. 10. Structure for 2 operations

After construction, the GAP-tree is used during on-the-fly generalisation based on the painter's algorithm idea in computer graphics. This means, once an important object is drawn, it will never be drawn again during refinement. Only additional details will be drawn on top of it, each time refining the 'image'. These details include the neighbour regions in aggregation and the divided parts in amalgamation. In this way, it can be compared with the progressive transfer of the raster images on the Internet, which are also displayed from coarse to fine as there is more and more time for sending details (refinements). In the case of joining disjoint neighbours in the GAP-tree construction, which have to be visually disconnected again, when using the GAP-tree, the 'neighbour region' will be drawn on top of this. This is an exception to the rule that the most important (and first drawn object) will never be drawn again. In this case, a small fragment will be 'redrawn'.

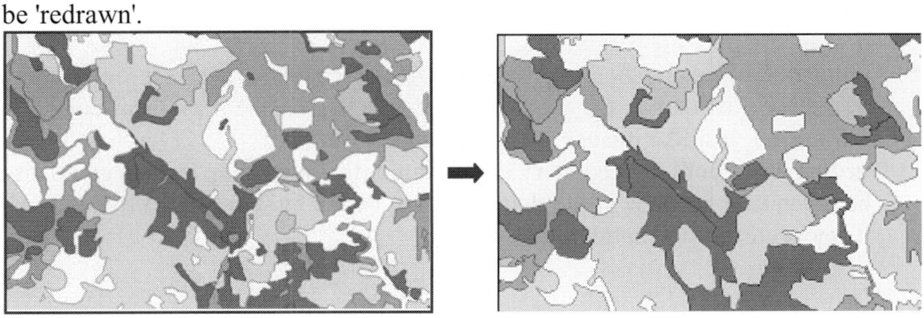

Fig. 11. The generalisation of land use objects based on the extended GAP-tree

The land use map, see Fig. 11, is a typical kind of categorical theme and has the properties covering the whole area with neither gaps nor overlapping areas. Usually it includes many categories in the classification hierarchy. Its generalisation has to consider geometric simplification, semantic hierarchy abstraction and statistical properties maintenance. In our generalization of land use data, we select one category of objects (resulting in a disjoint polygonal cluster) as operated objects, while others are classified as background objects. Fig. 11 gives an experimental illustration. Future research has to determine the proper parameters for compatibility formula computation in real land use data. For the time being, we just use the normal skeleton (equal partitioning) to separate insignificant objects in this experiment.

6 Conclusion

Based on skeleton handling, this paper extends the GAP-tree construction in two aspects. The first one is the introduction of aggregation of disjoint polygon objects. The second one is the improvement of amalgamation between different category objects. Both extensions use the Delaunay triangulation and skeleton, but in different ways. The idea of two operations can be unified in the GAP-tree. On-the-fly generalisation (continuous zoom-in/out) can be achieved by using the GAP-tree based on painters algorithm. The initial tests show good generalisation results. This strategy is also very suitable to progressively transfer polygonal map in Internet.

In the past other generalisation techniques have been based on the Delauney triangulation, Voronoi diagram or the skeleton (Poorten and Jones, 1999, Jones, Bundy and Ware, 1995, Ware and Jones, 1995, 1997, Ai, Guo and Liu, 2000, Ai and Guo, 2000). This paper has added two more generalisation techniques, which proves again how powerful these structures are.

Further research topics related to GAP-tree use in generalisation include:

a) Improvement of neighbourhood definition, simultaneously considering topological, geometric, semantic and Gestalt nature's impacts. The compatibility formula between two neighbours has been defined, especially how to compute the parameters.
b) Application of this method in a web transferring environment, in a server-client framework, to realise the transfer of vector map in the way from coarse to refinement, just like images are refined during transfer.
c) Uncertainty of both the geometry and the classification of the objects could also be taken into account in the determination of the least important feature and its most compatible neighbours.

Acknowledgements

Tinghua Ai's work is supported by the National Science Foundation, China under grant number 40101023. We would like to thank our colleague Tjeu Lemmes and the anonymous referees whose comments materially improved the paper. Special thanks go to Elfriede M. Fendel for editing the paper.

References

Ai T, Guo R, Liu Y (2000) A Binary Tree Representation of Bend Hierarchical Structure Based on Gestalt Principles. In: Forer P, Yeh AGO, He J (eds) Proceedings of the 9th International Symposium on Spatial Data Handling, Beijing, pp 2a30-2a43

Ai T, Guo R (2000) A Constrained Delaunay Partitioning of Areal Objects to Support Map Generalization. Journal of Wuhan Technical University of Surveying and Mapping 25(1):35-41 (in Chinese)

Bader M, Weibel R (1997) Detecting and Resolving Size and Proximity Conflicts in the Generalization of Polygonal Maps. In: Ottoson L (eds) Proceedings of the 18th International Cartographic Conference. Stockholm, pp 1525-1532

Hernandez D, Clementini E (1995) Qualitative Distance. In: Proceedings of COSIT'95, Semmering, Austria, pp 45-57

Jones CB, Bundy GL, Ware JM (1995) Map Generalization with a Triangulated Data Structure. Cartography and Geographic Information System 22(4): 317-331

Oosterom P Van (1995) The GAP-tree, An Approach to On-the-Fly Map Generalization of An Area Partitioning. In: Muller J C, Lagrange J P, Weibel R (eds) GIS and Generalization: Methodology and Practice. Taylor & Francis, London, pp 120-13

Oosterom P Van (1994) Reactive Data Structure for Geographic Information Systems. Oxford University Press, Oxford

Oosterom P Van (1991) The Reactive-Tree: A Storage Structure for a Seamless, Scaleless Geographic Database. In: Auto-Carto 10, pp 393-407

Poorten P, Jones C B(1999) Customisable Line Generalization Using Delaunay Triangulation, CD-Rom Proceedings of the 19th ICC, Ottawa, Section 8

Preparata FP, Shamos MI (1985) Computational Geometry An Introduction. Springers-Verlag

Putten J Van, Oosterom P Van (1998) New Results with Generalized Area Partitionings. In: Proceedings of Spatial Data Handling, Ottawa, Canada

Ware JM, Jones CB (1995) A Triangulated Spatial Model for Cartographic Generalization of Areal Objects. In: Proceedings COSIT'95. Semmering, Austria, pp 173-192

Ware JM, Jones CB(1997) A Spatial Model for Detecting (and Resolving) Conflict Caused by Scale Reduction. In: Kraak MJ, and Molenaar M (eds) In: Advance in GIS Research II (7th Int. Symposium on Spatial Data handling), Springer-Verlag

Improving Automated Generalisation for On-Demand Web Mapping by Multiscale Databases

Alessandro Cecconi, Robert Weibel and Mathieu Barrault

Department of Geography, University of Zurich,Winterthurerstrasse 190, CH-8057 Zurich, Switzerland, {acecconi, weibel, barrault}@geo.unizh.ch

Abstract

This paper describes the generation of maps on-demand with the use of a multi-scale database. It is based on an analysis of the requirements of on-demand mapping and points out the different requests and limits of on-demand cartography. The central idea is to combine two commonly used approaches in cartography: On the one hand the use of a multi-scale database which includes two or more levels of details, on the other hand the use of cartographic generalisation methods. For selected object classes the paper discusses and evaluates design and implementation options for the multi-scale database and the generalisation of parts of the framework. The importance lies in the optimal combination of these two methods – which tasks must be solved by the MSDB and which through the generalisation process.
Keywords: multi-scale database (MSDB), on-the-fly generalisation, level of detail (LoD), web mapping, on-demand mapping

1 Introduction

In recent years, a new field of application has opened for cartography: web mapping (Peterson 1999). More and more websites offer maps of various kinds (topographic maps, thematic maps) and topics (route planning, city guides, among many others.) over the net. Many of these websites are equipped with zooming capabilities, but as these are not true generalisation capabilities, the quality of most web maps is poor.

1.1 Maps on the Internet

Maps created for the Internet are based on different conditions than paper maps. The map provider cannot control many technical constraints and requirements. With respect to visualisation, several parameters cannot be defined, such as hardware (display resolution), system software (operating system) or application software (browser). Since the bandwidth is typically still narrow, it is also very important to take into account the amount of transmitted data in distributed systems. Despite technical restrictions it's possible to create maps for the web, witness the numerous examples such as route planning, location finders and others (*www.mapquest.com, www.map24.com*). In functional terms most of those services, however, are not flexible as they have been designed for a well-defined purpose, such as providing locator maps for user-specified street addresses. Hence, the user has no opportunity to change anything or to define his/her purposes and requirements for the map graphics. To remedy these drawbacks a new strategy must be found which allows more flexibility for web mapping.

1.2 On-demand Mapping

On-demand mapping is concerned with the generation of maps based on user request and according to user requirements. Users are able to produce their own maps and customise the process of generation. A short explanation of on-demand mapping can be found on *http://www.ngdc.noaa.gov/seg/tools/gis/ondemand.shtml.* It deals with the dynamic creation of digital cartographic products, like topographic or thematic maps. It is only concerned with displaying data and does not create a target data set (van Oosterom and Schenkelaars 1995). In principle, however, the approach could be extended to include on-demand creation of reduced databases too. For the flexible creation of maps on-demand at arbitrary scales and for arbitrary themes cartographic generalisation is a necessity. However, cartographic generalisation is also known to be a time-consuming process, which runs counter to user expectations for Internet services, where the time factor is critical. Based on possible parameters defined by the user, three scenarios can be determined (table 1). These differ in the strength of the generalisation process and hence in the time required for map creation.

Table 1. Description of different scenarios for the on-the-fly and on-demand mapping

	Description	Map generation
Scenario 1	The user wants to have a first map for an overview of the requested information. The compute time should be very short and generalisation is only partially applied.	On-the-fly
Scenario 2	The advantages of the other two scenarios are combined here. The user needs a map with good cartographic quality within a limited time.	On-the-fly and on-demand
Scenario 3	In this scenario, the time component plays a minor role, and the user is willing to wait for the desired map. The generation can thus encompass the entire process of generalisation to obtain a high-quality map.	On-demand

The generalisation process critically determines how long it takes to create a map. Because automated generalisation is only partially solved, other solutions must be found 1) to substitute for missing automated generalisation operations and 2) to speed up cartographically appropriate yet computationally intensive generalisation algorithms. Multi-scale databases offer a possible solution (Buttenfield 1993, Timpf and Devogele 1997). Using a multi-scale database, the automated generalisation process can be divided into two steps. First, time consuming generalisation algorithms can be precomputed in advance (offline) and the results stored as levels of detail (LoDs) in the multi-scale database. Non-existing generalisation algorithms can be substituted (i.e., simulated) by means of interactive operations used to build the multi-scale database. Second, computationally efficient generalisation algorithms can be computed on-the-fly and used to refine the nearest LoD to the requested map scale. By means of this combined process on-demand map creation can be optimised and made more flexible at the same time.

Section 2 explains how such a MSDB can help to speed up the generalisation process and, conversely, how generalisation can help to render a more flexible MSDB. Section 3 briefly presents an example. Section 4 studies the generalisation process for selected object classes and shows the use of the different generalisation operators and algorithms. The paper will finish off with some conclusions and an outlook (section 5) of future work.

2 Combination of Map Generalisation and a MSDB for On-demand Map Creation

This paper explores an approach that combines components of a MSDB (Devogele et al. 1997) with components of on-the-fly generalisation (van Oosterom and Schenkelaars 1995). On the basis of user specifications (map scale, map purpose, symbology, information density) the appropriate object classes are selected from the database LoD for the desired map scale and further refined at run-time by on-the-fly generalisation algorithms. The overall design of this combined approach was described in Cecconi and Weibel (2001). To implement this architecture the main issues to be resolved are:

- to study the generalization process in conjunction with multi-scale databases;
- to develop a schema and structure for the MSDB which includes hierarchical linking of corresponding objects between different levels.

This paper focuses on the discussion of the former issue, trying to develop cartographically sound generalisation processes that exploit the combined strength of existing generalisation algorithms and MSDB. For brevity, we refrain from the discussion of the second issue. A first schema based on a relational approach has been designed for the road network (Cecconi 2001). More work on schemas for other object classes will follow after a careful study of the generalisation process (parts of which are presented in the paper), which allows us to better define the structure of the MSDB. Jones et al. (2000), as a further example of a combined approach, devote more attention to MSDB design, spending less time on building generalisation processes. Glover and Mackaness (1999) experimented with on-the-fly generalisation from a single scale database.

2.1 Use of a Multi-Scale Database for Cartographic Generalisation

As explained in the previous section the MSDB is used as a base element for the generation of maps. It is defined as a composition of different data sets, in which corresponding object elements are linked (Kilpeläinen 1997).

The MSDB includes a minimum of two data sets of geographic objects represented at different levels of detail (LoDs), or scales. In our case study, the first level is equivalent to a 1:25’000 topographic database. The second level is equivalent to 1:200’000. Fig. 1 (left) shows this set-up. The corresponding objects of the two levels are linked together (shown by dashed lines). The definitions of the links (Fig. 1, right) are very important and needed for on-the-fly generalisation. Every object in one scale has to know its counterpart in the larger or smaller scale. This information can be used to simplify the generalisation process. Given the links between the different levels of scale, generalisation can be understood as an ‘interpolation’ (or morphing) process between two different geometries.

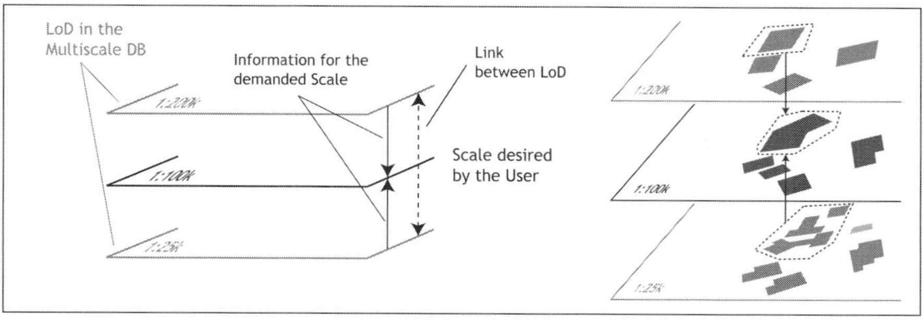

Fig. 1. Which data set should be selected for the desired scale (left)? How are the objects linked between the different data sets (right)?

For creating a map at the scale desired by the user (1:100'000 in our case) the appropriate LoD has to be determined. To this end, so-called limits of applicability are defined. These pre-defined limits can vary from object class to object class and depend on the given levels of detail. Fig. 2 shows the selection of the corresponding data set for the desired map scale based on these limits.

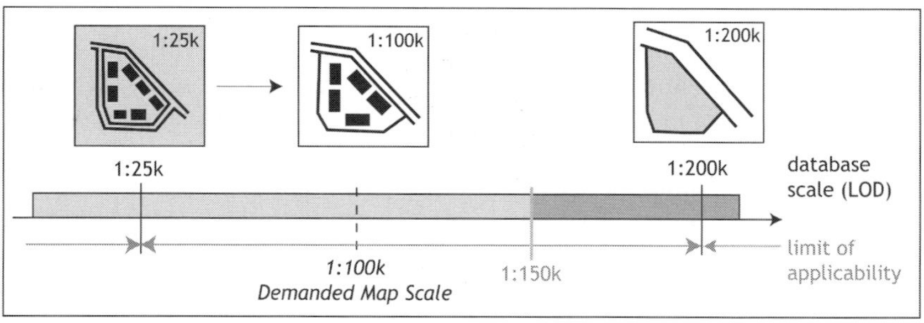

Fig. 2. Definition of the limits of applicability for the desired scale

Based on the assumption that the depiction of individual buildings is only applicable at scales of 1:150,000 or larger, the data set at1:25'000 was set to 150'000. Hence, in our example, the appropriate LoD is 1:25'000. For scales smaller than 150'000 the 1:200'000 LoD will be taken and built-up areas rather than individual buildings are shown.

2.2 Application of Generalisation Operators in different Scale Bands

The manual generalisation process is of an intrinsically holistic nature. For automated generalisation it is – at least for the time being – almost impossible to develop such a holistic solution. Hence, the generalisation process must be subdivided into a set of generalisation operators. McMaster and Shea (1992) have defined several generalisation operators, which – in an extended version – are building the basis of this research. It should be noted that other authors use different operators and may define them differently (see Weibel and Dutton 1999 for a review). The task of a generalisation operator is to solve a specific generalisation problem. Using such operators, it is possible to break down the generalisation process into smaller sub-processes; the combination of several operators can then be used to build an entire generalisation process or workflow.

The use of generalisation operators for the creation of maps is fundamentally dependent on the scale(s) and object classes involved (e.g., road network, hydrography). Depending on scale different operators (e.g. selection, simplification, typification and displacement) will be applied in different sequences. This knowledge of map generalisation can be used to design the contents of a multi-scale database. In particular aspects like 1) the levels of detail, 2) the limits of applicability for the LoDs of the MSDB, and 3) the generalisation operators used to transform between the LoDs can be addressed. Obviously, these decisions must be made separately for each object class and depend largely on the symbol specifications used (e.g. whether built-up areas are shown by individual buildings or simply as a tinted area). So, how can the above elements be defined in a meaningful way? Surely, we will seek to minimise the number of LoDs, and extend the limits of applicability as far as possible. Each additional LoD implies a significant additional cost, not only during database creation but even more importantly during database updates (which need to be consistently propagated across LoDs). In addition, it is rarely possible that data for the different LoDs can be gathered from existing sources, such as a topographic map series. We propose to analyse the 'generalisation complexity' as well as the application scope of the generalisation operators over the desired range of scales (Fig. 3). By 'generalisation complexity' we mean the combined 'cost' involved in generalisation; low values indicate simple operators can be used (e.g., selection and simplification) while high values hint at major modifications requiring complex contextual operators such as typification or displacement. The complexity values and generalisation operators can be established by analysis of maps and literature. While this is merely a qualitative method, it is nevertheless systematic and allows us to define the limits of applicability (i.e. the points where the 'regime' of generalisation operators changes) and hence the appropriate minimal number of LoDs. Note that this procedure bears some resemblance with the *points de généralisation* observed by Ratajski (1967).

Fig. 3 shows the behaviour of generalisation complexity and the relevant generalisation operators for a range of scales and for several selected object classes. For lack of space, we will only point out some essential features. The complexity curve for the "road network" class suggests three scale bands. In the first scale band (large scales down to 1:50'000) the selection and simplification operators are sufficient because there is usually enough room for displaying all road objects without much modification. In the intermediate scale band (1:50'00 to 1:300'000) important shape changes become noticeable. Because there is still enough space most road objects can be displayed, but must be transformed due to increased symbol width. This will be achieved by the operators typification (transformation of an initial set of objects into a new set, maintaining the typical arrangement) and displacement. Hence, the generalisation complexity increases.

In the third scale band the available space is so small that only the most important road objects appear; as the number of objects is low, typification and displacement are becoming less important and generalisation complexity decreases again.

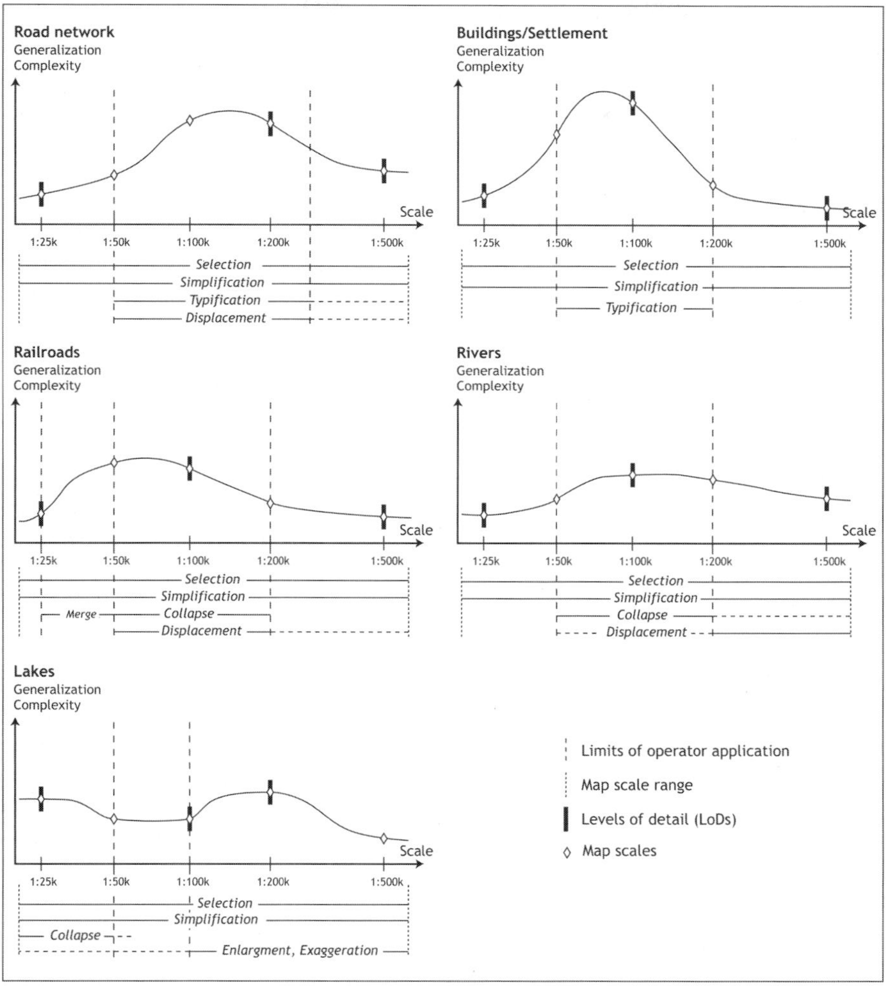

Fig. 3(a-e) The 'generalisation complexity' and the application scope of generalisation operators for different object classes

The three scale bands can be represented by three LoDs, 1:25'000, 1:200'000 and 1:500'000. The object class "buildings/settlement" again shows three scale bands. The main break point (between scale bands 2 and 3) is the transition from individual buildings to tinted polygons for the display of built-up areas. On Swiss topographic maps (which are used in our research) this occurs around 1:200'000. A further break point can be found around 1:50'000. Up to 1:50'000 , the main operators include elimination of small buildings (i.e. selection of large buildings) and simplification of building outlines. Typification of groups of buildings and object displacements only occurs later, and significantly increases the generalisation complexity.

The object class “railroads” shows a different behaviour with a narrow band at large scales. The large complexity of in the first scale band is due to merge and collapse operations in station zones and switchyards, while overland tracks do not pose a problem due to large curve radii. It is only later, in the intermediate scale band, that overland tracks need to be generalised by various operators. Finally, at small scales beyond 1:200’000 only a small number of objects remain, decreasing the generalisation complexity.

For the class “rivers” most objects can be drawn as polygons up to 1:50’000. At medium scales objects need to be collapsed to lines and must eventually be displaced if overlaps are detected. As the collapse operation actually creates space the complexity in the intermediate scale band is lower than for other object classes.

Finally, the object class “lakes” exhibits a special behaviour with high complexity at large scales, low complexity at medium scales and large complexity at small scales. The large complexity in the first scale band is due to collapse operations of small inlets of rivers leading into the lakes.

Clearly, the advantage of using MSDB in conjunction with generalisation is that the automated generalisation process must not be carried out from a single detailed data set. Rather, it should start from a data set that is closer to the target scale, thereby simplifying the generalisation process enormously.

3 Example – Description of a More Detailed Scenario

We are presently implementing an experimental platform that uses the following scenario, taking place in an Internet (Client/Server) environment.

Technical specifications: The information will be transferred via network from a server to a local computer (client). Bandwidth is presently not of concern, as it is very hard to predict. The following technical restrictions must be taken into account (Kraak and Brown 2000): The display resolution of 72 dpi and the colour depth of 16 bits influences the symbolisation process; the display area of 1024x768 limits the real estate available for map display.

Application scenario: The main aim of the work is to generate a user-defined map from a multi-scale database online with the help of different generalisation techniques. The focus is on the generalisation components of the map creation process, meaning that the quality of the maps must be adapted to the user request. The user is assumed to request a topographic map, where the main focus will be on the road network, while the other topographic elements are used for orientation purposes. The target scale is assumed at 1:100’000, the MSDB that is being used consists of two data sets at 1:25’000 and 1:200’000, respectively.

Test area: The data set represents an area of 210km^2 with different topographic feature elements. The elements, which will be used for map display, are road network, buildings and hydrography.

4 Generalisation Processes for Selected Object Classes

This section intends to define possible sequences of the generalisation process for the selected object classes and determine which generalisation operators and algorithms will be used. A generalisation operator conceptually defines the type of (generalisation) transformation that is to be achieved; a generalisation algorithm is then used to implement the particular transformation. Note, however, that we are currently building a MSDB that relies on data from Swiss topographic maps. Hence, the characteristics of the Swiss national map series apply.

4.1 Highways

At both scales of our MSDB, 1:25'000 and 1:200'000, all highways are drawn. Modifications in the transition from the large to the small scale are relatively small, mainly affecting interchanges and ramps (entries, exits). Table 2 shows which operators are used for the generalisation process and which algorithms will therefore be implemented.

Table 2. Use of generalisation operators for the object class "highways" and the corresponding algorithms

	Operators	Algorithms
Selection	Because every object has to be selected, the selection operation is trivial to handle.	-
Weeding Simplification	While weeding is necessary, changes should be very small since highways are very important features on topographic maps.	Implementation of the Douglas-Peucker algorithm for weeding. Simplification of slip roads is done by using the information from the 1:200'000 data set.
Smoothing	Weeding is followed by smoothing.	Algorithm for smoothing by Lowe (1988).
Displacement	Due to the overriding importance and small number of highways, displacement will only occur in few places.	Displacement algorithm using elastic beams (Bader 2001).

The proposed generalisation process using a MSDB for the object class „highways" is shown in Fig. 4. Shaded boxes denote use of a data set, while white boxes indicate procedures. Starting with the data set of 1:25 '000, the data set 1:100'000 for the requested map will be created. Because all highway elements will be selected, the first step (selection) is trivial. After that the symbolisation (i.e. line width, colour type) must be defined either by the user (as part of his/her map request at the beginning) or by default values. Next, the weeding operator will reduce the number of points of the line, followed by the smoothing operator. The entries and exits will be simplified by using the knowledge of the data set at 1:200'000. This information can help to decide which segment must be shown. The rendering step is responsible for the representation of the elements on the screen. If unsolved problems remain after rendering and subsequent displacement, the process may return to symbolisation and restart with smaller symbol sizes.

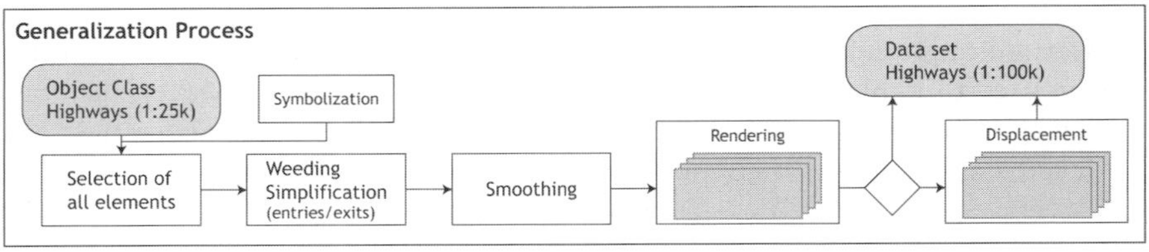

Fig. 4. The different steps of the generalisation process for the object class "highways"

The displacement step is optional. In the case of highways, it will only be needed in special situations, as highways are of key importance and hence will resist displacement.

4.2 Main and Minor Roads

The object class "major and minor roads" is more interesting because the selection operator plays an important role. At a scale of 1:25'000 usually all main and minor roads are represented, because sufficient space is available. At smaller scales, only main roads and important minor roads are displayed. Special road elements, such as "roundabouts" or squares will be collapsed into road sections.

Table 3. The use of generalisation operators for the object class "main and minor roads" and their corresponding algorithms

	Operators	Algorithms
Selection	Road selection is important. The roads are selected by importance and road class (comparison with LoD 1:200k).	Selection by using information of 1:200'000.
Morphing	The data sets 1:25'000 and 1:200'000 form the 'key frames'. The map scale 1:100'000 are generated from these data sets by a morphing algorithm.	Morphing algorithm by Sederberg and Greenwood 1992 (shape blending).
Weeding	If there is no possibility for morphing a road (the road segment is not conclude in both data sets) the weeding process must be done.	Douglas-Peucker algorithm.
Smoothing	Weeding is followed by smoothing.	Lowe (1988).

Displacement	Displacement plays an important part in this case. After the symbolisation process it could happen that road objects are located too close to each other or to other objects.	Displacement algorithm using elastic beams (Bader 2001).

Very complex elements such as hairpin bends must be simplified or replaced by information from the data set 1:200’000. Table 3 shows the selected generalisation operators for this object class and the corresponding algorithms. The step ”morphing” does not belong to the generalisation operators in the closer sense (McMaster and Shea 1992), because it needs two data sets to transform the selected features. Nevertheless, in our case – a MSDB is used – it can be utilised for the generalisation process.

Fig. 5 shows the proposed generalisation process for the object class ”major and minor roads. Road selection is carried out in a two-step sequence. The first selection step is based on the comparison with the smaller data set. Every road section that appears at both scales must appear. The second step decides on the basis of the road attribute data and which road section is to be shown (assuming that the data in the database has been semantically enriched by importance scores in advance).

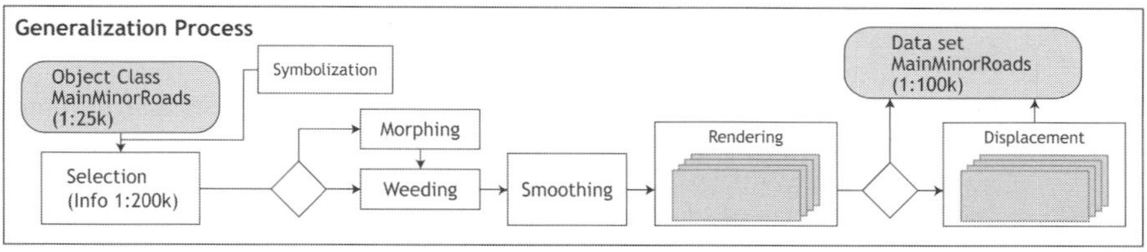

Fig. 5. Generalisation process for the object class “main and minor roads”

The symbolisation step is the same as for highways, though more classes need to be distinguished (major and minor roads are displayed differently). Next, it has to be decided whether a morphing transformation will take place or a weeding operator is directly applied. The difference between the two options depends on the data in the database. If the corresponding road section is contained in both data sets, a morphing process can be started; otherwise the weeding operator is used. The remainder of the process (rendering and displacement, as well as potential reiteration) is equivalent to the one explained for highways.

4.3 Buildings

Buildings are displayed in different shapes depending upon the scale. At a scale of 1:25’000 most objects are represented as polygon elements, at a scale of 1:200’000 as rectangle items, with the exception of large buildings (e.g. factory buildings). The smaller the map scale, the simpler the outlines of the buildings are. In densely settled areas (e.g. town centres) they may simply be represented as city blocks. At smaller scales the representation changes: the built-up area is now shown as tinted polygons instead of individual buildings. As we are using Swiss data in this work, however, this change of representation will not take place. Both scales use individual buildings.

Thus, the generalisation process is based on the assumptions of table 4. The main task of settlement generalisation for smaller scales using individual buildings is to maintain the overall pattern and arrangement of the large-scale representation. Hence, particularly in dense areas (like town centres), typical arrangements of buildings (clusters, alignments etc.) must be identified and defined in the MSDB. These specific groups of buildings must be linked across scales. This data enrichment process is typically carried out offline and prior to on-demand mapping process.

Table 4. Generalisation operators for the class "buildings" and the corresponding algorithms

	Operators	Algorithms
Selection	With the data set, 1:200'000 it can be decided which (groups of) objects are represented.	The 1:200'000 LoD provides the basis (data enrichment: building blocks and clusters in dense areas).
Typification	A reduced set of buildings that exhibits the typical arrangement is displayed at the small scale.	Interpolation-based typification (Hangouët 1996).
Rectification	Rectifies the geometry of objects, which are expected to have a rectangular shape.	Airault (1996), global algorithm for rectification.
Displacement	On overlap of single objects a displacement can be executed. In the case of groups of objects data from the 1:200'000 LoD can be used.	Displacement algorithm using elastic beams (Bader 2001).

The generalisation process for this object class is summarised in Fig. 6. As explained above using the previously defined cluster elements can alleviate the selection step. Also, the typification process can work with these clusters. The rectification step can be essential for the appearance of the buildings in the final map if the available data have not been previously rectified (e.g. if they were manually digitised). Displacement can be accomplished relatively easily since overlaps are easily detected.

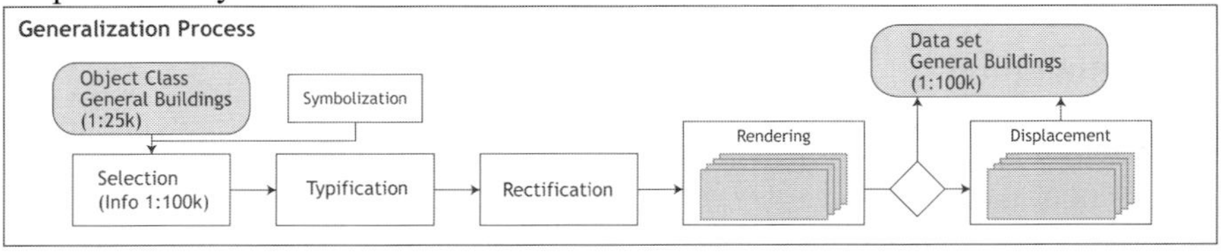

Fig. 6. Generalisation process for the object class "buildings"

Special buildings are buildings that are represented by a special symbol on large and small-scale maps. Examples include churches, castles, power plants etc.

The generalisation process can be executed easily in this case. In many cases all elements can be selected (compare with 1:200'000) and replaced by a previously defined symbol. In case of overlaps with other map objects the object under consideration can be eliminated or, alternatively, displaced if it is a very important building. More important is the definition of the symbol that represents the corresponding object. Typically, this information is stored in the MSDB. Due to the special symbols used, rendering also becomes more demanding.

4.4 Rivers

River items are represented at smaller map scales – in contrast to larger map scales – as linear elements. The change from surface objects to linear objects frees up map space (cf. section 2.2).

The selection can be simplified by the definition of stream orders for river sections (e.g. the Horton order). Table 5 shows the operators and algorithms used. Offline data enrichment can save precious time for on-the-fly generalisation. Two things can be done during the MSDB creation.

First, collapsing polygons to linear objects can be precomputed and added to the data in the MSDB as an alternative representation (i.e. a river object may both have a polygon and line representation). Second, stream ordering (in particular the so-called Horton order; Thomson and Brooks 2000) for the river network can be precomputed and stored as attributes for each river section. The Horton order can directly be used for river selection by network pruning.

Table 5. The use of generalisation operators for the object class "rivers" and the corresponding algorithms

	Operators	Algorithms
Collapse	The collapse process will be carried out offline and stored in the database.	Collapsing using a skeleton algorithm (Bader 1997). Takes place offline.
Selection	The rivers are selected on-the-fly depending upon the defined ordering scheme. LoD 1:200'000 can also be used to decide which elements should be displayed.	Selection by a predefined ordering scheme of river sections (Thomson and Brooks 2000).
Morphing	The LoDs 1:25'000 and 1:200'000 form the basis. Scale 1:100'000 will be generated from these by morphing.	Morphing algorithm by Sederberg and Greenwood 1992 (shape blending).
Weeding/ Smoothing	Weeding is followed by smoothing.	Douglas-Peucker; Lowe (1988).
Displacement	Due to polygon-to-line collapse, additional space is available, thus decreasing the need for displacement. On overlap, the same algorithms can be	Displacement using elastic beams (Bader 2001).

	used as for roads.

The selection can be simplified by the definition of stream orders for river sections (e.g. the Horton order). Table 6 shows the operators and algorithms used.

After having selected the objects that are retained the morphing or the weeding step will follow (Fig. 7). Morphing will be done if both data sets contain the corresponding object. Otherwise, simple weeding can be applied. Displacement will not be used often since the collapsing process can generate more space for the object.

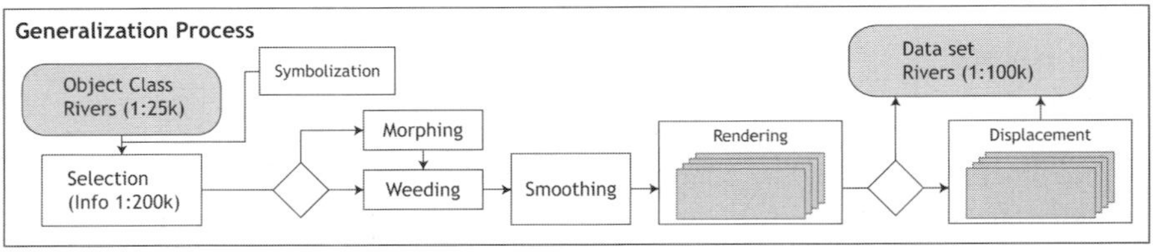

Fig. 7. Generalisation process for the object class "rivers"

4.5 Lakes

Lakes will be retained as long as possible through the scales. In most cases, they are represented as a polygon; if they are long and narrow they are collapsed to linear elements and treated as river object (stored as an alternative representation in the MSDB).

The lake surface generally determines which items will be displayed. As the map scale gets smaller, the outlines of lakes are increasingly simplified in each case. As lakes often exhibit elongated shapes, the minimum width of narrow passages and river inlets must be observed. Table 6 shows the generalisation operators and algorithms used for this object class.

Table 6. The use of generalisation operators for the object class "lakes" and their corresponding algorithms

	Operators	Algorithms
Selection	The lake surface (compared to 1:200'000) determines whether a lake will be retained.	Selection of the lakes in comparison with the data set 1:200'000.
Morphing	The LoDs 1:25'000 and 1:200'000 form the basis. Scale 1:100'000 will be generated from these by morphing.	Morphing algorithm by Sederberg and Greenwood 1992 (shape blending).
Weeding	Data reduction requires co-ordinate weeding.	Douglas-Peucker algorithm.
Smoothing	Weeding is followed by smoothing.	Lowe (1988).

Because lake objects are very important for orientation it makes sense to select as many as possible. If the surface of a lake falls below the minimum size limit, it is eliminated. Objects that appear in both data sets, however, should be shown (including those that are too small, in which case they will be enlarged).

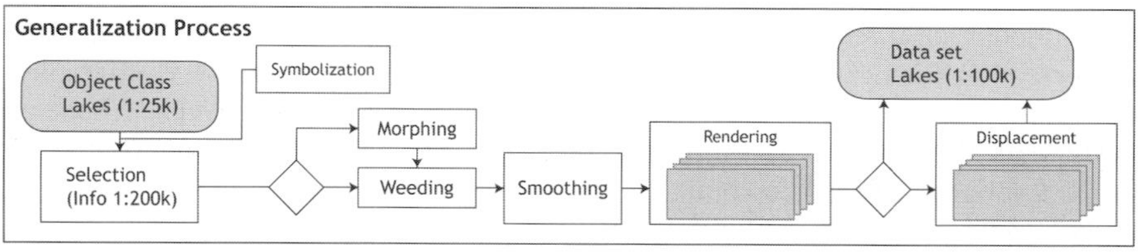

Fig. 8. Generalisation process for the object class "lakes"

Again, it must be decided if morphing or simple weeding will be used (Fig. 8). Since so many lakes will be retained at the small scale, it is important that the outline of the selected lakes is sufficiently simplified.

5 Conclusions and Outlook

This paper has proposed new ways of on-demand mapping in an Internet context. The creation of maps on-demand is a time-critical task. Yet, despite the fact that map generalisation is a computationally intensive task, capabilities for scale changing and generalisation should be added to on-demand mapping services on the web. Existing examples that offer zooming and scaling functions (e.g. www.mapquest.com) are however, usually defined for a single purpose and use multiple (independent) LoDs. As such, they do not allow flexible generation of maps at arbitrary scales and for arbitrary themes. Furthermore, multiple scale levels invariably mean more overhead during database creation and even more so during database updates (as updates need to be propagated across scales to maintain consistency). Fully automated map generalisation is not yet possible – some generalisation operations are still unresolved, and others are too time-consuming. Therefore, we propose to combine both multi-scale databases (MSDB) and map generalisation to exploit the strengths of both approaches in mutual support of each other. On the one hand, MSDB alleviates and speeds up the generalisation task and even substitutes for missing generalisation operators. On the other hand, generalisation renders MSDB more flexible and leads to better cartographic results.

Using our proposed approach, automated generalisation can be separated into two tasks and thus greatly simplified. The first task is carried out offline and consists of MSDB creation and data enrichment. The second task is computed online, adjusting the LoDs pre-specified in the MSDB to the required target scale, using on-the-fly generalisation and morphing. In this paper, we have presented the general approach and then concentrated primarily on the design of generalisation processes related to the second of the above tasks. To this end, we have discussed a method for the selection of appropriate LoDs for relevant object classes based on an analysis of the 'generalisation complexity'. Furthermore, we have presented in detail the structure and contents of the generalisation processes for several object classes. This information is now serving as a basis for the implementation of an experimental MSDB as well as an MSDB-based generalisation prototype. The experimental case study that we are pursuing focuses on on-demand topographic map creation (using Swiss maps as an example). Besides the more cartography-oriented aspects of the work reported here, an initial relational database schema design has already been accomplished (Cecconi 2001) and a rudimentary version of the test database has been built. However, we intend to extend the schema using object-oriented techniques and implement it based on a commercial system (Lamps2). Following that, the database for the test region in the Canton of Fribourg (Switzerland) will need to be populated, in particular by defining the links between corresponding objects of different scales. Next, the algorithms discussed in section 4 need to be adapted for our purposes and integrated into the MSDB context. The ultimate goal is to expand the specifications for the specific test scenario and build a generic module for on-demand mapping.

Acknowledgments

This research is part of the project "GENDEM: Map Generalisation for Thematic and On-demand Mapping in GIS" supported by the Swiss National Science Foundation under contract 20-59249.99.

References

Airault S (1996) De la base données à la carte: une approche globale pour l'équarrissage de bâtiment. Revue internationale de géomatique 6(2-3): 203-217

Bader M (2001) Energy Minimizing Methods for Feature Displacement in Map Generalisation. Ph.D. thesis, Department of Geography, University of Zurich

Buttenfield B (1993) Research Initiative 3: Multiple Representations. Closing Report, National Center for Geographic Information and Analysis, Buffalo

Cecconi A, Weibel R (2001) Map Generalisation for On-demand Mapping. GIM International, Magazine for Geomatics 15(5):12-15

Cecconi A (2001) Schema for a Multiscale Database for Selected Object Classes [online]. Internal Report, University of Zurich, Available from: www.geo.unizh.ch/~acecconi/gendem/msdb

Devogele T, Trevisan J, Raynal L (1997) Building a Multi-Scale Database with Scale-Transition Relationships. In: Kraak MJ, Molenaar M (eds) Advances in GIS Research II. Taylor & Francis, London pp, 559-570

Glover E, Mackaness WA (1999) Dynamic Generalisation from Single Detailed Database to Support Web Based Interaction. In: Proceedings of the 19th International Cartographic Conference. Ottawa, pp 1175-1184

Hangouët, JF (1996) Automated Generalisation Fed on latent Geographical Phenomena. In: Proceedings of InterCarto 2. Irkutsk, pp 125-128

Jones C, Abdelmoty AI, Lonergan ME, van der Poorten P, Zhou S (2000) Multi-Scale Spatial Database Design for Online Generalisation. In: Proceedings of the 9th Spatial Data Handling Conference. Beijing, pp 7b.34-7b.44

Kilpeläinen, T (1997) Multiple Representation and Generalisation of Geo-Databases for Topographic Maps. Ph.D. thesis, Publications of the Finnish Geodetic Institute, Helsinki

Kraak MJ, Brown A (2000) Web Cartography: Developments and Prospects. Taylor & Francis, New York

Lowe DG (1988) Organization of Smooth Image Curves at Multiple Scales. In: Proceedings of the 2nd International Conference on Computer Vision. Tampa pp 558-567

McMaster RB, Shea KS (1992) Generalisation in Digital Cartography. Association of American Geographers, Washington

Peterson MP (1999) Trends in Internet Map Use: A Second Look. In: Proceedings of the 19th International Cartographic Conference. Ottawa, pp 571-580

Ratajski L (1967) Phénomènes des points de generalisation. In: International Yearbook of Cartography, Vol. VII. Gütersloh Bertelsmann, Bonn-Bad Godesberg pp, 143-152

Sederberg TW, Greenwood E (1992) A Physically Based Approached to 2-D Shape Blending. Computer Graphics 26(2): 25-34

Timpf S, Devogele T (1997) New Tools for Multiple Representations. In: Proceedings of the 18th International Cartographic Conference. Stockholm, pp 1381-1386

Thomson RC, Brooks R (2000) Efficient Generalisation and Abstraction of Network Data Using Perceptual Grouping. In: Proceedings of the 5th International Conference on GeoComputation. Manchester (CD-Rom)

van Oosterom P, Schenkelaars V (1995) The Development of an Interactive Multi-Scale GIS. International Journal of Geographic Information Systems 9(5): 489-507

Weibel R, Dutton G (1999) Generalizing Spatial Data and Dealing with Multiple Representations. In: Longley P, Goodchild MF, Maguire DJ, Rhind DW (eds) Geographical Information Systems: Principles, Techniques, Management and Applications. John Wiley, Chichester, pp 125-155

Map Samples to Help GI Users Specify their Needs

Frédéric Hubert[1,2]

COGIT, IGN, 2-4 avenue Pasteur, 94165 Saint-Mandé Cedex , France, email: frederic.hubert@ign.fr[1]
GREYC, Université de Caen, Boulevard du Maréchal Juin, 14032 Caen Cedex, France, email: fhubert@info.unicaen.fr[2]

Abstract

Accessing the user interfaces proposed by GISs, G-DBMS or geographic Web sites shows how complex they may prove for novice users of geographic information. Such users have needs in geographic information, but they do not have the means to express them and thus to adequately fulfil them. The objective of this paper is to outline an approach that may help users specify their needs with geographic information. This approach consists in converging on one available solution, the closest possible to the user needs, by using new means of communication between user and computer: map samples and natural language. Map samples are geographical data, extracted from the database and possibly processed by available GIS treatments. A few map samples are proposed to the user, who makes the relevant choices. The system tries to "understand" the interactions to help in the choosing. A search engine navigates through the available map samples in order to identify the most appropriate map samples and proposes them to help the user identify the need more precisely. Natural language allows the expression of user needs to be captured in a manner more consistent with a dialogue and provides more freedom or versatility for example, for general queries and quantitative evaluation. A virtual scenario shows the combination between these two languages.
Keywords: geographic information, user needs, map samples, natural language, user interface

1 Introduction

To access geographic information, we should use both geographic applications and data. These applications are available through user interfaces included in

Geographic Information Systems (GIS), Geographic Database Management Systems (G-DBMS) or other software. Moreover, GI is mainly accessible as maps on the Web and we can only visualise them. The Web is the visible part of Internet and the majority of users are novices in geographic information. Some advantages of Internet explain the increase in the number of geographic Web sites, but there are limitations as well. Geographic Web sites offer little functionality compared to standard desktop GISs. Most of the time no tools are available to help users specify their needs, although they would be most welcome especially when the user is not an expert (Hubert 2001).

Contributing to this issue, we focus on spatial query languages as a way to help formulate queries without using expert languages such as SQL. Some approaches have been proposed to improve the expression of queries by extending the current interfaces of GIS. They propose new kinds of interfaces that include texts, graphics, icons, and so on. (Egenhofer 94) proposes an extension of the SQL language with *Spatial SQL*. *PICQUERY* is a textual language inviting users to enter words in display tables corresponding to spatial operators (Joseph and Cardenas 88). Visual languages like *CIGALES* (Calcinelli and Mainguenaud 94) or *LVIS* (Bonhomme *et al.* 99) offer a graphic interface where users can draw their request by using icons and metaphors. *Sketch!* (Meyer 92), *Spatial-Query-by-Sketch* (Egenhofer 96) and *VISCO* (Haarslev and Wessel 97) are sketch languages where users draw their requests and the system interprets them by trying to identify the spatial operators involved. Hybrid languages combine a few means of communication, for example (Lee and Chin 95) with icons and graphics, *PEGASE* (Proulx *et al.* 95) with graphics and text and the *Geographical Anteserver* (Szmurlo *et al.* 98) with graphics and labels in natural language. Even though these spatial query languages offer more freedom to express requests, there are three major limitations: we can only make simple queries, the interfaces are not always intuitive, and user needs are not taken specifically into account.

Our work gives the priority to the expression and the understanding of users' needs. Our purpose is to identify and to understand these needs before running geographic applications that may be complex and time consuming. With regard to works on spatial query languages, we do not only want to replace SQL languages. Our aim is, on the one hand, to treat queries that modify representations, and on the other hand to treat queries that create information (i.e. information that will be computed from existing data by means of expert GIS functions such as generalisation). To improve the expression of needs in geographic information, a link should be established between novice users and the stored and computed information on remote machines. Needs could be, for example, *"show city monuments by localising them with a chosen symbol"* or "*create a map with only roads and some buildings along a river*". Actually, the challenge is to relate two means of thinking: human thinking and computer "thinking". Users should not have to adapt themselves to the machine, the machine should "understand" users. To conceive this link to help users, we propose to introduce new means of communication to help users: map samples and natural language. Through an interaction with the users, the system can be made to understand and to specify their requests. Our approach allows the running of geographic applications without prior knowledge in geographic information from the users.

This paper discusses the principles of our interface combining map samples and natural language: Section 2 explains why we have chosen map samples and natural language. Section 3 gives definitions about map samples. Section 4 presents a scenario on the specification of queries by combining map samples and natural language. Section 5 proposes a method to find map samples from a "map samples" database. Section 6 explains briefly the general architecture of our system and its different modules, more precisely the Web interface and the dialogue module, which have been fully implemented.

2 Why Use Map Samples and Natural Language?

The purpose of our work is to make a system to identify user's needs in an effort to apply the relevant geographic treatments on appropriate data. We introduce a mechanism between users and server of geographic applications designed to help users to specify and fulfil their needs in geographic information (see Fig. 1). Geographic data are currently manipulated through specific applications like GISs. For an expert in geographic information, using a GIS may be easy. For a novice, however, many non-intuitive, even random manipulations are necessary before finding some more or less satisfactory treatment. We think that an automated identification of the user's need is necessary to efficiently produce the desired result.

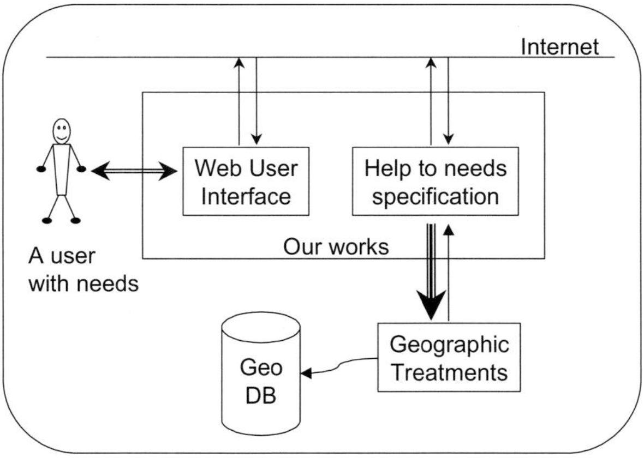

Fig. 1. General model

To obtain information on the user's needs, we propose a user interface to capture the needs. The interface's engine manages the events: input of user's needs and visualisation of the intermediate or final results. Hidden to the user is the "intelligent" engine, to receive the input data from the interface engine and to understand, validate and process the user's needs. As we focus on novice users, we must offer an intuitive means of communication. Our first choice is to use map samples. Map samples are like images. They are used to illustrate solutions as made possible by GIS treatments. The samples subdue the specificity of geographic information and of the precise treatments in favour of intuitive representation. Users indeed do not know the specific treatments and do not have expert background in geographic information. Moreover, users are in the habit of visualising and using maps, like paper maps.

We cannot only use map samples, however, since users cannot usually express their needs graphically (most of the communicability of a map indeed lies in its legend). Natural language is one an option for resolving this limitation.. It can be used to start the dialogue and to identify a global application concerning the need, for example "*I want to generalise buildings*". Further it can be used to validate a proposed map sample, for example "*I like that*" or "*I don't understand*". Finally, it allows the user to express refinements on a map sample, like "*This sample without roads*", "*More square*" or "*Less big*". The use of natural language is easy for users. They need no technical knowledge on interfaces and their mechanisms. Some freedom of expression is offered as well (Frank *et al.* 1991).

In combining these two means of communication, we obtain a harmonious dialogue with the users to help them specify their needs on geographic information: natural language corresponds to a facility for people, while map samples corresponds to a facility for machine. This article will focus on map samples.

3 What are Map Samples?

A map sample is a small extract or part of a geographic database. We distinguish two types of samples: basic samples and treated samples. Basic samples are a small subset of data from the GDB. Treated samples are results of geographic treatments on basic samples. Three views on map samples are presented and can be distinguished as follows: samples as a graphic object, samples as a discourse object and samples as a computer object. In fact, one sample corresponds to one graphic representation and one computer representation. The discourse representation is used to establish a relation between natural language and map samples.

3.1 Map Sample as Graphic Object

The graphic object representation of a map sample is a GIF, JPEG or TIFF image, which are easy formats to use in Web pages. This object can be easily integrated in Web pages and visualised by users. The main interests of using this representation are:

- Map sample is in the geographic information language with its cartographic representation;

- The user is in the habit of visualising maps without adaptation problems;
- The user has no specific knowledge in geographic information.

This last representation masks information on parameter's values or geographic treatments to the user.

3.2 Map Sample as Discourse Object

The graphic sample becomes a discourse object when the user can express reactions about the sample in natural language, when he can enter sentences, like "*I want that buildings are smaller*", "*I choose the third sample, but I want fewer details*", " *I don't like that*". In this case, it is necessary to establish the possibility of entering sentences, and the link between sentences and relevant samples. Some knowledge is necessary to translate words in context, like "*smaller*", "*less than details*", "*the third sample*": to associate a meaning to a word or a group of words in the sample's context, we use specific knowledge (see Section 5.3).

3.3 Map Sample as Computer Object

The computer object representation gives information on raw and treated samples. Replacing geographic applications by graphic and computer objects allows one to have some information on the creation of samples by GIS or other processes. The computer representation allows the handling of the graphic and discourse objects coherently. The user analyses the graphic properties of a map sample and criticises them in natural language. The system uses the computer representation of this sample to translate criticisms into expectations by using another knowledge defined in Section 5.3. The aim is to find another sample corresponding to the user's wishes by using the computer object representation of the current sample.

To understand the complementarity between map samples and natural language, a virtual scenario is given in the next Section. This scenario shows how the user may use our interface to interact with the system.

4 A virtual Scenario

Our user interface is organised into two major interaction areas (Fig. 2): (1) the dialogue box where the user interacts with the system in natural language and (2) the result area where the user interacts by mouse clicks.

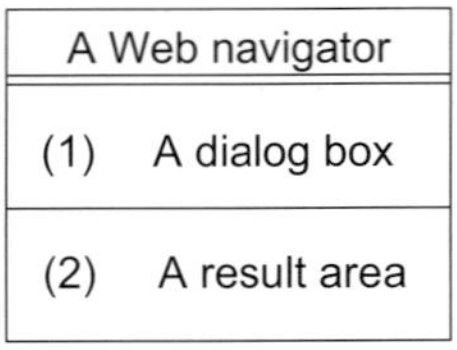

Fig. 2. The user interface

The following scenario, presents a simple query with few interactions between the user and the system, and provides an example of the interaction that is made by our interface between the user and the system. To start the dialogue, the user writes a request, in the form of a sentence: "*I want to generalise some buildings*" (Fig. 3).

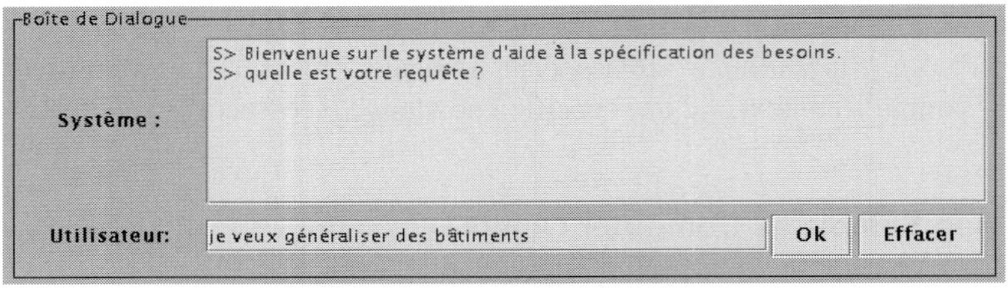

Fig. 3. The first query in the dialogue box

The system analyses this request by using a dictionary with subjective terms (e.g. "*I want to*", "*I like that*"), technique terms (e.g. "*generalise*", "*smooth*", "*zoom*"), and thematic terms (e.g. "*buildings*", "*house*", *roads*"). Then, the system proposes first result according to the buildings' nature (e.g. house, church, hospital, and school). The nature of the buildings is presented on a page by using checkboxes as graphic objects. The user can select an element or write a sentence in the dialogue box.

The system records the user's choice and displays another page with some images corresponding to map samples. It shows one or several basic samples and their treated samples. The user selects the building he wants (Fig. 4). For example, he chooses sample E2. In this case he is allowed to add a constraint by writing a sentence in the dialog box: "*I select this sample, but without the small square from the right wall*".

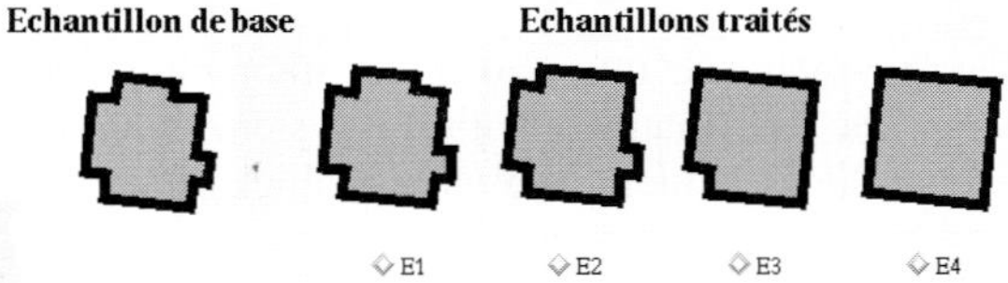

Fig. 4. A results Web page with map samples

The system records the user's choice and analyses the sentence. After the research of samples, it displays another page with one sample as close as possible to the user's request (Fig. 5).

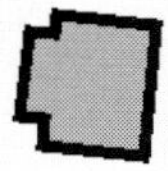

Fig. 5. A map sample as result

The user sees the map sample and reacts by writing: *"OK, for this sample"*. The system interprets the response as an order of validation of the results and can pursue the helping the user further specify his needs. The process is pursued by proposing additional map samples that correspond to the first selected sample. This is reasonable since the process should not be based on only one simple sample.

This virtual scenario shows that the system must be able to translate indications, from the users to find further map samples. We have defined some strategies to search one or several samples in a database of map samples.

5 Search Strategies on Map Samples

In the previous section, we have presented a simple virtual scenario, but the choice and the search of map samples are not trivial. To manage map samples, we connect our system with a "map samples" database. In this Section, we present this map samples database, and explain two different research strategies. Finally, we describe some knowledge to process these strategies.

5.1 The "Map Samples" Database

The map samples database has been created by geographic treatments on a GDB. These geographic treatments come from GISs, G-DBMS, and so on. This database stores computer objects of basic and treated map samples. An organisation of the database is necessary to improve the access to map samples. We distinguish some categories of samples according to their type (e.g. buildings, roads, urban blocks, map of city, and so on), their nature (e.g. house, school, church, administrative building, among other aspects). The database groups the basic and treated samples as well.

The current state of our map samples database contains only map samples resulting from cartographic generalisation processes. The computer properties are based on the use of constraints as defined in (Ruas 1999). The treated map samples are identified by different constraints such as "size", "width", "granularity", "squareness" and "concavity". These constraints have goal values and current values to identify the interest of the sample. In this way, and as a first prototype , samples are described by means of their properties and not by means of their GIS treatment. Our prototype database contains 63 basic map samples. Nicolas Regnauld from the University of Edinburgh has processed 32 different generalisation processes using these basic samples.

5.2 Search Strategies

Our two competitive strategies are based on the use of space criteria for map samples (Fig. 6). The axes represent the values of criteria. A criterion can be a parameter or a constraint. In the figure, we have 3 criteria: c1, c2 and c3. E1, E2, E3 and E4 are the treated map samples selected by a user at different steps. Each symbol on the figure corresponds to a map sample.

Strategy 1: Search of coherent values of criteria

Conditions: After several interactions, a set of map samples, or a selection of several samples has been selected by the user. **Actions**: According to the chosen samples, the set of relevant criteria is constituted, and a space of their possible values is defined. The search method then consists in building hypotheses in that space and in restricting them by proposing new samples. Fig. 6 shows a volume including the chosen samples E1, E2, E3 and E4. If the system finds 20 map samples in this volume, it proposes for example 4 samples to the user and so restricts the initial volume. The aim is to find the sample closest to the user's need in this volume.

Strategy 2: Queries in natural language on map sample

Conditions: A user has selected a map sample and written a sentence qualifying this sample (e.g. "*larger*", "*fewer details*", "*bigger*", "*smoother*" and so on.).

Actions: Expert linguistic and geographic knowledge is used to translate the previous terms into criteria. For example, "*fewer details*" corresponds to "*less granular*" which corresponds to a specific criterion in the search space, say c2 on Fig. 6. The system must find within the space one or several map samples whose value for criterion c2 is smaller than that of the selected sample.

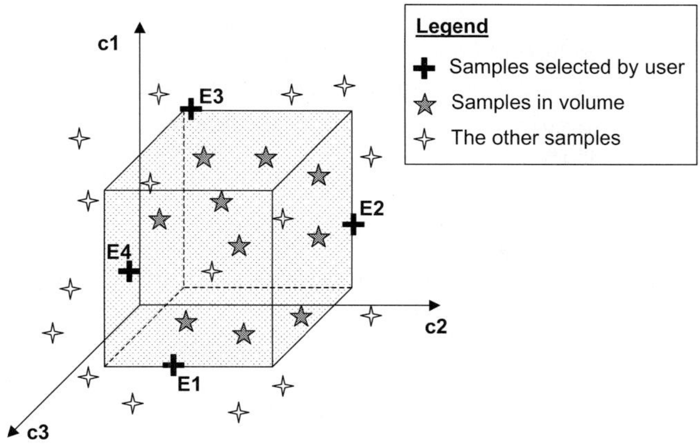

Fig. 6. A space of criteria for map samples

The aim is not to find the map sample which corresponds exactly to the request, but to converge on the must appropriate available solution. Some tools for proximity assessment should be included to answer terms such as "*close to*".

5.3 Necessary Knowledge

To run the research strategies, we need to use specific knowledge:

- Information on the map samples database: This information facilitates the research on available map samples, on possible parameters and on their possible values. For example, the "size" parameter of for building samples generally has values between 25 and 400 m2.
- Information on the generic terms of natural language: Allowing the use of terms in natural language implies their translation in parameter values for map samples. Generic terms include "*more*", "*less*", "*close to*", and so on. For example, the term "*more*" corresponds to an increase of one or several constraint values.
- Information between natural language and properties of map samples: This information is used to establish a mapping between a term in natural language and one or several parameters for map samples. Moreover, it depends on the context. For example, in the term "*bigger*", the system identifies the term "*big*" as corresponding to the size of a map sample. In fact, it must increase the "size" of the current map sample.
- Information on available treatments: Information on geographic treatments can make the searching easier. A geographic treatment can be associated with several constraints, where some variations are forbidden. For example, smoothing implies different values for some constraints. If the system "knows" the different parameters associated with a geographic treatment, it only searches map samples according to these parameters.

These strategies are integrated in a "map samples" manager. This manager is a module of the general architecture of our prototype presented in the next section.

6 General Architecture

Fig. 7 gives us an overview of our prototype. It is composed of 3 modules with different roles: the Web interface, the dialogue manager and the map samples manager.

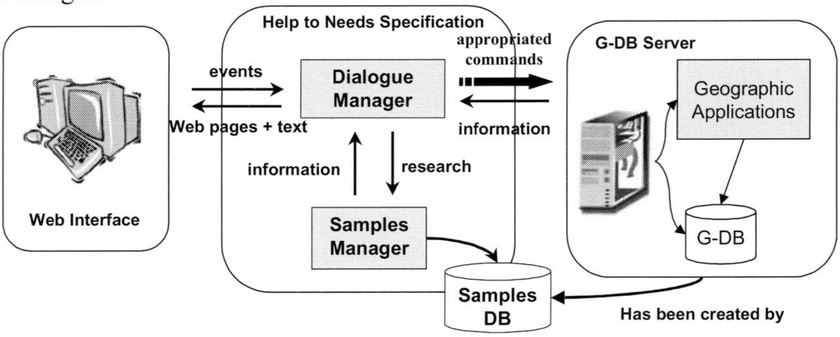

Fig. 7. General schema of our architecture

6.1 Web Interface

The Web user interface offers the user the means to interact with the system and to visualise results including graphics or texts. It is divided in 3 areas as shown on Fig. 8: (1) dialogue box, (2) result area and (3) "common ground" area.

The dialogue box (1) allows the input of text in natural language and displays the response from the system. This box is a Java applet and uses a RMI (Remote Method Invocation) connection with the dialogue manager. The result area (2) allows the display of intermediate and final results. The user can interact with the system by using graphic objects like checkboxes, radio boxes, buttons and so on. These pages are written in HTML and JavaScript languages. The connection with the dialogue manager is established with a JavaScript socket. The last area (3) is used to display the evolution of the "common ground". The "common ground" corresponds to knowledge resulting from an inter-understanding between the user and the system (Nicolle *et al.* 1999). In fact, the dialogue manager stores all interactions with the user by creating a reasoning tree. This tree is used to avoid the loops of reasoning and to describe the "common ground".

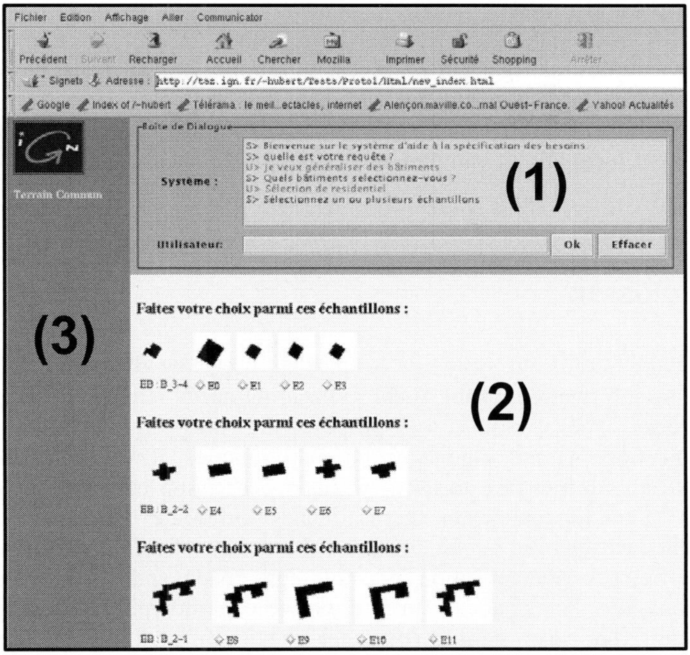

Fig. 8. The Web user interface

6.2 Dialogue Manager

The dialogue manager is the brain of our system. It establishes a link between the Web interface, and the map samples manager or the available GIS. This manager is divided into two modules: a Web server and a dialogue module.

The Web server intercepts events from the Web interface and sends them to the dialogue module. It has an "on-the-fly" generator of Web pages containing graphic objects, text, and images like map samples.

The dialogue module retrieves events and treats them by various actions. It analyses all inputs in natural language with a syntactic and semantic analyser. With these analyses, it can contact the map samples manager or a GIS software to run the relevant action with the relevant information such as constraints goal value. This module must be able to generate language acts, to report bugs or to manage a history of interactions. To conceive this module, we have integrated the Genedic prototype (Lemeunier 2000). Genedic manages a dialogue between human and machine in natural language and it includes functionalities such as the management of waits and the recognition of communication intentions.

6.3 Samples Manager

The map samples manager allows access to samples and analysis. Different kinds of knowledge corresponding to the different strategies (see Section 5) are integrated. Information from samples manager to dialogue manager must include the different locations of samples in the samples database and the results of the analyses

7 Conclusion

The primary role of our system is to translate user's needs in an effort to derive geographic information on the Web. Translating needs requires an expression of those needs. As a result, we propose the use of innovative means of communication: natural language and map samples. These choices allow the design of a Web interface to assist in the specification of needs for geographic information and to identify the necessary and appropriate data.

Current investigation is focussing on the design of a first prototype allowing the user-friendly generalisation of buildings, roads and urban blocks. Some adaptations are being added to the Genedic prototype to treat geographic information. The map samples manager is still to be designed. Further work includes adding the "common ground" facilities in the user interface. Additionally to validate both the interface and the dialogue on map samples, the application scope will be extended to address cartographic symbolisation and allow users to draw their own maps.

Acknowledgments

The author thanks Anne Nicolle of the AI laboratory at the University of Caen (France) for our discussions about dialogue issues in natural language. Thanks also to my reviewers: Jean-François Hangoët and Anne Ruas at the COGIT laboratory.

References

Bonhomme C, Trépied C, Aufaure MA, Laurini R. (1999) A Visual Language for Querying Spatio-Temporal Databases. In: Proceedings of the 7th International Symposium on Advances in Geographic Information Systems. 2-6 November, 1999, Kansas City, USA, pp 34-39

Egenhofer MJ (1994) Spatial SQL: A Query and Presentation Language. IEEE Transactions on Knowledge and Data Engineering 6(1):86-95

Egenhofer MJ (1996) Spatial-Query-by-Sketch. In: IEEE Symposium on Visual Language. 3-6 September, 1996, Boulder, Colorado, USA, pp 60-67

Frank A, Mark DM (1991) Language Issues for GIS. In: Maguire DJ, Goodchild, MF, Rhind DW (eds) Geographical Information Systems: Principles and Applications. Longman Scientific and Technical, London, pp 147-163

Hubert F (2001) Assistance Mechanisms Use for Needs Specifications in Geographical Information on the Web. In: Proceedings of the 20th International Cartographic Conference. 6-10 August, 2001. Beijing, China. 4. pp 2320-2329

Haarslev V, Wessel M (1997) Query GIS with Animated Spatial Sketches. In: IEEE Symposium on Visual Language. 23-26 September, 1997, Capri, Italy, pp 201-208

Joseph Th, Cardenas AF (1988) PICQUERY: A High Level Query Language for Pictorial Database Management. IEEE Transaction on Software Engineering 14(5):630-638

Lee YC, Chin FL (1995) An Iconic Query Language for Topological Relationships in GIS. International Journal of Geographical Information Systems 9(1): 25-46

Lemeunier T (2000) Modélisation Dialogique des Intentions de Communication en Dialogue Homme-Machine. Ph.D. thesis, Le Mans University, France

Mainguenaud M, Portier MA (1990) Cigales: A graphical query language for geographical information systems. In: Proceedings of 4th International Symposium on Spatial Data Handling. 22-28 July, 1990, Zurich, Switzerland, pp 393-404

Meyer B (1992) Beyond Icons: Towards New Metaphors for Visual Query Languages for Spatial Information Systems. In: Proceedings of the First International Workshop on Interfaces to Database Systems. Glasgow, UK, pp 113-135

Nicolle A, Saint-Dizier de Almeida V (1999) Vers un modèle des interactions langagières. In: Moulin B, Delisle S, Chaib-draa B (eds) Analyse et Simulation de Conversations : De la théorie des actes de discours aux systèmes multiagents. L'interdisciplinaire, Lyon, pp 133-169

Proulx MJ, Bédard Y, Moulin B (1995) Développement d'un nouveau langage d'interrogation de base de données spatio-temporelle. In: La 7ème Conférence internationale sur la géomatique. 11-15 June, 1995, Ottawa, Canada

Ruas A (1999) Modèle de Généralisation de Données Géographiques à Base de Constraintes et d'Autonomie. Ph.D. thesis, Marne La Vallée University, France

Szmurlo M, Gaio M, Madelaine J (1998) The geographical anteserver : a client/server architecture for gis. In: Proceedings of EODEO'98 Workshop. February, 1998, Salzburg, Austria

Representation of Map Objects with Semi-Structured Data Models

Emmanuel Stefanakis

Department of Geography, Harokopio University, 70 El. Venizelou Ave., 17671 Athens, Greece, e-mail: estef@hua.gr

Abstract

Mapping agencies world-wide produce a range of different types of maps, in efforts to meet different user requirements. These maps are usually designed in a digital environment and some of them are distributed through the World Wide Web. One important issue for those agents is the appropriate modelling and handling of map objects from which the digital maps are composed. , This paper focuses on this aspect and as such addresses two aspects. First, the semi-structured nature of data composing a map is discussed. Second, the Object Exchange Model (OEM), a popular model for semi-structured data, is adopted to represent a map. Several configurations regarding the representation of a map in OEM are proposed. Finally, it is shown how useful information can be extracted from those configurations using Lorel query language for semi-structured data.
Keywords: semi-structured data, OEM, Lorel, XML

1 Introduction

Semi-structured data has recently emerged as an important topic of study for a number of reasons (Buneman, 1997): (a) there's a need to handle data sources that cannot be constrained by a schema, through the web (b) there's a need for a flexible format for data exchange between remote databases, and (c) there's a need to assist the browsing on (structured or semi-structured) data based on a self-describing schema. The use of semi-structured data models has been investigated in the past mostly in more traditional data, such us data for bibliographic databases. Expressive data models for representing semi-structured data have been proposed and efficient languages to query semi-structured data have been developed (Abiteboul, 1997; Suciu, 1998).

This study focuses on the use of semi-structured data models for representing *map objects*. Currently there's an increased demand from mapping agencies to produce digital products and distribute them electronically through the web (Kraak and Brown, 2001). When applicable, map objects are modelled using proprietary models of software systems (CAD, GIS, etc.) which are applied to generate the digital maps. Although, some of those systems already use sophisticated models, such as the object-oriented model, several deficiencies emerge. First, map data is usually only semi-structured and traditional database models are not well suited to represent it, primarily because they assume a constrained schema. Second, the database schema is strict and when data is exchanged well described metadata should accompany the data so that the recipient is well advised of the detail.

The scope of this paper is twofold. First, to discuss the advantages of using semi-structured data for composing a map, and second, to show how semi-structured data models can be adapted to model map data. Taking into account the current developments on the web technology and the rapid spread of *XML* (eXtensible Mark-up Language) (W3C, 1998) and its extensions, the reader would expect the above specification to be adopted for modelling map objects. However, this is not the case in this paper, and *OEM* (Object Exchange Model) (Papakonstantinou *et al.*, 1995) has been chosen instead. There are several reasons for making this choice.

First, OEM is a pure database model for semi-structured data, in contrast to XML, which has been built for the exchange of structured data over the web. It is argued (Abiteboul *et al.*, 2000) that the XML specification and its extensions, although suitable for modelling semi-structured data, will not solve the problem of efficiently extracting the required portion of the underlying data, in other words, XML does not efficiently query large sets of data and extract the required information. The solution to this problem is in part through the development of parallel databases that have semi-structured data. It is therefore well recognised that the combination of XML with semi-structured data models (OEM is the most popular of them) will yield a new technology for web data (Abiteboul *et al.*, 2000) and generally for information exchange across agencies.

Second, considerable efforts have been given to advancing or improving the representation of geographic information with XML, and *GML* (Geographical Mark-up Language) (OpenGIS, 1999). On the other hand, the representation of geographic information with semi-structured data models does not, as yet, appear to have been fully investigated.

Third, it has already been shown that an OEM representation can be generally mapped to an XML file (Suciu, 1998; Abiteboul *et al.*, 2000). XML does not have an associated data model, OEM on the other hand, provides a more suitable and convenient environment to model data.

The discussion is organized as follows. Section 2 briefly presents the application requirements, i.e., the modelling of map objects, and provides a simplified example of a map object collection, to demonstrate the discussion. Section 3 briefly presents the *Object Exchange Model* (OEM), which is the most popular semi-structured data model found in the literature (Papakonstantinou *et al.*, 1995) and has been adopted in this study. Section 4 elaborates on the semi-structured nature of data involved in a map. Section 5 shows how an individual map can be modelled in OEM, while section 6 is focused on the operational issues. Specifically, it shows how *Lorel* language, designed to query semi-structured data (Abiteboul *et al.*, 1997; Quass *et al.*, 1995), can be used to extract information from an OEM representation of a map. Finally, Section 7 concludes the discussion by summarising the contribution of the paper and giving hints for future research.

2 The Application Domain

A map, either on paper or in digital form, can be seen as a database. It consists of a set of objects (or entities), the *map objects*, which represent real world entities, physical (e.g., a road, a river, etc.) or conceptual (e.g., poverty, criminality, among other aspects.), on the map. The representation of an entity on the map may take different forms, depending on the map scale, the map content, or the map scope (Keates, 1989).

Any map object, presented either in analogue or digital form, can be considered as the *representation instance* of some real world entity. A map object has several dimensions. First, it has a unique identifier. Second, it is assigned the identifier of the real world entity it represents. Third, it has a specific *cartographic symbol*. Fourth, it is positioned in a specific location on the medium, namely it has *geometry*. Both the symbol and geometry are characterised by the *symbol type*, which may be (Keates, 1989): (a) point, (b) line, (c) area, and (d) text, or combinations of them. Each symbol, depending on its type, is described by appropriate graphic *parameters*, such as form, dimension, colour, orientation, pattern, font, etc.

Fig. 1a presents a simplified example map. This map consists of a set of map objects, which represent real world entities on a paper medium. These objects are assigned unique identifiers (Fig. 1b). Letters *P*, *L*, *A*, and *T* refer to the type of the symbol, i.e., point, line, area and text, respectively. The bridge located at the crossing between main road *L2* and lake *A1* (LakeA) is a point symbol, with parameters: colour = “black”, orientation = 30deg, dimension = 8mm, form = “II-like” (a geometric description might be given instead). Main road *L2* is a line symbol, with parameters: colour = “red”, dimension (i.e., width) = 2mm, form = “solid line". LakeA, represented by *A1*, is an area symbol, with parameters: colour = “cyan”, pattern = “plain”. Finally, annotation “LakesB”, represented by *T2*, is a text symbol with parameters: colour = “black”, font = “Times bold”, dimension = “16pt”, orientation = -45deg.

The map objects geometry, is a function of symbol type. In other words, the map object geometry can be seen as an ordered set of primitive geometric elements. An element is the basic building block of geometry, i.e., point, line (or polyline) and area (or polygon). This schema, also adopted in spatial database management systems (DBMS), such as Oracle Spatial (Oracle, 2000), is capable of modelling simple and complex map object geometries. Text objects can be handled as objects of type either point or line. In the former case, text centroid is the point of reference. In the latter case, a polyline models text curvature (Keates, 1989).

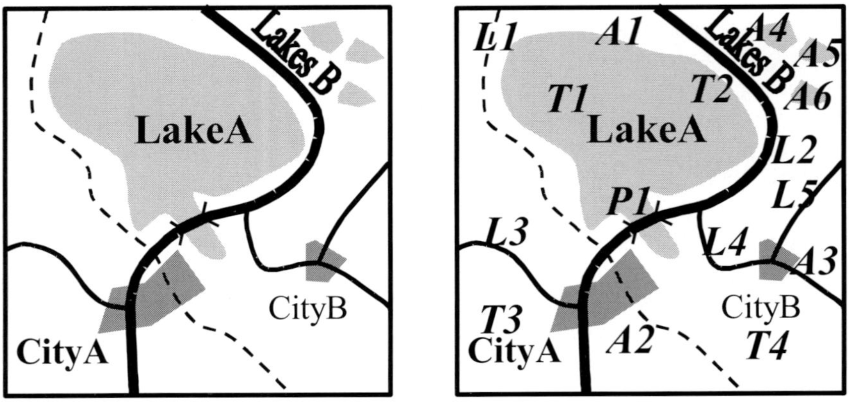

(a) (b)

Fig. 1(a) A simplified example map, **(b)** with map objects labelled

Although the map in Fig. 1 contains only a few simple symbols and by no means can be considered as a complex map representation, it is adequate to assist the discussion on the highlights of semi-structured data necessary for composing a map. Further, it will show how semi-structured data models can be used to efficiently handle map objects.

3 Semi-Structured Data and the Object Exchange Model

The observation that much of today's electronic data does not conform to traditional relational or object oriented models, combined with the need to integrate heterogeneous data, has led to the development of data models for handling semi-structured data. Additionally, query languages for handling this data have recently been proposed and investigated (Abiteboul, 1997; Buneman, 1997; Suciu 1998).

Semi-structured data is characterised by an irregular structure. Some objects may have missing attributes, while others may have multiple occurrences of the same attribute. Further, a single attribute may share different objects, and semantically related information may be represented differently for various objects.

The most popular model of semi-structured data is the *Object Exchange Model* (OEM) (Papakonstantinou *et al.*, 1995), originally introduced for the Tsimmis data integration project (Chawathe *et al.*, 1994). OEM is a simple, self-describing, model, where data is represented by a collection of objects. There is no need to define the structure of an object in advance, and there is no notion of a fixed schema or object class. In a sense, each object contains its own schema.

Every object in the OEM data model has an *identifier*, a *label*, a *type*, and a *value*. The *identifier* uniquely identifies the object among all objects in the domain of interest. The *label* is a string (the tag) presumably denoting the "meaning" of the object. Labels may be used to group objects by assigning the same label to related objects. Labels are usually drawn from a universe of label names. The *type* refers to the data type of object value. The *value* can be of a scalar type, such as an integer or string, or it can be a set of (sub)objects.

Each object may be *atomic* or *complex*. The value of an atomic object is of some base type (e.g., integer, string, image, sound, etc.). The value of a complex object is a set of subobjects, i.e., a set of object identifiers.

OEM is a collection of tagged values and is usually represented as a graph, where the nodes are the objects, the edges are labelled with object labels, leaf nodes are atomic objects associated with an atomic value, and non-leaf nodes are complex objects pointing to their subobjects. The graph also has a root, which is a distinguished object with access to all other objects.

OEM is much simpler than object-oriented models. It supports only *object nesting* and *object identity*, while other features such as classes, methods, and inheritance are omitted. When dealing with semi-structured data the OEM has several advantages as compared to relational or object-oriented models (Abiteboul, 1997; Buneman, 1997; Suciu, 1998).

A representative general-purpose data management system for semi-structured data, which uses OEM, is *Lore* (Lightweight Object Repository), which provides *Lorel* query language (Quass *et al.*, 1995). Notice that contrary to traditional data models, in OEM it is not required to be aware of the schema in order to pose a query. The schema can be discovered as queries are posed (Papakonstantinou *et al.*, 1995).

4 Map Objects are Semi-Structured in Nature

Data involved in a map forms a special category of semi-structured data. The following paragraphs explore this issue.

The symbols used to represent real world entities on the map may have a complex type described by many and variable parameters. Specifically, there are four basic symbol types, i.e., point, line, area or text, which can also be combined. Each type is described by different parameters (Keates, 1989). For instance, a point symbol is usually described by three parameters: form, colour, and dimension. There are, however, cases where more parameters are applied, such as orientation for non-symmetric symbols, or additions, extensions and iconic forms depending on the choice of the cartographer. The use of images, in place of point symbols, is also possible. Things are more complicated when dealing with line or area symbols, where additional parameters may be present, such as multiple lines, or area patterns with variable orientations and densities. Text symbols, used in annotations, are also hard to model, provided that a text is described by its colour, type (or font), size, rotation and curvature. The whole problem of modelling symbols becomes much more complex if the requirement calls for symbols that combine two or more types, e.g., point and area types, or all three point, line and area types. Hence, it is hard to generate a schema in advance that can describe every symbol used in the map, especially in a cartographic organisation, where symbology is a subject of continuous revision.

The granularity of space, i.e., the precision of coordinates describing geometry, is variable and depends on source data and map scale. Hence, the geometry of a map object is expressed with variable precision. Even further, in some cases the coordinate reference system for geometries may not be the same for all map objects. This situation may occur when geometric data is derived from different sources (e.g., land surveys, photogrammetric detail, and paper map digitisation, among others) and they are maintained in the database without being transformed to a common reference system. The simultaneous and integrated visualisation of this type of data can only be obtained by performing appropriate registration procedures.

The units used to express the parameters of symbols may not be unique. For instance, line width may be expressed in units on the map or on the ground. In addition, the units in each case may vary between symbols. For example, the use of mm/inches or meters as integers/meters as floats with two decimal figures, respectively. Another example concerns the colour value, which can be expressed as a triple in a RGB system, as another triple in a HSL system or simply as a standard string, such as "dark red".

All features above render objects involved in a map to form a category of semi-structured data. It is apparent that defining a schema, that uses traditional data models, and that will meet the future modelling requirements of the larger mapping agencies (e.g., national cartographic organisation) is a complex and demanding task. It seems that a more convenient configuration is the adoption of a *semi-structured database* (Suciu, 1998), which does not require an a-priory schema, and is based on a self-describing model that is readily extensible.

The following discussion is focused on how a semi-structured data model, and specifically OEM, can be applied to model map objects.

5 Representation of a Map in OEM

Consider the map in Fig. 1. This map can be modelled in OEM. The map consists of complex and atomic objects. An example of an atomic object is the dimension (i.e., width) of a linear symbol, e.g., "1.5 mm". Another example is a point location described by a string that has two double numbers, the coordinate values for X and Y in a reference system, separated by a comma; for example, the UTM coordinates "515456.67, 4125113.71".

An example of a complex object is the main road, the thick solid line, represented by *L2*, in the map. This object has four sub-objects (with labels) which are as follows:

(a)"oid", an atomic object with a value that the user identifier assigns to the representation instance, i.e., *L2*;
(b) "type", an atomic object that accommodates the value "line" and describes the symbol type used to represent the road;
(c) "geometry", an atomic object with a value of type string, e.g., "12, 515356.67, 4125003.71, 515358.23, 4125110.29, …, 515528.14, 4125998.18", where the first integer, i.e., 12, refers to the number of vertices composing the polyline, and the twelve pairs of numbers that follow are the UTM coordinates of polyline vertices, separated by commas, and;
(d) "parameters", a complex object, that points to three sub-objects, with labels "colour", "form" and "size".

Objects labelled as "form" and "size" are atomic, with values "solid line" (a string) and 2.0 (a float) respectively, while an object labelled as "colour" is a complex object pointing to three atomic objects labelled as "hue", saturation", and "lightness" with values 0, 240, and 120 respectively, representing in this case, red.. Fig. 2 presents the graphical and textual representation of the main road in OEM. Notice that *mo* stands for "map object".

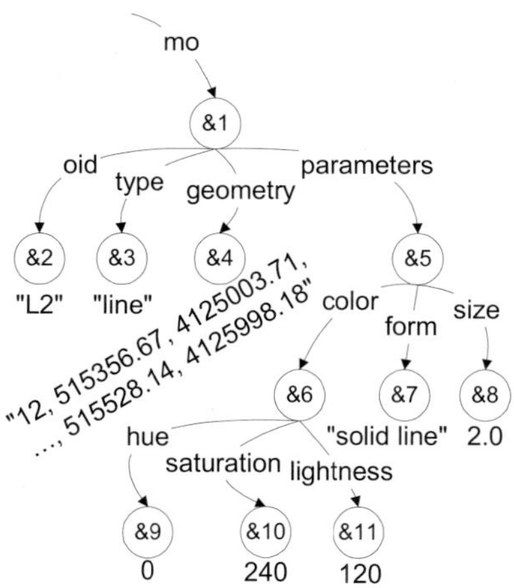

<&1, mo, set, {&2, &3, &4, &5}>
<&2, oid, string, "L2">
<&3, type, string, "line">
<&4, geometry, string, "12, 515356.67, 4125003.71,…,515528.14, 4125998.18">
<&5, parameters, set, {&6, &7, &8}>
<&6, colour, set, {&9, &10, &11}>
<&9, hue, integer, 0>
<&10, saturation, integer, 240>
<&11, lightness, integer, 120>
<&7, form, string, "solid line">
<&8, size, float, 2.0>

Fig. 2. Example representation of a map object in OEM

It is clear, that in the above case, the OEM representation is based on some hidden semantics. For example, the size of line symbol refers to its width and is expressed in millimetres. If units are not the same, the label may be used to convey this fact. For example, label "size" can be replaced by label "size-in-mm" (Papakonstantinou, 1995), and similar aspects apply to the "form", of the coordinate reference system, among other aspects.

In addition, the representation of polyline geometry as a string assumes that the parser is aware of the string syntax. An alternative way to represent this geometry is by a set of point objects, i.e., the vertices of the polyline. At this point the problem of ordering the vertices, in an effort to form the polyline, arises. One way to overcome this inefficiency is to use appropriate labels on point objects, which can depict the sequence of vertices. An alternative way is to represent each point as a complex object with two sub-objects, which consist of the point order in polyline and the point location. A third solution is to extend OEM, so that complex objects accommodate ordered lists of sub-objects instead of sets. This last solution has been already proposed for mapping an OEM to an extensible mark-up language (XML) document (Suciu, 1998b). The latter is by definition ordered, while the former is not. As for the point geometry representation, the "X,Y" string can be readily replaced by two atomic objects with labels "x" and "y", and of type double.

In the following, geometry in 2D space is represented according to the configuration shown in Fig. 3. Specifically, *point geometry* is a complex object, which points to two atomic sub-objects, with labels "x" and "y". These sub-objects accommodate projection or geographic coordinates of the point location in space. *Line geometry* is a complex object, which points to as many sub-objects (with labels "vertex") as the number of vertices composing the line. Each sub-object is in turn a complex object, which points to three atomic sub-objects, with labels "x", "y" and "order", where the first two accommodate the coordinates of a line vertex, whereas the third is the order of this vertex in the polyline. In a similar way, *area geometry* is represented. The only difference, which does not affect OEM representation, is that area outline has one more line segment, connecting the last vertex in the sequence with the first. Notice that this geometry representation scheme is compatible with original OEM definition, where no ordering of sub-objects is assumed.

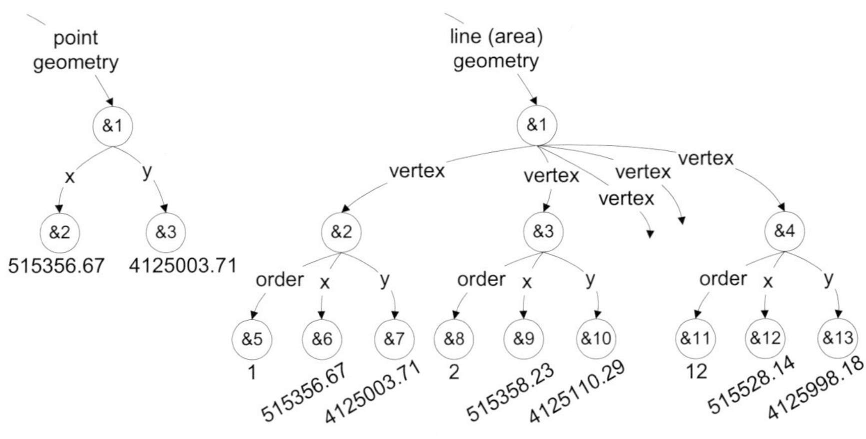

Fig. 3. Geometry representation in OEM

Coming back to the map in Fig. 1, it can also be seen as a complex object, consisting of a set of sub-objects, the representation instances of cartographic entities (i.e., the map objects). Hence, OEM can be used to model this map. In the graphical representation of OEM, the root of the graph is the map, while the leaves accommodate the user identifier (oid), the geometry, the symbol type and the symbol parameters of the map objects. Fig. 4 presents a portion of the OEM diagram used to represent the map. For simplicity, several symbol parameters are relatively abstract, e.g., a unique string represents colour (e.g., “red”).

As mentioned previously, the object label in an OEM can, by definition, be used to group related objects. Hence, objects may primarily be grouped based on their type, i.e., point, line, area, and text. Further, they may be grouped based on their properties. For example, all linear objects representing roads can be nested in a super-object labelled as “road network”. Obviously, this configuration approaches a cartographic database (or GIS database) organisation.

The use of symbols in representing cartographic entities is not arbitrary. It is usually based on some standards adopted by organisations or companies producing maps. In most cases, those standards define the symbol types and parameters to represent a cartographic entity at a specific scale. Such a standard forms a *symbol library* (Robinson *et al.*, 1995). It is a common practice to model a symbol library in a separate schema, where map objects point to.

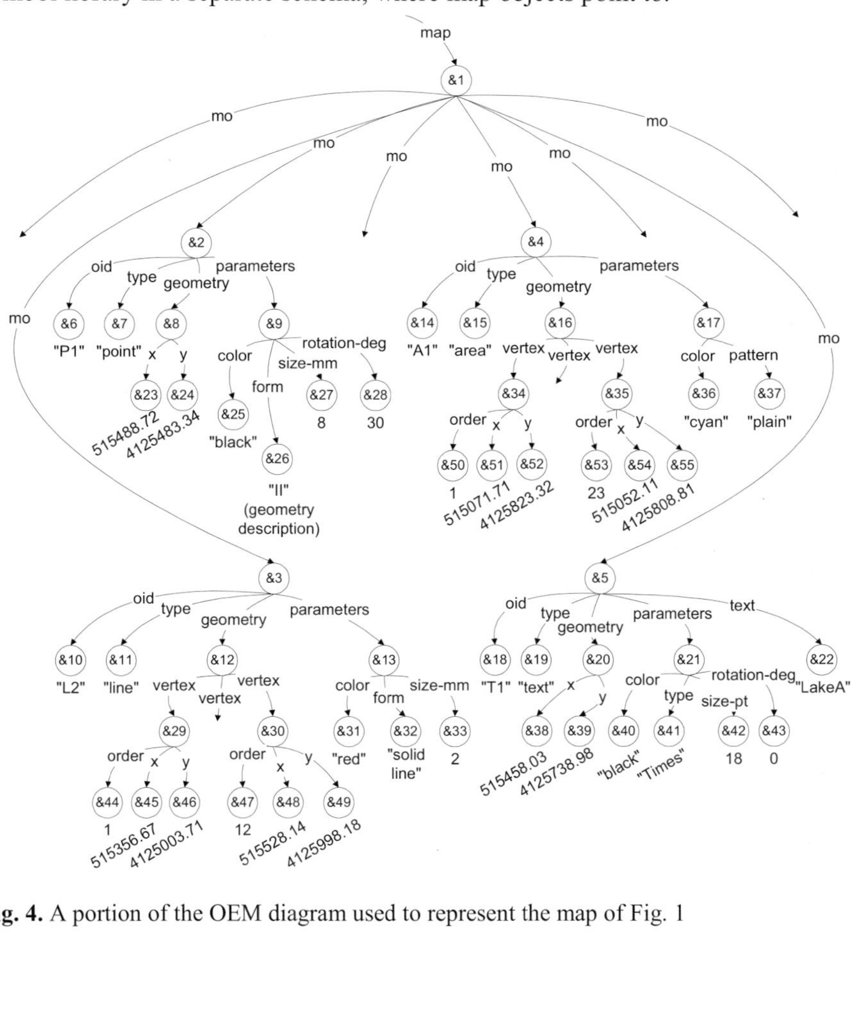

Fig. 4. A portion of the OEM diagram used to represent the map of Fig. 1

When an OEM is adopted to represent the maps of an organisation, a separate OEM can be adopted to accommodate the common symbols used in these maps. Fig. 5 presents a portion of an OEM diagram used to model the symbol library adopted in the map of Fig. 1. The same figure shows an alternative version of OEM in Fig. 4, whose map objects point to the symbol library. Geometry sub-objects are omitted for simplicity.

This configuration is flexible, when graphic limitations or other reasons require the representation of a cartographic entity and it's symbol, which does not conform to the symbol library standard or is not included in the symbol library. In this case, the specific symbol parameter object is attached to the map object directly as a sub-object. An example is given in Fig. 5, for the bridge map object *P1*.

6 Operational Issues

The discussion so far has focused on the representation of a map in OEM. It is argued that this is a convenient practice, due to the semi-structured nature of the map objects involved. The rest of this discussion is focused on handling issues such as extracting information from an OEM representation of a map. The task to be accomplished requires the existence of a query language, able to handle semi-structured data.

In the past, several languages have been proposed to query semi-structured data (Abiteboul, 1997). This study adopts *Lorel*, the query language supported by *Lore system* (Quass *et al.*, 1995), which makes use of the OEM data model. Lorel query language is a compatible extension to the OQL object-oriented query language (Cattell, 1994), with new features designed specifically for querying semi-structured data. Briefly, those features are: partially specified path expressions, wildcards, automatic type coercion in comparisons, and a special semantics for disjunction. Unlike OQL, Lorel does not enforce strong typing, and allows querying and schema browsing when the object structure is unknown or partially known. In other words, the uniqueness of Lorel includes: (a) the extensive use of *coercion*, and (b) the powerful *path expressions* (Abiteboul *et al.*, 1997).

The scope of this Section is to show how Lorel can be used to support some basic queries in the application domain of handling a map, which has been modelled using a semi-structured data model, specifically OEM. This task is accomplished through a series of examples. For brevity, it is assumed that the reader is familiar with Lorel query language syntax (Abiteboul *et al.*, 1997). All example queries refer to the database (fragment) in Fig. 6. The database refers to the map shown in Fig. 1. The graphical OEM representation of this database is given in Fig. 4.

The textual syntax adopted in Fig. 6 has been simplified according to Abiteboul *et al.* (1997). Tabs have been used to represent the nesting of objects. Each object has a unique *object identifier* (obj_id). Some objects are atomic and contain a value from one of the disjoint basic types, e.g., integer, real, string, gif, html,

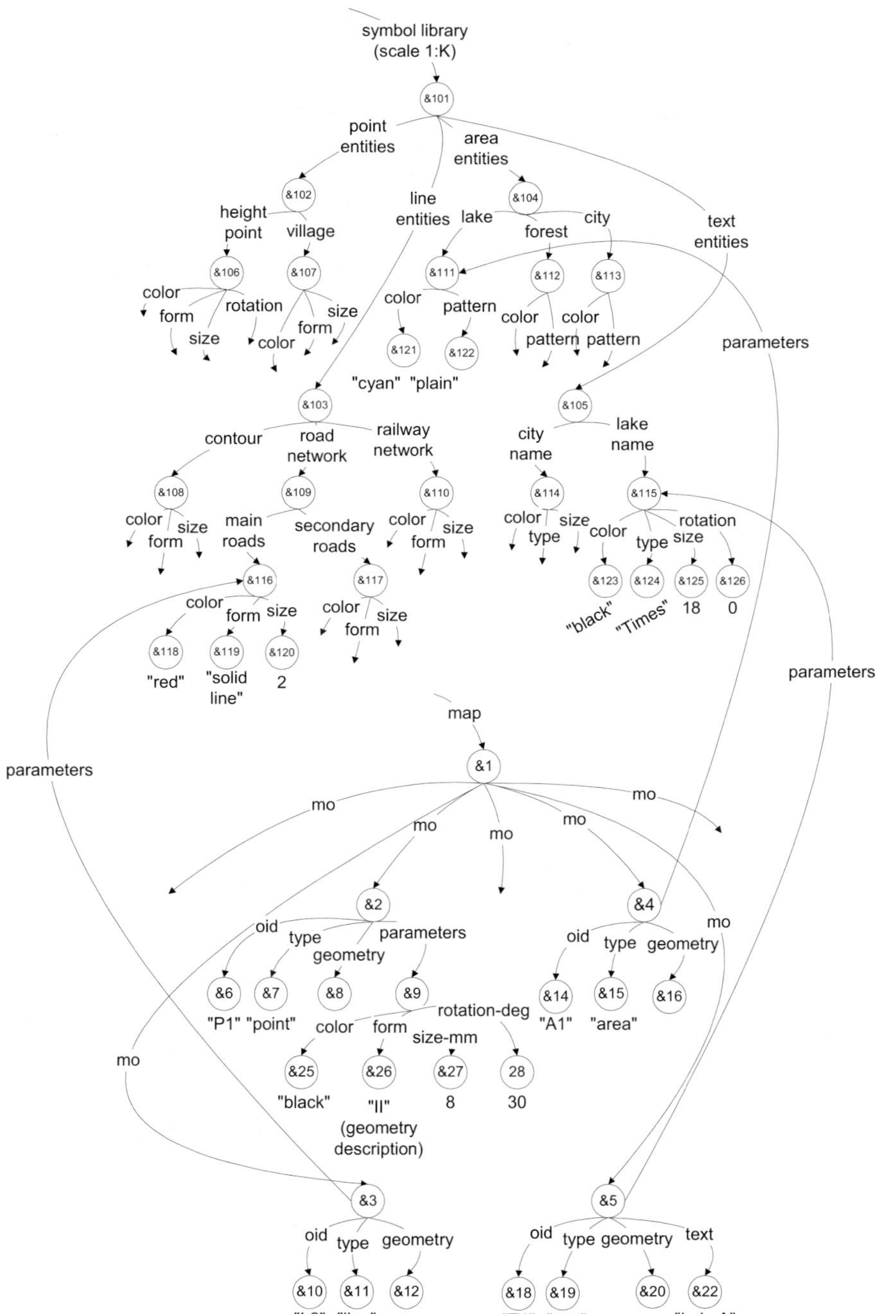

Fig. 5. The symbol library and an alternative version of OEM in Fig. 4

audio, java, etc. All other objects are complex; their value is a set of object references, denoted as a set of (*label, obj_id*) pairs. The labels are all taken from the atomic type string.

Notice that "object identifiers" retrieved in most of the query examples that follow and denoted by "oid" refer to the user-defined labels assigned to map objects in Fig. 1b (e.g., *P*1, *L*3, *A*5) and differ from those system-generated object identifiers (denoted by "obj_id") described in the previous paragraph.

```
map &1
   mo &2
      oid &6 "P1"
      type &7 "point"
      geometry &8
         x &23 515488.72
         y &24 4125483.34
      parameters &9
         color &25 "black"
         form &26 "II"
         size-mm &27 8
         rotation-deg &28 30
   mo &3
      oid &10 "L2"
      type &11 "line"
      geometry &12
         vertex &29
            order &44 1
            x &45 515356.67
            y &46 4125003.71
         ...
         vertex &30
            order &47 12
            x &45 515528.14
            y &46 4125998.18
      parameters &13
            color &31 "red"
            form &32 "solid line"
            size-mm &33 2
```

```
   mo &4
      oid &14 "A1"
      type &15 "area"
      geometry &16
         vertex &34
            order &50 1
            x &51 515071.71
            y &52 4125823.32
         ...
         vertex &35
            order &53 23
            x &54 515052.11
            y &55 4125808.81
      parameters &17
         color &36 "cyan"
         pattern &37 "plain"
   mo &5
      oid &18 "T1"
      type &19 "text"
      geometry &20
         x &38 515458.03
         y &39 4125738.98
      parameters &21
            color &40 "black"
            form &41 "Times"
            size-pt &42 18
            rotation-deg &43 0
      text &22 "LakeA"
```

Fig. 6. A textual OEM database (graphical representation in Fig. 4)

Query 1: "Find the identifiers (oid) of all line or area objects present in the map"

Lorel expression:	*Answer object:*
select A.oid	answer &101
from map.mo A	oid &10 "L2"
where A.type = "line"	oid &14 "A1"
or A.type = "area"	

This is a simple query example. It involves two conditions which are both included in a 'where clause'.

Query 2: "Find all black-colored map objects and list their type"

Lorel expression:
```
select A, A.type
from map.mo A
where A.parameters.color = "black"
```

Answer object:
```
answer &102
    mo &2
        type &7 "point"
    mo &5
        type &19 "text"
```

With the query above, map objects can be separated based on their colour. This operation is usually applied in the cartographic production process (Keates, 1989).

Query 3: "Find the identifiers (oid), along with coordinates, of all point objects that fall inside the rectangular window with lower left (south-west) corner (515000, 4125000) and upper right (north-east) corner (515500, 4125500)" (this is the south west quadrangle of the map in Fig. 1).

Lorel expression:
```
select B.oid, A.x, A.y
from map.mo{B}.geometry A
where B.type = "point"
and A.x >= 515000
and A.x <= 515500
and A.y >= 4125000
and A.y <= 4125500
```

Answer object:
```
answer &103
    oid &6 "P1"
        x &23 515488.72
        y &24 4125483.34
```

This is a simple *window query* example. This query is commonly applied in map browsers and GIS software packages and is well known as a *zoom-in* operation. The extension of this query to select line and area objects involves spatial operators, which are not provided by Lorel query language. Notice that this is one of the open issues for future research.

Query 4: "Find the identifiers (oid) of all point objects and line/area objects that have a vertex (for line and area objects) at position (515488.72, 4125483.34) (i.e., the position of the bridge in Fig. 1).

Lorel expression:
```
select B.oid
from map.mo{B}.geometry(.vertex)? A
where A.x = 515488.72
and A.y = 4125483.34
```

Answer object:
```
answer &104
    oid &6 "P1"
    oid &10 "L2"
```
(*L2* has a vertex on the bridge)

or alternatively,
```
select B.oid
from map.mo{B}.geometry A
where A.#@P.x = 515488.72
and A.#@Q.y = 4125483.34
and P == Q
```

This is a simple *point query* example. This query is also commonly applied in map browsers and GIS. Usually, the user is prompted to select an object by clicking with the mouse on the graphical screen. The query as expressed above restricts mouse clicks to coincide with a point object location or a line/area object vertex location. The extension of this query to select point, line and area objects with a tolerance of, e.g., 0.5mm (i.e., distance less than 0.5mm on the screen, from the mouse click position), involves spatial operators, which are not provided by the Lorel query language. The path expression given in the 'from' clause of the first alternative query guarantees a path to x and y coordinates of both point objects and line/area object vertices.

The second alternative query is a more complex expression, but highlights the flexibility provided by Lorel language such that :

(a) it involves the use of wildcard (#), which matches to any data path of length 0 or more (here, 0 for point objects and 1 for line/area objects); and
(b) the use of path variables P and Q, which assure that x and y coordinate values are examined each time and are assigned to the same line/area vertex.

The last condition in the 'where' clause eliminates from the answer object cases where a vertex of a line/area object has the desired X value and another vertex of the same object has the desired Y value.

Query 5: "Find identifiers (oid) of all line objects that are composed by less than 20 vertices".

<u>*Lorel expression:*</u>

```
select A.oid
from map.mo A
where A.type = "line"
and A SATISFIES
   20 > COUNT(select B.geometry.vertex
              from map.mo B)
```

<u>*Answer object:*</u>

```
answer &105
   oid &10 "L2"
```

This query uses a sub-query as operand to the aggregation operator COUNT. What the sub-query does is to count the number of vertices per line object. Notice that B already has its mapping fixed by the object assignment A (Quass *et al.*, 1995) in the enclosing query. The latter reports line objects, which satisfy the condition posed by the query.

Query 6: "Find the identifiers (oid) of all line objects intersecting area objects. Also report the corresponding area object identifier".

<u>*Lorel expression:*</u>

```
select distinct A.oid, B.oid
from map.mo A, map.mo B, A.geometry AG,
      B.geometry BG
where A.type = "line"
and B.type = "area"
and AG.x = BG.x
and AG.y = BG.y
```

<u>*Answer object:*</u>

```
answer &106
   oid &10 "L2"
   oid &14 "A1"
```

This query is a simple example of *join query*. Notice that the key word "distinct" eliminates duplicates from the answer object, e.g., *L*2 intersect twice *A*1, and however the pair appears once in the answer object. This query assumes the existence of coincident vertices on both objects at their intersections. Otherwise, no map objects will be included in the answer object (due to the lack of spatial operators).

7 Conclusions and Future Research

This paper argues that objects involved in a map form a category of semi-structured data. It adopts OEM, the most popular semi-structured data model that exists in the literature, and shows how objects of an individual map can be represented using alternative configurations. In addition, some basic application domain queries for handling these data are expressed in Lorel, also a popular query language for semi-structured data.

Several issues remain open for future research. One important issue is the integration of topological relations and constraints, which characterise cartographic entities of the real world, in the same model. This information is of significant value to the cartographer when performing several cartographic design and composition processes, such as the generalisation processes (Keates, 1989).

Another direction for future research is the mapping of the OEM representation of a map into XML variants (Suciu, 1998b) and specifically to GML (OpenGIS, 1999). Use of XML is expanding rapidly, and XML data will be exchanged freely in the near future between applications, belonging to the same or different organisations, in the same way HTML documents are currently exchanged. Hence, geographic and cartographic data exchange using XML must be seriously considered. Additionally, as already mentioned, the combination of XML with semi-structured data models will give as result a new technology for web data (Abiteboul *et al.*, 2000).

An issue that requires future examination is the extension of the Lorel language with spatial operators. This will enhance the functionality of the language and enrich the possibilities for spatial semi-structured data handling.

References

Abiteboul S (1997) Querying semi-structured data. In: Proceedings of the Sixth International Conference on Database Theory (ICDT'97). Delphi, Greece, pp 1-18

Abiteboul S, Quass D, McHugh J, Widom J, Wiener JL (1997) The Lorel query language for semistructured data. International Journal on Digital Libraries 1:68-88

Abiteboul S, Buneman P, Suciu D (2000) Data on the Web: From Relations to Semi-Structured Data and XML. Morgan-Kaufmann

Buneman P (1997) Semistructured data. In: Proceedings of the Sixteenth ACM SIGACT-SIGMOD-SIGART Symposium on Principles of Database Systems (PODS'97). Tucson, Arizona, pp 117-121

Cattell RGG (1994) The Object Database Standard ODMG-93. Morgan Kaufmann

Chawathe SS, Garcia-Molina H, Hammer J, Ireland K, Papakonstantinou Y, Ullman JD, Widom J (1994) The TSIMMIS project: integration of heterogeneous information sources. In: Proceedings of the Tenth Meeting of the Information Processing Society of Japan, Tokyo, Japan, pp 7-18

Keates JS (1989) Cartographic Design and Production, 2nd Edition. Longman

Kraak MJ, Brown A (2001) Web Cartography: Developments and Prospects. Taylor & Francis

OpenGIS (1999) Open GIS Consortium (OGC) [online]. Available from: http://www.opengis.org/

Oracle (2000) Oracle Spatial user's guide and reference. Oracle Corporation

Papakonstantinou Y, Garcia-Molina H, Widom J (1995) Object exchange across heterogeneous information sources. In: Proceedings of the Eleventh International Conference on Data Engineering (ICDE'95). Taipei, Taiwan, pp 251-260

Quass D, Rajaraman A, Sagiv Y, Ullman JD, Widom J (1995) Querying semistructured heterogeneous information. In: Proceedings of the Fourth International Conference on Deductive and Object-Oriented Databases. Singapore, pp 319-344

Robinson AH, Morrison JL, Muehrcke PC, Kimerling AJ, Guptill SC (1995) Elements of Cartography, 6th Edition. Wiley

Suciu D (1998) An overview of semistructured data. ACM SIGACT News 29:28-38

Suciu D (1998b) Semistructured data and XML. In: Proceedings of the 5th International Conference of Foundations of Data Organization (FODO'98). Kobe, Japan

W3C (1998) The World Wide Web Consortium (W3C) [online]. Available from: http://www.w3c.org/

Druck: Strauss Offsetdruck, Mörlenbach
Verarbeitung: Schäffer, Grünstadt